# THE FROZEN CLIMATE VIEWS OF THE IPCC

## An analysis of AR6

EDITED BY MARCEL CROK, ANDY MAY

clintel.org

## Colophon

Clintel Foundation
Zekeringstraat 41C
1014 BV, Amsterdam
The Netherlands
https://clintel.org
https://clintel.nl

Send feedback to office@clintel.org

ISBN: 979-8-89074-861-4 (ebook)
ISBN: 979-8-89074-862-1 (paperback)

Edited by Marcel Crok and Andy May
Graphic design by Maarten Bosch (Little Shop of Graphics)

Contributing authors:
Dr. Javier Vinós (molecular biologist, writer, Spain)
Dr. Ross McKitrick (Professor of Economics, University of Guelph, Canada)
Dr. Nicola Scafetta (Professor of Atmospheric Physics, University of Naples Federico II, Italy)
Kip Hansen (science research journalist, USA)
Dr. Fritz Vahrenholt (Professor, University of Hamburg, Germany)
Dr. Ole Humlum (Professor, University of Oslo, emeritus, Norway)
Marcel Crok (Director, Clintel, The Netherlands)
Andy May (Science writer and retired petrophysicist, USA).

Special discounts for bulk sales are available. Please contact office@clintel.org.

Publisher: Andy May Petrophysicist LLC, The Woodlands, Texas, USA

# Contents

# Introduction

BY MARCEL CROK AND ANDY MAY

n March 2023, with the publication of the so-called Synthesis report, the IPCC completed its Sixth Assessment (AR6) cycle. During this cycle, which started in 2015, the IPCC published three special reports: *Global Warming of 1.5°C* in October 2018; *Climate Change and Land* in August 2019; and *Special Report on the Ocean and Cryosphere in a Changing Climate* in September 2019. These reports were followed by reports of three Working Groups.

The Working Group 1 (WG1) contribution to AR6, *Climate Change 2021: the Physical Science Basis*, was released on 9 August 2021. The Working Group II (WG2) contribution, *Climate Change 2022: Impacts, Adaptation and Vulnerability*, was released on 28 February 2022. The Working Group III contribution, *Climate Change 2022: Mitigation of Climate Change*, was released on 4 April 2022. The cycle was then completed with the AR6 *Synthesis Report, Climate Change 2023*.[1]

## Big Panel

The Assessment cycle thus spanned 8 years and yielded 7 volumes. In their somewhat older but still interesting book *Taken by Storm*[2], Canadian scientists Ross McKitrick and Chris Essex, call the IPCC the "Big Panel". This is an apt description, except the IPCC is no longer a single entity, instead it now consists of numerous "Big Panels" which have less and less in common with one another. Each one produces large reports, sometimes thousands of pages, with contributions from hundreds of scientists and social scientists from all around the world. In this cycle, for example, the WG1 report was 2409 pages long, the WG2 report even longer with 3068 pages and the WG3 report also contained 2913 pages.

Apart from the sprawling nature of the reports, the IPCC is also a "Big Player" in the sense it dominates the narrative on climate change, although the nature of its influence is deceptive. People speak as if there is a single 'view' attributable to the IPCC which is the 'consensus' of the thousands of participating expert authors and reviewers. This picture of the IPCC makes it hard to criticize the claims of those who claim to be invoking the consensus position. Yet it is doubtful that any of the IPCC contributors or reviewers even read all seven volumes, nor were any asked to indicate their agreement with everything therein. Since, on many important topics, the chapters describe conflicting lines of evidence and admit to only limited levels of agreement or confidence, it is not plausible to suppose that the IPCC holds a single view on every given topic.

Yet the media routinely invokes the IPCC as a unified body putting forward a simple, clear (and dire) message, and politicians rely on this message to justify their climate policies. The IPCC has become a 'knowledge monopolist'[3] and this brings all kinds of dangers into play. Who is in a position to check the Big Player or challenge the way its authority gets invoked in political circles?

The IPCC describes its own work as follows: "An open and transparent review by experts and governments around the world is an essential part of the IPCC process, to ensure an objective and complete assessment and to reflect a diverse range of views and expertise." Do they succeed in this?

## Clintel Foundation

The Clintel Foundation[4], founded in The Netherlands in 2019, decided to analyse parts of the AR6 report, especially parts of the Working Group I and Working Group II report. We did that with an international group of scientists and experts, who, in general, have also signed the World Climate Declaration of Clintel and its central message "there is no climate emergency".[5] Some of us were

---

1   All these reports can be found here: https://www.ipcc.ch/ar6-syr/
2   Taken By Storm: The Troubled Science, Policy and Politics of Global Warming Paperback – May 1, 2003, Christopher Essex and Ross McKitrick, Key Porter Books, 320 pages.
3   Tol, Richard S J (2011) *Regulating knowledge monopolies: The case of the IPCC.* Climatic Change, 108 (4). pp. 827-839. ISSN 0165-0009
4   See Clintel.org and clintel.nl
5   https://clintel.org/world-climate-declaration/

also expert reviewers of the IPCC reports and commented on drafts of the report. The project was coordinated by Marcel Crok (co-founder of Clintel) and Andy May (retired petrophysicist and author of several climate books and frequent climate blogger[6]). The full list of authors can be found in the colophon of this report.

We didn't check all – almost 10.000 – pages of the AR6 report of course. That would be beyond the scope of our possibilities. We looked at topics that we know – based on our long experience with the climate debate – are highly relevant. Think of trends in extremes, disaster losses, sea level rise, climate sensitivity, scenarios etc. Even though we limited our effort to 13 topics, it turned out to be a very heavy project. It also generated very interesting internal discussion, some of which is reflected in the report.

## Frozen Climate Views

Our conclusions are quite harsh. We document biases and errors in almost every chapter we reviewed. In some cases, of course, one can quibble endlessly about our criticism and how relevant it is for the overall 'climate narrative' of the IPCC. In some cases, though, we document such blatant cherry picking by the IPCC, that even ardent supporters of the IPCC should feel embarrassed.

The IPCC seems obsessed with a few themes: the current warming is unique or their favourite word unprecedented, climate change is all bad and it's caused by $CO_2$. This attitude leads to tunnel vision and therefore we chose the title "The Frozen Climate Views of the IPCC". This doesn't mean that $CO_2$ is not having any effect. Of course, it has. But the evidence that $CO_2$ and other greenhouse gases are causing 'dangerous climate change' is, even after 30 years and 6 major IPCC reports, rather thin.

Roger Pielke Jr, a critic of the IPCC, who is mentioned several times in this report, often says: the IPCC is so important, that if it didn't exist, it should be invented. But given its importance and influence, the IPCC should also take criticism seriously. We really hope people involved in the IPCC will look seriously at our findings and draw lessons from them.

*Marcel Crok and Andy May*
*May 2023*

---

6    https://andymaypetrophysicist.com/

# Summary

BY MARCEL CROK AND ANDY MAY

f we have to summarize the IPCC-reports in one paragraph, it might sound like this: Climate change is happening at an increasingly rapid pace. Current warming is unprecedented in at least 125,000 years and the current $CO_2$ concentration is unprecedented in at least two million years. $CO_2$ and other greenhouse gases have caused all or most of the warming since 1850. As a result, some changes, like sea level rise, are already irreversible for centuries to come. Climate change is already making the weather more extreme. Around half of the global population is very vulnerable to climate change. Only urgent climate action, i.e., reducing $CO_2$, methane, and other greenhouse gases, can secure a liveable future for all. Luckily, renewable energy has become much cheaper in the past decade, so we can do it.

Some sentences here are paraphrased, but others are literally from IPCC text. An even shorter summary would be this: the current warming is unprecedented, is caused by us, is very dangerous, and we should stop it by reducing our $CO_2$ emissions, preferably by enhancing the production of renewable energy.

This is the 'science based' message that the IPCC has delivered after six assessment reports. Each report consists of three working group reports and a synthesis report. In March 2023, with the publication of the AR6 Synthesis Report, the IPCC finished its sixth assessment cycle.

What is the IPCC and what is its role? From the IPCC website:

Created in 1988 by the World Meteorological Organization (WMO) and the United Nations Environment Programme (UNEP), the objective of the IPCC is to provide governments at all levels with scientific information that they can use to develop climate policies. IPCC reports are also a key input into international climate change negotiations. The IPCC is an organization of governments that are members of the United Nations or WMO. The IPCC currently has 195 members. Thousands of people from all over the world contribute to the work of the IPCC.

The role of the IPCC is laid down in its procedures.[1] Here is the most relevant one (our bold):

The role of the IPCC is to assess on **a comprehensive, objective, open and transparent basis** the scientific, technical and socio-economic information relevant to understanding the scientific basis of risk of human-induced climate change, its potential impacts and options for adaptation and mitigation. IPCC reports should be neutral with respect to policy, although they may need to deal objectively with scientific, technical and socio-economic factors relevant to the application of particular policies.

The IPCC can also be seen as a "knowledge monopoly" and as such it suffers from the same dangers as any other monopoly. The well-known Dutch (climate) economist Richard Tol, who contributed to several IPCC reports, but was not invited to work on AR6, after he criticized and left the author team of the AR5 Working Group 2 (WG2) *Summary for Policy Makers* report in 2013.[2] He wondered how you could regulate such a knowledge monopoly.[3] In his abstract, Tol described the IPCC process in the following way (our bold):

The Intergovernmental Panel on Climate Change has a monopoly on the provision of climate policy advice at the international level and a strong market position in national policy advice. This may have been the intention of the founders of the IPCC. **I argue that the IPCC has a natural monopoly, as a new entrant would have to invest time and effort over a longer period to perhaps match the reputation, trust, goodwill, and network of the IPCC.** The IPCC is a not-for-profit organization, and it is run by nominal volunteers. It therefore cannot engage in the price-gouging that is typical of monopolies. However, the IPCC has certainly taken up tasks outside its mandate. The IPCC has been accused of haughtiness. Innovation is slow. Quality may have declined. And the IPCC may have used its power to hinder competitors. [These] are all

1 https://www.ipcc.ch/site/assets/uploads/2018/09/ipcc-principles.pdf
2 https://www.bbc.com/news/science-environment-26655779
3 Tol, Richard S J (2011) *Regulating knowledge monopolies: The case of the IPCC*. Climatic Change, 108 (4). pp. 827-839. ISSN 0165-0009

things that monopolies tend to do, against the public interest. The IPCC would perform better if it were regulated by an independent body which audits the IPCC procedures and assesses its performance; if outside organizations would be allowed to bid for the production of reports and the provision of services under the IPCC brand; and if policy makers would encourage potential competitors to the IPCC.

This was written by Tol in 2011, a year after the Interacademy Council (IAC) investigated the IPCC process, after errors in the IPCC AR4 report received a lot of attention in the media.[4] The most striking error was the claim in the AR4 WG2 report that Himalayan glaciers would be completely gone in 2035, a claim the IPCC later admitted was unfounded.[5]

The IAC made several recommendations. In our (i.e., Clintel's) view a key IPCC problem is group-think. The IPCC tends to invite only those scientists that strongly agree with claims in earlier IPCC reports, i.e., that current warming is unprecedented, caused by greenhouse gases, and is dangerous. Then they write the same conclusion in the next report. Big surprise.

The IAC review was quite clear about dealing with a range of views (page 17-18, our bold):

> ### Handling the full range of views
> An assessment is intended to arrive at a judgment of a topic, such as the best estimate of changes in average global surface temperature over a specified time frame and its impacts on the water cycle. Although all reasonable points of view should be considered, they need not be given equal weight or even described fully in an assessment report. Which alternative viewpoints warrant mention is a matter of professional judgment. Therefore, Coordinating Lead Authors and Lead Authors have considerable influence over which viewpoints will be discussed in the process. **Having author teams with diverse viewpoints is the first step toward ensuring that a full range of thoughtful views are considered.**
>
> **Equally important is combating confirmation bias—the tendency of authors to place too much weight on their own views relative to other views** (Jonas et al., 2001). As pointed out to the Committee by a presenter and some questionnaire respondents, alternative views are not always cited in a chapter if the Lead Authors do not agree with them. Getting the balance right is an ongoing struggle. However, concrete steps could also be taken. For example, **chapters could include references to all papers that were considered by the authoring team and describe the authors' rationale for arriving at their conclusions.**

## Investigation by Clintel

In this Clintel[6] report we will show that not only did the IPCC not follow this recommendation, it did the opposite. It went to great lengths to exclude "diverse viewpoints" to draw its often alarmist conclusions. We will show that one well-known scientist, Roger Pielke Jr., whose work is relevant for many chapters, is treated by the IPCC as a 'Voldemort', the Harry Potter villain 'whose name shall not be named'.[7] Indeed, as we document in several chapters of this report, the IPCC avoids mentioning his work, so they can draw opposite conclusions. Pielke told us that a U.S. IPCC contributor literally told him that "he would never be involved in IPCC".

Other well-known sceptical scientists, like Richard Lindzen, John Christy, and Roger Pielke Sr (yes, the father of Jr) have contributed or tried to contribute to earlier WG1 reports but were disappointed about the process and decided not to spend their energy on it anymore. A pity, because if the IPCC author teams would recruit scientists with diverse viewpoints, a lot of the shortcomings

---

4   https://archive.ipcc.ch/organization/organization_review.shtml

5   This became known as 'Himalayan Blunder' or 'glacier-gate': https://www.theguardian.com/environment/2010/jan/20/ipcc-himalayan-glaciers-mistake

6   Clintel.org

7   This is going on for a long time. Here an example by Andrew Revkin in 2012: The Superstorm and Humanity's Disaster Blind Spot - The New York Times (nytimes.com)

that we document in this report could have been prevented. The conclusions of the IPCC reports would be radically different though, and far less apocalyptic.

This Clintel report is written by scientists and experts who were not directly involved in the writing of the IPCC reports (although some have been "expert reviewer" of one or more IPCC reports) and who are experienced with the underlying climate science literature. We investigated if the IPCC followed its own principles. Are the reports and its claims (especially in the *Summary for Policy Makers*) really based on a comprehensive review of the literature? Are the conclusions unbiased, objective and the methods of reaching them transparent? The short answer to these questions unfortunately is a very clear "no".

The report is divided into four parts. Part 1 deals with observations, starting at the end of the last ice age (the start of the Holocene) all the way to the current modern warming period. Part 2 looks at causes of climate change, including the role of the sun and the effect of additional greenhouse gases. Part 3 examines the scenarios used by the IPCC especially the most extreme one, the so-called RCP8.5 or SSP5-8.5 scenario. In part 4 we delve into the impacts of climate change, mainly on humans. Parts 1 to 3 of the report discuss the Working Group 1 report (WG1) of AR6 while part 4 deals with the Working Group 2 report (WG2).

## Erasing climate history

In AR6, the IPCC makes the remarkable claim that "global surface temperatures are more likely than not unprecedented in the past 125,000 years." This claim erases the so-called Holocene Thermal Maximum, sometimes called the Holocene Climatic Optimum, terms that are avoided by the IPCC. The IPCC flattens our climate history thereby making the current warming look "unprecedented" and therefore "unique". But is this realistic?

The Holocene Thermal Maximum is well documented in the literature and can be considered a period that extended from c. 9800-5700 before present (BP[8]) when temperatures varied considerably in many parts of the globe and maximal Holocene temperatures were reached in many areas, but often at different times. As the Spanish scientist Javier Vinós, author of the recent book *Climate of the Past*[9], notes in Chapter 1:

> Multi-proxy reconstructions are useful, but biases and unavoidable limitations of the technique result in their inability to answer the IPCC question: Was the last decade the warmest the planet has been during the Holocene?

As Vinós explains a multi-proxy reconstruction is very dependent on researcher's choices, starting with the proxies included and excluded, whether land and marine proxies are representative of temperatures in the area, and what their respective weight should be in the mix. Attempting to measure the average temperature of the planet with a few hundred low-precision uncalibrated proxy thermometers that provide a reading once a decade to once a century or two at best is a laughable task. Comparing the resulting global average with our daily modern measurements, including satellites and thousands of high-precision calibrated thermometers distributed all over the world, including all the oceans, and then declaring we can trust that it is *more likely than not* that the past decade is warmer than any century during the past 12,000 years is an untenable claim.[10]

8    B.P. (Before the Present) is the number of years before the present. Because the present changes every year, archaeologists, by convention, use A.D. 1950 as their reference.

9    Vinós, J. (2022). Climate of the Past, Present and Future, A Scientific Debate. 2nd ed. Critical Science Press

10    Chapter 1: No confidence that the present is warmer than the Middle Holocene

# New hockey stick

A big surprise in the AR6 Working Group 1 report was the publication in the key *Summary for Policymakers* (SPM) of a new hockey stick graph. The first pronounced hockey stick graph was published by Michael Mann in 1998 and 1999[11] and it was heavily promoted in the 3rd IPCC (TAR) report in 2001.

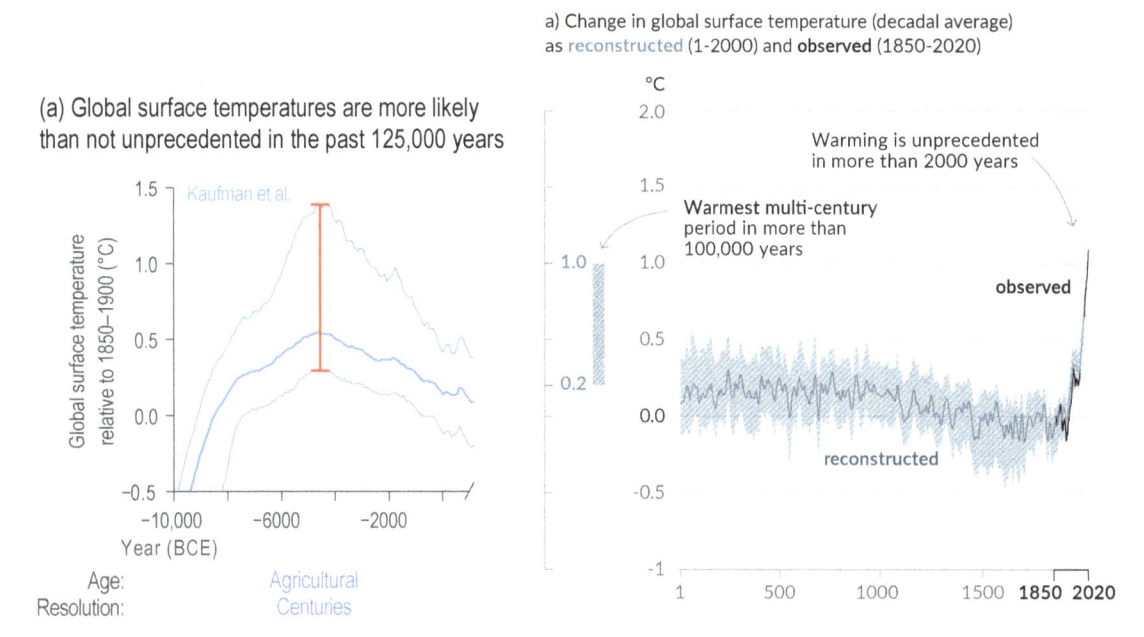

**Figure 1:** The new AR6 hockey stick graph (Figure SPM.1a) is shown on the right. The vertical bar on its left is stated to show the estimated temperature (*very likely* range) during the warmest multi-century period in at least the last 100,000 years. On the left is part of AR6 WG1 figure 2.11(a). It shows a temperature reconstruction by Kaufman et al. (2020) with the 90% uncertainty range noted with a red bar at the peak of the Holocene Thermal Maximum. The uncertainty is as large or larger than the total modern warming.

Hockey stick graphs have been used by the IPCC to claim that the current warming is unprecedented in the last 1000 or 2000 years. Both the current (AR6) hockey stick, and the first one by Mann et al. in 1998 attempt to erase the historically well-documented Medieval Warm Period and Little Ice Age. They are trying to send the message that these were only regional phenomena, with little consequence globally.

The earlier hockey stick was heavily criticized for major deficiencies in its paleoclimatic proxies, and the statistical methods used to construct it (Soon et al. 2003[12], McIntyre and McKitrick, 2003[13], 2005[14]; McShane and Wyner, 2011a, b[15]; Montford, 2010[16]).

According to Stephen McIntyre, the problem with all these reconstructions is more or less the same. Authors select proxies from thousands of available proxy series in international databases. Most proxies show little more than noise raising doubt about their validity as a temperature proxy. Authors then select their proxies, apply one or more statistical methods to them to end up with their hockey stick. The latest incarnation of the hockey stick is examined in more detail in Chapter 2.[17]

11   Mann, M. E., Bradley, R. S., and Hughes, M. K., 1999, Northern Hemisphere Temperatures during the past Millennium: Inferences, Uncertainties, and Limitations: Geophysical Research Letters, v. 26, no. 6, p. 759-762.

12   Soon, Willie, and Sallie Baliunas. "Proxy Climatic and Environmental Changes of the Past 1000 Years." *Climate Research*, vol. 23, no. 2, 2003, pp. 89–110. *JSTOR*, http://www.jstor.org/stable/24868339. Accessed 3 Apr. 2023.

13   McIntyre, S., and McKitrick, R., 2003, Corrections to the Mann et al. (1988) proxy data base and northern hemispheric average temperature series: Energy & Environment, v. 14, no. 6, p. 751-771.

14   McIntyre, Stephen and Ross McKitrick (2005a) "The M&M Critique of the MBH98 Northern Hemisphere Climate Index: Update and Implications." Energy and Environment 16(1) pp. 69-100; (2005b) "Hockey Sticks, Principal Components and Spurious Significance" Geophysical Research Letters Vol. 32, No. 3, L03710 10.1029/2004GL021750 12 February 2005.

15   McShane, B. B., and Wyner, A. J., 2011a, Rejoinder: The Annals of Applied Statistics, v. 5, no. 1, p. 99-123. -, 2011b, A statistical analysis of multiple temperature proxies: Are reconstructions of surface temperatures over the last 1000 years reliable?: The Annals of Applied Statistics, v. 5, no. 1, p. 5-44.

16   Montford, A. W., 2010, The Hockey Stick Illusion, London, Stacey International, 482 p.

17   Chapter 2: The Resurrection of the Hockey Stick

In summary, we find the claims by the IPCC, that current warming is unprecedented in the last 2000 or even the last 125,000 years, very unconvincing to say the least. There is good evidence that both in the last 2000 years as well during the Holocene Thermal Maximum, temperatures were broadly similar to, or perhaps higher than, during the current warming period. In this case, the IPCC seems to act like George Orwell's ministry of truth, by rewriting earth's climate history. Moreover, the IPCC failed to discuss these issues in a comprehensive and transparent way. Their bias is revealed in their choice of what studies they include in the report and what studies they ignore.

## Global temperature

The Global Mean Surface Temperature (GMST) has become the iconic parameter in the climate change debate. It is the measurement of choice when deciding international climate policy, as in are we to exceed either the 1.5°C or 2°C target. Even though these targets are arbitrary and political[18] rather than scientific. Unscientific as they are, these targets dominate the scientific discourse about climate change. But is that deserved? How reliable are these temperature measurements and are there 'better' alternatives?

Before Andy May delves into detailed discussions[19] about the many different temperature datasets and their uncertainties, he first puts the rise in global average temperature of one degree Celsius since 1850 into perspective.

Each year the globe experiences temperature swings that are much larger than the one degree rise in the annual average temperature seen in the past 170 years. The global average temperature of Earth varies over three degrees every year, it is just over 12 degrees in January and just under 16 degrees in July as shown in figure 2 from Phil Jones and colleagues at the UK Met Office. The Northern Hemisphere average temperature has a larger swing from eight degrees in January to over 21 degrees in July, a remarkable change of 13°C in only six months.

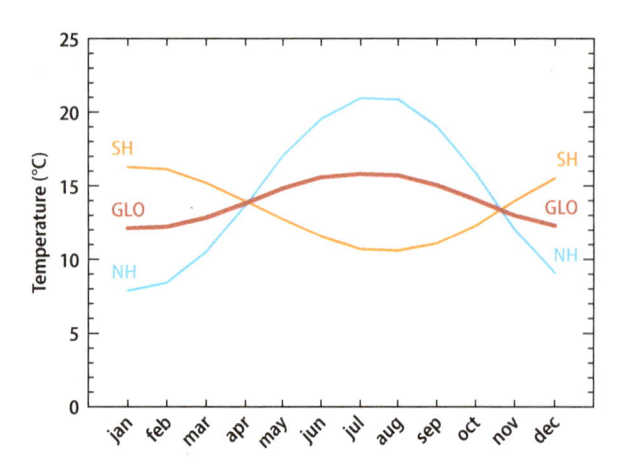

**Figure 2:** Average global surface temperatures from 1961-1990 for the globe (GLO), Northern Hemisphere (NH) and Southern Hemisphere (SH) by month. After: (Jones, New, Parker, Martin, & Rigor, 1999).[20]

The IPCC appears to agree that GMST is a poor measure of climate change and provides a plot of the change in ocean heat content in AR6 on page 350, it is shown below as Figure 3. The steep upward slope appears alarming as it moves from 0 to 500 zettajoules. Even the unit "zettajoules" sounds scary. But how many zettajoules of energy do the global oceans contain? A staggering 1,514,000! So, an increase of 500 zettajoules is a change of 0.03% in the global energy content, hardly an alarming change. The IPCC avoided giving this important background information.

---

18   https://rogerpielkejr.substack.com/p/the-two-degree-temperature-target

19   Chapter 3: Measuring global surface temperature

20   Jones, P. D., New, M., Parker, D. E., Martin, S., & Rigor, I. G. (1999). Surface Air Temperature and its Changes over the Past 150 years. Reviews of Geophysics, 37(2), 173-199.
     Retrieved from https://citeseerx.ist.psu.edu/viewdoc/download?doi=10.1.1.546.7420&rep=rep1&-type=pdf

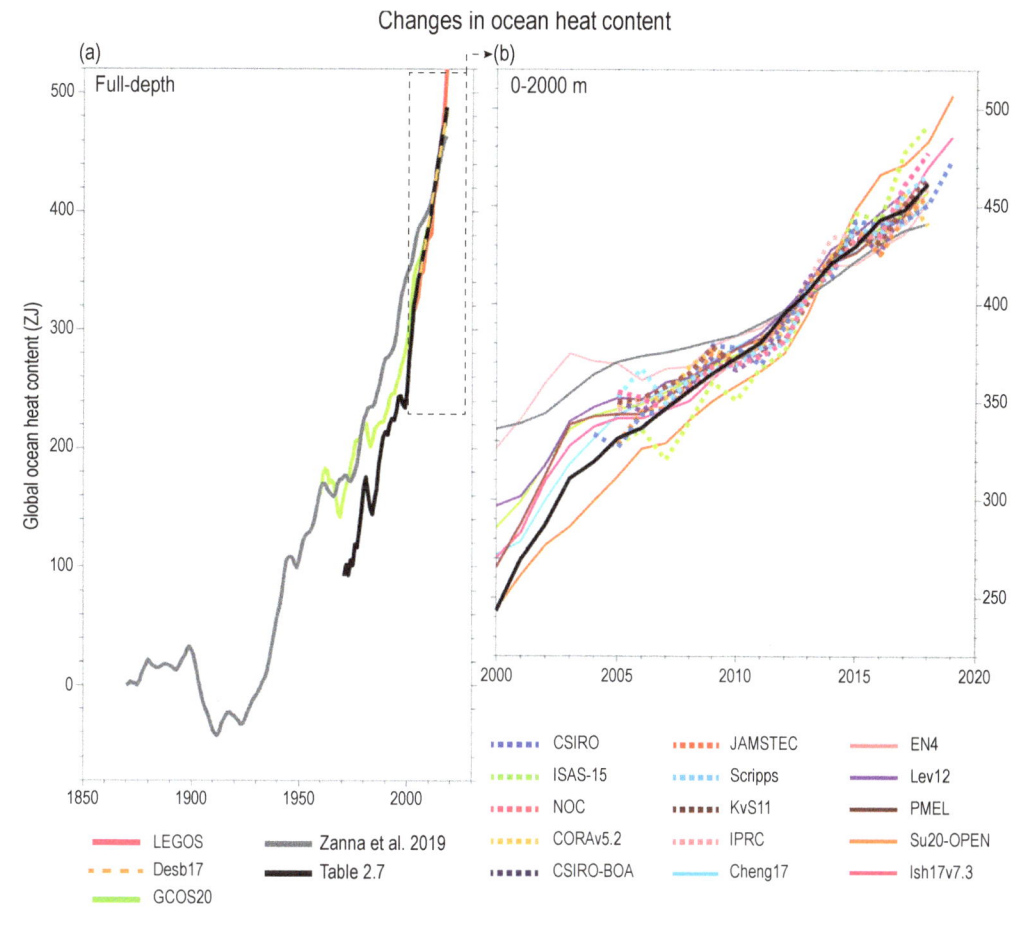

**Figure 3:** Changes in ocean heat content from AR6, Chapter 2, page 350.

The steep slopes are an artifact of the scales chosen and the starting point. Figure 4 is more meaningful and is roughly the same body of water as shown in the right graph in figure 3. The only difference is figure 3 is from the surface, and figure 4 is from 100 meters depth, in both cases to 2000 meters.

In figure 4 we see a rate of increase of about 0.4°C/century. This is less than half that reported for the surface over the past century or so. Reporting the change in ocean temperature is a more relevant and understandable way to show recent changes in the climate system, as Roger Pielke Sr. wrote in 2003.[21] When interested in global warming or cooling one should look at the ocean heat content.

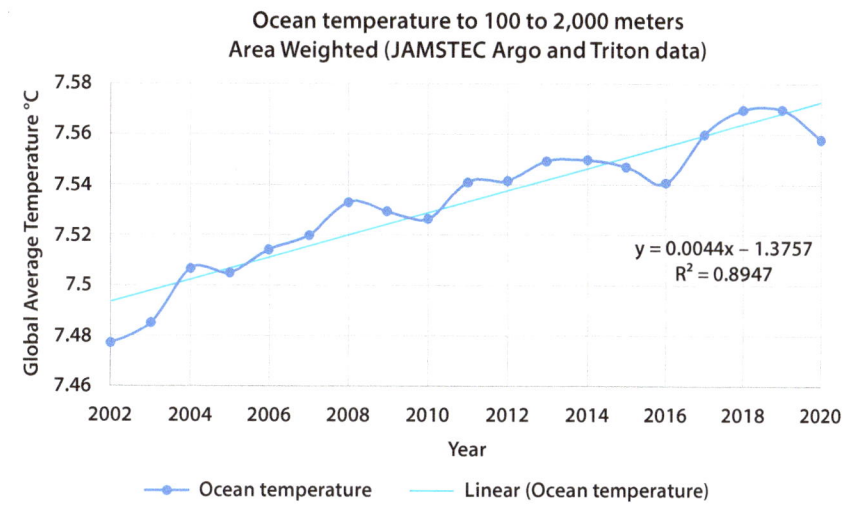

**Figure 4:** Average temperature for the world ocean from 100 to 2,000 meters. Data from JAMSTEC.

---

21   Pielke Sr., r. (2003, March). Heat Storage within the earth System. BAMS, 84(3), 331-335.
     Retrieved from https://journals.ametsoc.org/view/journals/bams/84/3/bams-84-3-331.xml

May concludes by answering questions he posed at the beginning of his chapter. Are the estimates of global temperature change since 1850 accurate and comprehensive enough to tell us how quickly Earth's entire surface, including the oceans, is warming? *No.* Is the global mean surface temperature a key indicator of climate change? *No*, the measurements used partly reflect local weather and environmental conditions and are affected by the chaotic conditions at the surface. Further, the total change recorded over the past century is quite small relative to the basic temperature measurement accuracy and natural climate variability.

## Snow cover

In 2000, Dr David Viner, a senior research scientist at the climatic research unit (CRU) of the University of East Anglia, said that within a few years winter snowfall in the UK would become "a very rare and exciting event". "Children just aren't going to know what snow is," he said.[22]

It's now 2023 and his prediction didn't come through. It's tempting to think that global warming will mean less snow. On the other hand, warming could mean more evaporation and more precipitation, including in the form of snow. There is no necessary relationship between global average temperature and snowfall.

The IPCC decided to show a snow trend graph only for the month of April (figure 5):

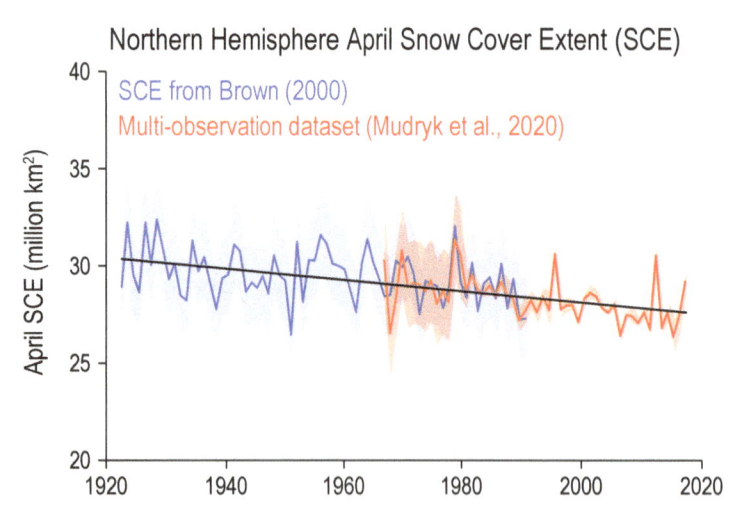

**Figure 5:** April snow cover extent (SCE) for the Northern Hemisphere. From: AR6, Figure 2.22.

The IPCC introduced a brand-new dataset that wasn't even published yet at the time of the Second Order Draft, the last version that is seen by reviewers. Not surprisingly, the scientist behind this new dataset, Lawrence Mudryk, was also a contributing author of the chapter in the IPCC report. This new dataset is a so-called hybrid dataset, it consists of seven different datasets. Some of these datasets use measurements, others use models or a combination of models and measurements. We tried unsuccessfully to download the different datasets behind this new dataset.[23]

The new dataset cited by the IPCC makes the claim that snow cover extent is in decline in all months of the year. This is remarkable, because until now, the well-known Rutgers Global Snow Map dataset showed increasing snow cover extent during the Fall and Winter.

The IPCC mentions a relevant paper by Connolly et al. but failed to mention its key conclusion, namely that climate models are unable to simulate the increasing trend in snow cover in the Fall and Winter:[24]

22  https://web.archive.org/web/20130422045937/http://www.independent.co.uk/environment/snowfalls-are-now-just-a-thing-of-the-past-724017.html

23  Chapter 4: Controversial Snow Trends, by the Clintel Team.

24  Connolly, et al., 2019, Northern Hemisphere Snow-Cover Trends (1967-2018): A Comparison between Climate Models and Observations, Geosciences, 9, 135, doi:10.3390/geosciences9030135

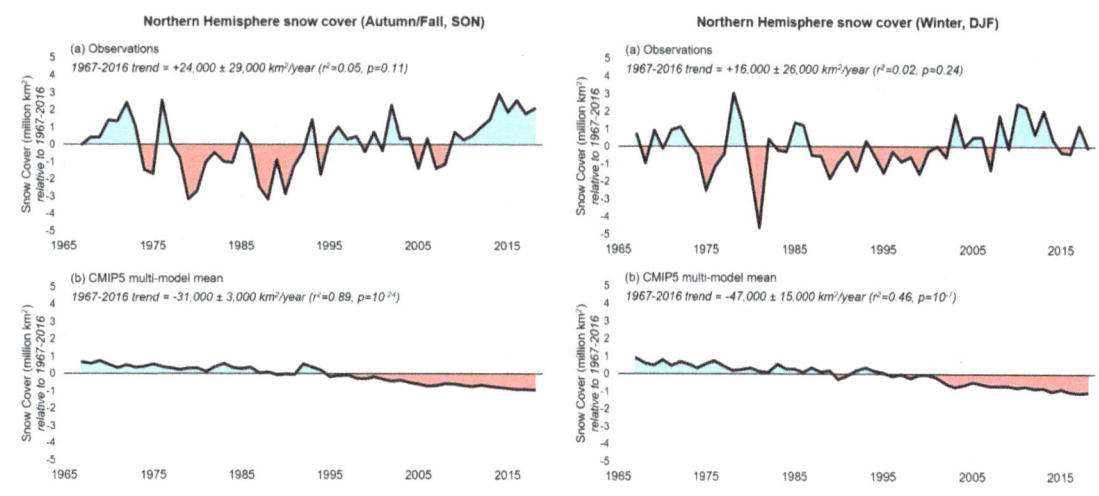

**Figure 6:** Northern Hemisphere snow cover in Autumn/Fall and winter. Top panel: observations based on the Rutgers Snow Lab data. Bottom panel: based on CMIP5 simulations. Source of the figure: Connolly et al. (2019).

The IPCC introduced a brand-new dataset far too late in the process. Reviewers were unable to check the validity of this radically new dataset.

The trends in Northern Hemisphere snow cover extent are just one of many examples of biased reporting in the IPCC AR6 report. The outcome of the assessment is largely decided when report authors are nominated. In this case, the lead author of a key paper, Lawrence Mudryk, was nominated as contributing IPCC author and most likely influenced the direction of the IPCC literature review in his own favour.

## Sea Level Rise

The IPCC's Sixth Assessment Report (AR6) claims that sea level rise is accelerating. However, the evidence for this is rather thin.

As Kip Hansen points out[25], the best available evidence for long-term sea level changes comes from tide gauge records. These records typically show remarkably linear behaviour for more than a century. The IPCC likes to use satellite sea level measurements combined with a blend of tide gauge records to show that sea level rise is accelerating. The IPCC ignores the fact that the rise in sea level shows multidecadal variability, probably related to the Atlantic Multidecadal Oscillation. A paper the IPCC itself frequently cited has this figure:

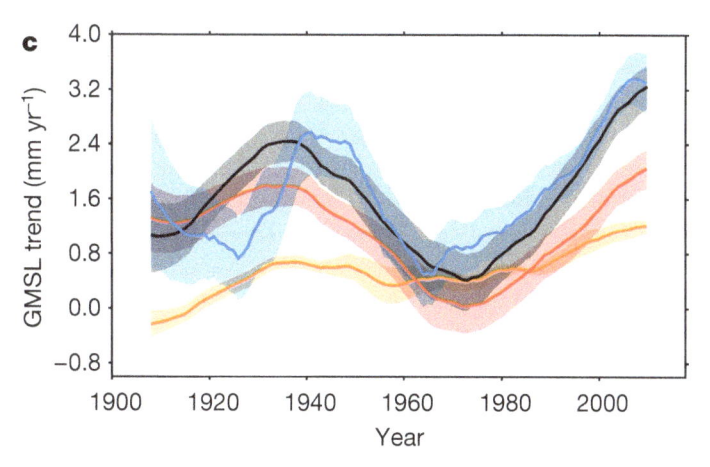

**Figure 7:** Sea level trend based on a so-called sea level budget method model as used in Frederikse et al 2020.[26] The thick blue line is the observed trend and black is the sum of different model components. Orange and red are two important components, they are thermosteric (thermal expansion) change and barystatic (the change in seawater mass) respectively.

---

25   Chapter 5: Accelerated Sea Level Rise: not so fast
26   Frederikse, T., Landerer, F. C., Caron, L., Adhikari, S., Parkes, D., & Humphrey, V. (2020). The causes of sea-level rise since 1900. Nature, 584, 393-397. Retrieved from https://www.nature.com/articles/s41586-020-2591-3

So, it is likely that the IPCC conflates their recent 'acceleration' of the sea level with this multidecadal variability. This should become clear in the next 10 to 20 years. Right now, it is very preliminary to claim there is an acceleration of the sea level rise.

In Chapter 10[27], Ole Humlum uses the IPCC sea level projection tool,[28] which the public can use to 'make' different sea level scenarios for tide gauge stations around the world, to project sea level for four Scandinavian capitals and shows us the surprising results. It seems that the IPCC projections contrast sharply with observations. Below, in figure 8, we compare the IPCC projections and observations for Stockholm, Sweden.

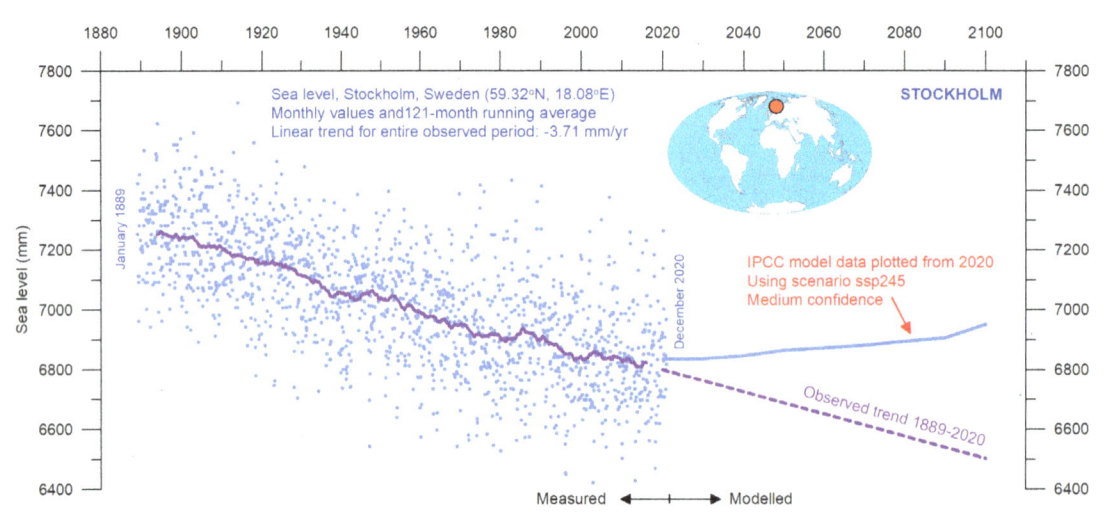

**Figure 8:** The IPCC projected sea level change versus observations for Stockholm, Sweden.

Only time will tell if the IPCC sea level projections are correct, but they do contrast strongly when compared to recent observations. Humlum observes:

> It is … extremely surprising that the modelled effect of this [change] should first appear in 2020 as a rather marked step change in the relative sea level. Had the modellers instead modelled their sea level data from an earlier date, e.g., 1950, which would have been entirely possible, the conflict between measured and modelled data would immediately have become apparent. Usually, model improvements would then have been initiated as the next scientific step. It is highly disappointing that such a simple quality- or sanity check was apparently never requested or performed by the IPCC.

It seems that this tool was not produced to test the validity of a scientific idea. It was instead an attempt to alarm the user.

## The Sun's Role in Climate Change

We start Part 2 on the causes of climate change with Nicola Scafetta's and Fritz Vahrenholt's chapter on the Sun.[29] They point out that the Medieval Warm Period (MWP) and the Little Ice Age (LIA) are historically well documented climatic anomalies in the peer-reviewed literature from around the world. Historical records of sunspots, auroras, and solar proxies, also document significant changes in solar activity. The climate changes and the solar activity changes correlate well.[30] Temperatures were relatively warmer during the MWP and solar activity was higher, temperatures were lower during the LIA and solar activity was lower. Common sense suggests there is probably some connection between the two.

---

27 Chapter 10: A miraculous sea level jump in 2020

28 https://sealevel.nasa.gov/data_tools/17

29 Chapter 6: Why does the IPCC downplay the Sun?

30 Connolly et al., R. (2021). How much has the Sun influenced Northern Hemisphere temperature trends? Research in Astronomy and Astrophysics, 21(6). Retrieved from https://iopscience.iop.org/article/10.1088/1674-4527/21/6/131?fbclid=IwAR0U5WARVnuGVjj2qeiiBYgGo0IlxXb9NNzUbeqqN-th2Zp1YU8rLOZkrMM

As Scafetta and Vahrenholt point out, these two well documented periods provide an excellent scientific blind test of the solar-climate connection. They list many peer-reviewed studies that show the close connection between climate changes and solar activity. They correlate well in Spain, Portugal, Slovakia, China, Bhutan, and the Canadian Rocky Mountains. During the Wolf, Spörer, and Maunder solar minima, the intermediate water layers of the North Atlantic cooled by 2-3°C, while the surface water in the tropical North Atlantic off Mauritania cooled by 1°C. On Sakhalin, Russia's largest island, the lowest temperatures were recorded during the Maunder Solar Minimum. In Tasmania, Australia, proxies show cold periods during the Spörer and Maunder Solar Minima. Even in Antarctica, climate proxies correlate with repeated drops in solar activity.

Similar evidence shows that the Medieval Warm Period, which coincides with a solar maximum, was unusually warm around the world. Further, historical records and climate proxies show that solar minima and maxima correlate with precipitation around the world, including in the USA, Tibet, South America, India, China, Egypt, and elsewhere.

The first page of Chapter 6 contains a quote from a review paper by Connolly et al. (including Scafetta) that was published after the IPCC deadline but came to a different conclusion about the potential role of the sun in the warming period since 1850. The paper discusses the current uncertainties regarding both solar and climate data, and concludes that the data on past solar activity and climate changes:

> "suggest everything from no role for the Sun in recent decades (implying that recent global warming is mostly human-caused) to most of the recent global warming being due to changes in solar activity (that is, that recent global warming is mostly natural)".[31]

Thus, it appears that the conclusions presented in IPCC AR6 are consistent only with a portion of the published scientific literature, the portion that minimizes the role of the sun so as to maximize the anthropogenic component.

The exact mechanisms for the climate/solar correlation are unclear and the chapter lists and discusses several possible mechanisms. However, the correlation exists and for the IPCC to ignore it, and claim that modern climate change is 100% anthropogenic, simply because the solar connection cannot be explained is unacceptable.

## Climate Sensitivity to $CO_2$

One of the most important conclusions of the AR6 report was to reduce the uncertainty in estimates of climate sensitivity to doubling the amount of carbon dioxide in the atmosphere. The IPCC calls climate sensitivity "ECS," which stands for equilbrium or effective climate sensitivity. Since the Charney Report 1979[32], the *likely* range (66% chance) of climate sensitivity has been between 1.5°C and 4.5°C. This range has remained stubbornly wide, until the IPCC AR6 narrowed the *likely* range to be between 2.5°C and 4.0°C.

The report discusses the AR6 estimate of climate sensitivity to $CO_2$ in Chapter 7.[33] It explains that AR6 relied heavily on Sherwood et al. (2020), an important paper (Chapter 7 mentions the Sherwood paper 22 times) that was written by the who's who of the professional climate sensitivity community.[34] In earlier IPCC reports estimates for climate sensitivity relied heavily on climate model calculations, but the 'good news' is that the Sherwood et al paper (that was the main basis for the AR6 report), which narrowed the likely range to almost the same as the 2.5°C to 4.0°C AR6 range, did not.

31   Connolly et al., R. (2021). How much has the Sun influenced Northern Hemisphere temperature trends? Research in Astronomy and Astrophysics, 21(6). Retrieved from https://iopscience.iop.org/article/10.1088/1674-4527/21/6/131?fbclid=IwAR0U5WARVnuGVjj2qeiiBYgGo0IlxXb9NNzUbeqqN-th2Zp1YU8rLOZkrMM

32   Charney, J., Arakawa, A., Baker, D., Bolin, B., Dickinson, R., Goody, R., . . . Wunsch, C. (1979). *Carbon Dioxide and Climate: A Scientific Assessment*. National Research Council. Washington DC: National Academies Press. doi:https://doi.org/10.17226/12181

33   Chapter 7: Misty climate sensitivity

34   Sherwood, S. C., Webb, M. J., Annan, J. D., Armour, K. C., J., P. M., Hargreaves, C., . . . Knutti, R. (2020, July 22). An Assessment of Earth's Climate Sensitivity Using Multiple Lines of Evidence. Reviews of Geophysics, 58. doi:https://doi.org/10.1029/2019RG000678

The 'bad news' is that when independent scientist Nic Lewis redid the analysis of Sherwood et al (after the deadline of AR6) he discovered flaws in the statistics and shortcomings in the input data.[35] Lewis remedied these flaws and shortcomings and also revised certain key input data, almost entirely to reflect more recent evidence. The results of Lewis' analysis determined a *likely* range of 1.75 to 2.7°C for climate sensitivity. The central estimate from Lewis' analysis is 2.16°C, which is well below the IPCC AR6 *likely* range. This large reduction relative to Sherwood et al. shows how sensitive climate sensitivity estimates are to input assumptions. Lewis' analysis implies that climate sensitivity is more likely to be below 2°C than it is to be above 2.5°C.

The lower estimates of climate sensitivity determined by Nic Lewis have profound implications for climate models and projections of warming for the 21st century. Climate models used in the IPCC AR6 had values of climate sensitivity ranging from 1.8°C to 5.6°C. The IPCC AR6 judged that some of the climate models had values of climate sensitivity that were too high. Hence the AR6 selected only the climate models with reasonable values of climate sensitivity to be used in projections of 21st century climate change. Lewis' analysis indicates that a majority of climate models used in the IPCC AR6 have values higher than the *likely* range.

The report discusses more evidence that the IPCC climate sensitivity is too high and speculates that the estimate is too high due to incorrect IPCC assumptions about cloud cover. The IPCC admits that a "multitude of studies" imply that the AR6 ECS is too high but ignores the "multitude of studies" without explaining why. Or rather their explanation is the IPCC is correct and everyone else is wrong.

## Are climate models unreliable?

Ross McKitrick shows that the IPCC climate models compute global and tropical tropospheric air temperatures too high relative to observations.[36] This error appears in the model results from every model at a statistically significant level and invalidates the climate models. Since climate model projections are used to compute the future impact of climate change, this result invalidates the future projections as well.

Surprisingly, McKitrick found that if the impact of anthropogenic greenhouse gas emissions are removed from the climate models, the results match observations in the tropical troposphere much more closely. McKitrick also found that the AR6 model results, which are higher than the previous (AR5) results, universally overestimate *global* average temperature as shown in figure 9 below for the lower troposphere. The red dots and error ranges (95% confidence intervals) are the model results for 38 models and the blue dots and error ranges are observations from three data sources.

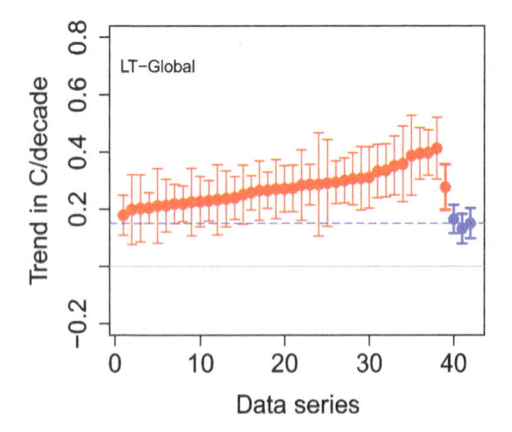

**Figure 9:** Model results in red, and observations in blue, with 95% confidence intervals indicated. The red dot and bar most right is the average of all the models. The Y axis is the warming trend in °C/decade for 1979-2014 for Earth's global lower troposphere.

---

35   climate sensitivity evidence. Climate Dynamics.
     Retrieved from https://link.springer.com/article/10.1007/s00382-022-06468-x?mc_cid=6760f55b0f&mc_eid=133f53df

36   Chapter 8: AR6: More confidence that models are unreliable

As McKitrick tells us in the conclusions of Chapter 8:

> If the discrepancies in the troposphere were evenly split across models between excess warming and cooling, we could chalk it up to noise and uncertainty. But that is not the case: it's all excess warming. The AR5/CMIP5 models warmed too much over the sea surface and too much in the tropical troposphere. Now the AR6/CMIP6 models warm too much throughout the global lower- and mid-troposphere. That's bias, not uncertainty, and until the modeling community finds a way to fix it, the economics and policy making communities are justified in assuming future warming projections are overstated, potentially by a great deal depending on the model.

## The Climate Change Scenarios

[...] Crok takes a close look at the $CO_2$ human emissions scenarios used by the IPCC to predict [...] temperatures and climate.[37] He finds that the IPCC admission that the higher emissions sce-[...], SSP5-8.5 and SSP3-7.0 are unlikely is deeply buried in the report, and unlikely to be read [...] policy makers. In addition, he finds that significant and important sections still emphasize [...] too-high unlikely scenarios, potentially invalidating those sections of the report.

[...]as serious implications from a policy standpoint. Figure 10 compares the various emissions [...]rios and the resulting projected temperatures.

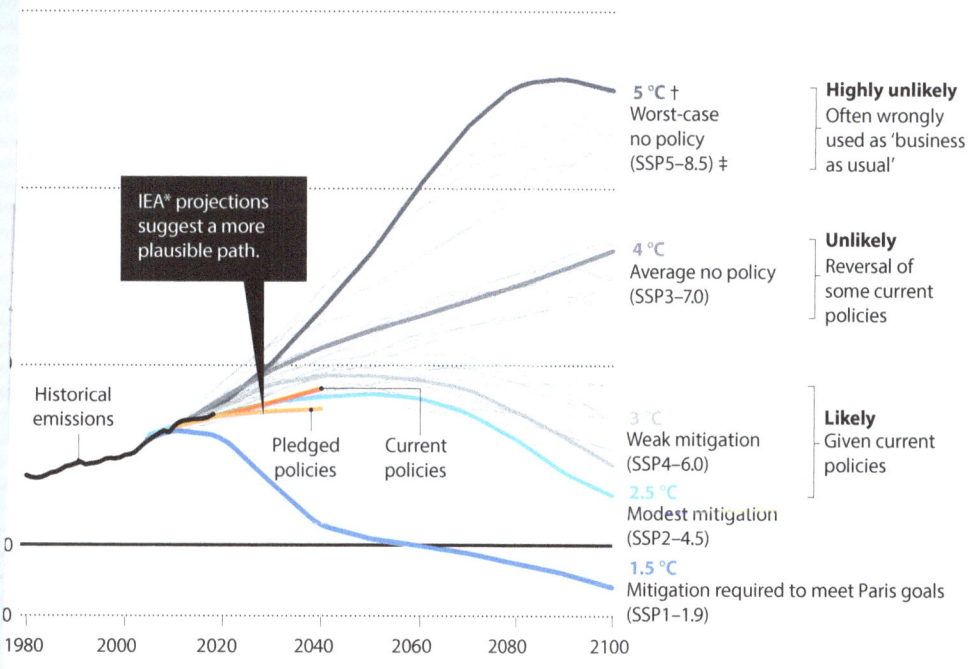

**e 10:** Various emissions scenarios projected into the future. This plot is from an article by Hausfather and Peters in *Nature* in [...] The assessments of likelihood on the right are from Hausfather and Peters.

[...]re 10 makes it clear that the extreme IPCC emissions scenarios are unlikely and should be [...]sidered unrealistic, academic extremes. Crok reports that 42% of the AR6 mentions of sce-[...]ios are of the most unlikely SSP5-8.5 scenario. Given that numerous authors have called this [...]nario "highly unlikely," this causes AR6 to lose credibility.

## [...]ding good news on extremes

The final part of the report is on the human impacts due to climate change. It starts with a chapter by Marcel Crok on hiding the good news.[38] He points out that AR6 claims that climate is becoming more extreme with time, but the data suggests this is not the case in most categories of climate (or

---

37   Chapter 9: Extreme scenarios

38   Chapter 11: Hiding the good news on hurricanes and floods.

more accurately "weather") events. For example, deep inside the report the IPCC acknowledges that there is no trend in tropical cyclones and floods. Such extreme events cause about 90% of the global disaster losses, so it should be regarded as 'good news' that they show no increasing trend.

The longest available time series about landfall hurricanes is from the US. Although the graph below has been published in a peer reviewed paper it is not shown in any of the IPCC reports:

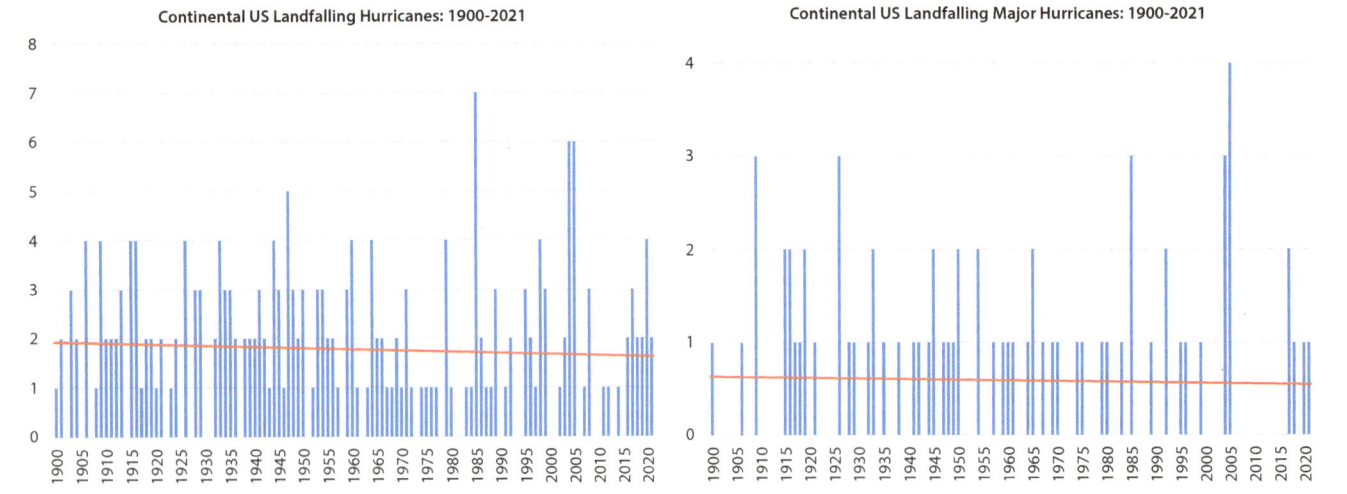

**Figure 11:** Number of US landfalling hurricanes and major hurricanes between 1900 and 2021. Updated graph from Klotzbach 2018.[39]

Global cyclones and U.S. hurricanes are decreasing in frequency and strength in recent decades, not increasing. On (hydrological and meteorological) droughts, deep in AR6 WG1 (pages 1578-1579) Crok finds that the authors have low confidence that human activities have contributed to those droughts at a regional scale. They do conclude that heat waves have increased globally since 1950, but the warmest years in the U.S. were in the 1930s, so this may be a result of the time period chosen.

Crok points out some serious contradictions between various parts of AR6 on this subject, especially between the WG1 and WG2 reports. The full WG1 report states that the IPCC has low confidence that humans have contributed to flooding, yet the *Summary for Policy Makers* says the opposite, they believe that human influence has increased "compound" flooding (WG1 SPM A.3.5).

Crok concludes that the full AR6 WG1 report, except for the SPM, did a reasonably good job reporting on trends in extreme weather, however all the good news is buried and only the bad news is brought forward to the *Summary for Policy Makers*. In WG2 things really get worse, the IPCC even contradicts many of its own claims in the WG1 report.

## Disaster losses

In the next chapter (Chapter 12[40]) Crok continues discussing weather disasters and the possible attribution, if any, to human activities or emissions. In Chapter 12 he focusses on comparing the results of past disasters to today.

During the past century, the global population has increased from 2 to 8 billion people and people today are much more affluent than they were 100 years ago. Buildings, roads, and other infrastructure destroyed by extreme weather are much more valuable and numerous than in the past. Thus, comparing the dollar-to-dollar nominal cost of destruction today to in the past is invalid. Present costs must be adjusted for inflation, population growth, economic growth, and affluence. The adjustment is called normalization.

39    Klotzbach, Philip J., et al. „Continental US hurricane landfall frequency and associated damage: Observations and future risks." Bulletin of the American Meteorological Society 99.7 (2018): 1359-1376

40    Chapter 12: Extreme views on disasters

"The **secret of joy** is found in **charity**"

Crok examines the peer-reviewed literature on normalization of disaster costs from their beginning in a landmark paper by Roger Pielke Jr. in 1998.[41] Since this paper was written, more than 50 normalization studies have been published, and the technique has become established and routine. All 54 papers, except one, conclude that the costs associated with the extreme weather events they studied could *not* be attributed to human activities. Guess which one of the 54 studies is cited in AR6?

The earlier AR5 report acknowledges Pielke Jr.'s conclusions and restates them as follows:

> The 2014 IPCC assessment reinforced the conclusions of the IPCC (2012) special report on extreme events, providing even stronger evidence: 'There is medium evidence and high agreement that long-term trends in normalised losses have not been attributed to natural or anthropogenic climate change' and 'Increasing exposure of people and economic assets has been the major cause of long-term increases in economic losses from weather- and climate-related disasters (high confidence)' (IPCC, 2014).

So, the earlier reports acknowledged that normalization of costs is needed and the increase in nominal costs was largely caused by increased exposure of people and assets. Crok concludes:

> with respect to the literature on disaster losses, the latest AR6 WG2 report was neither comprehensive, open and transparent (it ignored most of the published literature on the topic), nor objective (it cherry picked the few studies that claimed an increase of losses due to greenhouse gases while the majority of the published studies show the opposite, no increasing trend after normalisation of the data). This is very poor performance by the IPCC.

In this chapter it becomes clear that Roger Pielke Jr is a real 'Voldemort' for the IPCC. They do everything to ignore his work even though it is highly relevant. So, apart from his important review article about disaster losses, the IPCC also ignored another of his papers that shows an important graph of normalized global weather disaster losses as a percent of global GDP (figure 12, below):

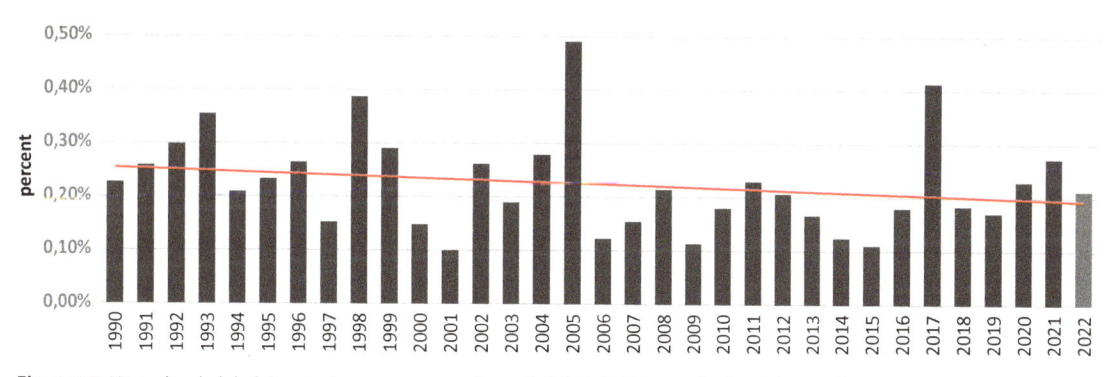

**Global Weather Losses as Percent of Global GDP: 1990-2022**
Sources: World Bank, IMF, Munich Re, Pielke 2019

**Figure 12:** Normalised global disaster losses as a percentage of global GDP. Source: Updated from Pielke (2019), from Pielke's website here.[42]

## Climate-related deaths

In Chapter 13[43], Crok examines UN Secretary-General Antonio Guterres statement that *"we are on a highway to climate hell with our foot on the accelerator"*. He made this claim during his speech to delegates of the COP27 conference in Egypt. Is there any truth in the statement?

41   Pielke Jr. & Landsea, (1998). Normalized hurricane damages in the United States: 1925–95, Weather and Forecasting, 13(3).

42   Pielke, R. (2019). Tracking progress on the economic costs of disasters under the indicators of the sustainable development goals. Environmental Hazards, 18(1), 1–6. https://doi.org/10.1080/ 17477891.2018.1540343

43   Chapter 13: Say goodbye to climate hell, welcome climate heaven

It seems there isn't. Crok explains that Bjorn Lomborg has shown that climate (strictly speaking "weather") related deaths have plummeted in the past 100 years from nearly half-a-million per decade in 1920 to a few thousand per decade today.[44] An astounding 96% decrease.

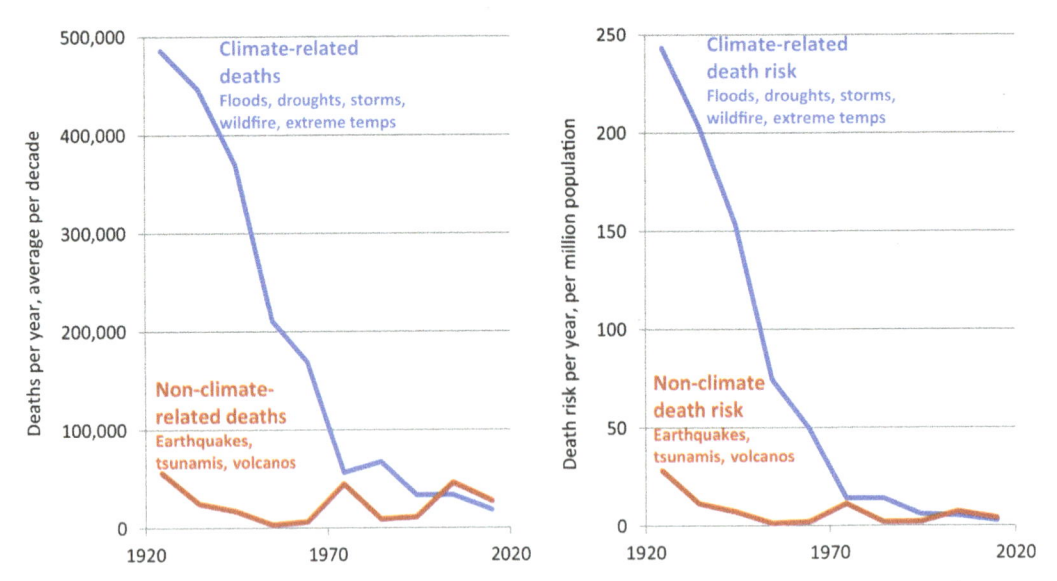

**Figure 13:** Climate and non-climate-related deaths and death risks from disasters 1920–2018, averaged over decades. Data comes from EM-DAT (2019), using floods, droughts, storms, wildfire, and extreme temperatures for climate-related deaths, and earthquakes, tsunamis, and volcanos for non-climate-related deaths. Source: Lomborg (2020).

Yet, this good news is not to be found in AR6, nor is there any mention of Lomborg's 2020 paper. Although most peer-reviewed papers cannot attribute any extreme weather events to human activities, the AR6 *Summary for Policy Makers* states:

> Increasingly since AR5, these observed impacts have been attributed to human-induced climate change particularly through increased frequency and severity of extreme events. These include increased heat-related human mortality (medium confidence) ... [AR6 WG2 SPM B.1.1]

This is the opposite of the conclusions reached by most researchers and previous IPCC reports. It is also opposite of what is written on page 2435 of the AR6 WG2 report, where we see:

> Formetta and Feyen (2019)[45] demonstrate declining global all-cause mortality and economic loss due to extreme weather events over the past four decades, with the greatest reductions in low-income countries, and with reductions correlated with wealth. (AR6 WG2 p 2435)

Although the main body of the WG1 reports has arguably improved since AR4, other IPCC reports (and the WG1 SPM) have continued to deteriorate in quality and increase in bias with time, as is evident to anyone who has read all of them. The first report (FAR) in 1990 was a reasonably fair assessment of climate science at the time, but the subsequent reports have become more biased with each passing year. No honest assessment of AR6 would conclude it is fair and unbiased, quite the opposite. The problems seem to be considerably worse in Working Group 2 than in the Working Group 1 report.

## Our summary

We started this summary with how you could summarize the IPCC view on climate change. In this report we have shown that many of the important claims of the IPCC – i.e., that current warming is unprecedented, that it is 100% caused by humans, that it is dangerous – are all questionable.

44  Bjorn Lomborg, (2020), Welfare in the 21st century: Increasing development, reducing inequality, the impact of climate change, and the cost of climate policies, Technological Forecasting and Social Change, Volume 156, 119981, ISSN 0040-1625, https://doi.org/10.1016/j.techfore.2020.119981.

45  Giuseppe Formetta, Luc Feyen, Empirical evidence of declining global vulnerability to climate-related hazards, Global environmental change, Volume 57, 2019, 101920, ISSN 0959-3780, https://doi.org/10.1016/j.gloenvcha.2019.05.004.

Based on the same available evidence the Clintel team would phrase a summary in the following way:

Warming in the Holocene likely peaked during the Holocene Thermal Maximum, when century-scale global temperatures were probably similar (within uncertainty) to those observed in the last decade. Decadal average temperatures are not available during the Holocene Thermal Maximum and proxy-temperatures, when averaged, reduce extremes. After which a slow cooling began, that follows the Milankovitch cycles. The cooling climaxed in the Little Ice Age, which was probably the coldest period of the Holocene. Greenhouse gases have likely contributed to the moderate modern warming since 1850. It is impossible to state with reasonable accuracy what percentage of the warming is due to greenhouse gases. Sea level started rising in the 19th century and there is no acceleration visible after 1950, the period in which the climate is supposedly dominated by greenhouse gases.

Moreover, most types of extreme weather have not become more frequent or more intense. This is especially true for tropical cyclones and floods, events that cause the most damage globally. Disaster losses, if normalized for economic development, show a slight decrease since the 1990s. Climate-related deaths show more than a 95% drop since the 1920s. This reflects increasing wealth and availability of technologies that better prepare humanity for disasters. In short, a prosperous humanity is largely prepared for climate change and can easily cope with it.

*Marcel Crok and Andy May*
*May 2023*

# A
## Observations

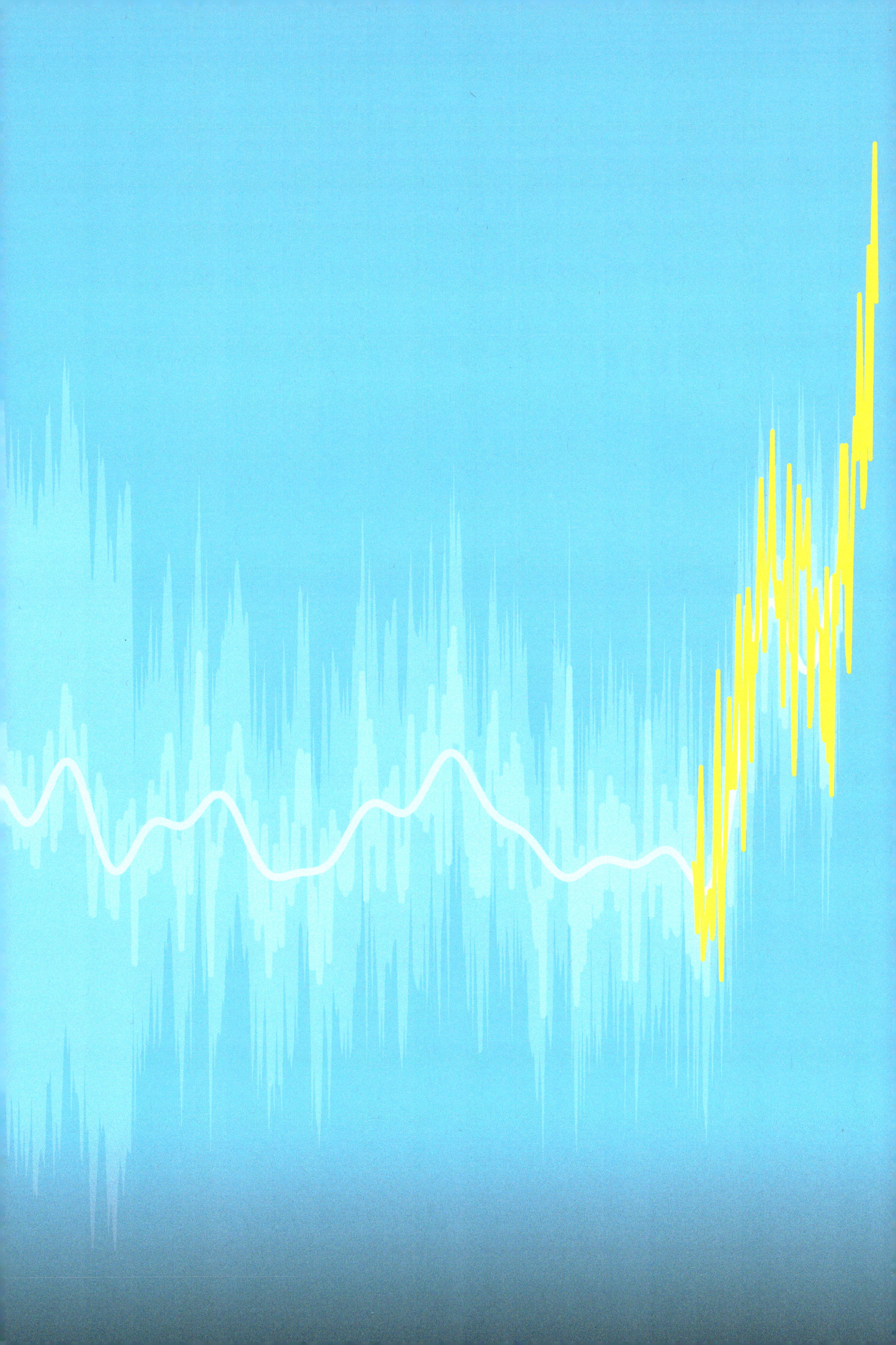

# 1 No confidence that the present is warmer than the Middle Holocene

BY JAVIER VINÓS

**The new IPCC report came with the remarkable claim that it is now warmer than in the past 125000 years. This would mean it is now warmer than during the Mid Holocene, a period that was also relatively warm. Javier Vinós investigates the evidence for this and shows that glaciers and treelines contradict this evidence. It is more likely that the Holocene Thermal Maximum was warmer than it is now.**

S everal of the IPCC's Assessment Reports have been preceded by controversial new past temperature reconstructions. The latest IPCC report, AR6 (WG1), includes the surprising new position that present "global surface temperatures are *more likely than not* unprecedented in the past 125,000 years" (IPCC AR6, Gulev et al. 2021; Fig. 1[1]), and that "it is therefore *more likely than not* that no multi-centennial interval during the post-glacial period was warmer globally than the most recent decade" (IPCC AR6, Gulev et al. 2021; 2.3.1.1.2). It is expected that such an extraordinary claim, that breaks the tradition of considering the Holocene Thermal Maximum (HTM, previously known as the Holocene Climatic Optimum, or Altithermal) the warmest period of the Holocene, must be based on extraordinary evidence. However, it is based on the work of a group of authors (Kaufman et al. 2020) that have performed a new multi-proxy reconstruction.

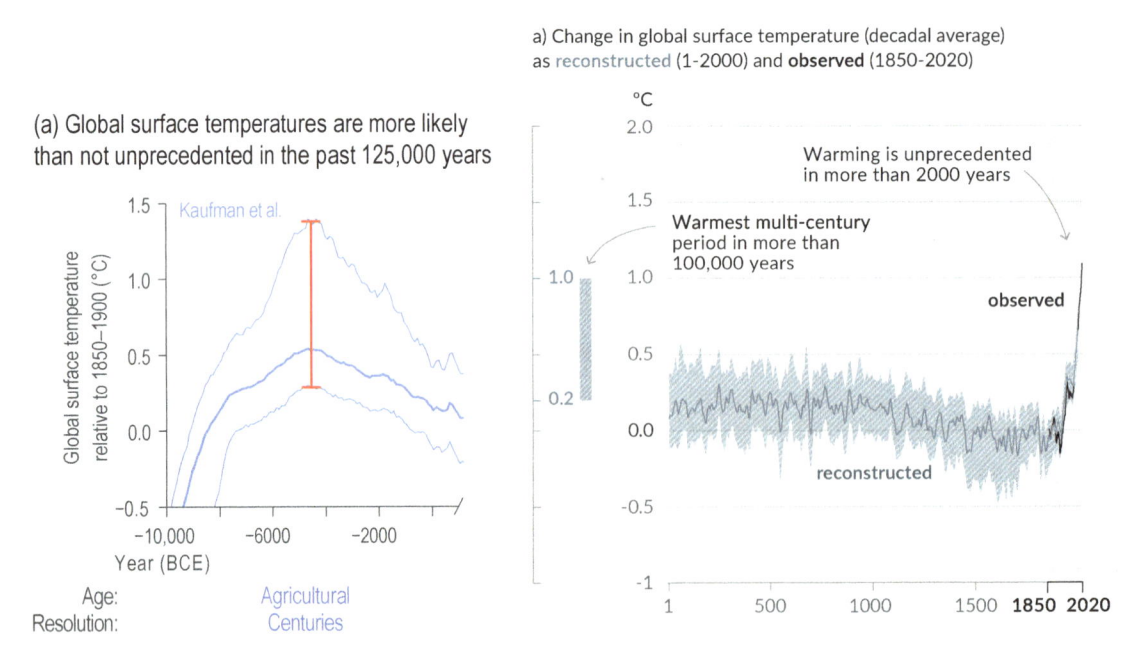

**Figure 1:** Evidence for the IPCC AR6 claim that present temperature is unprecedented in 125,000 years. Left, part of AR6 WG1 figure 2.11a based on Kaufman et al. 2020 with the uncertainty at the mid-Holocene highlighted with a red bar (added). Right, part of AR6 figure SPM.1 at the same temperature scale showing that present temperature is included in mid-Holocene uncertainty. The claim is based on a statistically weak medium confidence range (grey vertical bar at right figure) from a single multi-proxy reconstruction.

## Proxy-Based Temperature Reconstructions

Trying to gauge past global temperatures from proxy reconstructions is a task so full of uncertainties and potential problems that most authors will not attempt it. Proxies do not record temperatures but physical, chemical, or biological processes that are affected by temperature. When con-

1    Gulev SK, Thorne PW, Ahn J et al (2021) Changing State of the Climate System. In Masson-Delmotte V, Zhai P, Pirani SL et al (eds) Climate Change 2021: The Physical Science Basis. Contribution of Working Group I to the Sixth Assessment Report of the Intergovernmental Panel on Climate Change. Cambridge University Press.

verting a proxy into temperature there are many things that can go wrong, and the researcher won't know it. The proxy concentrates into a small interval (e.g., a few cm in a speleothem), non-uniformly, what has happened to the proxy at a single location over thousands of years. There are uncertainties in the dating of that interval, uncertainties about how other environmental changes have affected the proxy over that long time (e.g., how precipitation changes affect tree-ring width), uncertainties about a possible proxy non-linear response to temperature changes, the proxy might respond to temperature changes at some times and not others (e.g., growing season for plants). These are only some of the known unknowns, there are unknown unknowns affecting proxies.

Even if the researcher does his best there is a great uncertainty in the conversion of the changes the proxy has recorded at a specific location into a set of temperatures. Averaging several dozens or a few hundreds of such proxy-derived temperature records from different locations into a multi-proxy reconstruction adds new uncertainties. Some proxies record huge local changes that might seriously bias the average. For example, the change in the position of the Inter-Tropical Convergence Zone that took place in the Middle Holocene due to changes in insolation, resulted in an estimated change of 5°C over the course of a century in a proxy from the West African coast (core 658C; de Menocal et al. 2000[2]). This is a known case of an extreme local effect, but any multi-proxy average is bound to include such "locally biased proxies" that unknowingly bias the average at different times.

A multi-proxy reconstruction thus depends greatly on researcher choices. One outstanding example is the Holocene temperature reconstruction by Marcott et al. 2013[3]. In his doctoral dissertation (Marcott 2011[4]), Shaun Marcott reconstructed Holocene temperature changes using 73 proxies. His reconstruction displayed a weak uptick of +0.2°C after 1750, as his dissertation figure 4.3a shows (our Fig. 2a). However, when his reconstruction was published in *Science* just ahead of AR5 (Marcott et al. 2013), the reconstruction displayed a huge uptick of +0.8°C in the last 250 years, as can be appreciated in their figure 1c (our Fig. 2b). The difference was the result of decisions taken by Marcott et al. on how to perform the reconstruction from the same proxies. The press release by the National Science Foundation that funded the study highlighted the magnitude of the final warming resulting in the news media transmitting a misleading message to the public (Fig. 2c).

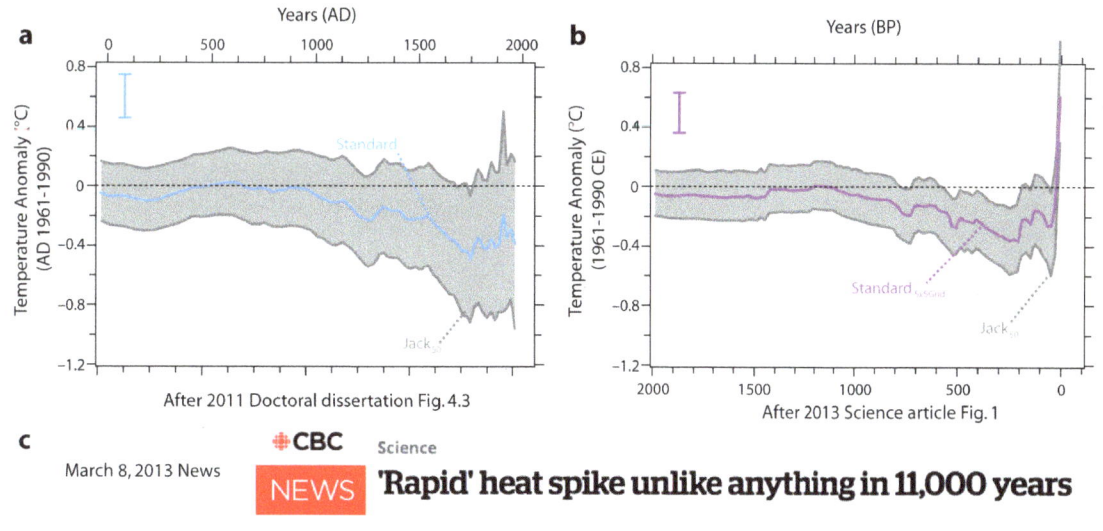

**Figure 2:** Researchers' decisions can greatly affect multi-proxy reconstructions. a) The final 2000 years of a multi-proxy Holocene global temperature reconstruction, after Marcott 2011 doctoral dissertation figure 4.3a. b) The final 2000 years of a reconstruction with the same proxies, after Marcott et al. 2013 figure 1c. c) An example of the media reaction to the press release ahead of Marcott et al. 2013 publication.

2   de Menocal P, Ortiz J, Guilderson T and Sarnthein M (2000) Coherent high-and low-latitude climate variability during the Holocene warm period. Science 288 (5474) 2198-2202

3   Marcott SA, Shakun JD, Clark PU and Mix AC (2013) A reconstruction of regional and global temperature for the past 11,300 years. Science 339 (6124) 1198-1201

4   Marcott SA (2011) Late Pleistocene and Holocene glacier and climate change. Dissertation, Oregon State University. https://ir.library.oregonstate.edu/concern/graduate_thesis_or_dissertations/3484zm26f Accessed 11 Oct 2021

Multi-proxy reconstructions are useful, but biases and unavoidable limitations of the technique result in their inability to answer the IPCC question: Was the last decade the warmest the planet has been during the Holocene? The final result of a multi-proxy reconstruction is very dependent on multiple researcher's choices, starting with the proxies included and excluded, whether land and marine proxies are representative, and what their respective weight should be in the mix. Proxies have big intrinsic uncertainties many of which cannot be properly estimated. On top of that, global coverage is very low.

Attempting to measure the average temperature of the planet with a few hundred low-precision uncalibrated thermometers that provide a reading once a decade at best would be a laughable task, and proxies are not even thermometers. Comparing the resulting global average with our modern measurements, including satellites and thousands of high-precision calibrated thermometers distributed all over the world, including all the oceans, and then declaring we can trust that it is *more likely than not* that the past decade is warmer than any century during the past 12,000 years is an untenable claim.

## The Holocene Temperature Conundrum

The Holocene Thermal Maximum (HTM) can be considered a period that extended from c. 9800-5700 BP when temperatures varied considerably in many parts of the globe and maximal Holocene temperatures were reached in many areas, but often at different times. For example, in the Baltic Sea region the highest temperatures occurred c. 6500 BP and are estimated at 1.5–2.5°C above present-day values in the north-west area and 1.0–1.5 °C in the north-east (Borzenkova et al. 2015[5]). The HTM was certainly not a time of uniformly warm temperatures and trying to determine if at a certain time there was a global average, higher than the global average of the past decade has little scientific value, but apparently great political value. Regionally, most parts of the globe were warmer sometime during the HTM than now. Determining the global average temperature at any time in the HTM is not possible, we can only roughly determine the temperature, relative to today, in special locations with crude proxies.

At the core of the problem is that climate models do a poor job of reproducing a warmer past during the Holocene, given that $CO_2$ levels were much lower. This is known as the Holocene temperature conundrum (Liu et al. 2014[6]). Samantha Bova and colleagues tried to explain it was because of a more marked seasonality during the HTM that multi-proxy reconstructions fail to capture (Bova et al. 2020[7]). However non-biogenic mean annual-temperature reconstructions also underscore the conundrum (Affolter et al. 2019[8]) and point instead to strong latitudinal temperature gradients that models are unable to reproduce.

Another significant problem is that the oceans, which respond more slowly to temperature changes, appear to have cooled significantly since the HTM. The fossil coral Sr/Ca record at the Great Barrier Reef, Australia, shows that the mean SST (sea surface temperature) c. 5350 BP was 1.2°C warmer than the mean SST for the early 1990s (Gagan et al. 1998[9]). At the Indo-Pacific Warm Pool, the warmest ocean region in the world, Stott et al. (2004[10]) find that SST has decreased by c. 0.5°C in the last 10,000 years, a finding confirmed by Rosenthal et al. (2013[11]), who show a

5    Borzenkova I, Zorita E, Borisova O, et al (2015) Climate change during the Holocene (past 12,000 years). In The BACC II Author Team (eds) Second assessment of climate change for the Baltic Sea basin 25-49. Springer Cham.

6    Liu Z, Zhu J, Rosenthal Y et al (2014) The Holocene temperature conundrum. Proceedings of the National Academy of Sciences 111 (34) E3501-E3505

7    Bova S, Rosenthal Y, Liu Z et al (2021) Seasonal origin of the thermal maxima at the Holocene and the last interglacial. Nature 589 (7843) 548-553

8    Affolter S, Häuselmann A, Fleitmann D, et al (2019) Central Europe temperature constrained by speleothem fluid inclusion water isotopes over the past 14,000 years. Science advances 5 (6) eaav3809

9    Gagan MK, Ayliffe LK, Hopley D et al (1998) Temperature and surface-ocean water balance of the mid-Holocene tropical western Pacific. Science 279 (5353) 1014-1018

10   Stott L, Cannariato K, Thunell R et al (2004) Decline of surface temperature and salinity in the western tropical Pacific Ocean in the Holocene epoch. Nature 431 (7004) 56-59

11   Rosenthal Y, Linsley BK and Oppo DW (2013) Pacific Ocean heat content during the past 10,000 years. Science 342 (6158) 617-621

decrease of 1.5-2°C for intermediate waters. The tropics display a very reduced climate change compared to the rest of the globe, and if the tropical oceans were warmer at the HTM it would be difficult to claim that the Earth is warmer globally now.

## Glacier Advances

To solve the problem of comparing HTM and present temperatures we should avoid uncertain proxy blends and try to find indicators less prone to artifacts, even if they will not give us a quantitative response. One of these indicators is glacier changes. We know that on average glaciers were at their most reduced state for the past 100,000 years during the HTM and at their most expanded state for the past 7,000 years during the LIA. In most regions of the mid-high latitudes of the NH (Northern Hemisphere), glaciers were smaller than now between 8000-4000 BP (Solomina et al. 2015[12]).

In their outstanding work, Solomina et al. divided 189 glacier timelines into 17 regions: 12 from the NH, 1 from low latitudes, and 4 from the SH. Then they studied the major glacier advances at each region for each of the 118 centuries of the Holocene. Figure 3a shows how many of those regions were experiencing advances at each century. It is very interesting that their result essentially reads as a negative print of a global temperature anomaly change reconstruction, despite being completely independent.

For comparison a reconstruction using the same 73 proxies of Marcott et al. (2013) is made, so a new selection bias is not introduced, but with their originally published dates, and averaging them after expressing them as differences to their individual means to convert them into anomalies. This reconstruction (Fig. 3b, inverted) ends in 1910 due to lack of sufficient proxies afterward, so it does not include 20th century warming. The temperature anomaly is expressed as a Z-score, or distance to the mean Holocene temperature anomaly (upper horizontal straight dashed line in Fig. 3b), to avoid making inferences about actual Holocene temperatures that we cannot possibly know.

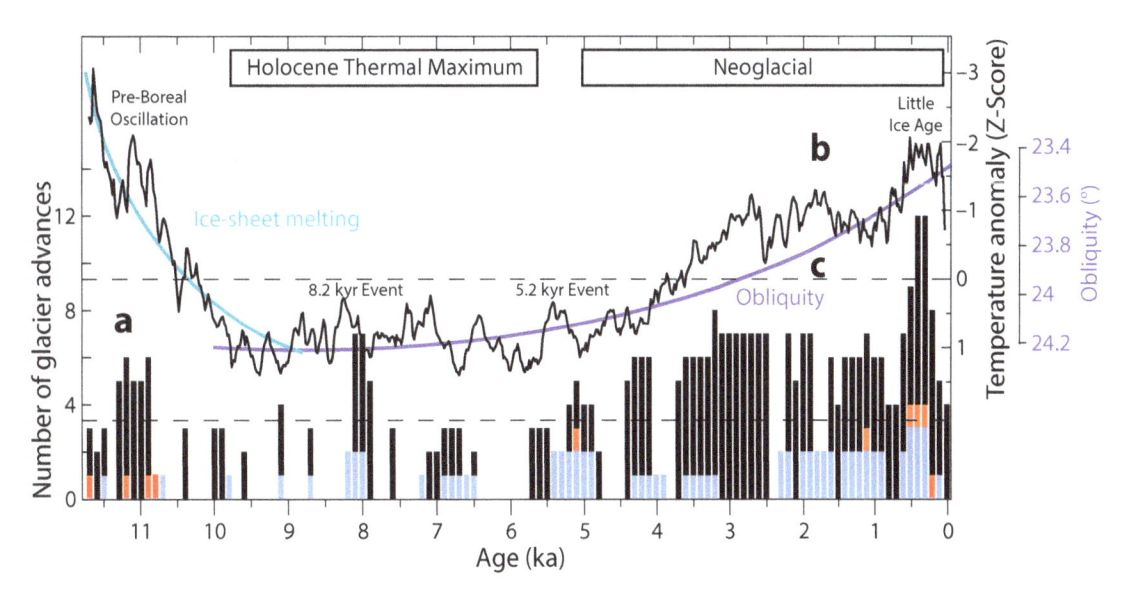

**Figure 3:** Glacial advances versus temperature (inverted) during the Holocene. a) Number of regions displaying glacier advances at each Holocene century. Black, NH 17 regions; orange, low latitudes single region; ice blue, SH 4 regions. Bottom dashed line, Holocene glacial position average. b) Inverted temperature reconstruction from the same 73 proxies used by Marcott et al. 2013. The reconstruction ends in 1910 and does not include modern warming. Temperature anomaly changes are expressed as Z-score (distance to the mean). Some well-known cooling periods or events are indicated by their accepted names. The period affected by the melting of the extra-Arctic ice sheets is indicated in aquamarine. c) Inverted changes in obliquity. Vinós, 2022, page 55.[13]

---

12   Solomina ON, Bradley RS, Hodgson DA et al (2015) Holocene glacier fluctuations. Quaternary Science Reviews 111 9-34

13   Vinós, J. (2022). Climate of the Past, Present and Future, A Scientific Debate. 2nd ed. Critical Science Press.

A comparison of centurial glacier advances to this multi-proxy reconstruction shows that both share a very similar general pattern. From the start of the HTM both follow the obliquity curve (Fig. 3c; Bosmans et al. 2015[14]), the main Milankovitch parameter for interglacials, as it is the only one that alters the insolation distribution between low and high latitudes. Not only that, but major Holocene cooling events show clear correspondence, like the Pre-Boreal Oscillation, the 8.2 kyr event, the 5.2 kyr event that initiated the Neoglaciation, or the LIA; they all coincide with increases in glacier advances. Less conspicuous are warming periods like the Roman or Medieval Warm Periods, that can also be detected in both records (Solomina et al. 2015). The 20th century has 4 glacier-advancing regions in the NH in Figure 3. This is the right-most bar in Figure 3, and it is slightly above the Holocene average of 3.35 (Fig. 3, bottom dashed line).

Centurial glacier advances appear to reconstruct the general temperature evolution during the Holocene, lending strong support to the Holocene temperature conundrum. Temperature changes during the Holocene do not appear to follow $CO_2$ changes. This is strongly supported by implied large perturbations in ocean heat content and Earth's energy budget at odds with the very small radiative forcing anomalies throughout the Holocene (Rosenthal et al. 2017[15]).

AR6 recognizes that most NH glaciers are larger now than at the HTM (IPCC AR6, Gulev et al. 2021; section 2.3.2.3) but points towards the relatively long adjustment time of glaciers. That is true for large slow continental glaciers, however 80% of world glaciers are very small (with area $\leq 1$ km$^2$; Li et al. 2019[16]) and glaciers respond to mean annual temperature and precipitation at their surface, not to global warming. The glaciers that have reduced the most since 1980 are tropical glaciers, where warming is less intense. In the mid-high latitudes, where warming has been more intense (Li et al. 2019), the glacial retreat is less.

The great worldwide glacier retreat started around 1850. By 1950, before the fast increase in anthropogenic $CO_2$ started, 169 glaciers from different parts of the world had already reduced their length by 70% of the year 2000 total (Oerlemans 2005[17]). Glaciers are reducing due to a combination of factors that includes anthropogenic black carbon and debris accumulation, not just temperature increases. Since the extra-tropical NH has warmed the most due to modern global warming, the presence of multiple glaciers and small permanent ice patches (Koch et al. 2014[18]) in the NH that did not exist during the HTM is a strong argument that the HTM was warmer than the present.

## Treelines

Another independent means of assessing whether the HTM was warmer than the present is through biology. Trees grow on the slope of mountains up to a certain altitude—the treeline—above which they are unable to survive. Temperature is the primary control on treeline formation and maintenance (Körner 2007[19]) and consequently the treeline has been moving higher over the past century at over half the locations studied, while receding at only 1% (Harsch et al., 2009[20]). The advances have taken place mainly in the extra-tropical NH, where more warming has been experienced, and particularly in locations where winter warming has been stronger. This aspect is also important as it indicates that winter tree survival might be a limiting factor for treeline altitude, and not only growth-season mean temperature.

14    Bosmans JH, Hilgen FJ, Tuenter E and Lourens LJ (2015) Obliquity forcing of low-latitude climate. Climate of the Past 11 (10) 1335-1346

15    Rosenthal Y, Kalansky J, Morley A and Linsley B (2017) A paleo-perspective on ocean heat content: Lessons from the Holocene and Common Era. Quaternary Science Reviews 155 1-12

16    Li YJ, Ding YJ, Shangguan DH and Wang RJ (2019) Regional differences in global glacier retreat from 1980 to 2015. Advances in Climate Change Research 10 (4) 203-213

17    Oerlemans J (2005) Extracting a climate signal from 169 glacier records. Science 308 (5722) 675-677

18    Koch J, Clague JJ and Osborn G (2014) Alpine glaciers and permanent ice and snow patches in western Canada approach their smallest sizes since the mid-Holocene, consistent with global trends. The Holocene 24 (12) 1639-1648

19    Körner C (2007) The use of 'altitude' in ecological research. Trends in ecology & evolution 22 (11) 569-574

20    Harsch MA, Hulme PE, McGlone MS and Duncan RP (2009) Are treelines advancing? A global meta-analysis of treeline response to climate warming. Ecology letters 12 (10) 1040-1049

There are a plethora of studies all over the NH showing that the treeline was much higher than present during the HTM. In the Italian Alps, it was 400 m higher than today from 8.4 and 4 ka (Badino et al. 2018[21]), and 150-200 m higher between 9 and 2.5 ka in the Swiss Central Alps (Tinner & Theurillat 2003; Fig. 4[22]). In the Pyrenees it was 400 m higher than the current treeline (Cunill et al. 2012[23]). In the Swedish Scandes 600-700 m higher between 9.5-6.5 ka (Kullman 2017[24]). In the British Columbia it was 235 m higher from 10.6 to 7.5 ka (Pisaric et al. 2003[25]). In New Zealand's South Island, where mean annual temperatures were at least 1.5°C warmer than present in the Early Holocene, treelines were lower however, suggesting shorter and cooler summers (McGlone et al. 2011[26]).

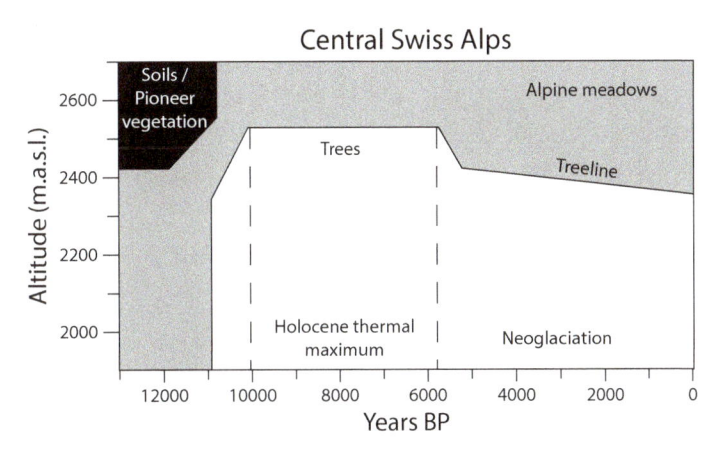

**Figure 4:** Fluctuations of the treeline in the Swiss Central Alps during the Holocene. The limits of the vegetational zones are placed between the sites according to the presence of the respective vegetation type as inferred by macrofossil analysis. Altitude in meters above sea-level. After Tinner & Theurillat 2003. Current treeline in the Swiss Central Alps is 150-200 m below Holocene Thermal Maximum treeline limit.

Randin et al. (2013)[27] showed that half of the 18 deciduous tree species they studied in Europe filled their thermal niche both at high latitude and high altitude, while 7 reached their latitudinal thermal limit, but not their elevational limit, where competition for space is stronger. The NH is the region that has experienced the strongest recent climate warming. Since so many NH species are at their thermal equilibrium, and yet at a great distance from their HTM limits, it is obvious that the planet cannot be warmer now, irrespective of conclusions reached via proxy reconstructions, temperature database kriging, and data homogenizing.

## Instrumental Temperature Changes Uncertainty

We do not know the HTM global average temperature, with enough confidence. Also, we cannot have much confidence in present global average temperature measurements. The continuous adjustments made to current global temperature datasets demonstrate how immature that data is.

This immaturity underscores a more worrisome problem. Figure 5 is an overlay of two GISS global temperature graphs, one from 2001 and the other from 2015. With the adjustments, the year 2000 became 0.4°C warmer with respect to 1880 than previously. Ole Humlum has been tracking these

21    Badino F, Ravazzi C, Valle F et al (2018) 8800 years of high-altitude vegetation and climate history at the Rutor Glacier forefield, Italian Alps. Evidence of middle Holocene timberline rise and glacier contraction. Quaternary Science Reviews 185 41-68

22    Tinner W and Theurillat JP (2003) Uppermost limit, extent, and fluctuations of the timberline and treeline ecocline in the Swiss Central Alps during the past 11,500 years. Arctic, Antarctic, and Alpine Research 35 (2) 158-169

23    Cunill R, Soriano JM, Bal MC et al (2012) Holocene treeline changes on the south slope of the Pyrenees: a pedoanthracological analysis. Vegetation history and archaeobotany 21 (4) 373-384

24    Kullman L (2017) Further details on holocene treeline, glacier/ice patch and climate history in Swedish Lapland. International Journal of Research in Geography 3 (4) 61-69

25    Pisaric MF, Holt C, Szeicz JM et al (2003) Holocene treeline dynamics in the mountains of northeastern British Columbia Canada inferred from fossil pollen and stomata. The Holocene 13 (2) 161-173

26    McGlone MS, Hall GM and Wilmshurst JM (2011) Seasonality in the early Holocene: Extending fossil-based estimates with a forest eco-system process model. The Holocene 21 (4) 517-526

27    Randin CF, Paulsen J, Vitasse Y et al (2013) Do the elevational limits of deciduous tree species match their thermal latitudinal limits?. Global Ecology and Biogeography 22 (8) 913-923

changes to GISS since 2008 (Climate4you 2021). At the very least these changes demonstrate that the confidence intervals of modern instrumental measurements mean very little and are only valid until the next major adjustment, therefore we do not really know how much the world has warmed since pre-industrial times with sufficient certainty.

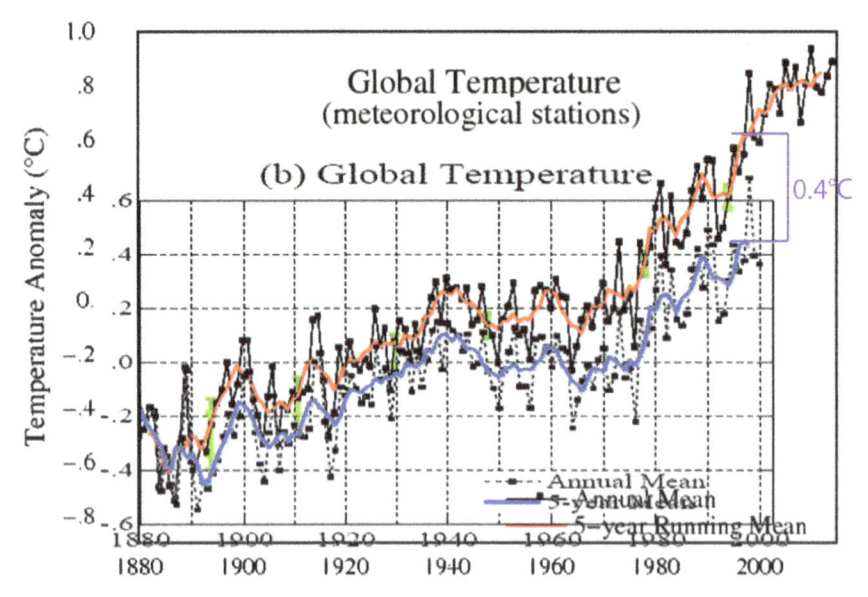

**Figure 5:** Overlay of GISS global temperature graphs from 2001 and 2015. Changes introduced in the dataset resulted in a 0.4°C increase in the 5-yr running mean from 1880 to 2000 between the 2001 data (blue curve) and the 2015 data (red curve).

In conclusion, there is too much uncertainty in proxy reconstructions and instrumental temperature datasets to sustain with any degree of confidence that the present is warmer than the Holocene Thermal Maximum, and independent evidence from glacier and treeline changes supports the opposite assessment.

# 2

# The Resurrection of the Hockey Stick

CLINTEL TEAM

**A big surprise in the new IPCC report is the publication of a brand new hockey stick. The IPCC once again has to cherry pick and massage proxy data in order to fabricate it. Studies that show larger natural climate variations are ignored.**

One of the big surprises of the IPCC's AR6 report was the comeback of the so-called "hockey stick". This term refers to the northern hemispheric and global temperature development of the past 1000-2000 years. More than two decades ago, Mann et al. (1999)[1] published a reconstruction in which the temperatures of the pre-industrial period 1000-1850 AD appear rather flat and uneventful (the "shaft" of the ice hockey stick), followed by a fast and allegedly unprecedented warming since 1850 (the "blade"). The hockey stick became world famous because it was featured prominently in the *Summary for Policymakers* (SPM) in the IPCC's 3rd Assessment report, TAR (Fig. 1). Subsequently, the work of Mann et al. (1999) was heavily criticized for major deficiencies in paleoclimatic proxies and statistical processing (McIntyre and McKitrick, 2003[2], 2005[3]; McShane and Wyner, 2011a, b[4]; Montford, 2010[5]).

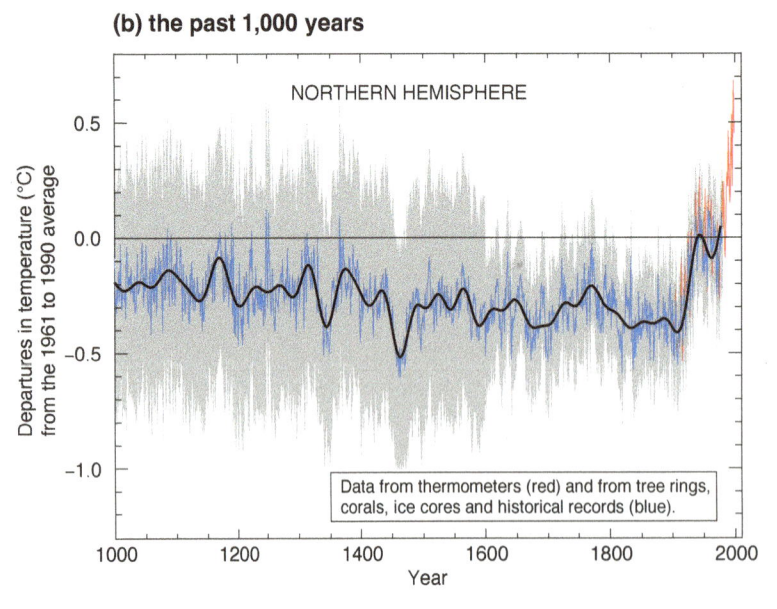

**(b) the past 1,000 years**

NORTHERN HEMISPHERE

Data from thermometers (red) and from tree rings, corals, ice cores and historical records (blue).

**Figure 1:** The original hockey stick by Mann et al. (1999) as illustrated in the Summary for Policymakers in the 3rd IPCC climate assessment report in 2001.

Interestingly, it was the group led by Michael E. Mann itself that partly corrected the hockey stick nearly a decade later (Mann et al., 2008).[6] In this new version the Medieval Warm Period (MWP, 800-1200 AD) was warmer than in the version of Mann et al. (1999). Two years later, Ljungqvist

1   Mann, M. E., Bradley, R. S., and Hughes, M. K., 1999, Northern Hemisphere Temperatures during the past Millennium: Inferences, Uncertainties, and Limitations: Geophysical Research Letters, v. 26, no. 6, p. 759-762.

2   McIntyre, S., and McKitrick, R., 2003, Corrections to the Mann et al. (1988) proxy data base and northern hemispheric average temperature series: Energy & Environment, v. 14, no. 6, p. 751-771.

3   McIntyre, Stephen and Ross McKitrick (2005a) "The M&M Critique of the MBH98 Northern Hemisphere Climate Index: Update and Implications." Energy and Environment 16(1) pp. 69-100; (2005b) "Hockey Sticks, Principal Components and Spurious Significance" Geophysical Research Letters Vol. 32, No. 3, L03710 10.1029/2004GL021750 12 February 2005.

4   McShane, B. B., and Wyner, A. J., 2011a, Rejoinder: The Annals of Applied Statistics, v. 5, no. 1, p. 99-123. -, 2011b, A statistical analysis of multiple temperature proxies: Are reconstructions of surface temperatures over the last 1000 years reliable?: The Annals of Applied Statistics, v. 5, no. 1, p. 5-44.

5   Montford, A. W., 2010, The Hockey Stick Illusion, London, Stacey International, 482 p.

6   Mann, M. E., Zhang, Z., Hughes, M. K., Bradley, R. S., Miller, S. K., Rutherford, S., and Ni, F., 2008, Proxy-based reconstructions of hemispheric and global surface temperature variations over the past two millennia: PNAS, v. 105, no. 36, p. 13252-13257.

(2010)[7] published another reconstruction in which the Little Ice Age (LIA, 1400-1850 AD) was colder than in the papers of the Mann group. This further increased the temperature difference between MWP and LIA, essentially eliminating the hockey stick shape by significantly deforming the "shaft". Another few years later, the PAGES 2k Consortium (2013)[8] published a reconstruction in which parts of the first millennium were occasionally as warm as present-day (Fig. 2).

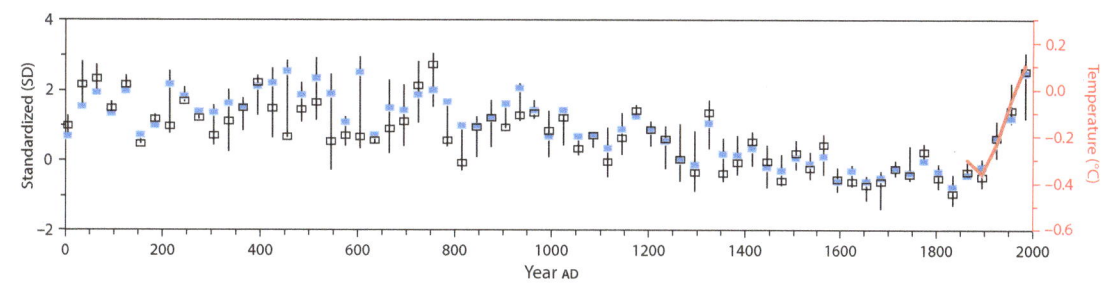

**Figure 2:** Global temperature reconstruction by PAGES 2k Consortium (2013)

The documented systematic pre-industrial warming and cooling presented a major challenge for the climate modellers because their simulations do not contain any powerful natural forcings that could produce such natural temperature changes. Hence, the climate models were not able to reproduce the real-world, reconstructed climate evolution.[9] This was a problem because the same, apparently deficient models are used for future temperature projections that form the basis for far-reaching political decisions and the costly transformation of global energy systems. Something was clearly wrong.

## Are humans 100% responsible for Modern Warming?

The discrepancy could be resolved by two possible solutions. In the first case, climate modelers could have added stronger natural forcings to their simulations, in order to replicate the documented pre-industrial climate change. However, this would mean that the warming effect of greenhouse gases likely would be reduced. That is because the temperature rise of the past 170 years would have to be shared with anthropogenic and natural causes. However, this was complicated, because in its special report on the 1.5°C target the IPCC had claimed in 2018 that 100% (!) of the observed modern warming was anthropogenic (IPCC, 2018).[10] Natural climate factors play no significant role, says the IPCC nowadays. This was a major shift for the IPCC because only five years earlier in its AR5 report, the organization still found it reasonable that "more than half" of the observed warming was man-made, leaving theoretically up to 49% to natural causes (IPCC, 2013). As natural contributions to modern warming have been essentially excluded, the IPCC opted for

7   Ljungqvist, F. C., 2010, A new reconstruction of temperature variability in the extra-tropical northern hemisphere during the last two millennia: Geografiska Annaler: Series A, v. 92, no. 3, p. 339-351.

8   PAGES 2k Consortium, 2013, Continental-scale temperature variability during the past two millennia: Nature Geosci, v. 6, no. 5, p. 339-346.

9   Büntgen, U., Krusic, P. J., Verstege, A., Sangüesa-Barreda, G., Wagner, S., Camarero, J. J., Ljungqvist, F. C., Zorita, E., Oppenheimer, C., Konter, O., Tegel, W., Gärtner, H., Cherubini, P., Reinig, F., Esper, J. (2017): New Tree-Ring Evidence from the Pyrenees Reveals Western Mediterranean Climate Variability since Medieval Times: Journal of Climate 30 (14), 5295-5318.
    Wilson, R., Anchukaitis, K., Briffa, K. R., Büntgen, U., Cook, E., D'Arrigo, R., Davi, N., Esper, J., Frank, D., Gunnarson, B., Hegerl, G., Helama, S., Klesse, S., Krusic, P. J., Linderholm, H. W., Myglan, V., Osborn, T. J., Rydval, M., Schneider, L., Schurer, A., Wiles, G., Zhang, P., Zorita, E. (2016): Last millennium northern hemisphere summer temperatures from tree rings: Part I: The long term context: Quaternary Science Reviews 134, 1-18.
    Luterbacher, J., Werner, J. P., Smerdon, J. E., Fernández-Donado, L., González-Rouco, F. J., Barriopedro, D., Ljungqvist, F. C., Büntgen, U., Zorita, E., Wagner, S., Esper, J., McCarroll, D., Toreti, A., Frank, D., Jungclaus, J. H., Barriendos, M., Bertolin, C., Bothe, O., Brázdil, R., Camuffo, D., Dobrovolný, P., Gagen, M., García-Bustamante, E., Ge, Q., Gómez-Navarro, J. J., Guiot, J., Hao, Z., Hegerl, G. C., Holmgren, K., Klimenko, V. V., Martín-Chivelet, J., Pfister, C., Roberts, N., Schindler, A., Schurer, A., Solomina, O., Gunten, L. v., Wahl, E., Wanner, H., Wetter, O., Xoplaki, E., Yuan, N., Zanchettin, D., Zhang, H., Zerefos, C. (2016): European summer temperatures since Roman times: Environmental Research Letters 11 (2), 024001.
    Fernández-Donado, L., González-Rouco, J. F., Raible, C. C., Ammann, C. M., Barriopedro, D., García-Bustamante, E., Jungclaus, J. H., Lorenz, S. J., Luterbacher, J., Phipps, S. J., Servonnat, J., Swingedouw, D., Tett, S. F. B., Wagner, S., Yiou, P., Zorita, E. (2013): Large-scale temperature response to external forcing in simulations and reconstructions of the last millennium: Clim. Past 9 (1), 393-421.

10  IPCC, 2018, Special Report on global warming of 1.5 °C above pre-industrial levels and related global greenhouse gas emission pathways: http://www.ipcc.ch/report/sr15/.

the second option to solve the dilemma of the model vs. reality mismatch. Flattened pre-industrial temperatures would provide a much better fit with the modelling results.

The PAGES 2k group is specialised in climate reconstructions and back in 2013 was comprised of the majority of all active paleoclimatologists. In 2019, PAGES 2k published a new version of the temperature development of the past 2000 years (PAGES 2k Consortium, 2019)[11]. Surprisingly, it differed greatly from the predecessor version. Even though the database had only mildly changed, the pre-industrial part was now suddenly nearly flat again. The hockey stick was reborn, and the modelling discrepancy conveniently solved. At least it seemed so.

The IPCC must have been delighted to get rid of this problem. The new hockey stick was immediately incorporated into the AR6 report (IPCC, 2021). Oddly, it was included in the first order draft (FOD) of AR6 in May 2019, even though the paper by the PAGES 2k Consortium (2019) was published in July 2019 (Fig. 3). Reference in the FOD was made to "PAGES 2k Consortium, in revision". Clearly, some of the IPCC authors were already aware of the manuscript prior to publication and used it in the IPCC report, even though it had not fully passed the journal review.

Among the lead authors of AR6 chapter 2 is Darrell S. Kaufman who is a co-author of the new hockey stick in the PAGES 2k Consortium (2019). This is probably not a coincidence.

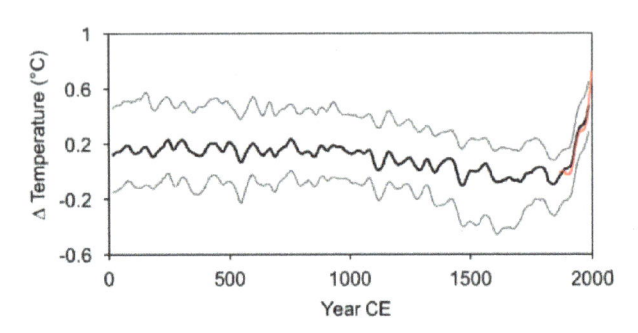

**Figure 3:** Global temperature development (the new hockey stick) as illustrated in the first order draft of chapter 2, working group 1, AR6, page 2-155. Reference is made to "PAGES 2k Consortium, in revision". Graph only shows palaeoclimatologiocally reconstructed, not instrumentally measured data.

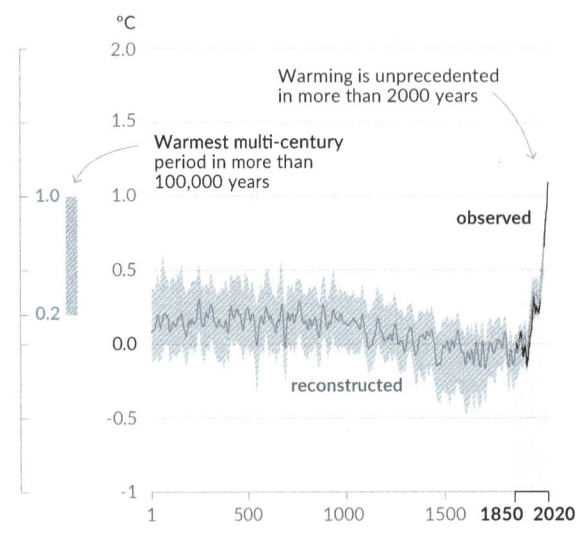

**Figure 4:** Global temperature development of the past 2000 years as illustrated in the final version of the Summary for Policymakers (SPM) of AR6, WG1, page SPM-7. Graph mixes reconstructed and instrumentally measured data and is based on PAGES 2k Consortium (2019).

---

11  PAGES 2k Consortium, 2019, Consistent multidecadal variability in global temperature reconstructions and simulations over the Common Era: Nature Geoscience, v. 12, no. 8, p. 643-649.

## University of Bern and the Hockey Stick

Another interesting role in the hockey stick saga may have been played by the climate researcher and manager Thomas Stocker of the University of Bern. Stocker has contributed to IPCC reports since 1998, and in 2015 he even ran for the IPCC's chairmanship, but was defeated by South Korean Hoesung Lee. Stocker appears to have never been far from the hockey stick, both the original one and the new one. Notably, Stocker co-authored the *Summary for Policymakers* of the IPCC's 3rd report in 2001, in which the Hockey Stick played a central role. Twenty years later, the resurrected hockey stick comes from PAGES 2k (Fig. 4), a group that is headquartered at the University of Bern, where Stocker chairs the climate and environmental physics department.

It cannot be ruled out that the new hockey stick was particularly commissioned for the 6th IPCC report. Five of the 19 authors of the new field hockey stick curve are from Bern (PAGES 2k Consortium, 2019).

Evidence suggests that a significant part of the original PAGES 2k researchers could not technically support the new hockey stick and seem to have left the group in dispute. Meanwhile, the dropouts published a competing temperature curve with significant pre-industrial temperature variability (Büntgen et al., 2020)[12] (EA and EA+ in Fig. 5). On the basis of thoroughly verified tree rings, the specialists were able to prove that summer temperatures had already reached today's levels several times in the pre-industrial past. However, the work of Ulf Büntgen and colleagues was not included in the IPCC report, although it was published well before the editorial deadline.

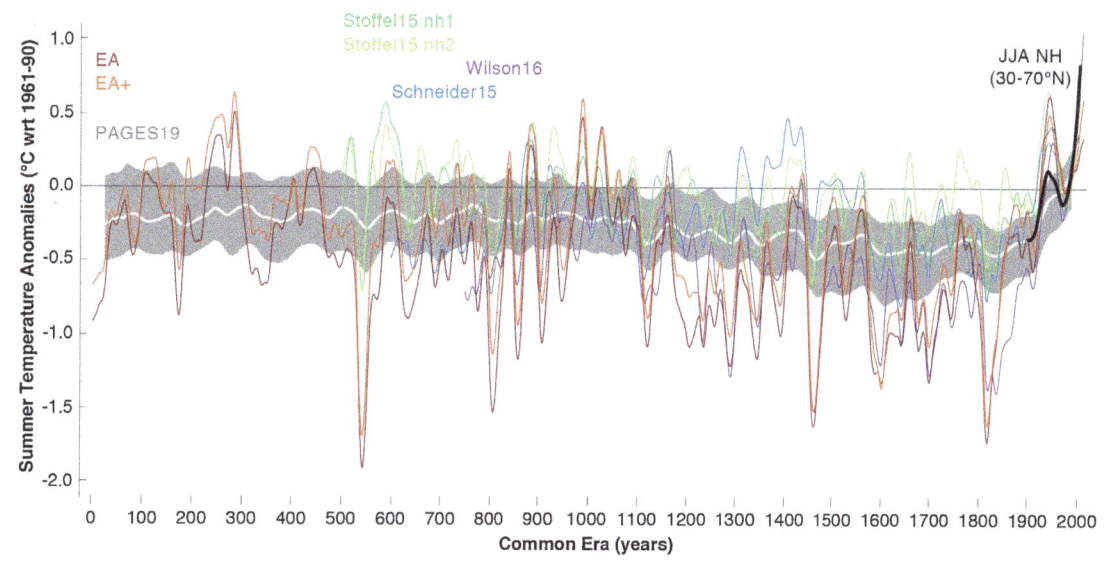

**Figure 5:** Temperature reconstruction for the extra-tropical northern hemisphere of the past 2000 years by Büntgen et al. (2020). The curves EA and EA+ in dark red and orange are from the Büntgen group, the other curves in green, blue and purple from other authors. The white line encased in grey is the new hockey stick by PAGES2k 2019 for comparison.

## How the Medieval Warm Period disappeared from AR6

The arbitrariness of the IPCC also becomes clear in another example. The IPCC explicitly listed the Medieval Climate Anomaly and the Little Ice Age in a summary table in chapter 1 of the first order draft of the AR6 report. (Fig. 6).

---

12 Büntgen, U., Arseneault, D., Boucher, É., Churakova, O. V., Gennaretti, F., Crivellaro, A., Hughes, M. K., Kirdyanov, A. V., Klippel, L., Krusic, P. J., Linderholm, H. W., Ljungqvist, F. C., Ludescher, J., McCormick, M., Myglan, V. S., Nicolussi, K., Piermattei, A., Oppenheimer, C., Reinig, F., Sigl, M., Vaganov, E. A., and Esper, J., 2020, Prominent role of volcanism in Common Era climate variability and human history: Dendrochronologia, v. 64, p. 125757.

| Period | Age/year* | Significance of climate state |
|---|---|---|
| Little Ice Age (LIA) | 1450–1850 CE (defined by AR5) | Series of globally heterogeneous cold periods lasting decades to centuries and including some of the lowest temperatures of the post-glacial period. |
| Medieval Climate anomaly (MCA) | 950–1250 CE (defined by AR5) | Loosely defined interval of relative warmth, especially prevalent in the circum North Atlantic region that preceded the LIA. |
| Last Millennium | 850–1850 CE (PMIP) or 1000-1999 CE | PMIP interval for transient climate model experiments. Encompasses the MCA and LIA, with demonstrable effects of volcanic and solar forcing. |

**Figure 6:** Explanations of Medieval Climate Anomaly and Little Ice Age in the first order draft of AR6, WG1 (chapter 1, page 1-71)

In the second order draft (SOD), the neutral term "Medieval Climate Anomaly" was even highgraded to the more classical "Medieval Warm Period", MWP (Fig. 7). The FOD claim that the MWP was a local circum—Atlantic phenomenon was dropped in the SOD, as a response to reviewer criticism. The description was now much improved compared to the FOD. The version in SOD was the last one that reviewers have seen and could comment on.

| Medieval Warm Period (MWP) | 950–1250 CE | Series of globally heterogeneous relatively warm periods lasting decades to centuries. Also known as, Medieval Climate Anomaly. | |
|---|---|---|---|
| Little Ice Age (LIA) | 1450–1850 CE | Series of globally heterogeneous relatively cold periods lasting decades to centuries and including some of the lowest temperatures of the post-glacial period. Coldest decades generally coincide with more frequent volcanic activity and low total solar irradiance. | |

**Figure 7:** Explanations of Medieval Warm Period and Little Ice Age in the second order draft of AR6, WG1 (chapter 2, page 2-10)

What followed was a big surprise. In the finally published version of the table all reference to the Medieval warming and Little Ice Age cooling was silently removed (Fig. 8). Instead the collective term "the last millennium" was introduced which downplays the significance of pre-industrial climate change. Three small asterisks explain to the reader equipped with reading glasses that one does not want to use, that the terms "Medieval Warm Period" and "Little Ice Age" were removed from the report because they are allegedly too poorly defined and too regionally variable. Once the manuscript moved beyond the review stage, the IPCC has apparently gone into full reverse gear. This is how the IPCC rewrites climate history behind closed doors, ignoring reviewers' comments. And hardly anyone in the public notices.

| Last millennium*** | 850-1850 CE | Climate variability during this period is better documented on annual to centennial scales than during previous reference periods. Climate changes were driven by solar, volcanic, land cover, and anthropogenic forcings, including strong increases in | 2.3.1.1.2 2.3.2.3 8.3.1.6 |
|---|---|---|---|
| | | greenhouse gasses since 1750. *PMIP4 past1000*, 850–1849 CE (Jungclaus et al., 2017) | 8.5.2.1 Box 11.3 |

*** The terms "Little Ice Age" and "Medieval Warm Period" (or "Medieval Climate Anomaly") are not used extensively in this report because the timing of these episodes is not well defined and varies regionally. Since AR5, new proxy records have improved climate reconstructions at decadal scale across the last millennium. Therefore, the dates of events within these two roughly defined periods are stated explicitly when possible.

**Figure 8:** Explanations of the "Last Millennium" in the final version of AR6, WG1 (chapter 2, pages 2-10 and 2-11)

The term "medieval" indeed no longer appears in chapter 2, in contrast to the FOD and SOD. Exceptions are the triple asterisk explanation and the titles of three recent MWP papers that are cited in the chapter (Lüning et al., 2019a; Lüning et al., 2018; Lüning et al., 2019b) (Fig. 9):

Lüning, S., M. Ga, I.B. Danladi, T.A. Adagunodo, and F. Vahrenholt, 2018a: Hydroclimate in Africa during the Medieval Climate Anomaly. *Palaeogeography, Palaeoclimatology, Palaeoecology*, **495**, 309–322, doi:10.1016/j.palaeo.2018.01.025.

Lüning, S., M. Ga, I.B. Danladi, T.A. Adagunodo, and F. Vahrenholt, 2018b: Hydroclimate in Africa during the Medieval Climate Anomaly. *Palaeogeography, Palaeoclimatology, Palaeoecology*, **495**, 309–322, doi:10.1016/j.palaeo.2018.01.025.

Lüning, S., M. Gałka, F.P. Bamonte, F.G. Rodríguez, and F. Vahrenholt, 2019a: The Medieval Climate Anomaly in South America. *Quaternary International*, **508**, 70–87, doi:10.1016/j.quaint.2018.10.041.

Lüning, S., L. Schulte, S. Garcés-Pastor, I.B. Danladi, and M. Gałka, 2019b: The Medieval Climate Anomaly in the Mediterranean Region. *Paleoceanography and Paleoclimatology*, **34(10)**, 1625–1649, doi:10.1029/2019pa003734.

**Figure 9:** The final version of chapter 2 cites four recent papers on the Medieval Climate Anomaly. Source: AR6, WG1, chapter 2, page 2-140.

## How robust is the new hockey stick?

Like its predecessor, the new hockey stick by PAGES 2k 2019 is based on a large variety of proxy types and includes a large number of poorly documented tree ring data. In many cases, the tree rings' temperature sensitivity is uncertain. For example, both PAGES 2k Consortium (2013) and PAGES 2k Consortium (2019) used tree ring series from the French Maritime Alps, even though tree ring specialists had previously cautioned that they are too complex to be used as overall temperature proxies (Büntgen et al. 2012[13]; Seim et al., 2012[14]).

In contrast, Büntgen et al. (2020) were more selective, relied on one type of proxy (in this case tree rings) and validated every tree ring data set individually. Their temperature composite for the extra-tropical northern hemisphere differs greatly from the studies that use bulk tree ring input.

In some cases, PAGES 2k composites have erroneously included proxies that later turned out to reflect hydroclimate and not temperature. In other cases, outlier studies have been selected in which the proxies exhibit an anomalous evolution that cannot be reproduced in neighbouring sites (e.g. MWP data from Pyrenees and Alboran Sea in PA13) (Lüning et al., 2019b[15]). Outliers can have several reasons, e.g. a different local development, invalid or unstable temperature proxies, or sample contamination.

Steve McIntyre has studied the PAGES 2k proxy data base in great detail and summarized his criticism in a series of blog posts on his website Climate Audit (McIntyre, 2021).[16] For example, the PAGES 2k Consortium (2019) integrated a tree ring chronology from northern Pakistan near Gilgit ("Asia_207") which shows an extreme closing uptick (Fig. 10). Incorporation of data series like this strongly promote the hockey stick geometry of the resulting temperature composite.

---

13 Büntgen, U., Frank, D., Neuenschwander, T., and Esper, J., 2012, Fading temperature sensitivity of Alpine tree growth at its Mediterranean margin and associated effects on large-scale climate reconstructions: Climatic Change, v. 114, no. 3, p. 651-666.

14 Seim, A., Büntgen, U., Fonti, P., Haska, H., Herzig, F., Tegel, W., Trouet, V., and Treydte, K., 2012, Climate sensitivity of a millennium-long pine chronology from Albania: Climate Research, v. 51, no. 3, p. 217-228.

15 Lüning, S., Schulte, L., Garcés-Pastor, S., Danladi, I. B., and Gałka, M., 2019b, The Medieval Climate Anomaly in the Mediterranean Region: Paleoceanography and Paleoclimatology, v. 34, no. 10, p. 1625-1649.

16 McIntyre, S., 2021,
https://climateaudit.org/2021/11/02/the-decline-and-the-stick/
https://climateaudit.org/2021/09/15/pages-2019-0-30n-proxies/
https://climateaudit.org/2021/09/02/pages19-0-30s/
https://climateaudit.org/2021/08/26/pages2019-30-60s/
https://climateaudit.org/2021/08/15/pages19-asian-tree-ring-chronologies/
https://climateaudit.org/2021/08/11/the-ipcc-ar6-hockeystick/.

PAGES2K Asia_207

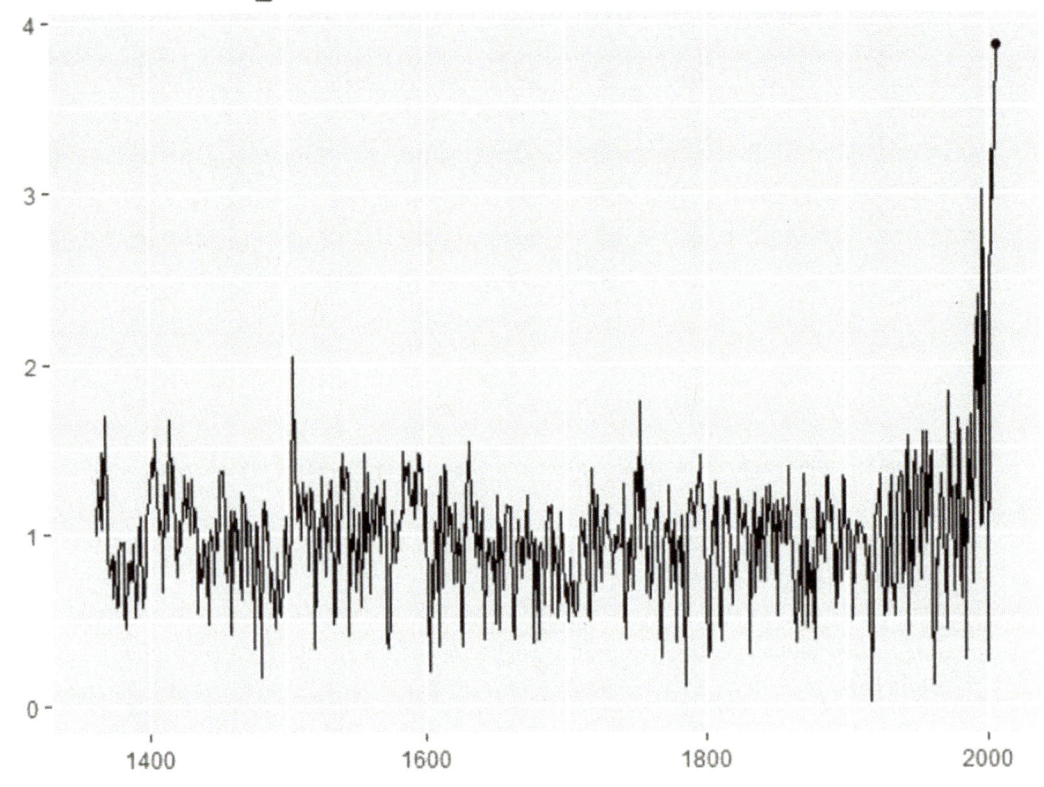

**Figure 10:** Hockey-stick-like tree ring chronology "Asia_207" as used by PAGES 2k Consortium (2019). Figure from McIntyre (2021).

McIntyre analysed the original tree ring data and found that the steep uptick in the Asia_207 chronology is the result of questionable data processing. When calculating the site chronology using the rcs function from Andy Bunn's dplR package, the uptick surprisingly disappears. In fact, the series declines over the 20th century (Fig. 11).

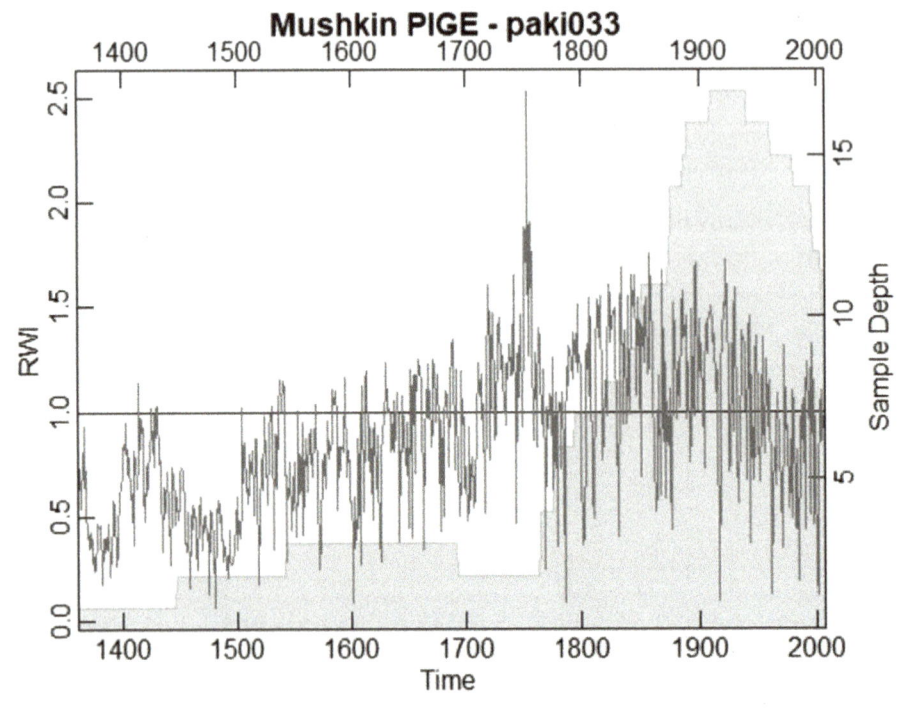

**Figure 11:** Tree ring chronology "Asia_207" calculated using the dplR function. Figure from McIntyre (2021).

An almost identical chronology to Figure 11 is also achieved by fitting a single Hugershoff curve to allow for growth prior to chronology calculation (Fig. 12).

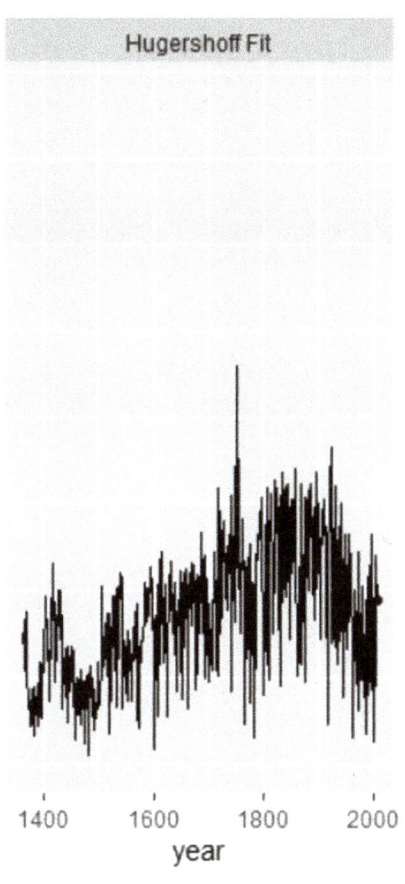

**Figure 12:** Tree ring chronology "Asia_207" calculated by fitting a single Hugershoff curve. Figure from McIntyre (2021).

## Conclusion

The resurrected hockey stick of AR6 shows how vulnerable the IPCC process is to scientific bias. Cherry picking, misuse of the peer review process, lack of transparency, and likely political interference have led to a gross misrepresentation of the pre-industrial temperature evolution. Neutrality, scientific robustness and reliability of the IPCC and the organization's quality assurance process has to be questioned.

# 3

# Measuring Global Surface Temperatures

BY ANDY MAY

**The Global Mean Surface Temperature has become the iconic parameter/ indicator in the climate change debate. Political climate targets – like the Paris targets of 2 and 1.5 degree Celsius – are determined by it. Is this deserved, how reliable are these temperature measurements and are there 'better' alternatives? Andy May takes a deep dive into the temperature records.**

C hapter 2 of the sixth IPCC assessment report on climate change (AR6) asserts that "Global Mean Surface Temperature (GMST) is a key indicator of the changing state of the climate system."[1] While surface temperature helps define a climate state globally, it also helps define local and regional climates that are arguably more relevant to the people in those areas. Historically, climate is a term used to describe the long-term weather trend for a specific region. One might say, for example, that northern Europe, is wetter and warmer now than previously. In recent decades, though, we have begun to talk about a "global" climate.

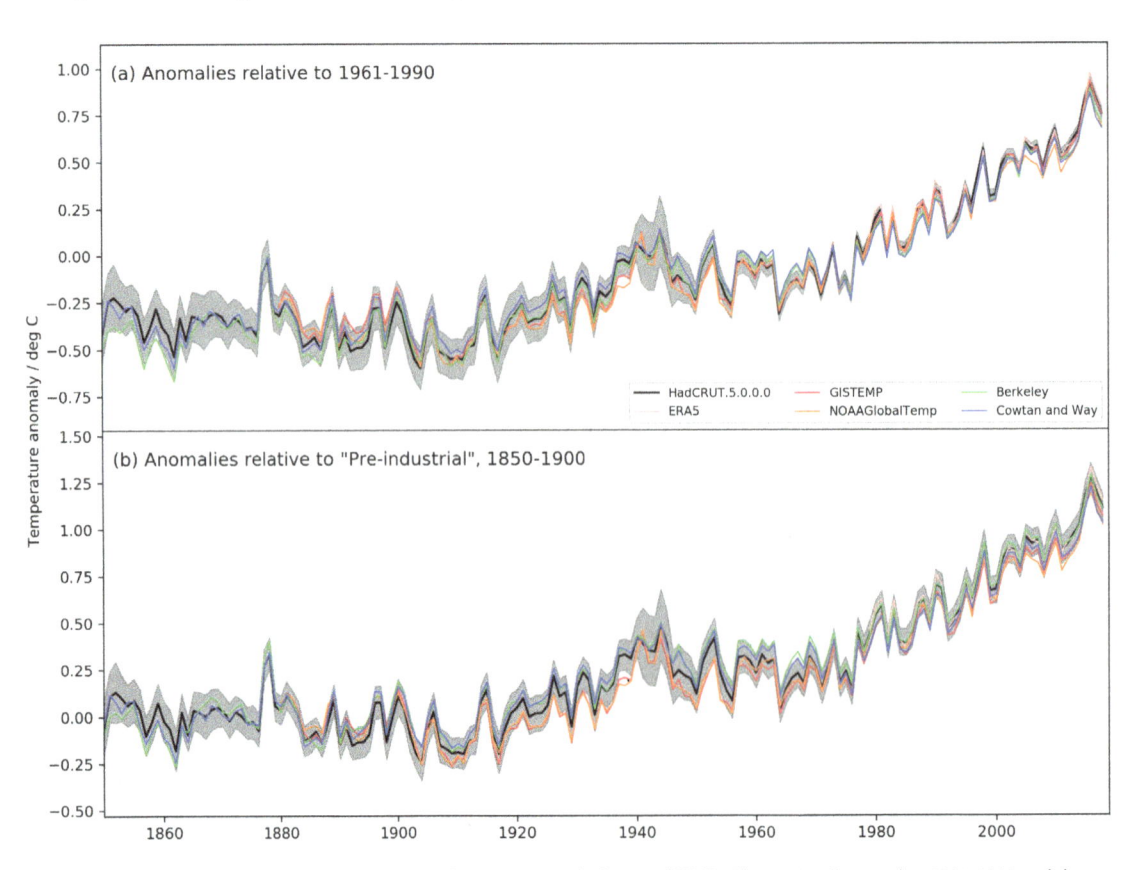

**Figure 1:** Various global surface temperature anomaly estimates, including HadCRUT5. The top is referenced to 1961-1990 and the bottom to the "pre-industrial period," also called the Little Ice Age. These estimates suggest about one degree of warming since 1900. Plot is from the Met Office Hadley Center web page.[2]

GMST is one indicator of climate change, whether global or local. But climate change is a long-term thing, and it has varied a lot over time. Figure 1 shows six estimates of GMST, and the estimates vary little from 1960 to the present day. The measured temperatures have been converted to "anomalies" by subtracting them from a chosen reference period. The top graph subtracts the local average from 1960-1990 from each temperature, the bottom uses an assumed "pre-industrial" or Little Ice Age

---

1  AR6, p. 294
2  Met Office Hadley Centre observations datasets

temperature. This is done to make the temperatures comparable and averageable; it is an attempt to make an intrinsically intensive local temperature measurement into an approximation of an extensive property. The actual measured global average surface temperature is not very meaningful, different parts of the Earth have different trends and elevations, some are warming, some are cooling, and all at different rates. As Figure 1 shows, the rate of global surface temperature change today, around one degree per century, is small relative to local temperature swings. In July, the global average surface temperature might be changing a few hundredths of a degree per annum, but the average low temperature in July at Vostok Station, Antarctica is −70°C and in Doha, Qatar the average high is 41°C. What does an average of −70 and 41 tell us about July climate change? Not much.

Figure 1 tries to suggest our estimates of global warming are accurate since all the lines are close together, but they share the same raw data and use very similar methods to correct and "homogenize" the data. While it is possible to calculate a reasonable GMST anomaly today, with thousands of ocean weather buoys[3] and ARGO floats,[4] as well as many thousands of land-based weather stations,[5] we've only had sufficient data to do so somewhat accurately for the past twenty to forty years or so.[6] While it is true that surface temperature is an indicator of climate change, are the estimates of global temperature change in Figure 1 accurate and comprehensive enough to tell us, with any precision, how quickly Earth's entire surface, including the oceans, are warming? Is the two meters of atmosphere, just above the solid or liquid surface, a *key* indicator of change in the entire climate system? We will examine just how *key* this measurement is.

The IPCC likes to frame the issue of climate change in terms of surface temperature change per volume of $CO_2$ and other greenhouse gas emissions. Implicit in this framing of the issues is the assumption that natural variability is insignificant. If all, or nearly all, of global warming is due to greenhouse gases and other human activities, then climate sensitivity is easily calculated. But the accuracy of the calculation is dependent upon the accuracy of the warming estimate. It also assumes that measuring surface temperature accurately reflects changes in the entire climate system. Here we examine the assumptions that the global surface temperature record is accurate and that the record reflects changes to the whole climate system.

The global average temperature of Earth varies over three degrees[7] every year, it is just over 12 degrees in January and just under 16 degrees in July as shown in Figure 2. The Northern Hemi-

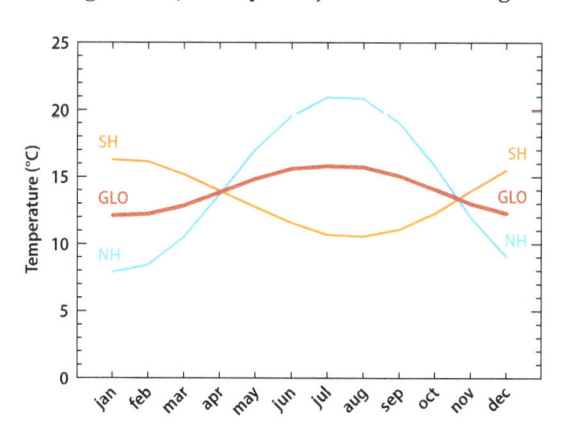

**Figure 2:** Average global surface temperatures from 1961-1990 for the globe (GLO), Northern Hemisphere (NH) and Southern Hemisphere (SH) by month. Source: (Jones, New, Parker, Martin, & Rigor, 1999)

3   National Data Buoy Center (noaa.gov)

4   Argo (ucsd.edu)

5   The U.S. National Temperature Index, is it based on data? Or corrections? | Andy May Petrophysicist

6   May, A. (2020e, November 27). *Ocean Temperature Update*. Retrieved from andymaypetrophysicist.com: https://andymaypetrophysicist.com/2020/11/27/ocean-temperature-update/, Kennedy, J. J., Rayner, N. A., Smith, R. O., Parker, D. E., & Saunby, M. (2011). Reassessing biases and other uncertainties in sea surface temperature observations measured in situ since 1850; 1. Measurement and sampling uncertainties. *Journal of Geophysical Research, 116*. Retrieved from https://agupubs.onlinelibrary.wiley.com/doi/full/10.1029/2010JD015218, Hosada, S., Ohira, T., Sato, K., & Suga, T. (2010). Improved description of global mixed-layer depth using Argo profiling floats. *Journal of Oceanography, 66*, 773-787. doi:10.1007/s10872-010-0063-3

7   Jones, P. D., New, M., Parker, D. E., Martin, S., & Rigor, I. G. (1999). Surface Air Temperature and its Changes over the Past 150 years. *Reviews of Geophysics, 37*(2), 173-199.
Retrieved from https://citeseerx.ist.psu.edu/viewdoc/download?doi=10.1.1.546.7420&rep=rep1&type=pdf

sphere has a larger swing from eight degrees in January to over 21 degrees in July, a remarkable change of 13°C in only six months. Compare this monthly change in global temperatures to the HadCRUT4 global change of one degree since 1850 shown in Figure 3 or the HadCRUT5 change of over 1°C since 1850 in Figures 1 and 4.

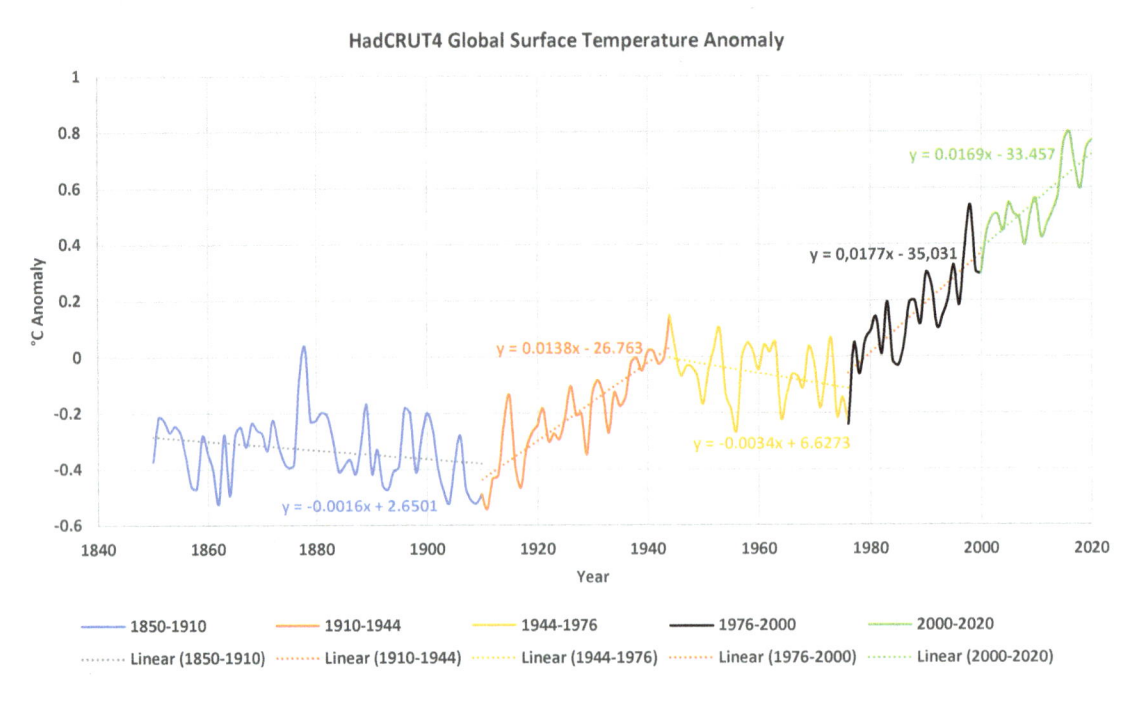

**Figure 3:** HadCRUT4 temperature record since 1850.

Figure 4 breaks down the HadCRUT5 temperature record into the same segments as Figure 3. Comparing Figure 2 to Figures 3 and 4 leads us the conclusion that the impact of the warming of the past 170 years is not very significant. Everyone experiences a larger global or hemispheric change every year from July to January.

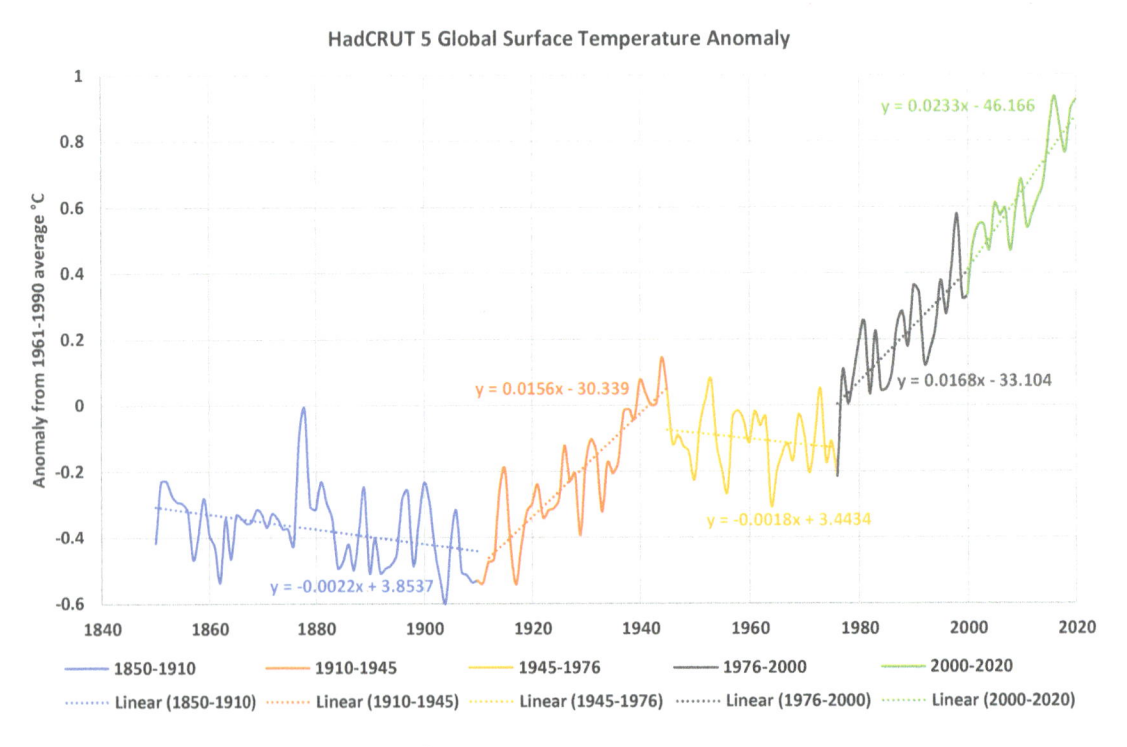

**Figure 4:** The HadCRUT5 temperature record.

# How significant is the global warming since the 19th century?

Comparing Figure 3 to Figure 4, we can see that the newly released HadCRUT5[8] dataset shows 0.2°C more warming than its predecessor, HadCRUT4[9] over the past 170 years. Table 1 breaks the 170-year period into warming and cooling periods and computes the difference for each period.

**Table 1:** HadCRUT 4 and 5 warming rates.

| WARMING RATES (PER DECADE) | | | |
|---|---|---|---|
| PERIOD | HadCRUT 4 | HadCRUT 5 | |
|  | °C/decade | °C/decade | Diff |
| 1850 - 1910 | –0.016 | –0.022 | –0.006 |
| 1910 - 1945 | 0.138 | 0.156 | 0.018 |
| 1945 - 1976 | –0.034 | –0.018 | 0.016 |
| 1976 - 2000 | 0.177 | 0.168 | –0.009 |
| 2000 - 2020 | 0.169 | 0.233 | 0.064 |

The cooling from 1850 to 1910 increased 37%, the warming from 1910 to 1976 increased, and the warming from 1976 to 2000 has decreased a little, but the warming from 2000 to 2020 has increased 38%! Considering this is the period with the best data, this is surprising. With swings such as these from one version of the HadCRUT record to another, just how accurate can their estimates of global surface warming be?

The most significant difference between the datasets is that HadCRUT4 is not infilled, that is if a grid cell has insufficient data, it is not included in the average. The HadCRUT5 dataset is infilled via interpolation and extrapolation. In the critical period from 2000 to 2020 the HadCRUT5 dataset has 99% to 100% of its cells filled and the HadCRUT4 dataset has 85% filled.[10] It seems unlikely that interpolating or extrapolating the existing data into 14-15% of the HadCRUT5 cells could cause a 38% change in the surface warming rate.

Another difference between HadCRUT4 and HadCRUT5 is that HadCRUT5 uses a new SST (sea surface temperature) dataset, HadSST4. HadSST4 has a much higher warming trend from 2000 to 2012 than HadSST3, and the estimated uncertainty in the estimate is high, relative to earlier periods.[11] This is odd, considering that the newer data is better than the older data. Kennedy, et al. note that the warming rates, from 2000 to 2012, of HadSST4, COBE-SST-2, and ERSSTv4 are very similar, but the raw unadjusted data for the period has a warming rate *near zero*. A map of the difference between HadSST4 and the raw unadjusted data for 1995-2018 is large, but still smaller than the applied adjustments to the raw data.[12] HadSST4 is not only warmer from 2000 to 2012 than HadSST3, it is also warmer than all the other SST datasets studied by Kennedy, et al. in their 2019 paper.

When compared to the UAH v6.0[13] global satellite temperature dataset, HadCRUT5 is very anomalous, as shown in Figure 5. The UAH dataset is not a surface temperature dataset, instead it is a global average temperature of the lower troposphere, and completely independent of the surface

---

8    Morice, C. P., Kennedy, J., Rayner, N., Winn, J., Hogan, E., Killick, R., ... Simpson, I. (2021, Feb. 16). An updated assessment of near-surface temperature change from 1850: the HadCRUT5 dataset. *Journal of Geophysical Research (Atmospheres), 126*(3). Retrieved from https://agupubs.onlinelibrary.wiley.com/doi/full/10.1029/2019JD032361

9    Morice, Kennedy, Rayner, & Jones. (2012, April). Quantifying uncertainties in global and regional temperature change using an ensemble of observational estimates: The HadCRUT4 dataset. *J Geophysical Research: Atmospheres, 117*(D8). Retrieved from https://agupubs.onlinelibrary.wiley.com/doi/full/10.1029/2011JD017187

10   Temperature data (HadCRUT, CRUTEM,, HadCRUT5, CRUTEM5) Climatic Research Unit global temperature (uea.ac.uk)

11   Kennedy, J., Rayner, N., Atkinson, C., & Killick, R. (2019). An ensemble data set of sea-surface temperature change from 1850: the Met Office Hadley Centre HadSST4 dataset. *JGR Atmospheres, 124*(14), 7719-7763. Retrieved from https://agupubs.onlinelibrary.wiley.com/doi/full/10.1029/2018JD029867, see figure 16.

12   Kennedy, J., Rayner, N., Atkinson, C., & Killick, R. (2019). An ensemble data set of sea-surface temperature change from 1850: the Met Office Hadley Centre HadSST4 dataset. *JGR Atmospheres, 124*(14), 7719-7763. Retrieved from https://agupubs.onlinelibrary.wiley.com/doi/full/10.1029/2018JD029867, see Figure 12 and the text.

13   Spencer, R., Christy, J., & Braswell, W. (2017). UAH Version 6 global satellite temperature products: Methodology and results. *Asia-Pacific J Atmos Sci, 53*, 121-130. Retrieved from https://link.springer.com/article/10.1007/s13143-017-0010-y

datasets plotted in Figure 1, which all share the same raw data. The satellite signal that is used to build the UAH lower troposphere average temperature is centered on about 600 hPa (same as millibars) or an altitude of about 4.5 km. Nearly all the signal used is captured from between 900 to 300 hPa (roughly 1 km to 9 km altitude).

All climate models[14] and logic suggest that, if greenhouse gases are causing our current surface warming, the temperature should be increasing in the lower to middle troposphere faster than at the surface. This is because the increased surface warming should cause more evaporation and the water vapor will condense between 2 and 12 km releasing latent heat that warms the surrounding air. In the tropics the extra warming extends much higher, up to 18 km in some extreme cases.

Since both theory and the models predict a higher warming rate in the lower and middle troposphere than we see at the surface, Figure 5 is surprising. It shows that the HadCRUT5 infilled land and ocean surface dataset is warming 36% faster than either the lower troposphere (per UAH v6.0) or the sea surface. The HadCRUT5 land and ocean surface temperature is constructed from the CRUTEM5 land data and the HadSST4 data. The same SST (Sea Surface Temperature) data that is plotted in Figure 5 and discussed above. The lower troposphere is warming at the same rate, to two decimals, as the world ocean surface HadSST4 data. UAH v6.0 shows a slower warming rate than the other lower troposphere satellite temperature datasets analyzed in AR6, but UAH 6.0 matches observations much better than the other satellite datasets.[15] The HadCRUT5 surface warming trend is also higher than the average shown in AR6.[16]

The lack of much tropospheric excess warming, over surface warming, suggests that changes in greenhouse gases are likely not a significant factor in current warming.[17] Further, since Earth's surface is 71% water and only 29% land, it seems unlikely that the land can be warming fast enough to increase the total surface temperature warming rate 36%. The UAH warming trend for global land, since 1979, is 50% larger than for the oceans, but land is only 29% of the surface,

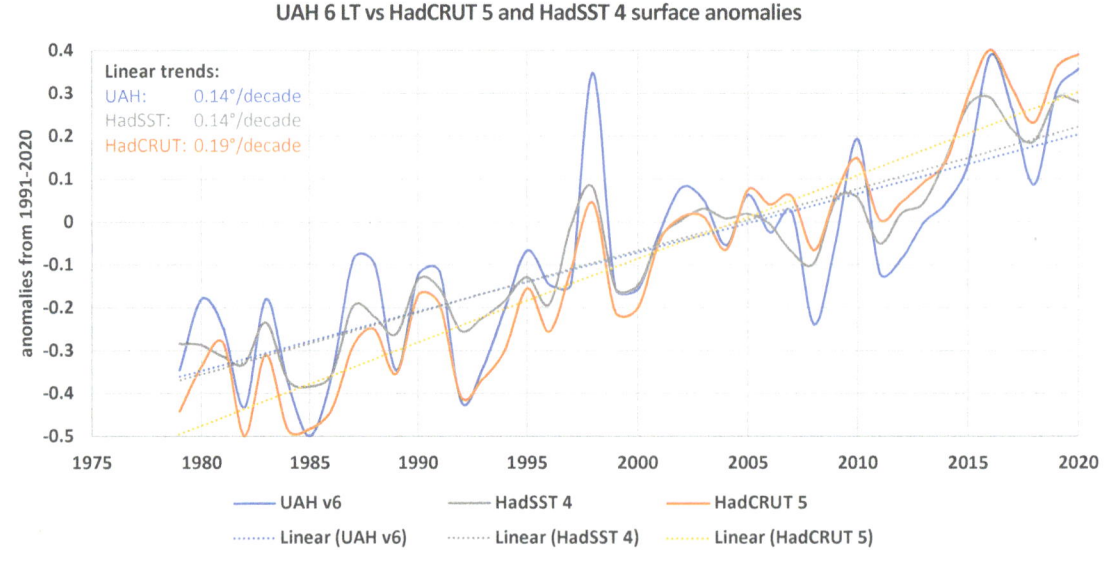

**UAH 6 LT vs HadCRUT 5 and HadSST 4 surface anomalies**

Linear trends:
UAH:       0.14°/decade
HadSST:    0.14°/decade
HadCRUT:   0.19°/decade

**Figure 5:** UAH lower troposphere temperatures compared to the Hadley SST v4 and the Hadley land plus SST surface temperatures. The plot spans 1979 (the first year of UAH data) to 2020.

---

14 The AR4, AR5, and AR6 climate models unanimously predict that the lower to middle troposphere will warm faster than the surface in response to greenhouse gas forcing. Natural (non-greenhouse gas forcing) also predicts some increase in warming rate, but much smaller than modeled. The observed difference is about 15-20% in the tropics and the modeled difference, with greenhouse gases, is over 30%. McKitrick, R., & Christy, J. (2018, July 6). A Test of the Tropical 200- to 300-hPa Warming Rate in Climate Models, Earth and Space Science. *Earth and Space Science, 5*(9), 529-536. Retrieved from https://agupubs.onlinelibrary.wiley.com/doi/full/10.1029/2018EA000401 and Blunden, J., & Arndt, D. S. (2020). *State of the Climate in 2019.* BAMS. Retrieved from https://www.ametsoc.org/index.cfm/ams/publications/bulletin-of-the-american-meteorological-society-bams/state-of-the-climate/

15 Christy, J. R., Spencer, R. W., Braswell, W. D., & Junod, R. (2018). Examination of space-based bulk atmospheric temperatures used in climate research. *International Journal of Remote Sensing,* 3580-3607. doi:10.1080/01431161.2018.1444293, see Figure 6

16 AR6 Table 2.5 shows an average trend warming of 0.70°C over the comparable 1980-2019 period, which compares with 0.75°C for HadCRUT5,

17 https://andymaypetrophysicist.com/2022/03/13/comparing-ar5-to-ar6/

so 29% of 50% is only 15%. The data in Figure 5 suggest that there are problems with the land-based temperature data or the processing of it.

## Ocean temperatures

The ocean mixed layer is a turbulent layer just below the ocean surface. Turbulence, due to the wind and weather in the atmosphere just above the ocean surface, keeps it well mixed and thus it has a nearly constant vertical temperature throughout. The thickness of the layer varies by location, but the global average thickness is roughly 72 meters according to data gathered and mapped by JAMSTEC.[18] Shigeki Hosada and his colleagues use Argo float data and ocean buoys to grid and map the mixed layer temperature across much of the world ocean.

The mixed layer is in constant communication with the surface, with a maximum delay of a few days to a few weeks. Because the mixed layer has over 27 times the heat capacity of the entire atmosphere, it moderates the speed of temperature changes in the overlying atmosphere.

Heat capacity tells us how much energy it takes to raise the temperature of a body one degree, thus if the atmospheric temperature increased 27 degrees, and all that thermal energy ("heat") were transferred quickly to the mixed layer, the temperature would only increase one degree. The mixed layer and the atmosphere are always trying to come to equilibrium, but the enormous heat capacity of the mixed layer means that its temperature fluctuates less rapidly than the more chaotic and active atmosphere. Atmospheric temperatures reflect the day-to-day weather, mixed layer temperatures reflect month-to-month and year-to-year climatic changes. Figure 6 is a plot of JAMSTEC global mixed layer temperatures, from 2002 to 2020.

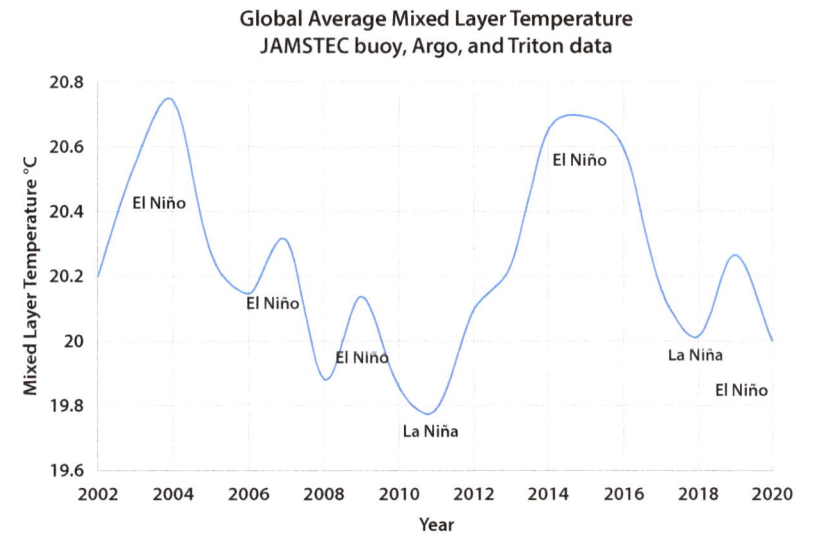

**Figure 6:** JAMSTEC area-weighted global average one-degree grid mixed layer temperatures. Only populated grid cells are averaged. Yearly averages are plotted. Some recent ENSO events are labeled in the graph.

We only have good mixed layer data for the past 19 years, some would say only since 2004 or 2005, but either way the data reflects recent ENSO (La Niña and El Niño) events and trends slightly downward with time so far. The plot is of actual temperature measurements, *not* anomalies from the mean. Various sea surface and mixed layer temperature measurement datasets have different trends, some up, and some down. When the measurements are converted to anomalies from a mean and "corrected" the trend is always slightly upward, this brings the conversion and corrections into question.[19]

---

18   The Japanese Agency for Marine-Earth Science and Technology (Hosada, Ohira, Sato, & Suga, 2010). Access to the JAMSTEC gridded data has been suspended, but more information is available here and here.

19   For a full discussion of mixed layer and sea surface measurements versus anomaly problems see these essays: (May, Ocean Temperatures, what do we really know?, 2020f), (May, Sea-Surface Temperatures: Hadley Centre v. NOAA, 2020g), (May, The Ocean Mixed Layer, SST, and Climate Change, 2020h), and (May, Ocean Temperature Update, 2020e)

While the mixed layer trends reflect climatic trends on a monthly or yearly scale, the deeper ocean looks even longer-term. Figure 7 shows the JAMSTEC grid temperatures for 100 to 2,000 meters below the ocean surface. The mixed layer rarely reaches as deep as 100 meters, so the temperatures shown in Figure 7 are partially insulated from the surface. In both Figure 6 and Figure 7 only populated cells are averaged, neither grid has values in all cells for every year. The values averaged for both graphs are area weighted since the grid cells in the higher latitudes are smaller than those at the equator.

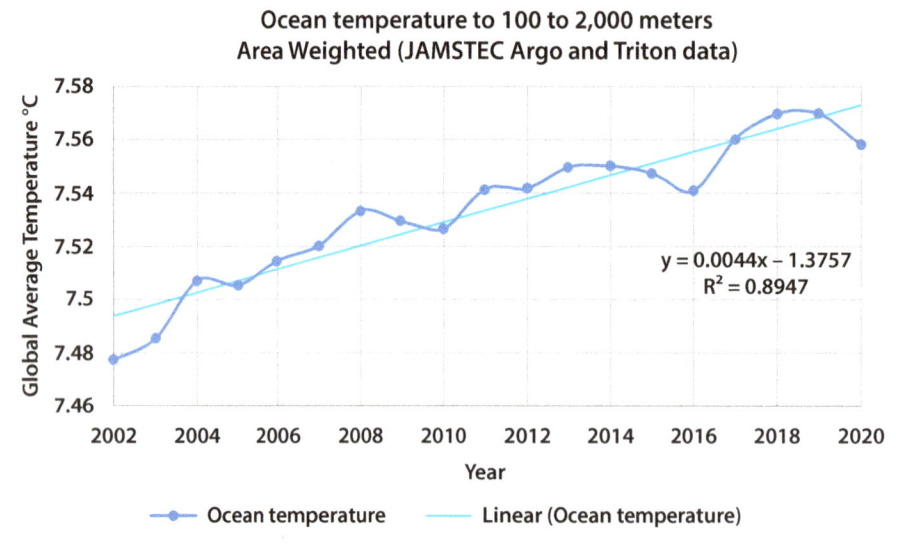

**Figure 7:** Average temperature for the world ocean from 100 to 2,000 meters. Data from JAMSTEC.

While recent atmospheric temperatures are reflected in Figure 6, with a delay of less than a month, they are reflected on a much longer time frame in Figure 7, perhaps centuries. In Figure 7 we see a rate of increase of about 0.4°C/century. This is less than half that reported for the surface over the past century or so. The ENSO features seen in Figure 6 are absent from Figure 7. With no ENSO related features, Figure 7 is remarkably linear, with an $R^2$ of 0.9. There is no sign of acceleration. Simply averaging the ocean temperatures from 100 meters to 2,000 meters is very crude, but it does make the point that a lot of historical temperature data probably exists in the deeper ocean water.

An earlier study of available ocean heat content by Roger Pielke Sr., in 2003,[20] shows no significant trend in heat storage in the upper three kilometers of the world ocean from 1958 to 1993. Pielke Sr. writes in the same paper, "Since the surface temperature is a two-dimensional global field, while heat content involves volume integrals, ... the utilization of surface temperature as a monitor of the earth system climate change is not particularly useful in evaluating the heat storage changes to the earth system." This emphasizes our point that surface temperature is not very useful as a measure of climate change.

Compare Figure 7 and Roger Pielke Sr.'s study to the IPCC AR6 Chapter 2 plot of ocean heat content shown in Figure 8. When reading the plot consider that the world ocean contains about 1,514,000 zettajoules of heat using an average ocean temperature of eight degrees C (see Figure 7). The increases shown in Figure 8 are very tiny.

Patrick Frank[21] reported that the warming from 1880 to 2000, as estimated by NOAA, is statistically indistinguishable from zero. Frank uses an estimate of the uncertainty in each land-based temperature reading of ± 0.35°C around the globe in his calculations. This estimate is derived from an error analysis study of MMTS (Minimum-Maximum Temperature System) weather stations by

20   Pielke Sr., R. (2003, March). Heat Storage within the Earth System. *BAMS, 84*(3), 331-335. Retrieved from https://journals.ametsoc.org/view/journals/bams/84/3/bams-84-3-331.xml

21   Frank, P. (2010). Uncertainty in the Global Average Surface Air Temperature Index: A Representative Lower Limit. *Energy and Environment, 21*(8), 968-989. Retrieved from https://meteo.lcd.lu/globalwarming/Frank/uncertainty_in%20global_average_temperature_2010.pdf

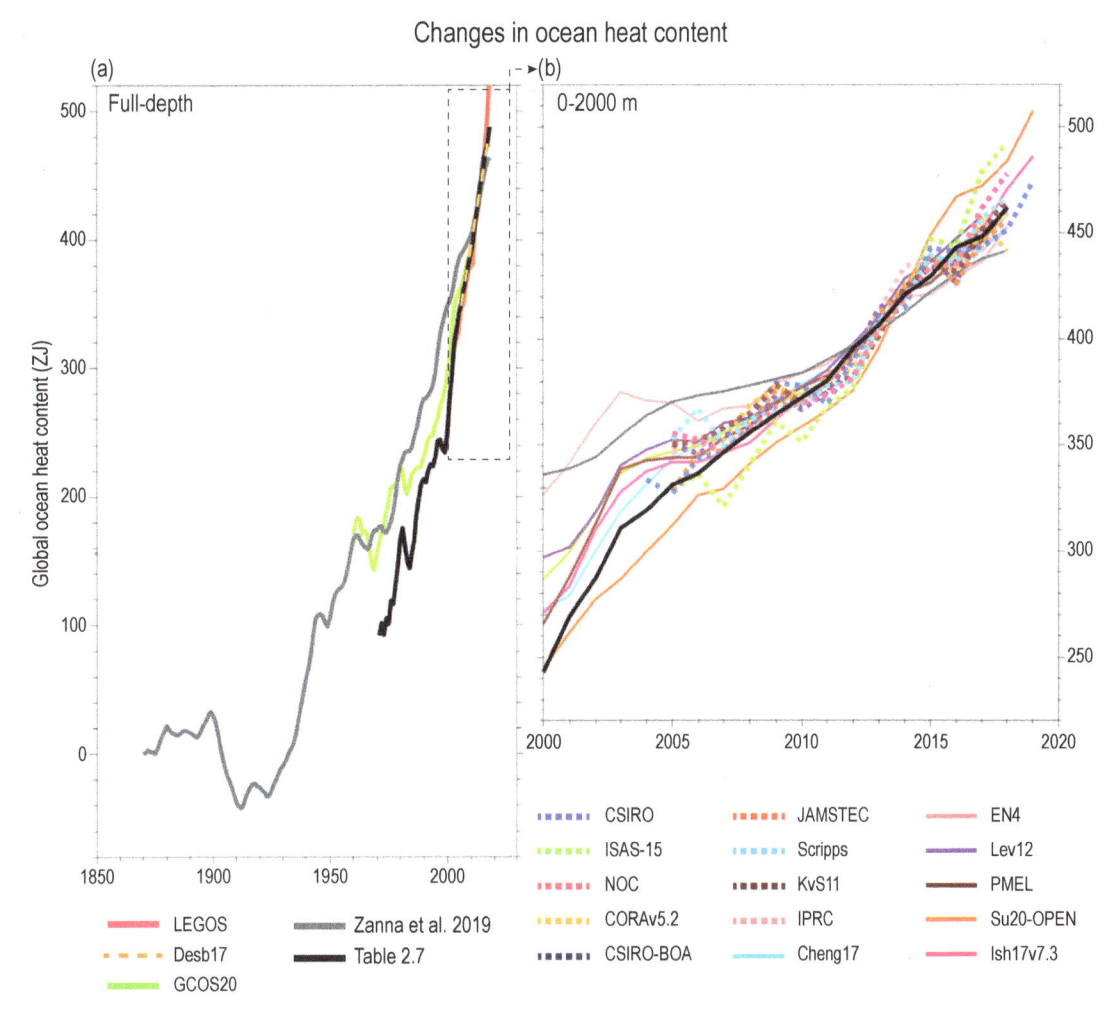

**Figure 8:** Changes in ocean heat content from AR6, Chapter 2, page 350. The graphs are in zettajoules change since ~1900, a relatively meaningless number given the huge heat capacity of seawater. The steep slopes are an artifact of the scales chosen. Figure 7 is more meaningful and is about the same body of water as in Figure 8, the only difference is Figure 8 is to the surface and Figure 7 is from 100 meters depth to 2000 meters.

Lin and Hubbard.[22] This error is not necessarily random. Even if the error is random, no extensive overall survey of station sensor variance has been published, and the precise accuracy of our global temperature record is unknown. Yet, we must consider the estimated warming of 0.87°C from 1875 to 2005 considering possible errors, and when we do, we see just how small the possible warming might be. Frank estimates that the total possible land-measured temperature error over this entire period is likely more than ± 0.46°C. There is also considerable uncertainty in ocean surface temperatures[23] so Frank concludes that the lower limit of uncertainty globally can credibly be set at ± 0.46°C. The various ocean SST datasets do not agree very well. Kennedy, et al. report that:

> *"The estimated uncertainties in the global and hemispheric averages are for the most part larger in HadSST4 than HadSST3 prior to around 1970." (Kennedy J. , Rayner, Atkinson, & Killick, 2019).*

It is very hard to interpret the various estimates of uncertainty in the large number of published hemispheric and global temperature records. This is particularly true of the more recent records, such as HadCRUT5 and HadSST4, where uncertainty is apparently increasing, as researchers uncover more problems in the raw data.

22  Lin, X., & Hubbard, K. (2004). Sensor and Electronic Biases/Errors in Air Temperature Measurements in Common Weather Station Networks. *J. Atmos. Ocean. Tech.*
Retrieved from https://journals.ametsoc.org/view/journals/atot/21/7/1520-0426_2004_021_1025_saeeia_2_0_co_2.xml

23  Kennedy, J. J., Rayner, N. A., Smith, R. O., Parker, D. E., & Saunby, M. (2011). Reassessing biases and other uncertainties in sea surface temperature observations measured in situ since 1850; 1. Measurement and sampling uncertainties. *Journal of Geophysical Research, 116.* Retrieved from https://agupubs.onlinelibrary.wiley.com/doi/full/10.1029/2010JD015218 and Kennedy, J. J., Rayner, N. A., Smith, R. O., Parker, D. E., & Saunby, M. (2011b). Reassessing biases and other uncertainties in sea surface temperature observations measured in situ since 1850: 2. Biases and homogenization. *J. Geophys. Res., 116.* doi:10.1029/2010JD015220

# GSAT, the Global Surface Air Temperature

Chapter 2 in AR6 has an extensive discussion of a new global average surface air temperature called "GSAT." GSAT stands for Global Surface Air Temperature and is different from GMST, or the global mean surface temperature. GSAT is the average air temperature over the ocean surface combined with the land-based surface air temperatures (LSAT), both sets of temperatures are meant to be from a two-meter altitude. GMST uses the same LSAT dataset but combines them with sea surface temperatures. SSTs are optimally at a 20 cm water depth.[24]

In his comments on the second order draft of AR6, Jim O'Brien commented that GSAT is probably artificially inflated by the Urban Heat Island (UHI) effect near coastal cities. The UHI is not addressed directly by either NOAA or the Met Office Hadley Centre in their respective estimates of global average surface temperature. Instead, they smooth through the excess warming in cities with a "homogenization" algorithm. This algorithm can smear warmer urban temperatures over large areas (Scafetta, 2021). The homogenization technique generally used, is best explained by Matthew Menne and Claude Williams in a 2009 *Journal of Climate* paper (Menne & Williams, 2009a). The basic technique described by Menne and Williams, has been updated several times, as described by several authors, but is still used for all the datasets in AR6.[25]

AR6 reports that GSAT is based upon nighttime marine air temperatures (NMAT) and weather reanalysis datasets. Weather reanalysis uses meteorological computer models to compute worldwide maps of meteorological conditions after the fact. The calculations are constrained by surface measurements, satellite measurements, and weather balloon data. The IPCC are interested in measuring this value because their climate models compute the global average surface air temperature at two meters, and they want to compare their model calculations to a comparable observed value.

GSAT and GMST are physically different, especially over sea ice. John Christy and colleagues examined the existing data and determined that GSAT and GMST are different measurements and there is no valid way to compute one from the other.[26] An interesting result of Christy, et al.'s study was that the difference between the air temperatures above the sea surface and the SST (as measured at about one-meter depth) is declining (going from positive to negative) from 1979 to 2000. That is, the lower-mid tropospheric air temperature of the tropical air over the sea surface cooled slightly over the period, and the SST warmed, this difference is statistically significant, but not constant. Further the NMAT, from buoys, were intermediate between the SST and the lower-mid tropospheric marine air temperature trends.

Christy and other researchers have tried to compare GSAT and GMST, but the results are confusing. Sometimes they found that GSAT warms faster than GMST and sometimes the reverse. They seem to differ, in warming rate, by less than 10%, but it can be in either direction. In other words, which one is warming faster globally, is unknown (AR6, p 192). Unable to model the difference between the two led the IPCC to throw up their hands and decided the warming rates would be "assessed to be identical."[27] This is despite the evidence presented by Christy and his colleagues that marine air temperatures (MAT) and SSTs can have different multidecadal trends and patterns of variability (AR6, p 318). Comparing the AR6 Technical Summary to the Chapter 2 text, suggests that there is a conflict within the AR6 team over the issue. AR6 Chapter 2 emphasizes the

---

24   Kennedy, J., Rayner, N., Atkinson, C., & Killick, R. (2019). An ensemble data set of sea-surface temperature change from 1850: the Met Office Hadley Centre HadSST4 dataset. *JGR Atmospheres, 124*(14), 7719-7763. Retrieved from https://agupubs.onlinelibrary.wiley.com/doi/full/10.1029/2018JD029867

25   Morice, C. P., Kennedy, J., Rayner, N., Winn, J., Hogan, E., Killick, R., . . . Simpson, I. (2021, Feb. 16). An updated assessment of near-surface temperature change from 1850: the HadCRUT5 dataset. *Journal of Geophysical Research (Atmospheres), 126*(3). Retrieved from https://agupubs.onlinelibrary.wiley.com/doi/full/10.1029/2019JD032361, for HadCRUT5 and Menne, m., Williams, C., & Gleason, B. (2018). The Global Historical Climatology Network Monthly Temperature Dataset, Version 4. *J of Climate, 31*(24). Retrieved from https://journals.ametsoc.org/view/journals/clim/31/24/jcli-d-18-0094.1.xml for GHCN version 4

26   Christy, J., Parker, D., Brown, S., Macadam, I., Stendel, M., & Norris, W. (2001, January). Differential Trends in Tropical Sea Surface and Atmospheric Temperatures since 1979. *Geophysical Research Letters, 28*(1), 183-186. Retrieved from https://agupubs.onlinelibrary.wiley.com/doi/abs/10.1029/2000GL011167

27   AR6, p 59

differences between the two measures and how important the differences are, and the Technical Summary "assesses" that they have the same long-term trend, and the two measures of temperature changes can be used interchangeably. This is a point of controversy; it is very unlikely GSAT and GMST trends are truly interchangeable.

GMST is grounded in observations, but there are no GSAT datasets as such, and none is expected for decades.[28] Thus, NMAT datasets and weather reanalysis models are used to construct current estimates of GSAT. Nighttime marine temperatures are used to estimate the difference between GSAT and SST because the daytime heating of ship superstructures and instruments creates a bias, and spurious trends in GSAT measurements.[29]

According to AR6, the importance of the difference in GSAT and GMST warming rates was raised in SR1.5,[30] but that assessment report still used GMST for their observation-based work. The AR6 second order draft[31] indicated that they had agreed to switch to GSAT as the primary metric of surface temperature changes, but this was removed from the near final draft we are discussing here. In the final draft, GSAT and GMST trends are treated as if they are interchangeable.

All CMIP simulations imply that GSAT increases faster than GMST, which is the reverse of what is seen in most observations.[32] A simple model was considered during the drafting of AR6 that applied a 4% global change in the SST warming rate. No data or observations were used, they just increased the GMST warming rate by 4% and called the result GSAT (SOD, page 2-35). Thankfully, this arbitrary model was abandoned in the final draft.

One of the datasets they examined while looking for evidence, was the HadNMAT2 dataset.[33] The dataset provided no evidence of a systematic difference between nighttime marine air temperatures (NMAT) and SSTs from 1920 to 1990, but SST warmed faster than NMAT during the 1990s. In contrast, Robert Junod and John Christy found that UAHNMATv1 warming trends were faster than SST trends from 1900 to 2010.[34] UAHNMATv1 also shows that the relative warming trends vary by region, as well as by timeframe. The interested reader may want to look at Figures 11 and 12 in Junod and Christy's *International Journal of Climatology* article to see the lack of coherence in the difference between the SST warming rate and the GSAT warming rate. AR6 points out that Nighttime Marine Atmospheric Temperature measurements are used to correct SSTs, so after this process is completed, using them to detect a difference between GSAT and GMST is partially circular.[35]

The complexity of the relationship in the real world and the uniformity of the climate model results suggests that the models are oversimplifying a complicated problem. Or, perhaps, the data simply aren't accurate enough to resolve the two temperature trends. Either way, simply assuming trends in the two temperatures are the same—as they did in the final version of AR6—was the only sensible option the IPCC had. But, given observed differences between the two values have an uncertainty of ± 10%, and it varies from GSAT > GMST to GSAT < GMST, a huge uncertainty between the modeled surface temperature and observations is introduced. Further, oceans cover 71% of Earth's surface.

Global average temperatures, in general, have little meaning unless there is a forcing agent that acts globally. $CO_2$ disperses rapidly, so if it is the dominant factor in global warming, we might ex-

---

28 AR6, p 319

29 AR6, p 319

30 IPCC. (2018). *Global Warming of 1.5 degrees C.* (Masson-Delmotte, V., P. Zhai, H.-O. Pörtner, D. Roberts, J. Skea, . a. T. Waterfield, Eds.) Geneva: World Meteorological Organization. Retrieved from https://www.ipcc.ch/sr15/

31 AR6 SOD, 2-35

32 AR6, p 319 and Christy, J., Parker, D., Brown, S., Macadam, I., Stendel, M., & Norris, W. (2001, January). Differential Trends in Tropical Sea Surface and Atmospheric Temperatures since 1979. *Geophysical Research Letters, 28*(1), 183-186. Retrieved from https://agupubs.onlinelibrary.wiley.com/doi/abs/10.1029/2000GL011167

33 The Met Office Night Marine Atmospheric Temperature dataset (Met Office , 2021)

34 Junod, R., & Christy, J. (2019, October 10). A new compilation of globally gridded night-time marine air temperatures: The UAHNMATv1 dataset. *RMetS, 40*(5). Retrieved from https://rmets.onlinelibrary.wiley.com/doi/full/10.1002/joc.6354

35 AR6, p 319

pect warming to occur over the globe approximately uniformly, if the sun and other natural factors are not changing. But this is not what has happened recently or in the distant past. The UAH satellite global temperature records show that the Northern Hemisphere is warming 30% faster than the tropics and 49% faster than the Southern Hemisphere since 1979. Figure 9 shows the warming rates for the Northern Hemisphere (NH), the tropics, the globe, and the Southern Hemisphere (SH). All are relative to the respective average from 1990 to 2020, which makes them bunch up a bit, but the differences in warming rates are significant.

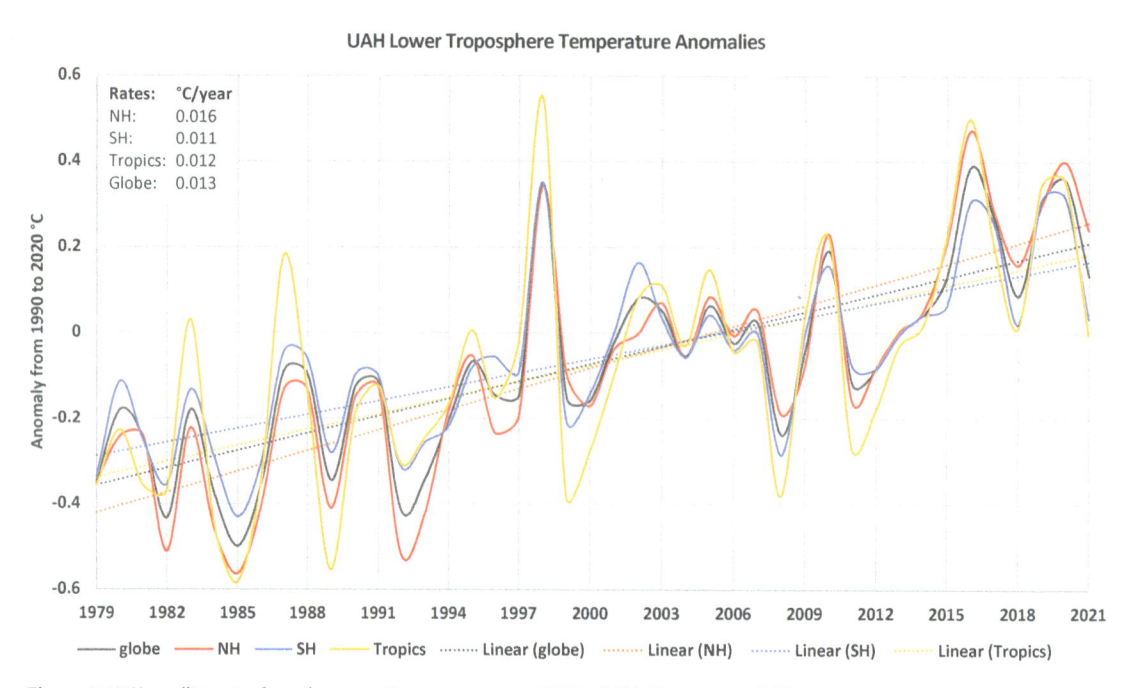

**Figure 9:** UAH satellite rates from the respective averages over 1990 to 2020. Data source: UAH.

The IPCC does their global temperature comparisons from 1850, the end of what they call the preindustrial era. The year 1850 is also close to the end of the Little Ice Age (LIA). This implies they consider the LIA an optimum temperature, but historian Wolfgang Behringer disagrees. He describes the Little Ice Age as a cold and miserable time for humanity, our modern climate is much better.[36] The LIA was a time when mountain glaciers advanced and swallowed entire towns; plagues, severe droughts, and famine due to poor crops were frequent. Geoffrey Parker estimates that a third of the population in Europe and Asia died during the mid-seventeenth century, the coldest portion of the LIA.[37] Very few informed people would want to return to the climate of the Little Ice Age.

## Discussion and Conclusions

There are no observational data to support shifting the GMST warming rate up to compute GSAT, thus the long discussion of GSAT versus GMST in Chapter 2 of AR6 is not necessary. Fortunately, this problem was recognized and GMST was not replaced by an estimated GSAT, we support this decision. However, the confusing differences between GMST and GSAT should be investigated, and not simply dismissed as irrelevant, the differences might be important.

Only a small portion of the surface (defined as the sea floor to the top of the atmosphere and excluding the land surface) heat content is in the atmosphere.[38] The lower atmosphere is very chaot-

36   Behringer, W. (2010). *A Cultural History of Climate*. Cambridge, UK: Polity Press. Retrieved from https://www.amazon.com/Cultural-History-Climate-Wolfgang-Behringer/dp/0745645291

37   Parker, G. (2012). *Global Crisis: War, Climate Change, and Catastrophe in the Seventeenth Century*. Yale University Press. Retrieved from https://www.google.com/books/edition/Global_Crisis/gjdDP15N4FkC?hl=en

38   May, A. (2020e, November 27). *Ocean Temperature Update*. Retrieved from andymaypetrophysicist.com: https://andymaypetrophysicist.com/2020/11/27/ocean-temperature-update/

ic, especially over land areas, and has a yearly range of surface temperatures that exceeds 110°C. The global average surface temperature varies three degrees from January to July every year, and the Northern Hemisphere average temperature varies over 12 degrees.[39] These yearly changes are much more dramatic than any decadal changes discussed in AR6 and have had no adverse effects on humanity.

The atmosphere is a good place to measure weather changes, but not a good place to measure longer term climatic changes. Temperatures measured in the atmosphere, close to the surface, require large corrections, as described by Matthew Menne,[40] these "corrections" introduce uncertainty.[41] In many ways, sea surface temperature measurements have worse problems than land-based temperatures, as described by John Kennedy and colleagues.[42] Problems measuring nighttime marine temperatures are described by Robert Junod and John Christy.[43]

Measurement of changes in bulk ocean temperatures provide a better indicator of the extent of disequilibrium in the climate system. Unfortunately, only since about 2002-2005 have we had good data on ocean temperatures at the surface and at depth.[44] But, going forward, the ocean interior will be the best place to look for a long-term stable record of the radiative disequilibrium that drives climate change.

In answer to the questions posed at the beginning of the chapter, are the estimates of global temperature change in Figure 1 accurate and comprehensive enough to tell us how quickly Earth's entire surface, including the oceans, are warming? *No.* Is the global mean surface temperature a key indicator of climate change? *No,* the measurements used simply reflect local weather and environmental conditions and are affected by the chaotic conditions at the surface. And the total change recorded over the past century is too small relative to the basic measurement accuracy and natural climate variability.

39  Jones, P. D., New, M., Parker, D. E., Martin, S., & Rigor, I. G. (1999). Surface Air Temperature and its Changes over the Past 150 years. *Reviews of Geophysics, 37*(2), 173-199.
    Retrieved from https://citeseerx.ist.psu.edu/viewdoc/download?doi=10.1.1.546.7420&rep=rep1&type=pdf

40  Menne, M., & Williams, C. (2009a). Homogenization of Temperature Series via Pairwise Comparisons. *Journal of Climate, 22*(7), 1700-1717. Retrieved from https://journals.ametsoc.org/jcli/article/22/7/1700/32422

41  Scafetta, N. (2021). Detection of non-climatic biases in land surface temperature records by comparing climatic data and their model simulations. *Climate Dynamics 56*, 2959–2982. Retrieved from https://doi.org/10.1007/s00382-021-05626-x

42  Kennedy, J. J., Rayner, N. A., Smith, R. O., Parker, D. E., & Saunby, M. (2011). Reassessing biases and other uncertainties in sea surface temperature observations measured in situ since 1850; 1. Measurement and sampling uncertainties. *Journal of Geophysical Research, 116.* Retrieved from https://agupubs.onlinelibrary.wiley.com/doi/full/10.1029/2010JD015218 and Kennedy, J. J., Rayner, N. A., Smith, R. O., Parker, D. E., & Saunby, M. (2011b). Reassessing biases and other uncertainties in sea surface temperature observations measured in situ since 1850: 2. Biases and homogenization. *J. Geophys. Res., 116.* doi:10.1029/2010JD015220

43  Junod, R., & Christy, J. (2019, October 10). A new compilation of globally gridded night-time marine air temperatures: The UAHNMATv1 dataset. *RMetS, 40*(5). Retrieved from https://rmets.onlinelibrary.wiley.com/doi/full/10.1002/joc.6354

44  May, A. (2020e, November 27). *Ocean Temperature Update.* Retrieved from andymaypetrophysicist.com: https://andymaypetrophysicist.com/2020/11/27/ocean-temperature-update/

# 4
# Controversial Snow Trends

BY CLINTEL TEAM

**SUMMARY: Is there less and less snow due to global warming?**
**The general public probably thinks this is the case. However, snow cover data in the Northern Hemisphere show a conflicting picture. In spring and summer a decline is visible, but the well-known Rutgers Data Lab shows an increase in autumn and winter. IPCC introduced a fresh new blended dataset in AR6 that changed the positive trend in the Rutgers dataset into a negative trend all year round. Is this new picture really the best available science or does it mainly demonstrate bias in the IPCC process?**

Global temperature has increased by more than 1°C over the past 170 years. Intuitively one might think that a warmer climate would automatically lead to a reduced snow cover on the planet. Several climate scientists have therefore predicted that in some parts of the world, due to global warming, snow will be a thing of the past. But this idea is too simple, because presence or absence of snow on the ground is not just dependent on temperature but also on precipitation, wind and cloud cover. For example, it is little known that Antarctica and large parts of the Arctic are actually classified as so-called 'polar deserts', i.e. regions with limited precipitation (Fig. 1). Especially in the Arctic, an increase in precipitation may therefore easily boost the snow cover extent in this region and in the Northern Hemisphere.

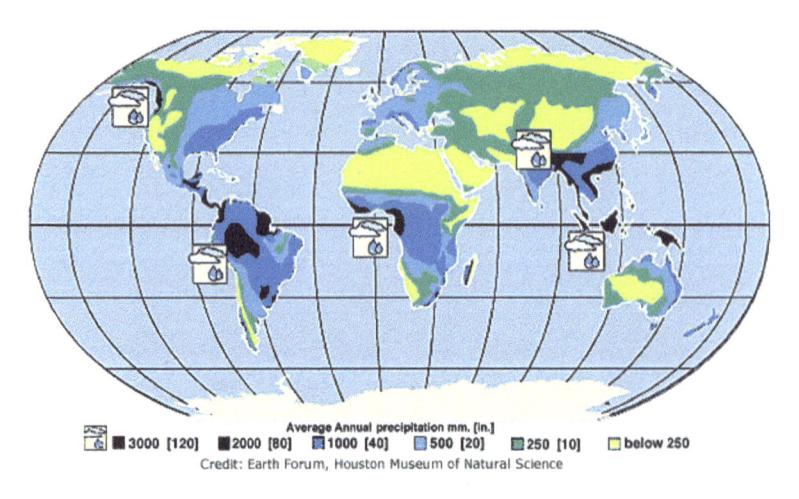

Average Annual precipitation mm. [in.]
■ 3000 [120]   ■ 2000 [80]   ▨ 1000 [40]   ☐ 500 [20]   ▨ 250 [10]   ☐ below 250
Credit: Earth Forum, Houston Museum of Natural Science

**Figure 1:** Average annual precipitation. Source: USGS, https://www.usgs.gov/media/images/generalized-world-precipitation-map

## Snow Cover Extent in AR6

Chapter 2 of the IPCC's 6th climate assessment report (AR6)[1] reports the 'Changing state of the climate system' (IPCC, 2021). This includes a summary of changes in snow cover extent (SCE) for the Northern Hemisphere (NH). The chapter reveals that SCE trends for autumn and winter are unclear. Whilst NOAA data suggests an increase in SCE (e.g. Hernández-Henríquez et al., 2015), composite ensemble data claims a decrease (e.g. Mudryk et al., 2020). Quote from the AR6 report, sub-chapter 2.3.2.2, page 2-67 (our bold):

---

1 IPCC. *Climate Change 2021: The Physical Science Basis. Contribution of Working Group I to the Sixth Assessment Report of the Intergovernmental Panel on Climate Change.* Cambridge University Press, 2021

*"Analysis of the combined in situ observations (Brown, 2002) and the multi-observation product (Mudryk et al. 2020) indicates that since 1922, April SCE in the NH has declined by 0.29 million km² per decade, with significant interannual variability (Figure 2.22) and regional differences (Section 9.5.3.1). [...]* **Analysis using the NOAA Climate Data Record shows an increase in October to February SCE** *(Hernández-Henríquez et al., 2015; Kunkel et al., 2016) while analyses based on satellite borne optical sensors (Hori et al., 2017) or* **multi observation products (Mudryk et al., 2020) show a negative trend for all seasons** *(section 9.5.3.1, Figure 9.23). The greatest declines in SCE have occurred during boreal spring and summer, although the estimated magnitude is dataset dependent* **(Rupp et al., 2013; Estilow et al., 2015**; *Bokhorst et al., 2016; Thackeray et al., 2016;* **Connolly et al., 2019***)."*

Interestingly, the IPCC in chapter 2 only illustrates the decreasing trend for April (i.e. spring), where full agreement of all authors exist:

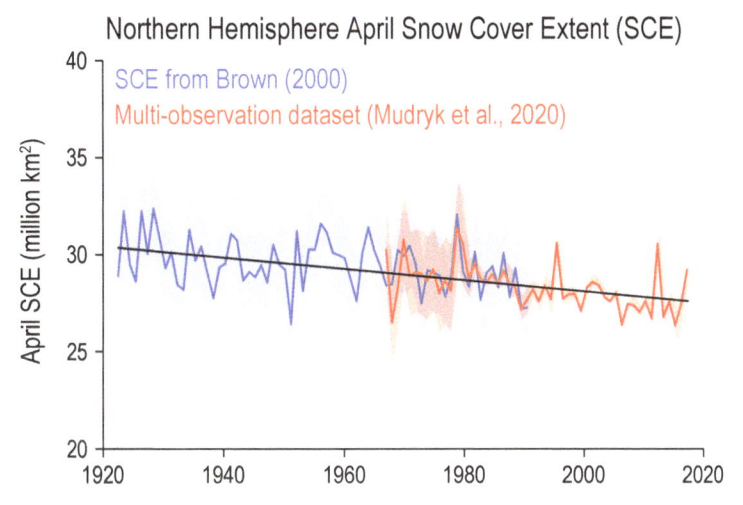

**Figure 2:** April snow cover extent (SCE) for the Northern Hemisphere. From: AR6, chapter 2.

The theme is being picked up again in Chapter 9 of AR6 (Ocean, cryosphere and sea level change). Here the IPCC authors add a judgment and explicitly favour the decreasing autumn SCE trend, because the increasing SCE trend could not be 'replicated' and is seen as somewhat 'anomalous'. The IPCC concludes: "There is therefore medium confidence that the NH SCE trend for the 1981-2016 period was also negative during these two months [October, November]". Quote from the AR6 report, sub-chapter 9.5.3.1, page 1281 (our bold):

*"Compared to numerous studies on spring SCE changes, less attention has been paid to* **changes in NH snow cover during the onset period in the autumn,** *a challenging period to retrieve snow information from optical satellite imagery due to persistent clouds and decreased solar illumination at higher latitudes.* **Positive trends in October and November SCE in the NOAA-CDR** *(Hernández-Henríquez et al., 2015)* **are not replicated in other surface, satellite, and model datasets** *(Brown and Derksen, 2013; Peng et al., 2013; Hori et al., 2017; Mudryk et al., 2017).* **The positive trends from the NOAA-CDR are also inconsistent with later autumn snow-on dates since 1980** *(-0.6 to -1.4 days per decade), based on historical surface observations, model-derived analyses and independent satellite datasets (updated from Derksen et al., 2017). Furthermore, the SCE trend sensitivity to surface temperature forcing in the NOAA-CDR* **is anomalous** *compared to other datasets during October and November (Mudryk et al., 2017). There is therefore* **medium confidence that the NH SCE trend for the 1981-2016 period was also negative during these two months** *(Mudryk et al., 2020)."*

The IPCC visually emphasizes their preference by replicating a figure from Mudryk et al. (2020) in the AR6 report:

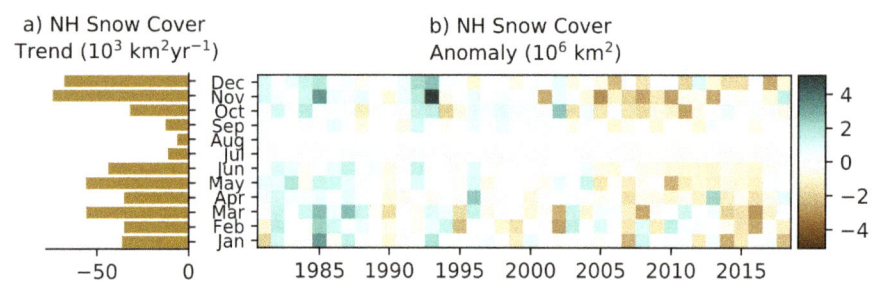

**Figure 3:** Snow cover extent (SCE) for the Northern Hemisphere. From: AR6, chapter 9, based on Mudryk et al. (2020).

The work of Mudryk et al. (2020)[2] is being explicitly praised by the AR6 authors:

> "*Since the SROCC, **progress in characterizing seasonal NH snow cover changes has been made through the combined analysis of datasets from multiple sources** (surface observations, remote sensing, land surface models and reanalysis products). A recent combined dataset (**Mudryk et al., 2020**) identified negative NH SCE trends in all months between 1981 and 2018, exceeding -50 × 10³ km² yr⁻¹ in November, December, March and May (Figure 9.23a,b).*"

## Discrepancy with other studies

The AR6 report clearly favours studies that suggest a *reduction* in NH autumn and winter snow cover extent (SCE) trend. This is surprising because the world authority group on the topic arrives at a very different conclusion. On its website the Rutgers Global Snow Lab[3] provides time series for Northern Hemisphere snow cover extent, separately plotted for autumn, winter and spring. Whilst SCE is decreasing in spring (and summer), it is clearly increasing in autumn and winter in this dataset (Fig. 4).

Rutgers generates their time series based on the Northern Hemisphere SCE CDR v01r01 from the National Centers for Environmental Information (NCEI). The Rutgers Snow Lab data formed also the basis for a NH SCE analysis by Connolly et al. (2019).[4] Not surprisingly, the study found similar results as Rutgers themselves, i.e. an increase in SCE for the NH during autumn and winter (though statistically not significant), and a decrease in spring and summer.

The Rutgers SCE time series is 54 years long, and the one used by Mudryk et al. (2020) only 37 years. The drivers and attribution of the documented fluctuations in SCE are therefore hard to interpret. It is well known that the Northern Hemisphere climate is strongly influenced by Atlantic multidecadal variability (Wyatt et al., 2012).[5] The Atlantic Multidecadal Oscillation (AMO) has a cycle duration of 60-70 years, which is longer than the time series of the available data. For example, it cannot be ruled out that the increase in NH snow cover extent during autumn (Fig. 4) actually reflects the transition of a negative AMO (1965-1995) to a positive AMO (prevailing since late 1990s) (Fig. 5). Notably, multidecadal climate variability is not even addressed in Chapter 3 (Human influence on the climate system) of the AR6 (sub-chapter 3.4.2, pages 470-471).

It is unclear why the IPCC favours the more dramatic version of the NH SCE development, with a reduction in all four seasons. In part this may be related to the fact that the first author of Mudryk et al. (2020), Lawrence Mudryk, is a Contributing Author to chapter 2 of AR6, in which the SCE trends

2   Mudryk, L., Santolaria-Otín, M., Krinner, G., Ménégoz, M., Derksen, C., Brutel-Vuilmet, C., Brady, M., and Essery, R., 2020, Historical Northern Hemisphere snow cover trends and projected changes in the CMIP6 multi-model ensemble: The Cryosphere, v. 14, no. 7, p. 2495-2514.

3   Rutgers Global Snow Lab. „Northern Hemisphere Seasonal Snow Cover Extent." *https://climate.rutgers.edu/snowcover/chart_seasonal.php?ui_set=nhland&ui_season=1* (2022).

4   Connolly, R., Connolly, M., Soon, W., Legates, D. R., Cionco, R. G., and Velasco Herrera, V. M., 2019, Northern Hemisphere Snow-Cover Trends (1967–2018): A Comparison between Climate Models and Observations: Geosciences, v. 9, no. 3, p. 135.

5   Wyatt, M. G., Kravtsov, S., and Tsonis, A. A., 2012, Atlantic Multidecadal Oscillation and Northern Hemisphere's climate variability: Climate Dynamics, v. 38, no. 5-6, p. 929-949.

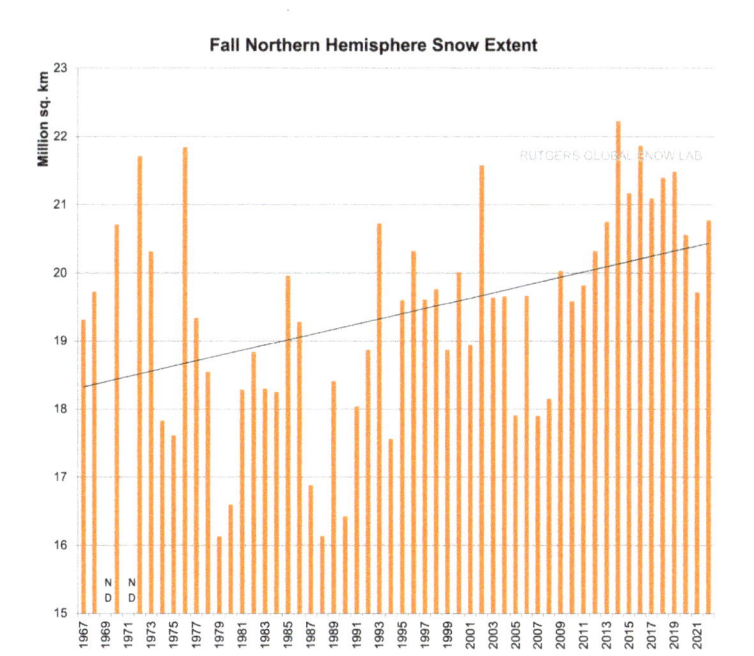

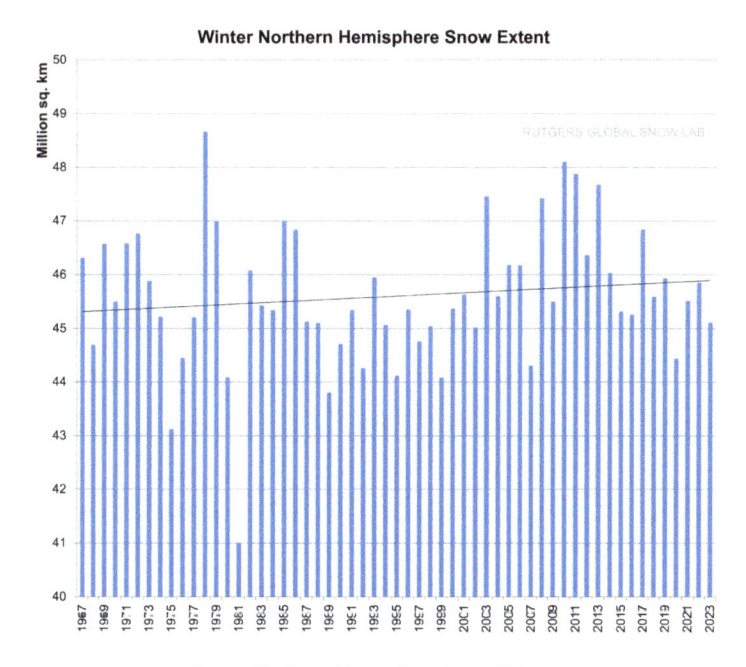

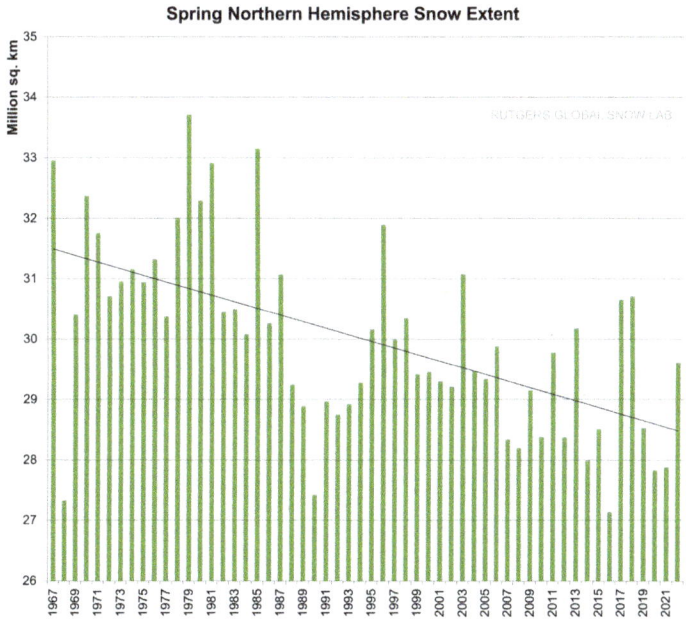

**Figure 4:** Snow cover extent (SCE) for the Northern Hemisphere for autumn, winter and spring. From: Rutgers Global Snow Lab (2023)

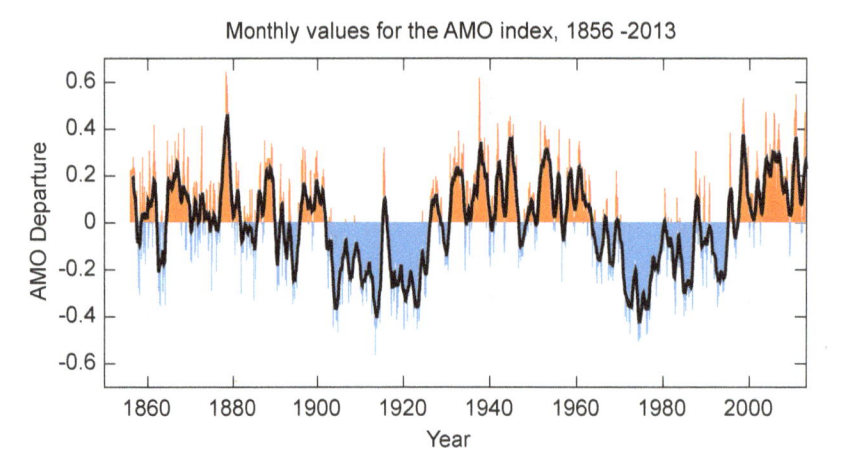

**Figure 5:** Development of the Atlantic Multidecadal Oscillation (AMO). Source: Wikipedia.

are evaluated. It appears that Mudryk may have favoured his own work over that of Rutgers Global Snow Lab. The lead researcher of Rutgers, David A. Robinson, has been co-author in two papers briefly cited in Chapter 2 of AR6, namely Rupp et al. (2013)[6] and Estilow et al. (2015).[7] However, the two papers are only mentioned in passing in AR6, without a serious attempt to discuss the different results of Rutgers. David A. Robinson was *not* among the authors of AR6, hence did not have the chance to influence the decision on which school of thought to promote in AR6.. On the other hand, Lawrence Mudryk is employed by Environment and Climate Change Canada, the department of the Government of Canada responsible for coordinating environmental policies and programs. A political influence can therefore not be excluded in his research. The author of this article contacted Lawrence Mudryk to better understand the differences in NH SCE trends. Unfortunately, the email remained unanswered by the time of the editorial deadline for this Clintel report.

## Statistical issues in Mudryk et al. 2020

Mudryk et al. (2020) produced a new time series of historical Northern Hemisphere snow extent anomalies and trends based on an ensemble of 'six observation-based products'. However, in their table 1, Mudryk et al. (2020) show seven, not six, products. Three are actual observational data. Two are models driven by reanalysis model output, one is an index based on gridded, observed, and reconstructed daily snow depth back to 1922, and one is a reanalysis model. So the composite contains a significant model component that is not purely observational.

Mudryk et al. (2020) do not illustrate the 6 (or 7) individual data products separately, which should have been done in light of transparency. It is therefore not possible to evaluate which of the various datasets actually dominates the final ensemble composite, whether they all agree, or differ greatly from each other. As a consequence, the origin of the new NH SCE composite time series remains a black box. Whilst data are provided by Mudryk et al. (2020) for download[8], few researchers will have the time and motivation to thoroughly evaluate this.

Mudryk et al. (2020) also fail to address the hot topic of autocorrelation, which is a huge and generally overlooked issue in natural datasets. Such datasets tend to be strongly autocorrelated, with major effects on the calculation of uncertainty. Mudryk et al. (2020) made no attempt to adjust for that. Typically, a probability is calculated for the case of a random time series showing a trend. A usual threshold of trends becoming statistically significant is $p<0.05$. Hence, probability values greater than 0.05 bear a greater risk of autocorrelation. Due to the high number of 12 monthly individual data series, multiple comparisons are made. The statistical significance criteria for

---

6    Rupp, D. E., Mote, P. W., Bindoff, N. L., Stott, P. A., and Robinson, D. A., 2013, Detection and Attribution of Observed Changes in Northern Hemisphere Spring Snow Cover: Journal of Climate, v. 26, no. 18, p. 6904-6914.

7    Estilow, T. W., Young, A. H., and Robinson, D. A., 2015, A long-term Northern Hemisphere snow cover extent data record for climate studies and monitoring: Earth Syst. Sci. Data, v. 7, no. 1, p. 137-142.

8    https://doi.org/10.18164/cc133287-1a07-4588-b3b8-40d714edd90e

such a group of 12 data series can be further tightened. The rational: The greater the number of data series, the greater the chance of trends occurring by chance. It would have been beneficial if Mudryk et al. (2020) had considered the Bonferroni Correction (BC), a method that corrects for the increased error rates when hypotheses are tested with multiple comparisons.

Figure 7 was provided by Willis Eschenbach and shows the NH SCE trends of the blended dataset of Mudryk et al. (2020) separately for all 12 months. The figure also shows the p values taking both autocorrelation and BC into account. The relevant threshold here is 0.05/12 = 0.004. Only five out of the 12 months of the Rutgers dataset have p values below the combined threshold (marked in red in Fig. 6). These are the months in which the trends can be safely considered statistically significant. This, however, does not rule out that other real trends exist, as this test is on the conservative side. Reviewers of Mudryk et al. (2020) should have picked up the autocorrelation analysis issues, which could have led to the rejection of the paper.

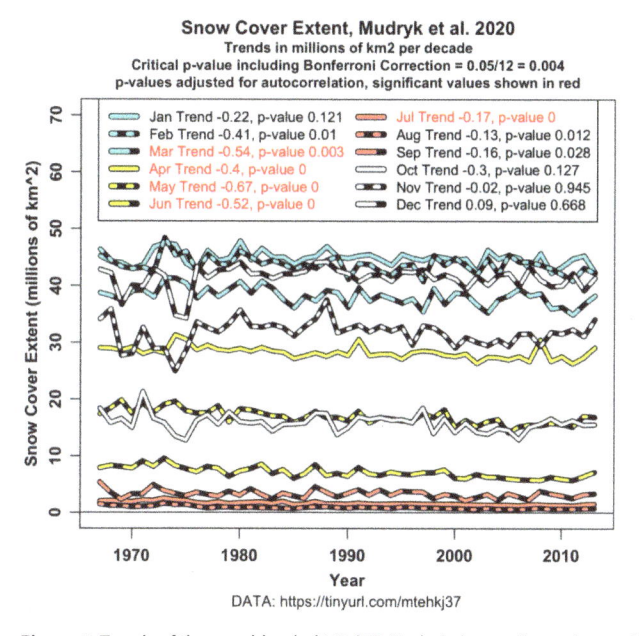

**Figure 6:** Trends of the new blended NH SCE Mudryk dataset for each month with p-values of trends (autocorrelation and Bonferroni Correction). The figure was prepared by Willis Eschenbach.

To further appreciate how small the differences are, it is also insightful to look at the full year round data from the Rutgers Snow Lab.

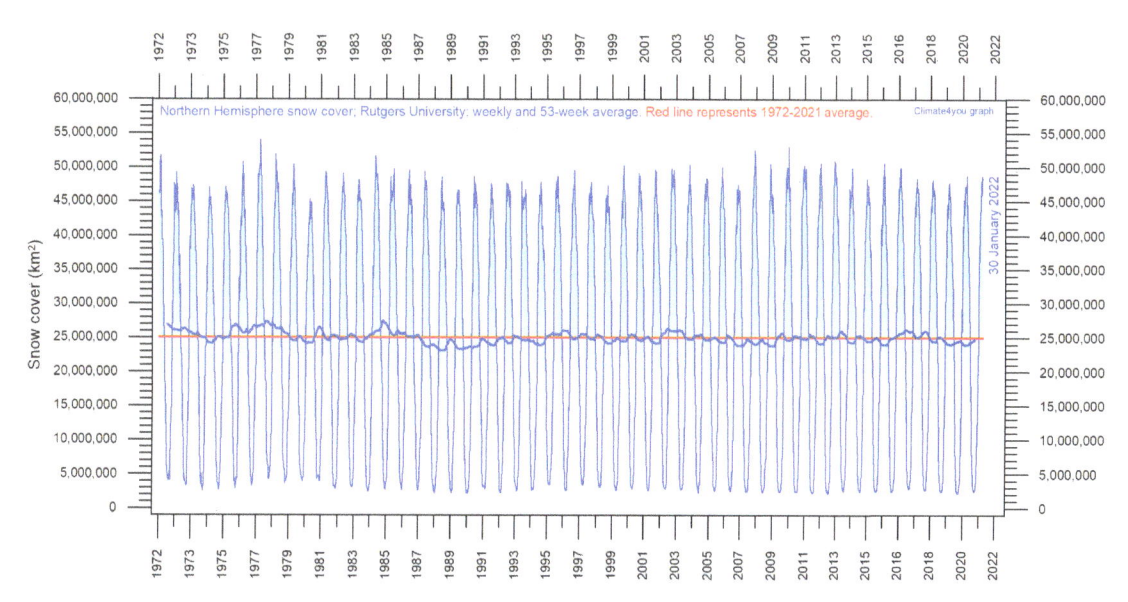

**Figure 7:** Northern hemisphere weekly snow cover since January 1972 according to the Rutgers University Global Snow Lab (http://climate.rutgers.edu/snowcover), the thin blue line is the weekly data, and the thick blue line is the running 53 week average (approximately 1 year). The horizontal red line is the 1972-2021 average. Last week shown: week 1 in 2022. Last figure update: 11 January 2022. Source: Ole Humlum, climate4you.com

In winter, Northern Hemisphere snow cover can reach values of 50 million km². Mudryk et al. (2020) now claims a decline in winter in the order of 50 thousand km²/year. That is a change of 0.1%/year. Given the apparent variability between different datasets you wonder if such changes can really be detected, let alone attributed to human causes.

## Comparison with climate models

In Chapter 9 (Ocean, cryosphere and sea level change) the IPCC suggests that the snow cover extent that was simulated by climate models generally matches well with 'observations' as published by IPCC chapter 2 co-author Lawrence Mudryk and colleagues. Quote from sub-chapter 9.5.3.2, page 1286 (our bold):

> *"Analysis of the available CMIP6 historical simulations for the 1981-2014 shows that on average,* ***CMIP6 models simulate well the observed SCE (Mudryk et al., 2020), except for outliers and a median low bias during the winter months*** *(Figure 9.24a). This is an improvement over CMIP5 (Mudryk et al., 2020), in which many snow-related biases were linked to inadequacies of the vegetation masking of snow cover over the boreal forests (Thackeray et al., 2015). A comparison between CMIP5 and CMIP6 results (Mudryk et al., 2020) shows that there is no notable progress in the quality of the representation of the observed 1981-2014 monthly snow cover trends."*

However, the conclusion would be very different, if the purely observational data (without major modeling input) of Rutgers Global Snow Lab was considered as reference. Connolly et al.[9] showed that climate models cannot replicate the increasing trend in the Rutgers dataset in the fall and winter.

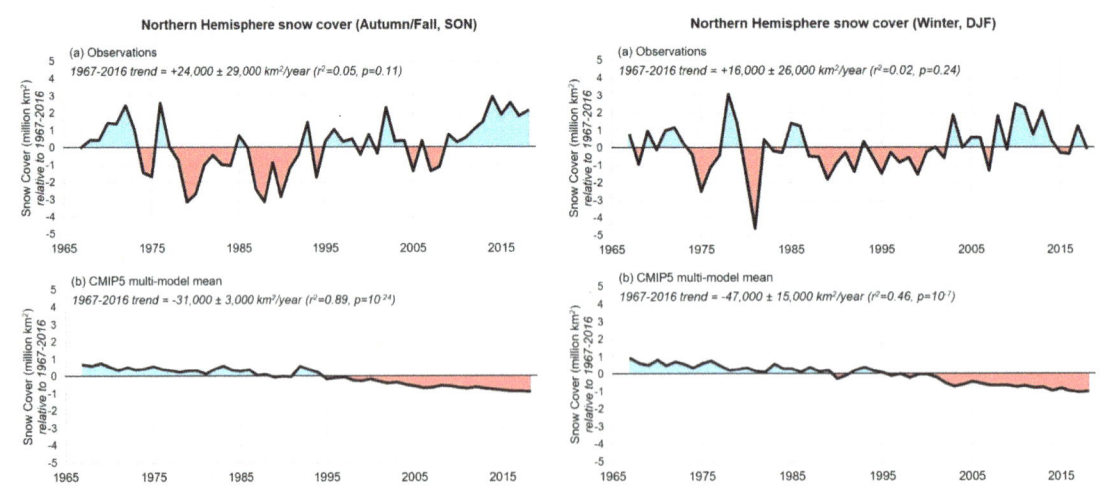

**Figure 8:** Northern Hemisphere snow cover in Autumn/Fall and winter. Top panel: observations based on the Rutgers Snow Lab data. Bottom panel: based on CMIP5 simulations. Source of the figure: Connolly et al. (2019).

The authors observe that:

> *"None of the current computer models can explain why snow cover might have increased in recent years."*

We note that the Connolly paper is mentioned in the chapter in the following way:

> *"The greatest declines in SCE have occurred during boreal spring and summer, although the estimated magnitude is dataset dependent (Rupp et al., 2013; Estilow et al., 2015; Bokhorst et al., 2016; Thackeray et al., 2016;* ***Connolly et al., 2019****)."*

To highlight Connolly et al. (2019) in this way is misleading. Yes, it also showed the decline in SCE in spring and summer, but the main message of the paper was that climate models can't reproduce

---

9    Connolly, et al., 2019, Northern Hemisphere Snow-Cover Trends (1967-2018): A Comparison between Climate Models and Observations, Geosciences, 9, 135, doi:10.3390/geosciences9030135

the increasing trend in Autumn/Fall and Winter. AR6 failed to acknowledge this. Are the AR6 authors favouring the results of Mudryk et al. (2020) over those of Rutgers Global Snow Lab because of a better model fit?

CMIP6 climate models also fail to correctly reproduce Northern Hemisphere snow depth trends (Zhong et al., 2022).[10] The simulations suggest decreasing snow depth trends for the last 70 years that contradict the observations. The study of Zhong et al. (2022) revealed that the simulated snow depths are insensitive to precipitation but too sensitive to air temperature. These inaccurate sensitivities could explain the discrepancies between the observed and simulated snow depth trends. Based on these findings, they recommend caution when using and interpreting simulated changes in snow depth and associated impacts. The CMIP6 models may require more detailed and comprehensive treatments of snow physics to more accurately project snow cover.

## Discussion

What physical processes could explain the positive NH SCE trends of Rutgers for autumn and winter? Allchin and Déry (2020)[11] have suggested that the changes might be a result of alterations in atmospheric patterns. Today, these patterns during autumn are delivering additional moisture northward to high latitude interior continental areas, where it is cold enough for snow to form. In previous decades, however, these "Arctic deserts" appear to have received less moisture, resulting in reduced snow cover.

### IPCC REPORTING

The trends in Northern Hemisphere snow cover extent are just one of many examples of biased reporting in the IPCC AR6 report. The outcome of the assessment is largely decided at the time when report authors are nominated. In this case, the lead author of a key paper, Lawrence Mudryk, was nominated as contributing IPCC author and most likely influenced the direction of the IPCC literature review in his own favour. The Second Order Draft (the last draft that is seen by external/ expert reviewers) already showed the results of the Mudryk et al. (2020) paper although that paper at the time had only been submitted to the journal. This is not forbidden under the IPCC rules, however it is clear that this favours the promotion of authors' and lead authors' own work. Also, no (or very few) expert reviewers will go to the effort to ask for the submitted paper so one can safely conclude that these results were not reviewed before they entered as the key claim surrounding snow cover trends in the AR6 report.

IPCC author nominations are typically initiated (at the upper levels, then cascading down) by the IPCC Bureau, a politically-controlled body of all IPCC member states. Lawrence Mudryk is employed by a Canadian government-related institute responsible for coordinating environmental policies and programs. The current Canadian Minister of Environment and Climate Change, Steven Guilbeault, was previously a Director and Campaign Manager for Greenpeace. The case demonstrates the importance of strictly separating science from policy. IPCC author nominations need to be done by politically-independent *scientific* bodies, not by *government*-related panels.

## Acknowledgements

We want to thank Willis Eschenbach for contributions on the statistical processing. David Robinson and Nicholas Pepin are thanked for valuable discussions. We are grateful for valuable comments received from two anonymous reviewers.

10  Zhong, X., Zhang, T., Kang, S., and Wang, J., 2022, Snow Depth Trends from CMIP6 Models Conflict with Observational Evidence: Journal of Climate, v. 35, no. 4, p. 1293-1307

11  Allchin, M. I., and Déry, S. J., 2020, The Climatological Context of Trends in the Onset of Northern Hemisphere Seasonal Snow Cover, 1972–2017: Journal of Geophysical Research: Atmospheres, v. 125, no. 17, p. e2019JD032367

# 5

# Accelerated Sea Level Rise: not so fast

BY KIP HANSEN

**The IPCC's Sixth Assessment Report (AR6) claims that sea level rise is accelerating. However, the evidence for this is rather thin. The best available evidence for long-term sea level changes comes from tide gauge records. These records typically show remarkably linear behavior for more than a century.**

Tide gauges around the world on average show a long term rise of about 1.7 mm/yr while satellite records since 1993 indicate double that rate, around 3.4 mm/yr. Tide gauges directly measure local sea surface height whereas satellite telemetry calculations measure something different, the eustatic sea level. IPCC's accelerating sea level rise seems to rely on "hybrid reconstructions" that combine these disparate datasets and often include modeled data.[1,2]

The IPCC's AR6 makes the following specific claims about present and future global mean sea level (GMSL) rise:

*"Global mean sea level increased by 0.20 [0.15 to 0.25] m between 1901 and 2018.* ***The average rate of sea level rise was 1.3 [0.6 to 2.1] mm yr$^{-1}$ between 1901 and 1971, increasing to 1.9 [0.8 to 2.9] mm yr$^{-1}$ between 1971 and 2006, and further increasing to 3.7 [3.2 to 4.2] mm yr$^{-1}$ between 2006 and 2018*** *(high confidence)...."*
*A.1.7, page SPM-5, Summary for Policymakers IPCC AR6 WGI*

*(Final Version)[3]*

*"The SROCC found that four of the five available tide gauge reconstructions that extend back to at least 1902 showed a robust acceleration (high confidence) of GMSL rise over the 20th century, with estimates for the period 1902-2010 (-0.002 to 0.019 mm yr$^{-2}$) that were consistent with the AR5."*

*Chapter 9, page 1287 (Final Version)[4]*

The increase of sea level during the past 200 years does not come as a surprise. Sea level was falling in the transition from the Medieval Warm Period (MWP, 850-1250 AD) to the Little Ice Age (LIA, 1450-1850 AD) as large amounts of water were taken up by glaciers and ice caps (Fig. 1).[5] The LIA represents one of the coldest phases of the entire last 10,000 years. After the LIA ended, glaciers and ice caps began to melt again and released water that ended up in the oceans, resulting in rising sea level.[6]

However, the recent further acceleration claimed by the AR6 coincides largely with the switch from pure tide gauge data to satellite data that became available only since 1992. Uncertainties in the calibration of the satellite results and discrepancies with synchronously recording tide

1    Sea Level Research Group University of Colorado https://sealevel.colorado.edu/presentation/what-definition-global-mean-sea-level-gmsl-and-its-rate

2    Rovere, A., Stocchi, P. & Vacchi, M. Eustatic and Relative Sea Level Changes. *Curr Clim Change Rep* 2, 221–231 (2016). https://doi.org/10.1007/s40641-016-0045-7

3    IPCC, 2021: Summary for Policymakers. In: Climate Change 2021: The Physical Science Basis. Contribution of Working Group I to the Sixth Assessment Report of the Intergovernmental Panel on Climate Change

4    IPCC AR6 Chapter 9 p. 1287 https://www.ipcc.ch/report/ar6/wg1/

5    Kopp, R. E., Kemp, A. C., Bittermann, K., Horton, B. P., Donnelly, J. P., Gehrels, W. R., Hay, C. C., Mitrovica, J. X., Morrow, E. D., Rahmstorf, S. (2016): Temperature-driven global sea-level variability in the Common Era: Proceedings of the National Academy of Sciences 113 (11), E1434-E1441.

6    Grinsted, A., Moore, J. C., Jevrejeva, S. (2010): Reconstructing sea level from paleo and projected temperatures 200 to 2100 AD: Climate Dynamics 34 (4), 461-472

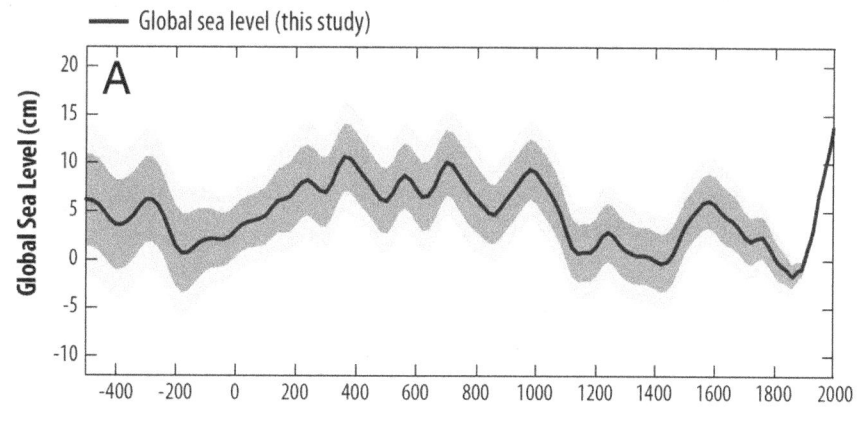

**Figure 1:** Global sea level reconstruction by Kopp et al. 2016.[5] X-axis shows years BC (negative)/AD (positive values).

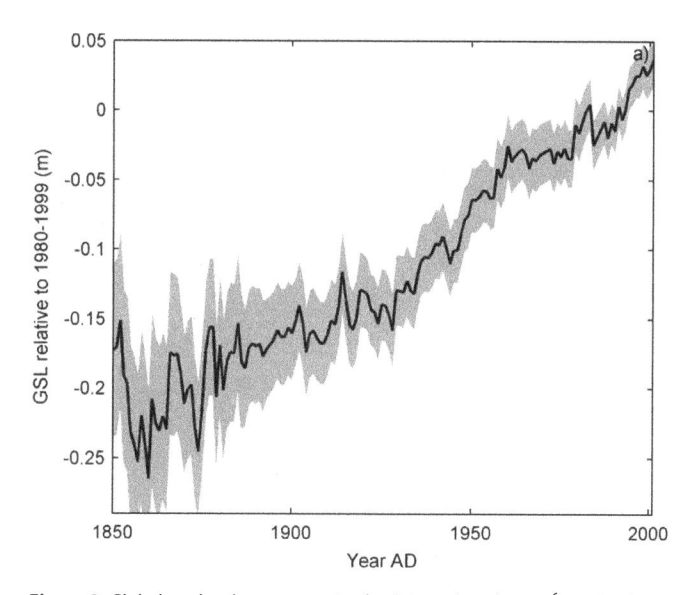

**Figure 2:** Global sea level reconstruction by Grinsted et al. 2010.[6] Sea level rose more slowly in the decades after the LIA from 1860 to 1940. As the subsequent warm period developed, sea level rose more quicky from mid-20th century onwards.

gauges suggests that at least part of the acceleration may in fact be an artefact due to the change in methodology.

During the 20th century, global sea level rose about 20 cm, at a rate, according to U.S. NOAA, of 1.7 mm/yr (+/- 0.4 mm). Other estimates and reconstructions are all just under 2.0 mm/yr. NOAA's rate is based on tide gauges[7] that directly measure local Relative Sea Level (RSL) – the height of the surface of the ocean relative to the land at the tide gauge location. "RSL is a combination of the sea level rise and the local Vertical Land Motion (VLM)."[8] In order to discover how much of RSL is *actual rise* in the height of the sea surface, which is known as *absolute* sea level (ASL)[9], and not sinking (or subsidence) of the land, tide gauges must be coupled to Continuously Operating Reference Stations (CORS) mounted on the same structure as the tide gauge. This is called Continuous Global Positioning System at Tide Gauges (CGPS@TG).

It is not possible to simply average global tide gauge records and determine a global figure for sea level rise unless these records have all been corrected for vertical land motion, which will not be possible until there are an adequate number of widespread CGPS@TG stations. U.S. NASA records a different metric for global mean sea level which it obtains through remotely sensed satellite telemetry.[10] Since 1993, NASA finds the change in their global mean sea level (GMSL) metric rising

7   Manual on Sea Level Measurement and Interpretation. https://library.wmo.int/doc_num.php?explnum_id=932

8   NOAA Tides and Currents https://tidesandcurrents.noaa.gov/sltrends/

9   CGPS@TG Working Group of the Sea Level Center at the University of Hawaii https://imina.soest.hawaii.edu/cgps_tg/introduction/index.html

10  NASA Earth Observatory https://earthobservatory.nasa.gov/images/147435/taking-a-measure-of-sea-level-rise-ocean-altimetry

at a rate of 3.4 mm/yr. Satellite GMSL however is not a measurement of the height of the surface of the oceans, but rather:

> "It can also be thought of as the 'eustatic sea level.' The Eustatic Sea Level [ESL] is not a physical sea level (since the sea levels relative to local land surfaces vary depending on land motion and other factors), but it represents the level **if all of the water in the oceans were contained in a single basin.**"[11] (emphasis added)

This is important because all the oceans are not at the same level or height, for example at the Panama Canal, sea level is 20 cm higher on the Pacific side than on the Atlantic side.[12]

The IPCC's AR6 claims that sea level rise has been accelerating, or rising faster and faster.[13] More specifically, the IPCC says that:

> **"four of the five available tide gauge reconstructions that extend back to at least 1902 showed a robust acceleration (high confidence) of GMSL rise over the 20th century**, with estimates for the period 1902-2010 (-0.002 to 0.019 mm yr $^{-2}$)" (AR6, page 1287, emphasis added)

Five sea level research groups did *reconstructions* of the uncorrected global tide gauge data. One of the five groups found **no acceleration**. Four of the five groups found some acceleration of SLR over the 20$^{th}$ century. Their estimates of SLR acceleration range from 2/1000ths to 2/100ths of a millimeter/yr$^2$. An acceleration of 2/1000ths of a mm/yr$^2$, for a century, raises sea level by less than an inch (2.54 cm), not a very "robust" finding. An acceleration of 2/100ths of a mm/yr$^2$ over a century results in an additional 10 cm of sea level rise. Here is what a sea level graph would look with and without these two rates of acceleration over a century:

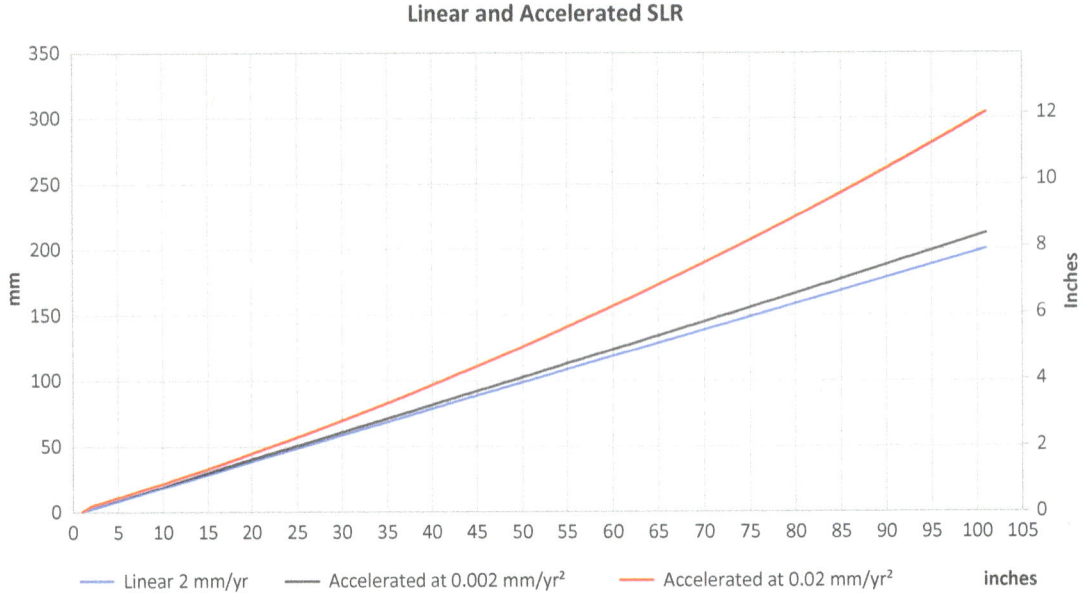

**Figure 3:** Linear increase (blue trace) at 2 mm/yr produces a straight line. Adding 0.002 mm/yr$^2$ (black trace) and 0.02 mm/yr$^2$ (orange trace) produce upward curving lines. End points on right show resulting increase in sea level after 100 years.

We see that the accelerating sea level produces a curved trend, while the steady sea level rise produces a trend that is straight. Here are the two graphs shown in the AR6 report:

11   Sea Level Research Group University of Colorado What is the definition of global mean sea level (GMSL) and its rate? | Sea Level Research Group (colorado.edu). Rovere, A., Stocchi, P. & Vacchi, M. Eustatic and Relative Sea Level Changes. *Curr Clim Change Rep* 2, 221–231 (2016). https://doi.org/10.1007/s40641-016-0045-7

12   Reid, Joseph, 1961, On the temperature, salinity, and density differences between the Atlantic and Pacific oceans in the upper kilometre - ScienceDirect

13   IPCC, 2021: Summary for Policymakers. In: Climate Change 2021: The Physical Science Basis. Contribution of Working Group I to the Sixth Assessment Report of the Intergovernmental Panel on Climate Change

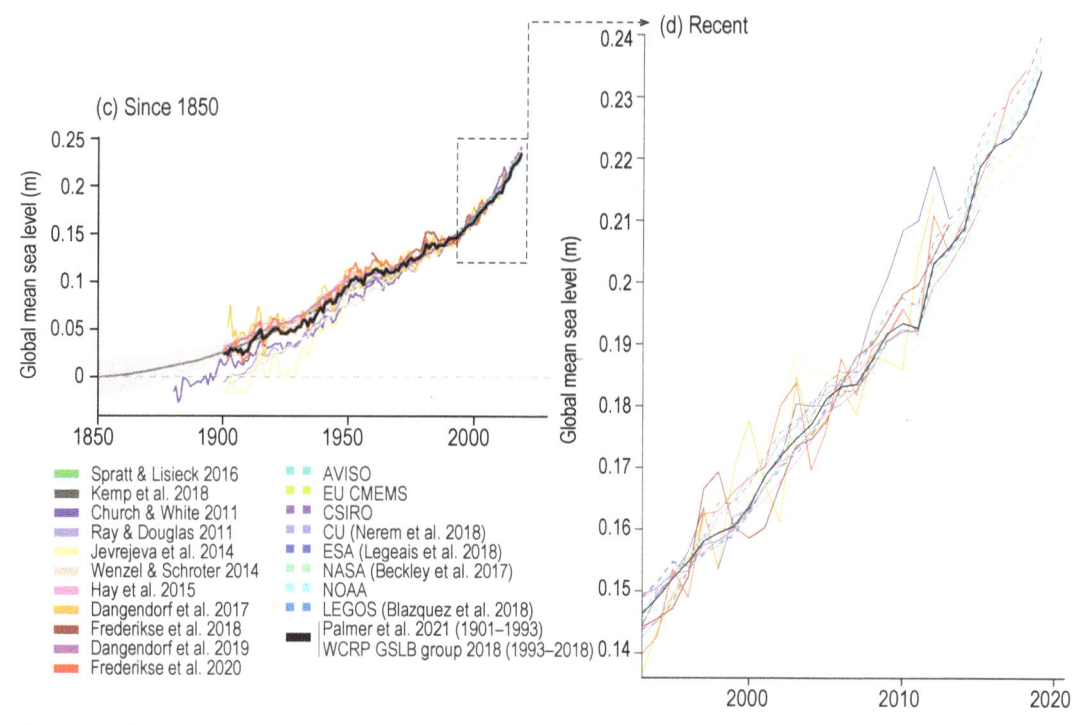

**Figure 4:** Sea level rise since 1850 and over the satellite era – 1993 to present. Figure 2.28 C and D in AR6, page 354.

This graph reminds us of the so-called spaghetti graphs that are used in millennial proxy reconstructions. The graph shows different data sets and mixes them together. Especially since the satellite era started, the different lines are pretty close to one another, strongly suggesting that reconstructions based on tide gauges confirm the higher satellite trend of 3 mm/yr. In the supplementary material the IPCC is apparently making the data available. But when we checked recently only two out of the twenty datasets were available. Only a few datasets are based on tide gauges (Church & White 2011, Ray & Douglas 2011, Jevrejeva et al 2014) and these datasets end around 2010 in the IPCC graph. But what is not shown is that these tide gauge measurements typically show multidecadal phases of acceleration and deceleration as is shown in the following figure:

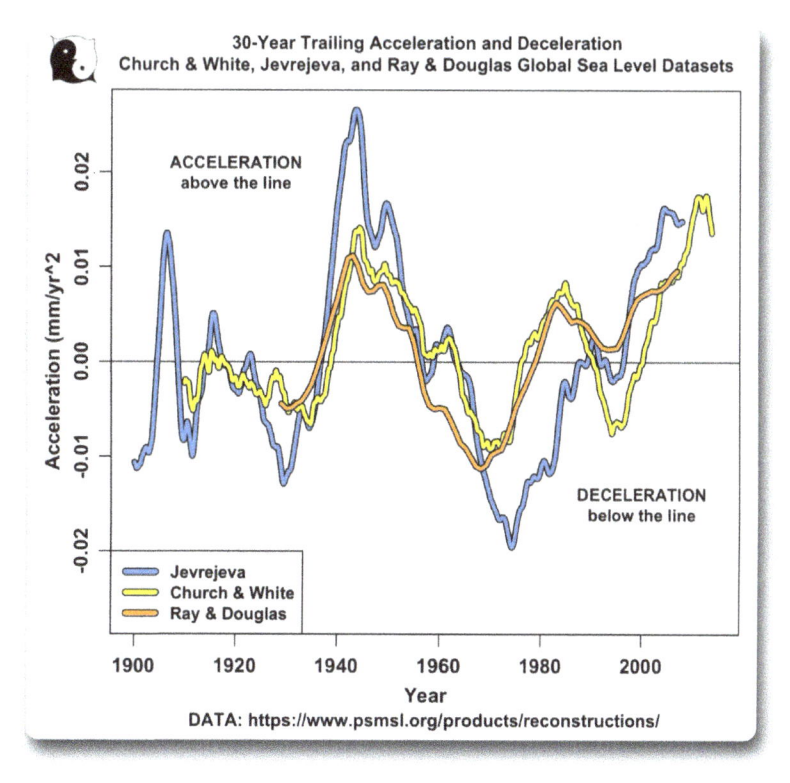

**Figure 5:** sea level acceleration and deceleration in three tide gauges datasets. Source: Willis Eschenbach

If the IPCC had mentioned the period 1940 to 1980 separately, they then should have mentioned a decreasing rate of sea level rise. The yellow and blue reconstructions show early signs of another period of deceleration, which you would expect based on the historical patterns. The Frederikse et al 2020 paper[14] that is part of the IPCC graph above did show these decadal fluctuations:

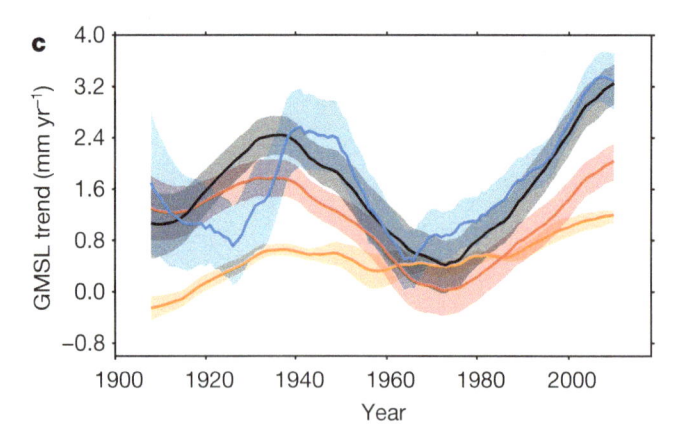

**Figure 6:** sea level trend based on a so-called sea level budget method as used in Frederikse et al 2020. The thick blue line is the observed trend and black is based on the sum of different components.

Many researchers have cautioned that sea level is associated with natural cycles with a duration of ~60 years.[15] It is possible that the acceleration is actually part of this natural cyclicity.

## Relative sea level

As mentioned, tide gauges directly measure relative sea level at a single location. The illustration in Figure 7 shows the three components of local relative sea level: vertical land motion of the regional land mass, subsidence of the tide gauge structure – such as a pier or a dock – and absolute sea level rise, the actual rising of the height of the surface of the sea. Added together they produce

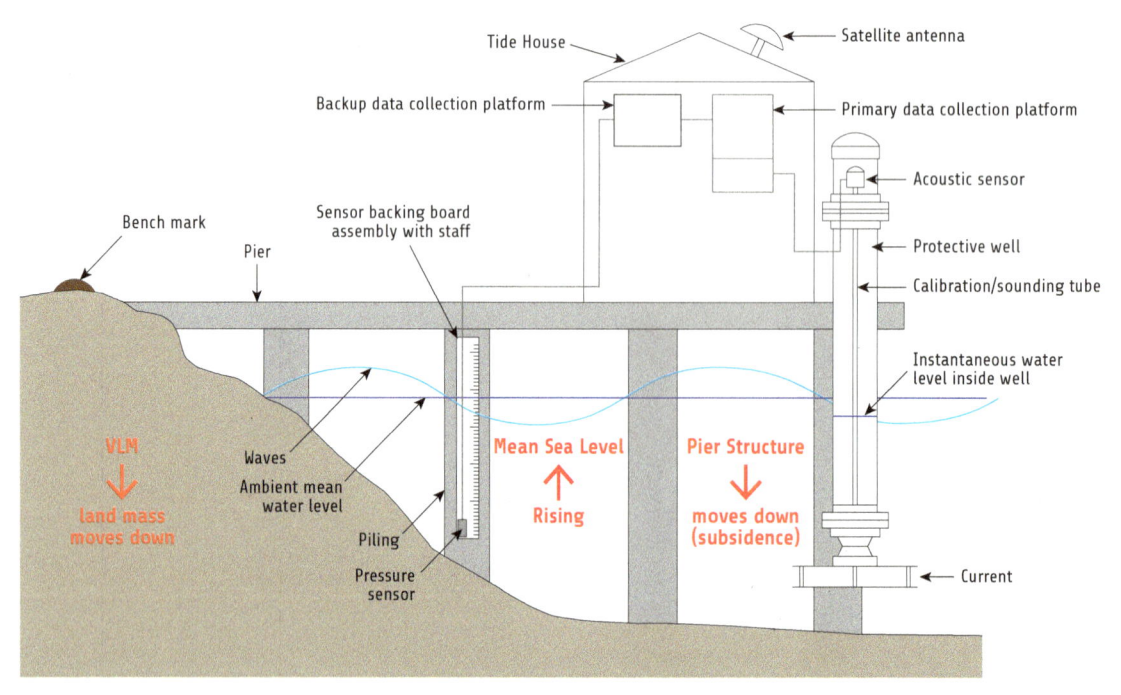

**Figure 7:** Typical modern tide gauge installation with components which contribute to measured Relative Sea Level rise (RSL).

14   Frederikse, T., Landerer, F. C., Caron, L., Adhikari, S., Parkes, D., & Humphrey, V. (2020). The causes of sea-level rise since 1900. Nature, 584, 393-397. Retrieved from https://www.nature.com/articles/s41586-020-2591-3

15   Ding, H., Jin, T., Li, J., & Jiang, W. (2021). The contribution of a newly unraveled 64 years common oscillation on the estimate of present-day global mean sea level rise. Journal of Geophysical Research: Solid Earth, 126, e2021JB022147. https://doi.org/10.1029/2021JB022147

the change measured by the tide gauge. The local tides, usually two highs and lows a day, can easily be averaged out to find a mean sea level for each day, week, month, and year.

The values are the mean relative sea level for that location. Figure 7 shows the graph of The Battery at New York City, USA.

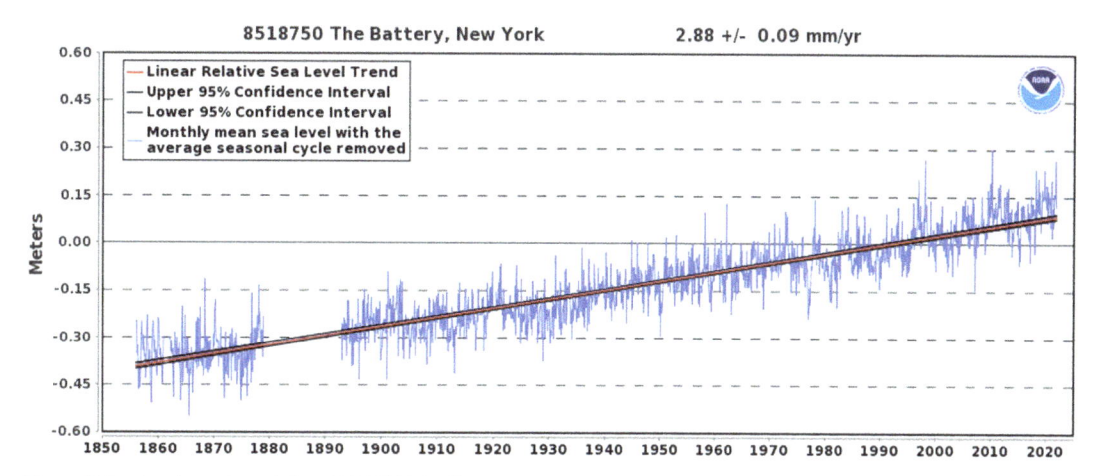

**Figure 8:** Sea Level trend graph from U.S. NOAA for The Battery at New York City. Linear across the entire 165-year record showing no acceleration.

Though it is highly variable, the graph is clearly linear – showing no acceleration over the entire 165-year record. This is directly measured data, not remotely sensed, and calculated.

Global sea level rise trends are highly variable, changing rates, and even direction decade by decade as the graph from Frederikse et al. shows. And, while it is easy to pick out decades with faster rise and fall, overall, the picture presented by long-term tide gauge records is clear.

The tide gauge record for NY City exhibits a steady rise of just under 3 mm/yr, including the downward vertical land motion which accounts for more than half of the relative sea level.[16]

Countering the IPCC's opinion that they have calculated "a robust acceleration (high confidence) of GMSL rise over the 20th century", are the directly measured, long-term (>50 years) tide gauge records all over the world which show the same linear trends, unaffected by acceleration, neither slowing down nor speeding up. Graphs of these tide gauge records can be found for U.S. and global tide gauges at NOAA's Tides and Currents web site.[17]

In another paper in 2016, Thompson et al.[18] provide the chart illustrating the consistency of relative sea level trends at tide gauges internationally, shown in Figure 9.

The mean of relative sea level trends before correction for vertical land motion is 1.69 mm/yr, very close to NOAA's 1.7 mm/yr, and when corrected, gives an estimate of absolute sea level rise trends, of 1.57 mm/yr.

The only sea level rise of any real concern for mankind is relative sea level at our shorelines, our ports and our cities. Relative sea level, averaged for all 149 U.S. NOAA tide stations and uncorrected for vertical land motion, is 2.01 mm/yr and for all of NOAA's global tide stations, it is 1.4 mm/yr.

---

16 Snay, Richard, et al. "Using global positioning system-derived crustal velocities to estimate rates of absolute sea level change from North American tide gauge records." Journal of Geophysical Research: Solid Earth 112.B4 (2007). https://agupubs.onlinelibrary.wiley.com/doi/full/10.1029/2006JB004606

17 For U.S. stations: https://tidesandcurrents.noaa.gov/sltrends/sltrends_us.html and for global stations: https://tidesandcurrents.noaa.gov/sltrends/sltrends_global.html

18 Thompson, P. R., et al. "Are long tide gauge records in the wrong place to measure global mean sea level rise?" Geophysical Research Letters 43.19 (2016): 10-403. https://doi.org/10.1002/2016GL070552

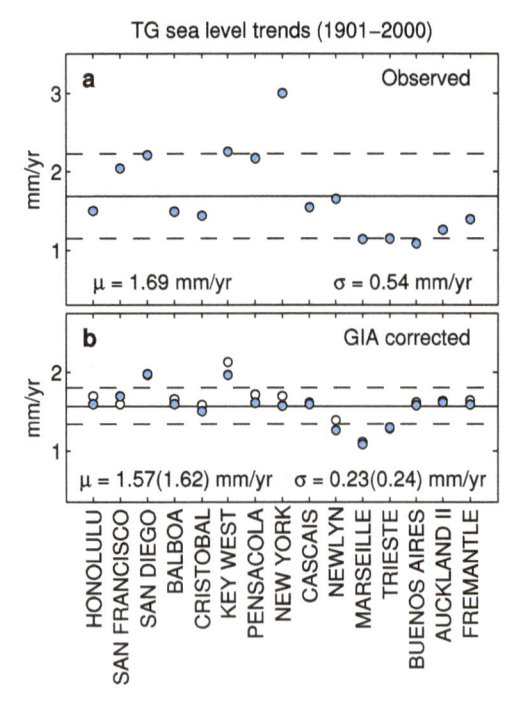

**Figure 1.** Least squares linear sea level trends during 1901–2000. (a) Observed. (b) Corrected for glacial isostatic adjustment (GIA) using the ICE-6G VM5a (blue) and ICE-5G VM2 (white). Solid and dashed lines in both panels represent the mean and 1σ range, respectively. In Figure 1b, the lines correspond to the ICE-6G VM5a correction; values for the mean and standard deviation using the ICE-5G VM2 correction are given in parentheses.

**Figure 9:** Tide gauge records from around the world show an uncorrected average rate of SLR of 1.69 mm/yr and a rate of 1.57 mm/yr when corrected for VLM. Contradicting the 3.4 mm/yr found by satellites.

## Absolute sea level rise

Absolute sea level rise, the actual increase in height of the sea surface, is much smaller than the value derived from tide gauge records corrected for vertical land motion by NOAA's network.[19] Snay et al. (2007)[20] found the average absolute sea level for its 37 corrected U.S. tide stations to be 1.28 mm/yr. Snay et al. (2016)[21] gives an average of vertical land motion for all 1289 CORS stations of minus 1.68 mm/yr – that is downward land movement.

All-in-all, vertical land movement is of the same magnitude as the actual increase in absolute sea level when directly measured by CGPS-corrected tide gauges[22] at places of interest to mankind. That measured increase in the height of the sea surface is found to be far less than 2 mm/yr, and not the 3.4 mm/yr reported for the conceptual eustatic sea level from satellite altimetry. Tide gauge records do not show any acceleration in the rate of either relative sea level or absolute sea level.

Mankind's ports, cities, and seashores have thrived despite the sea level rise of the last century, which in many cases has been ignored. That same slow and steady aspect of global sea level rise and continuing advances in technology make adaptation to future sea level rise eminently possible.

19  NOAA CORS Network (NCN) https://geodesy.noaa.gov/CORS/index.shtml

20  Snay, Richard, et al. "Using global positioning system-derived crustal velocities to estimate rates of absolute sea level change from North American tide gauge records." Journal of Geophysical Research: Solid Earth 112.B4 (2007). https://agupubs.onlinelibrary.wiley.com/doi/full/10.1029/2006JB004606

21  Saleh, Jarir, Yoon, Sungpil, Choi, Kevin, Sun, Lijuan, Snay, Richard, McFarland, Phillip, Williams, Simon, Haw, Don and Coloma, Francine. "1996–2017 GPS position time series, velocities and quality measures for the CORS Network" Journal of Applied Geodesy, vol. 15, no. 2, 2021, pp. 105-115. https://doi.org/10.1515/jag-2020-0041

22  CGPS@TG Working Group of the Sea Level Center at the University of Hawaii https://imina.soest.hawaii.edu/cgps_tg/introduction/index.html

# B

# Causes of Climate Change

# 6

# Why does the IPCC downplay the Sun?

NICOLA SCAFETTA AND FRITZ VAHRENHOLT

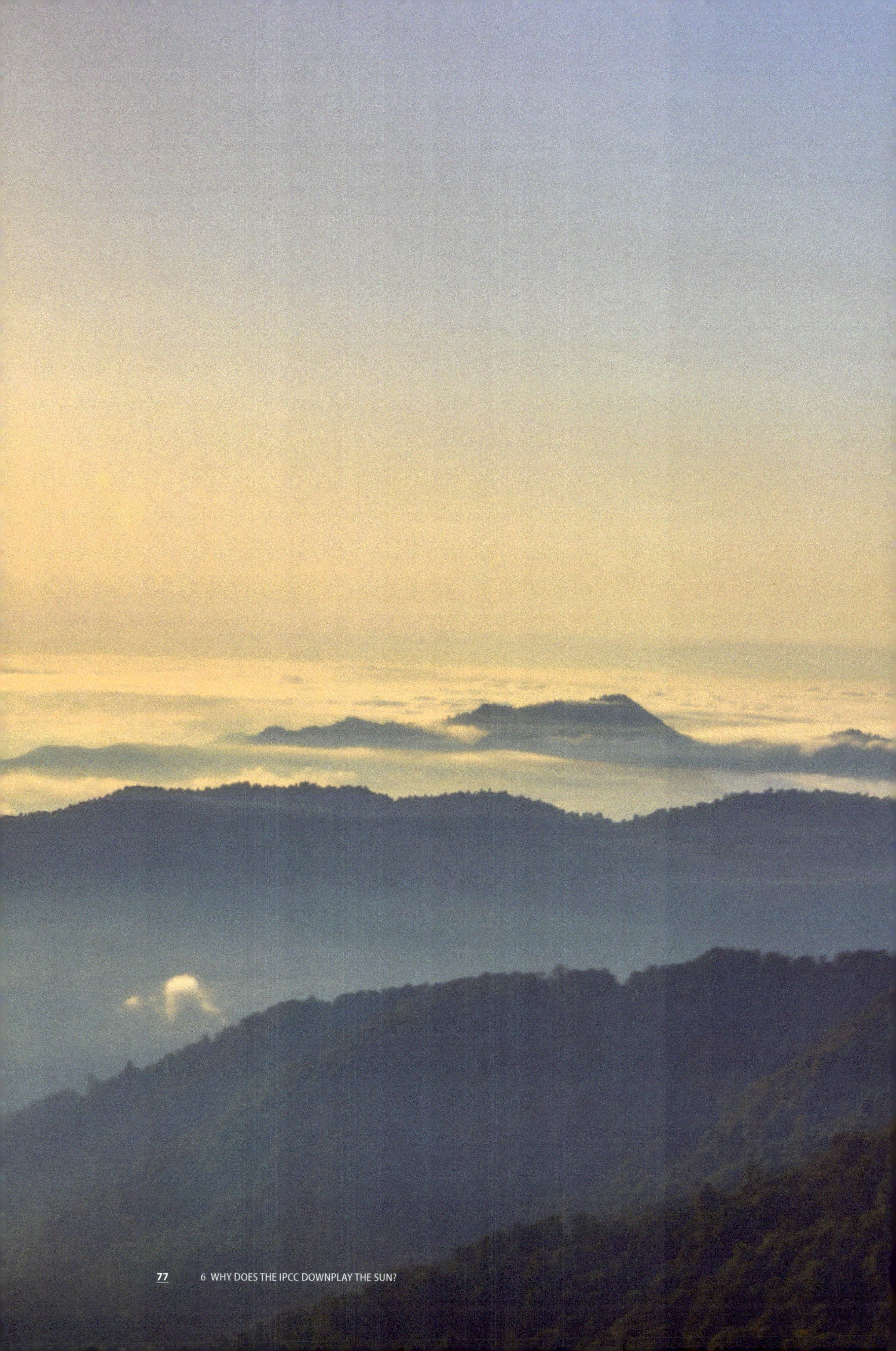

**H**ardly any other topic in the climate sciences is as controversial as the climate effect of the Sun. Whilst some are firmly convinced that the solar influence on climate is negligible, others attribute all climate change to the Sun. The debate is somewhat reminiscent of the geologists' dispute of the early 19th Century when scholars fervently argued about the origin of rocks. The "Neptunists" were quite certain that all rocks were formed as sedimentary deposits in water, while the "Plutonists" saw only volcanic forces at work. Today we know: the truth is in the middle, there are different ways rocks are formed. We smile today about this episode of scientific history, but can we be sure that history is not repeating itself right in front of our eyes? In this chapter we look at the 6th Climate Assessment Report (AR6) from the IPCC (available here) and compare its statements about the solar influence on climate to the wide spectrum of peer-reviewed scientific literature. It is expected that the regular IPCC reports are a thorough and balanced summary of climate publications where uncertainties and their implications are clearly discussed. How well did the IPCC do its job when it comes to the Sun and its potential role in climate change?

It is undisputed that a very large body of scientific publications from all over the world support the claim that variations in solar activity influence local and global climate. However, there is a heated debate as to whether solar related forcings are sufficiently relevant for interpreting global climatic changes and, in particular, the global warming trend of about 1 °C observed since 1900. More specifically the AR6 report mainly advocates the CMIP6 global circulation model climate change attribution results which conclude that the role of the Sun in 20th century global warming is negligible. Moreover, solar forcing is claimed to have been slightly "negative" since the 1980s, which would exclude any solar contribution to the warming observed over the last 40 years.

Let us briefly discuss this topic and highlight a number of issues that were not properly addressed in the IPCC AR6 report that suggest a significant solar contribution to climate change including the past 40 years. The interested reader can find a more detailed analysis of the ongoing debate about the influence of solar activity on climate in the recent review paper by Connolly et al. (2021)[1], which was co-authored by 23 experts in solar-climate interactions and cites 545 works. The review discusses the current uncertainties regarding both solar and climate data, and concludes that different solar forcings and climatic indicators:

> *"suggest everything from no role for the Sun in recent decades (implying that recent global warming is mostly human-caused) to most of the recent global warming being due to changes in solar activity (that is, that recent global warming is mostly natural)".*

Thus, it appears that the conclusions presented in IPCC-AR6 are consistent only with a portion of the published scientific literature, the portion that minimizes the role of the sun while maximizing the anthropogenic component.

## Numerous case studies support solar participation in the climate equation

Let's look at a recent example from Great Britain to illustrate the intriguing relationships found so far.[2] Figure 1 compares the British September temperatures with solar activity. Over many decades a stunning synchronicity of the parameters were observed. Between 1940 and 2000 the match is so good the relationship could have been used to forecast the weather. At the turn of the millennium, however, the correlation broke down, as it did several times in the first half of the 20th century. Quite likely, the two parameters will correlate better again at some point in the future, as the match seems to come and go every few decades. A possible solution to this apparent alternat-

1    Connolly R, Soon W, Connolly M, Baliunas S, Berglund J, Butler CJ, Cionco RG, Elias AG, Fedorov VM, Harde H, et al. How much has the Sun influenced Northern Hemisphere temperature trends? An ongoing debate. *Research in Astronomy and Astrophysics* 2021, 21:131.

2    Lüdecke H-J, Cina R, Dammschneider H-J, Lüning S. Decadal and multidecadal natural variability in European temperature. *Journal of Atmospheric and Solar-Terrestrial Physics* 2020, 205:105294.

ing correlation between the 11-year solar cycle and the corresponding decadal climatic oscillation was proposed by Scafetta[3,4] showing, for example, that globally at the decadal scale the climate system is regulated by two beating oscillations at about 9 and 11 years. The 9-year one appears to be related to long soli-lunar cycles while the 11-year one is related to the solar cycles. The beats between these two (and possible other) close oscillations may produce alternating periods of high and low correlation between solar and climate cycles.

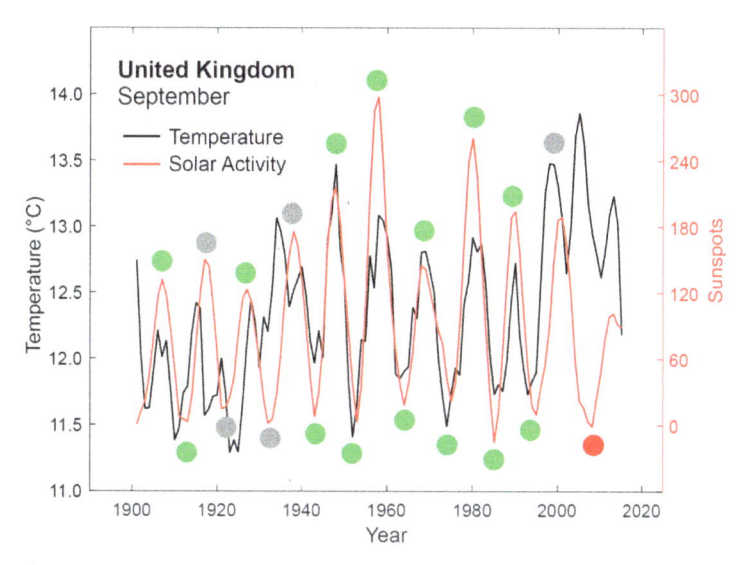

**Figure 1:** Comparison of September temperatures in the UK (black) and variations in solar activity (red) during the last 120 years. Green dots mark phases of good synchronicity, the red dot shows a contrary development. Graphic modified after Lüdecke, Cina[2].

The result shown in Figure 1 is not unique. There is an extensive literature with empirical case studies on solar influence on climate that shows that a good correlation between temperature development and solar activity exists in various regions of the world.[5,6,7,8,9]

The 11-year Schwabe solar cycle has been documented in numerous climate data series, for example in the westerly winds of Central Europe,[10] in tree rings in southern Germany[11] and Japan,[12] in sedimentary deposits of the Ionian Sea[13] and the Bering Sea,[14] in temperatures of Portugal[15] and

3   IPCC. Special Report on global warming of 1.5 °C above pre-industrial levels and related global greenhouse gas emission pathways. http://www.ipcc.ch/report/sr15/ 2018.

4   Scafetta, N.: 2010. Empirical evidence for a celestial origin of the climate oscillations and its implications. *Journal of Atmospheric and Solar-Terrestrial Physics*, 72, 951–970.

5   Usoskin IG, Schüssler M, Solanki SK, Mursula K. Solar activity, cosmic rays, and Earth's temperature: A millennium-scale comparison. *Journal of Geophysical Research: Space Physics* 2005, 110.

6   Zherebtsov GA, Kovalenko VA, Molodykh SI, Kirichenko KE. Solar variability manifestations in weather and climate characteristics. *Journal of Atmospheric and Solar-Terrestrial Physics* 2019, 182:217-222.

7   Lüning S, Vahrenholt F. The Sun's Role in Climate. In: *Chapter 16 in "Evidence-Based Climate Science" (Second Edition)*: Elsevier; 2016, 283-305.

8   Engels S, van Geel B. The effects of changing solar activity on climate: contributions from palaeoclimatological studies. *J. Space Weather Space Clim.* 2012, 2:A09.

9   Ibid 1

10  Schwander M, Rohrer M, Brönnimann S, Malik A. Influence of solar variability on the occurrence of central European weather types from 1763 to 2009. *Clim. Past* 2017, 13:1199-1212.

11  Güttler D, Wacker L, Kromer B, Friedrich M, Synal HA. Evidence of 11-year solar cycles in tree rings from 1010 to 1110 AD – Progress on high precision AMS measurements. *Nuclear Instruments and Methods in Physics Research Section B: Beam Interactions with Materials and Atoms* 2013, 294:459-463.

12  Miyahara H, Yokoyama Y, Masuda K. Possible link between multi-decadal climate cycles and periodic reversals of solar magnetic field polarity. *Earth and Planetary Science Letters* 2008, 272:290-295.

13  Taricco C, Vivaldo G, Alessio S, Rubinetti S, Mancuso S. A high-resolution δ18O record and Mediterranean climate variability. *Clim. Past* 2015, 11:509-522.

14  Katsuki K, Itaki T, Khim B-K, Uchida M, Tada R. Response of the Bering Sea to 11-year solar irradiance cycles during the Bølling-Allerød. *Geophysical Research Letters* 2014, 41:2892-2898.

15  Morozova AL, Barlyaeva TV. The role of climatic forcings in variations of Portuguese temperature: A comparison of spectral and statistical methods. *Journal of Atmospheric and Solar-Terrestrial Physics* 2016, 149:240-257.

Vancouver Island,[16] in the monsoons of East Asia[17,18] and India,[19] in the North American,[20] European,[21] and Arctic[22] winter climate, in the cyclone frequency of the Pacific coast of Mexico,[23] in the Pacific Walker Circulation,[24] in the oceanic heat content of the upper 700 m water column of the Pacific,[25] in the thunderstorms of Brazil[26] and in the water vapour concentration over the Arabian Peninsula.[27] According to mammal researcher Klaus Hackländer (University of Natural Resources and Applied Life Sciences, Vienna), even the mouse population in the Austrian Weinviertel region is controlled by the solar Schwabe cycle.[28] Due to the varying intensity of the winters, a so-called "mouse peak" occurs there about once per decade.

Sometimes it helps to look at things from a distance. For example, what is the situation in other parts of the solar system? Are other planets and their moons affected by fluctuations in solar activity of any kind? This is indeed the case. The 11-year solar cycle leads to systematic changes in other planetary atmospheres as well. On Uranus and Neptune, the brightness of the atmosphere changes, caused by fluctuations of UV and cosmic radiation.[29,30] In the atmosphere of Saturn's moon Titan the methane concentration changes with the rhythm of the solar cycle.[31]

However, solar variability is characterized by longer cycles as well and these too have been found in the climate system by numerous authors. For example, the solar Gleissberg cycles (90 years)[32] and Suess-DeVries cycles (210 years) were observed in the Atlantic deep water circulation,[33] in the westerly winds of the Falkland Islands,[34] in the climate of the North-East Pacific,[35] in the South American monsoon of Northeast Brazil,[36] in the temperatures of Tibet,[37] in the precipita-

16  Patterson RT, Chang AS, Prokoph A, Roe HM, Swindles GT. Influence of the Pacific Decadal Oscillation, El Niño-Southern Oscillation and solar forcing on climate and primary productivity changes in the northeast Pacific. *Quaternary International* 2013, 310:124-139.

17  Zhao L, Wang J-S. Robust Response of the East Asian Monsoon Rainband to Solar Variability. *Journal of Climate* 2014, 27:3043-3051.

18  Wang J-S, Zhao L. Statistical tests for a correlation between decadal variation in June precipitation in China and sunspot number. *Journal of Geophysical Research: Atmospheres* 2012, 117.

19  van Loon H, Meehl GA. The Indian summer monsoon during peaks in the 11 year sunspot cycle. *Geophys. Res. Lett.* 2012, 39:L13701.

20  Liu Z, Yoshimura K, Buenning NH, He X. Solar cycle modulation of the Pacific–North American teleconnection influence on North American winter climate. *Environmental Research Letters* 2014, 9:024004.

21  Brugnara Y, Brönnimann S, Luterbacher J, Rozanov E. Influence of the sunspot cycle on the Northern Hemisphere wintertime circulation from long upper-air data sets. *Atmos. Chem. Phys.* 2013, 13:6275-6288.

22  Roy I. Solar cyclic variability can modulate winter Arctic climate. *Scientific Reports* 2018, 8:4864.

23  Pazos M, Mendoza B. Landfalling Tropical Cyclones along the Eastern Pacific Coast between the Sixteenth and Twentieth Centuries. *Journal of Climate* 2013, 26:4219-4230.

24  Misios S, Gray LJ, Knudsen MF, Karoff C, Schmidt H, Haigh JD. Slowdown of the Walker circulation at solar cycle maximum. *Proceedings of the National Academy of Sciences* 2019, 116:7186-7191.

25  Wang G, Yan S, Qiao F. Decadal variability of upper ocean heat content in the Pacific: Responding to the 11-year solar cycle. *Journal of Atmospheric and Solar-Terrestrial Physics* 2015, 135:101-106.

26  Pinto Neto O, Pinto IRCA, Pinto O. The relationship between thunderstorm and solar activity for Brazil from 1951 to 2009. *Journal of Atmospheric and Solar-Terrestrial Physics* 2013, 98:12-21.

27  Maghrabi AH. Multi- decadal variations and periodicities of the precipitable water vapour (PWV) and their possible association with solar activity: Arabian Peninsula. *Journal of Atmospheric and Solar-Terrestrial Physics* 2019, 185:22-28.

28  Brickner I. Weinviertel: Von der Klimaerwärmung zur Mäuseplage. *Der Standard, 2.8.2019,* https://www.derstandard.at/story/2000106952286/von-der-klimaerwaermung-zur-maeuseplage 2019.

29  Aplin KL, Harrison RG. Determining solar effects in Neptune's atmosphere. *Nature Communications* 2016, 7:11976.

30  Aplin KL, Harrison RG. Solar-Driven Variation in the Atmosphere of Uranus. *Geophysical Research Letters* 2017, 44:12,083-012,090.

31  Westlake JH, Waite JH, Bell JM, Perryman R. Observed decline in Titan's thermospheric methane due to solar cycle drivers. *Journal of Geophysical Research: Space Physics* 2014, 119:8586-8599.

32  Feynman J, Ruzmaikin A. The Centennial Gleissberg Cycle and its association with extended minima. *Journal of Geophysical Research: Space Physics* 2014, 119:6027-6041.

33  Seidenglanz A, Prange M, Varma V, Schulz M. Ocean temperature response to idealized Gleissberg and de Vries solar cycles in a comprehensive climate model. *Geophys. Res. Lett.* 2012, 39:L22602.

34  Turney CSM, Jones RT, Fogwill C, Hatton J, Williams AN, Hogg A, Thomas ZA, Palmer J, Mooney S, Reimer RW. A 250-year periodicity in Southern Hemisphere westerly winds over the last 2600 years. *Clim. Past* 2016, 12:189-200.

35  Galloway JM, Wigston A, Patterson RT, Swindles GT, Reinhardt E, Roe HM. Climate change and decadal to centennial-scale periodicities recorded in a late Holocene NE Pacific marine record: Examining the role of solar forcing. *Palaeogeography, Palaeoclimatology, Palaeoecology* 2013, 386:669-689.

36  Novello VF, Cruz FW, Karmann I, Burns SJ, Stríkis NM, Vuille M, Cheng H, Lawrence Edwards R, Santos RV, Frigo E, et al. Multidecadal climate variability in Brazil's Nordeste during the last 3000 years based on speleothem isotope records. *Geophysical Research Letters* 2012, 39.

37  Li X, Liang J, Hou J, Zhang W. Centennial-scale climate variability during the past 2000 years on the central Tibetan Plateau. *The Holocene* 2015, 25:892-899.

tion of northwest,[38] northeast,[39] south,[40] and central[41] China, in the nitrate content of the polar ice caps,[42] in the growing season of the northern hemisphere,[43] in the subtropical monsoon of the Northern Hemisphere[44] and in global tree ring data.[45]

There is strong evidence that climatic millennium-scale cycles with periods of 1000-2500 years were caused by the Sun. Actually, the Sun appears to beat with a quasi-millennial cycle (known as the Eddy cycle)[46,47,48] and a quasi-2300-year cycle known as the Bray-Hallstatt cycle.[49] Gerard Bond and colleagues first described the cycles from the North Atlantic and explicitly stated that they were synchronous with solar activity.[50] Since then, the Millennium cycles have been described from all over the world.[51] In many cases, the respective study authors established a connection to solar activity, for example in the USA,[52,53] in Brazil,[54] Patagonia,[55] Peru,[56] Antarctica,[57] South Africa,[58] Morocco,[59] Oman,[60] India,[61] China,[62] Australia,[63] Spain,[64] Austria[65] and Finland.[66]

38   Tiwari RK, Rajesh R. Imprint of long-term solar signal in groundwater recharge fluctuation rates from Northwest China. *Geophysical Research Letters* 2014, 41:3103-3109.

39   Chu G, Sun Q, Xie M, Lin Y, Shang W, Zhu Q, Shan Y, Xu D, Rioual P, Wang L, et al. Holocene cyclic climatic variations and the role of the Pacific Ocean as recorded in varved sediments from northeastern China. *Quaternary Science Reviews* 2014, 102:85-95.

40   Zhao K, Wang Y, Edwards RL, Cheng H, Liu D, Kong X. A high-resolved record of the Asian Summer Monsoon from Dongge Cave, China for the past 1200 years. *Quaternary Science Reviews* 2015, 122:250-257.

41   Liu D, Wang Y, Cheng H, Edwards RL, Kong X. Cyclic changes of Asian monsoon intensity during the early mid-Holocene from annually-laminated stalagmites, central China. *Quaternary Science Reviews* 2015, 121:1-10.

42   Ogurtsov MG, Oinonen M. Evidence of the solar Gleissberg cycle in the nitrate concentration in polar ice. *Journal of Atmospheric and Solar-Terrestrial Physics* 2014, 109:37-42.

43   Ogurtsov M, Lindholm M, Jalkanen R, Veretenenko S. Evidence for the Gleissberg solar cycle at the high-latitudes of the Northern Hemisphere. *Advances in Space Research* 2015, 55:1285-1290.

44   Knudsen MF, Jacobsen BH, Riisager P, Olsen J, Seidenkrantz M-S. Evidence of Suess solar-cycle bursts in subtropical Holocene speleothem δ18O records. *The Holocene* 2012, 22:597-602.

45   Breitenmoser P, Beer J, Brönnimann S, Frank D, Steinhilber F, Wanner H. Solar and volcanic fingerprints in tree-ring chronologies over the past 2000 years. *Palaeogeography, Palaeoclimatology, Palaeoecology* 2012, 313–314:127-139.

46   Kerr, R.A.: 2001, A variable Sun paces millennial climate. *Science* 294, 1431.

47   Neff, U., Burns, S.J., Mangini, A., Mudelsee, M., Fleitmann, D., Matter, A.: 2001, Strong coherence between solar variability and the monsoon in Oman between 9 and 6 kyr ago. *Nature* 411, 290.

48   Ogurtsov, M.G., Nagovitsyn, Y.A., Kocharov, G.E., Jungner, H.: 2002, Long-period cycles of the Sun's activityrecorded in direct solar data and proxies. *Solar Phys.* 211, 371.

49   McCracken, K.G., Beer, J., Steinhilber, F., Abreu, J.: 2013, A phenomenological study of the cosmic ray variations over the past 9400 years, and their implications regarding solar activity and the solar dynamo. *Solar Phys.* 286, 609.

50   Bond G, Kromer B, Beer J, Muscheler R, Evans MN, Showers W, Hoffmann S, Lotti-Bond R, Hajdas I, Bonani G. Persistent Solar Influence on North Atlantic Climate During the Holocene. *Science* 2001, 294:2130-2136.

51   Ibid 7

52   Willard DA, Bernhardt CE, Korejwo DA, Meyers SR. Impact of millennial-scale Holocene climate variability on eastern North American terrestrial ecosystems: pollen-based climatic reconstruction. *Global and Planetary Change* 2005, 47:17-35.

53   Springer GS, Rowe HD, Hardt B, Edwards RL, Cheng H. Solar forcing of Holocene droughts in a stalagmite record from West Virginia in east-central North America. *Geophysical Research Letters* 2008, 35:1-5.

54   Bernal JP, Cruz FW, Stríkis NM, Wang X, Deininger M, Catunda MCA, Ortega-Obregón C, Cheng H, Edwards RL, Auler AS. High-resolution Holocene South American monsoon history recorded by a speleothem from Botuverá Cave, Brazil. *Earth and Planetary Science Letters* 2016, 450:186-196.

55   Kilian R, Lamy F. A review of Glacial and Holocene paleoclimate records from southernmost Patagonia (49–55°S). *Quaternary Science Reviews* 2012, 53:1-23.

56   Bush MB, Hansen BCS, Rodbell DT, Seltzer GO, Young KR, León B, Abbott MB, Silman MR, Gosling WD. A 17 000-year history of Andean climate and vegetation change from Laguna de Chochos, Peru. *Journal of Quaternary Science* 2005, 20:703-714.

57   Crosta X, Debret M, Denis D, Courty MA, Ther O. Holocene long- and short-term climate changes off Adélie Land, East Antarctica. *Geochem. Geophys. Geosyst.* 2007, 8:1-15.

58   outhern Africa. Quaternary *Science Reviews* 2003, 22:2311-2326.

59   Zielhofer C, Köhler A, Mischke S, Benkaddour A, Mikdad A, Fletcher WJ. Western Mediterranean hydro-climatic consequences of Holocene ice-rafted debris (Bond) events. *Clim. Past* 2019, 15:463-475.

60   Fleitmann D, Burns SJ, Mudelsee M, Neff U, Kramers J, Mangini A, Matter A. Holocene Forcing of the Indian Monsoon Recorded in a Stalagmite from Southern Oman. *Science* 2003, 300:1737-1739.

61   Thamban M, Kawahata H, Rao V. Indian summer monsoon variability during the holocene as recorded in sediments of the Arabian Sea: Timing and implications. *Journal of Oceanography* 2007, 63:1009-1020.

62   Wang Y, Cheng H, Edwards RL, He Y, Kong X, An Z, Wu J, Kelly MJ, Dykoski CA, Li X. The Holocene Asian Monsoon: Links to Solar Changes and North Atlantic Climate. *Science* 2005, 308:854-857.

63   McGowan HA, Marx SK, Soderholm J, Denholm J. Evidence of solar and tropical-ocean forcing of hydroclimate cycles in southeastern Australia for the past 6500 years. *Geophysical Research Letters* 2010, 37.

64   Fletcher WJ, Debret M, Goñi MFS. Mid-Holocene emergence of a low-frequency millennial oscillation in western Mediterranean climate: Implications for past dynamics of the North Atlantic atmospheric westerlies. *The Holocene* 2013, 23:153-166.

65   Mangini A, Verdes P, Spötl C, Scholz D, Vollweiler N, Kromer B. Persistent influence of the North Atlantic hydrography on central European winter termperature during the last 9000 years. *Geophysical Research Letters* 2007, 34:n/a-n/a.

66   Ojala AEK, Launonen I, Holmström L, Tiljander M. Effects of solar forcing and North Atlantic oscillation on the climate of continental Scandinavia during the Holocene. *Quaternary Science Reviews* 2015, 112:153-171.

The longer Bray-Hallstatt cycle was also found in a number of climate records.[67,68,69,70] In other cases, the connection with solar development was not checked, so that no statement was made about this for the time being,[71,72,73,74] but a solar climate driver is quite likely.

Despite the uncertainties appropriate to the various climatic records, which for the past are based on proxies, the observed solar-climate oscillations are likely real and not coincidental common patterns produced by complex but independent dynamics. In fact, it has been demonstrated that the typical observed frequencies correspond to the natural gravitational oscillations of the solar system (which are labelled as "invariant inequalities") that appear to simultaneously synchronize both solar activity and climate change from the decadal to the multimillennial scales.[75,76,77]

Why are these natural oscillations so important for accurately interpreting climate change during the last century? Because they reveal a natural climatic variability driven by solar or astronomical forcings that are not fully understood physically. Even so, they highlight some of the serious limitations of the present IPCC climate models (such as the CMIP5 and CMIP6 GCMs) that cannot reproduce them.[78,79,80] Let us discuss this point.

## The Past as a Plausibility Check

Let us take a brief look at the Medieval Warm Period (MWP), the last pre-industrial warm period, which in its wider interpretation is dated from 800-1300 AD. Solar activity started increasing around 700 AD and remained high until 1250 AD. Warm MWP and a strong sun - is this again just one of these "coincidences" (Figure 2)? Did the sun spend the first 100 years gradually revving up an inert climate system? Several researchers see the sun as the cause of the MWP warming, for example the authors of a study on the Tibet Plateau, where the MWP was warmer than it is today.[81] Even one of the hockey stick co-authors noted a connection to the sun, one year after the legendary temperature curve was published. In an e-mail (which was discovered in the "Climategate" collection from the University of East Anglia's climate research centre in Norwich, England in 2009) Raymond Bradley acknowledged to his hockey stick supporters and other colleagues that the MWP may have had a similar temperature level as today.[82] He suspected the Sun was the trigger:

67    Bray, J.R.: 1968, Glaciation and solar activity since the fifth century BC and the solar cycle. *Nature* 220, 672.

68    O'Brien, S.R., Mayewski, P.A., Meeker, L.D., et al.: 1995, Complexity of Holocene climate as reconstructed from a Greenland ice core. *Science* 270, 1962.

69    Pestiaux, P., Van Der Mersch, I., Berger, A., Duplessy, J.C.: 1988, Paleoclimatic variability at frequencies ranging from 1 cycle per 10,000 years to 1 cycle per 1000 years: Evidence for nonlinear behavior of the climate system. *Clim. Change* 12, 9.

70    Vasiliev, S.S., Dergachev, V.A.: 2002, The 2400-year cycle in atmospheric radiocarbon concentration: bispectrum of 14C data over the last 8000 years. *Ann. Geophys.* 20, 115.

71    Evangelista H, Gurgel M, Sifeddine A, Rigozo NR, Boussafir M. South Tropical Atlantic anti-phase response to Holocene Bond Events. *Palaeogeography, Palaeoclimatology, Palaeoecology* 2014, 415:21-27.

72    Voigt I, Chiessi CM, Prange M, Mulitza S, Groeneveld J, Varma V, Henrich R. Holocene shifts of the southern westerlies across the South Atlantic. *Paleoceanography* 2015, 30:39-51.

73    Arz HW, Gerhardt S, Pätzold J, Röhl U. Millennial-scale changes of surface- and deep-water flow in the western tropical Atlantic linked to Northern Hemisphere high-latitude climate during the Holocene. *Geology* 2001, 29:239-242.

74    Kemp J, Radke LC, Olley J, Juggins S, De Deckker P. Holocene lake salinity changes in the Wimmera, southeastern Australia, provide evidence for millennial-scale climate variability. *Quaternary Research* 2012, 77:65-76.

75    Scafetta, N.: 2012. Multi-scale harmonic model for solar and climate cyclical variation throughout the Holocene based on Jupiter-Saturn tidal frequencies plus the 11-year solar dynamo cycle. *Journal of Atmospheric and Solar-Terrestrial Physics*, 80, 296–311.

76    Scafetta, N., Milani, F., Bianchini, A., Ortolani, S.: 2016. On the astronomical origin of the Hallstatt oscillation found in radiocarbon and climate records throughout the Holocene. *Earth-Science Reviews*, 162, 24–43.

77    Scafetta, N.: 2020. Solar Oscillations and the Orbital Invariant Inequalities of the Solar System. *Solar Physics*, 295(2), 33.

78    Scafetta, N.: 2013. Discussion on climate oscillations: CMIP5 general circulation models versus a semiempirical harmonic model based on astronomical cycles. *Earth-Science Reviews*, 126, 321–357.

79    Scafetta, N.: 2021. Reconstruction of the Interannual to Millennial Scale Patterns of the Global Surface Temperature. *Atmosphere*, 12, 147.

80    Scafetta, N.: 2021. Testing the CMIP6 GCM Simulations versus Surface Temperature Records from 1980–1990 to 2011–2021: High ECS Is Not Supported. *Climate* 9, 161.

81    He Y, Liu W, Zhao C, Wang Z, Wang H, Liu Y, Qin X, Hu Q, An Z, Liu Z. Solar influenced late Holocene temperature changes on the northern Tibetan Plateau. *Chinese Science Bulletin* 2013, 58:1053-1059.

82    Bradley R. Email. *10.7.2000*, http://di2.nu/foia/foia2011/mail/0207.txt 2000.

*"[...] it may be that Mann et al simply don't have the long-term trend right [...] which of course begs the question as to what the likely forcing was 1,000 years ago. (My money is firmly on an increase in solar irradiance...)".*

The strong sun during the MWP also reduced the Aleutian Low system,[83] which is otherwise linked to the Pacific Decadal Oscillation (PDO) on shorter time scales.[84] Between 1010 and 1040 AD, during the solar Oort Minimum, the sun weakened briefly for three decades. In several local temperature reconstructions, the temperature also dropped, e.g., in Kenya,[85] Morocco,[86] and Antarctica.[87]

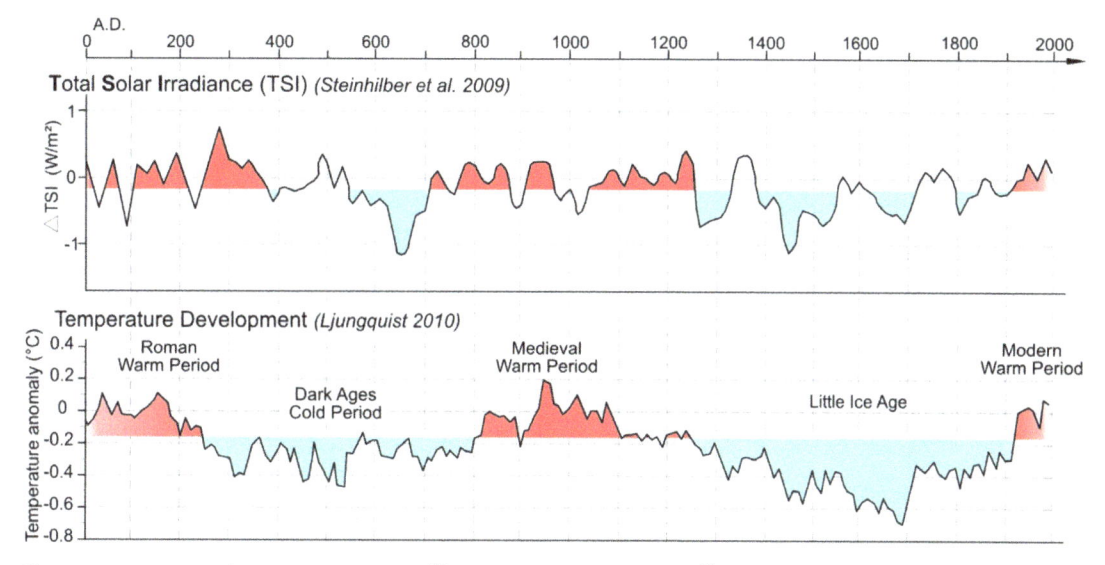

**Figure 2:** Long-term synchronicity of solar activity[88] and temperature development[89] (non-tropical Northern Hemisphere) over the last 2000 years.

The Little Ice Age (1300-1850 AD) was one of the coldest periods of the last 10,000 years. Solar activity was at a very low level, a level only rarely seen in the last ten millennia.[90] Is this really just a coincidence? Then it would also be purely coincidental that modern warming had its strongest warming impulse in the second half of the 20th century, when solar activity reached one of the highest values of the last 10,000 years.[91] Coincidences are unforeseen events that have a meaning, said the ancient Greek philosopher Diogenes of Sinope.

The Little Ice Age (LIA) offers an exciting research laboratory for solar-climate effects. On several occasions, solar activity fell sharply for several decades and then recovered rapidly. These are the Wolf Minimum (1280-1350), Spörer Minimum (1460-1550), Maunder Minimum (1645-1715), and the Dalton Minimum (1790-1830). Empirical paleoclimatology can be used to investigate how these phases of solar weakness led to climate change. If the sun really does not have a significant effect on the climate—as the AR6 assumes—the solar minima must have had no effect on the cli-

83  Osterberg EC, Mayewski PA, Fisher DA, Kreutz KJ, Maasch KA, Sneed SB, Kelsey E. Mount Logan ice core record of tropical and solar influences on Aleutian Low variability: 500–1998 A.D. *Journal of Geophysical Research: Atmospheres* 2014, 119:11,189-111,204.

84  Dong X, Su T-H, Wang J, Lin R-P. Decadal Variation of the Aleutian Low-Icelandic Low Seesaw Simulated by a Climate System Model (CAS-ESM-C). *Atmospheric and Oceanic Science Letters* 2014, 7:110-114.

85  Berke MA, Johnson TC, Werne JP, Schouten S, Sinninghe Damsté JS. A mid-Holocene thermal maximum at the end of the African Humid Period. *Earth and Planetary Science Letters* 2012, 351–352:95-104.

86  Zielhofer C, Fletcher WJ, Mischke S, De Batist M, Campbell JFE, Joannin S, Tjallingii R, El Hamouti N, Junginger A, Stele A, et al. Atlantic forcing of Western Mediterranean winter rain minima during the last 12,000 years. *Quaternary Science Reviews* 2017, 157:29-51.

87  Noon PE, Leng MJ, Jones VJ. Oxygen-isotope (δ18O) evidence of Holocene hydrological changes at Signy Island, maritime Antarctica. *The Holocene* 2003, 13:251-263.

88  Steinhilber F, Beer J, Fröhlich C. Total solar irradiance during the Holocene. *Geophysical Research Letters* 2009, 36.

89  Ljungqvist FC. A new reconstruction of temperature variability in the extra-tropical northern hemisphere during the last two millennia. *Geografiska Annaler: Series A* 2010, 92:339-351.

90  Steinhilber F, Abreu JA, Beer J, Brunner I, Christl M, Fischer H, Heikkilä U, Kubik PW, Mann M, McCracken KG, et al. 9,400 years of cosmic radiation and solar activity from ice cores and tree rings. *Proceedings of the National Academy of Sciences* 2012, 109:5967-5971.

91  Ibid 5

mate. The pronounced solar minima of the LIA therefore provide an excellent scientific blind test of the solar-climate connection.

The result is surprising. Contrary to expectations, the solar minima caused enormous climatic changes. Studies from Spain[92] and Portugal[93] document significant cooling during the Maunder and Dalton minima. Studies from Slovakia,[94] China,[95] Bhutan[96] and the Canadian Rocky Mountains[97] reported cold periods that occurred at the same time as the Spörer, Maunder, and Dalton periods of solar weakness. During the Wolf, Spörer and Maunder minima, the intermediate water layers of the North Atlantic cooled by 2-3 °C,[98] while the surface water in the tropical North Atlantic off Mauritania cooled by 1 °C.[99] On Sakhalin, Russia's largest island, the lowest temperatures were recorded during the Maunder Minimum.[100] Tree ring surveys in Tasmania, Australia, found cold phases during the Spörer and Maunder Minima.[101]

Even in Antarctica, the repeated drop in solar activity was very noticeable. A Chinese research group led by Yuesong Gao reconstructed the population of Adelie penguins in a bay in East Antarctica for the past 1000 years.[102] They took a sediment core from the seabed near the coast and examined it for penguin droppings. The researchers found that the population was strongly influenced by solar activity. Whenever solar activity decreased, the population collapsed, especially during the Spörer, Maunder, and Dalton minima. In between, the penguins recovered. Lower solar activity decreased the local phytoplankton, reducing the krill population, and that decreased the penguin population.

On a global scale, these alternating warm and cold periods are correlated with secular trends of solar activity.[103,104,105] Moreover, as Figure 2 shows, it appears that such common solar and climate variations are regulated mostly by a millennial cycle[106,107] and by a quasi-115-year cycle. Both cycles are accurately predicted by a combination of the 11-year solar cycle with two of the main gravitational oscillations of the solar system driven by the combined effects of the orbits of Jupiter and Saturn.[108,109]

Regarding the current warm period, as Figure 2 suggests, solar activity has increased from the 18th to the 20th century. The new level may be at a higher level than during the Medieval Warm

92  Tejedor E, Saz MÁ, Cuadrat JM, Esper J, de Luis M. Temperature variability in the Iberian Range since 1602 inferred from tree-ring records. *Clim. Past* 2017, 13:93-105.

93  Santos JA, Carneiro MF, Correia A, Alcoforado MJ, Zorita E, Gómez-Navarro JJ. New insights into the reconstructed temperature in Portugal over the last 400 years. *Clim. Past* 2015, 11:825-834.

94  Büntgen U, Kyncl T, Ginzler C, Jacks DS, Esper J, Tegel W, Heussner K-U, Kyncl J. Filling the Eastern European gap in millennium-long temperature reconstructions. *Proceedings of the National Academy of Sciences* 2013, 110:1773-1778.

95  Shi F, Yang B, Von Gunten L. Preliminary multiproxy surface air temperature field reconstruction for China over the past millennium. *Science China Earth Sciences* 2012, 55:2058-2067.

96  Krusic PJ, Cook ER, Dukpa D, Putnam AE, Rupper S, Schaefer J. Six hundred thirty-eight years of summer temperature variability over the Bhutanese Himalaya. *Geophysical Research Letters* 2015, 42:2988-2994.

97  Luckman BH, Wilson RJS. Summer temperatures in the Canadian Rockies during the last millennium: a revised record. *Climate Dynamics* 2005, 24:131-144.

98  Moffa-Sanchez P, Born A, Hall IR, Thornalley DJR, Barker S. Solar forcing of North Atlantic surface temperature and salinity over the past millennium. *Nature Geosci* 2014, 7:275-278.

99  Kuhnert H, Mulitza S. Multidecadal variability and late medieval cooling of near-coastal sea surface temperatures in the eastern tropical North Atlantic. *Paleoceanography* 2011, 26:PA4224.

100  Wiles GC, Solomina O, D'Arrigo R, Anchukaitis KJ, Gensiarovsky YV, Wiesenberg N. Reconstructed summer temperatures over the last 400 years based on larch ring widths: Sakhalin Island, Russian Far East. *Climate Dynamics* 2015, 45:397-405.

101  Cook E, Bird T, Peterson M, Barbetti M, Buckley B, D'Arrigo R, Francey R. Climatic change over the last millennium in Tasmania reconstructed from tree-rings. *The Holocene* 1992, 2:205-217.

102  Gao Y, Yang L, Yang W, Wang Y, Xie Z, Sun L. Dynamics of penguin population size and food availability at Prydz Bay, East Antarctica, during the last millennium: A solar control. *Palaeogeography, Palaeoclimatology, Palaeoecology* 2019, 516:220-231.

103  Hoyt, D.V.,Schatten,K.H.,1997.TheRoleoftheSunintheClimateChange.Oxford UniversityPress,NewYork.

104  Scafetta, N.,West, B.J.: 2007. Phenomenological reconstructions of the solar signature in the NH surface temperature records since 1600. *Journal of Geophysical Research*, 112, D24S03.

105  Scafetta, N.: 2009. Empirical analysis of the solar contribution to global mean air surface temperature change. *Journal of Atmospheric and Solar-Terrestrial Physics*, 71, 1916–1923.

106  Scafetta, N.: 2014. Discussion on the spectral coherence between planetary, solar and climate oscillations: a reply to some critiques. *Astrophysics and Space Science*, 354, 275–299.

107  Ibid 46

108  Ibid 75

109  Ibid 78

Period[110]. The increase in solar activity correlates well with the current global climate warming.[111,112,113] This is discussed in detail in Connolly et al. (2021)[114].

## Sun influences rain

The sun influences not only temperatures, but also rain. In Europe, the solar fingerprint is found in the precipitation during the months of February, April, June, and July.[115] Several studies report on flood phases, which occur mainly at times of low solar activity.[116,117] A solar signature also exists in rainfall in the USA,[118,119] of the Tibet Plateau,[120] the monsoon of South America,[121] the Indian monsoon[122], and the Asian summer monsoon[123]. A solar imprint is also seen in the water flow of rivers in the USA,[124] in Egypt[125], and Brazil.[126] The 11-year solar cycle shapes the flow rates of the Amazon[127] as well as the water levels of the Great Lakes in North America,[128] the Caspian Sea[129,130], and Lake Victoria in East Africa.[131] The influence of the sun on rain is at least as strong as on temperatures and is based upon shifts in wind patterns and clouds.

## IPCC's AR6 downplays the Sun

Considering the large number of publications on solar effects in the climate system, one would expect the IPCC to review this subject thoroughly and in great detail. Which climate elements in which parts of the world, during which season show the greatest link to solar activity changes? What are the potential physical processes behind these links? Have climate models been able to reproduce these empirically well-established relationships? Successful model "hindcasts" should reproduce the solar impact on climate correctly. Failure to replicate the observed relationship in

110 Ibid 5

111 Ibid 103

112 Ibid 104

113 Ibid 105

114 Ibid 1

115 Laurenz L, Lüdecke H-J, Lüning S. Influence of solar activity changes on European rainfall. *Journal of Atmospheric and Solar-Terrestrial Physics* 2019, 185:29-42.

116 Czymzik M, Muscheler R, Brauer A. Solar modulation of flood frequency in central Europe during spring and summer on interannual to multi-centennial timescales. *Clim. Past* 2016, 12:799-805.

117 Peña JC, Schulte L, Badoux A, Barriendos M, Barrera-Escoda A. Influence of solar forcing, climate variability and modes of low-frequency atmospheric variability on summer floods in Switzerland. *Hydrol. Earth Syst. Sci.* 2015, 19:3807-3827.

118 Nitka W, Burnecki K. Impact of solar activity on precipitation in the United States. *Physica A: Statistical Mechanics and its Applications* 2019, 527:121387.

119 Jones MD, Metcalfe SE, Davies SJ, Noren A. Late Holocene climate reorganisation and the North American Monsoon. *Quaternary Science Reviews* 2015, 124:290-295.

120 Sun J, Liu Y. Tree ring based precipitation reconstruction in the south slope of the middle Qilian Mountains, northeastern Tibetan Plateau, over the last millennium. *J. Geophys. Res.* 2012, 117:D08108.

121 Vuille M, Burns SJ, Taylor BL, Cruz FW, Bird BW, Abbott MB, Kanner LC, Cheng H, Novello VF. A review of the South American monsoon history as recorded in stable isotopic proxies over the past two millennia. *Climate of the Past* 2012, 8:1309-1321.

122 Kodera K. Solar influence on the Indian Ocean Monsoon through dynamical processes. *Geophysical Research Letters* 2004, 31.

123 Liu D, Wang Y, Cheng H, Edwards RL, Kong X. Remote vs. local control on the Preboreal Asian hydroclimate and soil processes recorded by an annually-laminated stalagmite from Daoguan Cave, southern China. *Quaternary International* 2017, 452:79-90.

124 Wallace MG. Application of lagged correlations between solar cycles and hydrosphere components towards sub-decadal forecasts of streamflows in the Western USA. *Hydrological Sciences Journal* 2019, 64:137-164.

125 Hennekam R, Jilbert T, Schnetger B, de Lange GJ. Solar forcing of Nile discharge and sapropel S1 formation in the early to middle Holocene eastern Mediterranean. *Paleoceanography* 2014, 29:343-356.

126 Mauas PJD, Buccino AP, Flamenco E. Long-term solar activity influences on South American rivers. *Journal of Atmospheric and Solar-Terrestrial Physics* 2011, 73:377-382.

127 Antico A, Torres ME. Evidence of a decadal solar signal in the Amazon River: 1903 to 2013. *Geophysical Research Letters* 2015, 42:10,782-710,787.

128 Watras CJ, Read JS, Holman KD, Liu Z, Song Y-Y, Watras AJ, Morgan S, Stanley EH. Decadal oscillation of lakes and aquifers in the upper Great Lakes region of North America: Hydroclimatic implications. *Geophysical Research Letters* 2014, 41:456-462.

129 Kaftan V, Komitov B, Lebedev S. Analysis of sea level changes in the Caspian Sea related to Cosmo-geophysical processes based on satellite and terrestrial data. *Geodesy and Geodynamics* 2018, 9:449-455.

130 Naderi Beni A, Lahijani H, Mousavi Harami R, Arpe K, Leroy SAG, Marriner N, Berberian M, Andrieu-Ponel V, Djamali M, Mahboubi A, et al. Caspian sea-level changes during the last millennium: historical and geological evidence from the south Caspian Sea. *Clim. Past* 2013, 9:1645-1665.

131 Stager CJ, Ryves D, Cumming FB, Meeker DL, Beer J. Solar variability and the levels of Lake Victoria, East Africa, during the last millenium. *Journal of Paleolimnology* 2005, 33:243-251.

climate models would imply that the solar effect is not understood well enough. In that case, quantification of the solar contribution to climate change of the past 170 years would not be possible.

Surprisingly, none of these topics are adequately addressed in AR6. The Sun appears in only three of the chapters of the AR6 *Climate Change 2021: The Physical Science Basis* volume. Chapter 2 describes changes in solar activity over the past century and millennia, concluding that the variability was much too small to impact climate in a significant way (subchapter 2.2.1). Chapter 7 acknowledges that solar activity changes in the ultraviolet (UV) part of the spectrum are much greater than in the visible part (subchapter 7.3.4.4). Nevertheless, according to the IPCC, even this does not imply a meaningful climate impact. In the same chapter, recent work by Henrik Svensmark and his team on galactic cosmic ray amplifiers of the solar climatic effect are rejected (subchapter 7.3.4.5). Yet, Svensmark and colleagues have recently published a significant new work supporting their hypothesis that cosmic rays have a significant effect on cloud formation.[132] Chapter 10 cites a few case studies in which solar influence on climate is well documented (subchapter 10.1.3.1). However, the AR6 authors do not follow up on these observations and investigate whether climate models are capturing these historical relationships.

Moreover, whilst the IPCC takes for granted strong amplifiers that boost the $CO_2$ warming effect from 1.1 to 4.0 °C per doubling of $CO_2$ so that the equilibrium climate sensitivity of the CMIP6 GCMs can vary from 1.8 to 5.7 °C, it denies similar amplifiers to solar forcing. Climate history shows that the sun has a considerable influence on the climate, both in pre-industrial and industrial times. Therefore, amplifiers and/or alternative reconstructions showing larger solar activity variations are clearly needed to explain the empirical data. In any case, the mechanism probably does not work solely through total solar irradiance (TSI). The TSI changes appear insufficient, but the TSI forcing model adopted by the CMIP5 and CMIP6 GCMs only considers the small, short-term, changes in total solar output. The solar magnetic field strength, cosmic rays, and UV radiation vary much more and are good candidates to amplify the solar TSI trends. Let us briefly discuss the possible amplifiers.

## The UV Amplifier

UV radiation increases during solar activity maxima, boosting ozone formation in the stratosphere at an altitude of 15 to 50 kilometres. The additional UV energy input converts a larger number of oxygen molecules ($O_2$) into ozone ($O_3$). A higher ozone concentration, in turn, intercepts more UV rays and converts their energy into heat, which causes the ozone layer or the stratosphere to warm. The search is now on for a process that combines the strong stratospheric fluctuations with the tropospheric climatic events below an altitude of around 15 kilometres.[133] Some researchers suggest that UV heating of the ozone layer creates anomalies in the atmospheric temperature gradient which are propagated to Earth's surface via intermediate steps.[134,135,136,137,138] Changes in wind patterns and atmospheric circulation apparently play a major role here.[139] For example, in times of low solar activity, westerly winds in the southern hemisphere shift toward the equator.[140]

132  Svensmark, H., Svensmark, J., Enghoff, M.B. et al. Atmospheric ionization and cloud radiative forcing. *Sci Rep* 11, 19668 (2021).

133  Niranjankumar K, Ramkumar TK, Krishnaiah M. Vertical and lateral propagation characteristics of intraseasonal oscillation from the tropical lower troposphere to upper mesosphere. *Journal of Geophysical Research* 2011, 116:1-10.

134  Meehl GA, Arblaster JM, Matthes K, Sassi F, Loon Hv. Amplifying the Pacific Climate System Response to a Small 11-Year Solar Cycle Forcing. *Science* 2009, 325:1114-1118.

135  Ineson S, Scaife AA, Knight JR, Manners JC, Dunstone NJ, Gray LJ, Haigh JD. Solar forcing of winter climate variability in the Northern Hemisphere. *Nature Geoscience* 2011, 4:753-757.

136  Kodera K. The role of dynamics in solar forcing. *Space Science Reviews* 2006, 125:319-330.

137  Gray LJ, Ball W, Misios S. Solar influences on climate over the Atlantic / European sector. *AIP Conference Proceedings* 2017, 1810:020002.

138  Wang W, Matthes K, Tian W, Park W, Shangguan M, Ding A. Solar impacts on decadal variability of tropopause temperature and lower stratospheric (LS) water vapour: a mechanism through ocean–atmosphere coupling. *Climate Dynamics* 2019, 52:5585-5604.

139  Kodera K, Thiéblemont R, Yukimoto S, Matthes K. How can we understand the global distribution of the solar cycle signal on the Earth's surface? *Atmos. Chem. Phys.* 2016, 16:12925-12944

140  Varma V, Prange M, Lamy F, Merkel U, Schulz M. Solar-forced shifts of the Southern Hemisphere Westerlies during the Holocene. *Clim. Past* 2011, 7:339-347.

# The Cosmic Ray Amplifier

The basic principle of this amplifier is the influence of fluctuations in the solar magnetic field on global or regional cloud cover. The possible path of action comprises several steps:

1) The strength of the solar magnetic field is coupled to solar activity.
2) The solar magnetic field shields the Earth from cosmic radiation coming from outer space.
3) The cosmic rays create condensation nuclei that help form clouds in the lowest three kilometres of the Earth's atmosphere. Similar to a cloud chamber, the particles charged by cosmic rays become condensation nuclei and attract water vapor.
4) Clouds limit the solar energy hitting the ground and thus the temperature.

In short: The stronger the sun, the stronger the solar magnetic field and the better protected Earth is from cosmic rays. The fewer cosmic rays that penetrate into the Earth's atmosphere, the less condensation and fewer clouds, which leads to warming. The result: a strong sun leads to global warming.

The model of the cosmic ray amplifier has been developed since the late 1990s by the Danish physicist Henrik Svensmark in collaboration with Eigil Friis-Christensen.[141,142,143,144,145,146] As might be expected, Svensmark's model met with fierce resistance in parts of the scientific community, because it was in competition with the dominance of $CO_2$ postulated by the IPCC. But Svensmark had the empirical data initially clearly on his side. During the period 1983-2002 global cloud cover developed synchronously with the eleven-year solar cycle (see Figure 3). After then, however, the relationship broke down, which led to criticism from Svensmark's scientific opponents. Notably, the temporary divergence of solar and climate trends could well be a consequence of non-linear and time-delayed processes, that have been reported for solar effects in the literature.[147,148]

At the turn of the millennium, the coupling between stratosphere and troposphere changed according to the 60-year cycle of atmospheric circulation, and the stratospheric polar vortex weakened, as Russian researchers have shown.[149,150] This reversed the solar effect on the cloud-generating low-pressure areas. While solar minima with more intense cosmic radiation used to bring more clouds, the cloud cover has now decreased with weak solar activity.[151,152]

The AR6 authors ignore the fact that other researchers have confirmed the Svensmark effect in general. However, the mechanism is probably not as simple and global as originally thought. It became clear that a much stronger differentiation into atmospheric altitudes, latitudes and seasons would be necessary.[153,154] The phase relationships were not uniform either. In some areas a direct

141 Svensmark H. Cosmic rays and earth's climate. *Space Science Reviews* 2000, 93:155-166.

142 Svensmark H, Friis-Christensen E. Variation of cosmic ray flux and global cloud coverage - a missing link in solar-climate relationships. *Journal of Atmospheric and Solar-Terrestrial Physics* 1997, 59:1225-1232.

143 Svensmark H, Pedersen JOP, Marsh ND, Enghoff MB. Experimental evidence for the role of ions in particle nucleation under atmospheric conditions. *Proc. R. Soc. A* 2007, 463:385-396.

144 Svensmark H, Friis-Christensen E. Reply to Lockwood and Fröhlich – The persistent role of the Sun in climate forcing. *Danish National Space Center, Scientific Report 2007*, 3 (2007).

145 Svensmark H. Cosmoclimatology: A New Theory Emerges. *Astronomy & Geophysics* 2007, 48:1.18-11.24.

146 Svensmark J, Enghoff MB, Shaviv NJ, Svensmark H. The response of clouds and aerosols to cosmic ray decreases. *Journal of Geophysical Research: Space Physics* 2016, 121:8152-8181.

147 van Loon H, Brown J, Milliff RF. Trends in sunspots and North Atlantic sea level pressure. *J. Geophys. Res.* 2012, 117:D07106.

148 Gusev AA, Martin IM. Possible evidence of the resonant influence of solar forcing on the climate system. *Journal of Atmospheric and Solar-Terrestrial Physics* 2012, 80:173-178.

149 Veretenenko S, Ogurtsov M. Cloud cover anomalies at middle latitudes: Links to troposphere dynamics and solar variability. *Journal of Atmospheric and Solar-Terrestrial Physics* 2016, 149:207-218.

150 Veretenenko S, Ogurtsov M, Lindholm M, Jalkanen R. Galactic Cosmic Rays and Low Clouds: Possible Reasons for Correlation Reversal. In: Szadkowski Z, ed. *Cosmic Rays*: IntechOpen, https://www.intechopen.com/books/cosmic-rays/galactic-cosmic-rays-and-low-clouds-possible-reasons-for-correlation-reversal; 2018, 79-98.

151 Ibid 149

152 Ibid 150

153 Ibid 48

154 Ibid 67

correlation was found, but in others an inverse one. AR6 decided to reject the simplistic version of the cosmic ray amplifier and remain silent about the complexity. The debate is, however, open as Svensmark and colleagues have recently published a new work further supporting their hypothesis of a significant effect of cosmic rays on cloud formation.[155]

It is also possible that the existence of an additional astronomical forcing of the climate system is related to interplanetary dust falling on Earth. In fact, Scafetta et al. (2020)[156] found that historical records of meteorites present a quasi-60-year oscillation that correlates with the quasi-60-year cycle observed in the climate record. A 60-year cycle is also present in the eccentricity variation of the orbit of Jupiter which may regulate the dust and comets moving toward the inner planets of the solar system. Consequently, it was proposed that an interplanetary-dust forcing of interplanetary ions might modify Earth's cloud system and regulate some climate changes.

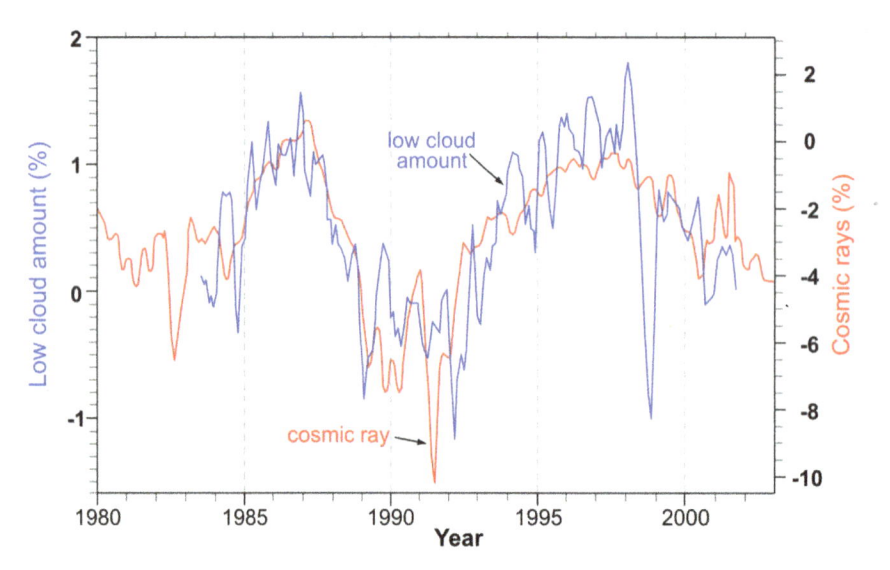

**Figure 3:** Correlation between cosmic radiation and global low cloud cover from 1980-2003 since the beginning of systematic cloud data collection.[157]

## Climate models cannot capture the sun

In addition to the clearly discernible linear climate effects of the sun, an even greater solar influence on the climate is probably achieved via non-linear effects. A growing number of scientists are pointing out the importance of non-linear relationships.[158,159,160,161,162] The complexity of the interaction between the sun (and other astronomical bodies) and Earth's climate is not even remotely considered in the climate models. At present, neither the linear nor the non-linear processes can be formulated physically, because the solar amplifiers are still being explored. In most cases, time delays, phase shifts and climate dipoles also cannot be reproduced by the models. The inconvenient truth is that climate simulations currently have no chance of reproducing, let alone quantifying, the solar influence on Earth's climate. Strong evidence supporting this claim is that climate models fail to properly reconstruct historical warm periods known to have occurred in

155  Ibid 132

156  Scafetta, N.; Milani, F.; Bianchini, A. A 60-year cycle in the Meteorite fall frequency suggests a possible interplanetary dust forcing of the Earth's climate driven by planetary oscillations. *Geophys. Res. Lett.* 2020, 47, e2020GL089954.

157  Ibid 145

158  Ratnam MV, Santhi YD, Kishore P, Rao SVB. Solar cycle effects on Indian summer monsoon dynamics. *Journal of Atmospheric and Solar-Terrestrial Physics* 2014, 121, Part B:145-156.

159  Wurtzel JB, Black DE, Thunell RC, Peterson LC, Tappa EJ, Rahman S. Mechanisms of southern Caribbean SST variability over the last two millennia. *Geophysical Research Letters* 2013, 40:5954-5958.

160  Lu H, Gray LJ, White IP, Bracegirdle TJ. Stratospheric Response to the 11-Yr Solar Cycle: Breaking Planetary Waves, Internal Reflection, and Resonance. *Journal of Climate* 2017, 30:7169-7190.

161  Kossobokov V, Le Mouël J, Courtillot V. On the Diversity of Long-Term Temperature Responses to Varying Levels of Solar Activity at Ten European Observatories. *Atmospheric and Climate Sciences* 2019, 9:498-526.

162  Le Mouël J-L, Lopes F, Courtillot V. A Solar Signature in Many Climate Indices. *Journal of Geophysical Research: Atmospheres* 2019, 124:2600-2619.

the Holocene such as the Medieval Warm Period.[163,164] It sounds like a joke when these historical warm periods are denied using the results of these deficient models.[165,166,167]

The situation is somewhat reminiscent of the debate on plate tectonics more than half a century ago. For a long time, scholars were reluctant to believe that the continents could be mobile and that they would constantly regroup and separate in the course of Earth's history. However, when more and more supporting evidence was found from 1960 onwards, the thinking quickly began to change. Alfred Wegener's idea had posthumously prevailed against all odds.

## IPCC has progressively downgraded the sun

IPCC reports have been issued since 1990. In the second report, called SAR, the IPCC attributed a radiative forcing value of +0.30 $W/m^2$ to solar variability. The greater this value, the larger the solar contribution to warming over the past 170 years. The value remained the same in the third report in 2001. The 4th Assessment report, however, reduced the value to +0.12 $W/m^2$. In the 5th report, the radiative forcing dropped to +0.05 $W/m^2$. AR6 also uses a very small value. The value is dwarfed by the suggested warming potential of $CO_2$ which the AR6 sets at +2.16 $W/m^2$ since 1750 (see AR6 p 959).

The attribution of the possible one degree C of global warming that occurred since the end of the Little Ice Age in 1850 AD is far from trivial. AR6 acknowledges *"that solar activity during the second half of the 20th century was in the upper decile of the range"* (chapter 2.2.1). Theoretically, this recent boost in solar activity might have contributed to warming if suitable amplifiers were considered. Another AR6 statement, however, is strongly misleading. The IPCC claims: *"New reconstructions of TSI over the 20th century [...] support previous results that the TSI averaged over the solar cycle very likely increased during the first seven decades of the 20th century and decreased thereafter."* Based on sunspots the strongest solar cycle may have indeed occurred around 1960 (Figure 4). However, this heating pulse of a few years is too short to have a long-term effect. Whilst the subsequent solar cycle in the 1970s was weak, the following three sunspot cycles in the late 20th and early 21st century were rather strong (orange lines in Figure 4). Their *cumulative* effect could have helped to warm Earth's climate, if the IPCC allowed suitable solar amplifiers in their equations.

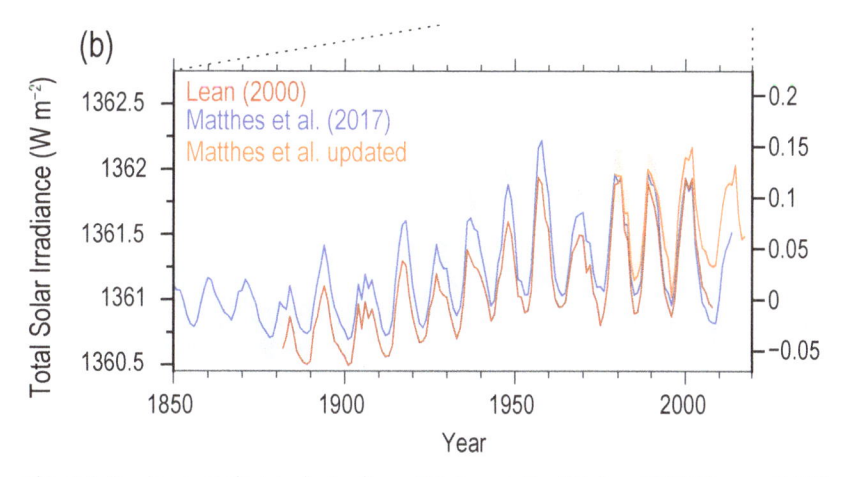

**Figure 4:** Development of total solar irradiance. This figure is Fig. 2.2b from AR6 (WG1, page 2-172).

163  Ibid 78

164  Ibid 79

165  Feulner G. Are the most recent estimates for Maunder Minimum solar irradiance in agreement with temperature reconstructions? *Geophysical Research Letters* 2011, 38:1-4.

166  IPCC. *Climate Change 2021: The Physical Science Basis. Contribution of Working Group I to the Sixth Assessment Report of the Intergovernmental Panel on Climate Change:* Cambridge University Press; 2021.

167  Ibid 3

A major problem is that the TSI forcing chosen for the CMIP6 GCMs is that proposed by Matthes et al. (2017)[168]. This TSI record is a combination of two TSI records (NRLTSI2 and SATIRE) that show a very small secular variability while many other TSI reconstructions show a much larger, up to about 10 times, larger secular variability and also slightly different patterns.[169,170,171] Moreover, the TSI recommended for the CMIP6 GCMs also decreases slightly since 1980, which roughly agrees with the TSI satellite composite proposed by PMOD. This decreasing pattern is important because it is usually cited as excluding solar forcing as a driver for the warming observed since the 1980s.

In AR6, the IPCC again fails to properly address the uncertainty and controversies regarding the competing TSI satellite composites from the ACRIM and PMOD science teams. Indeed, the TSI satellite composite proposed by the ACRIM team suggests that TSI has increased from 1980 to 2000 and slightly decreased afterward. The ACRIM TSI decadal pattern correlates well with the global surface temperature records available today. The main difference between the ACRIM and PMOD TSI satellite composites is that while the former uses the original raw satellite TSI records, the latter is based on TSI satellite records modified with a model. However, the original experimental science teams responsible for the measurements have consistently rejected the PMOD model-based modifications as physically unjustified. Moreover, Scafetta et al. (2019)[172] have recently reviewed the controversy and concluded that even the original NRLTSI2 and SATIRE data, when properly analysed, contradict the data modifications proposed by PMOD. This result calls into question the validity of all three records used by the CMIP6 GCMs to justify their claim that the solar effect on climate is negligible. Connolly et al. (2021)[173] provide more detail on these issues.

## Conclusions

The IPCC has once again missed an opportunity to address natural climate drivers in a balanced and fair way. Notably, AR5 argued cautiously that "more than half" of the observed warming of the past 170 years was of anthropogenic origin. This left up to 49% for natural climate drivers. The IPCC Special Report on the 1.5 °C goal (SR15, 2018) then jumped to 100% anthropogenic causes, leaving no room for natural climate change contributions. AR6 stuck to this narrative from three years earlier.

It is very clear from pre-industrial times that significant natural climate change similar to observed modern climate change has existed in the past and probably still exists. In many of the case studies, a solid solar influence was detected. Yet, the AR6 report presents a severely biased view of the solar role in the climate equation: it cherry-picks solar forcing that minimizes the solar effect without properly addressing the uncertainties and controversies surrounding this topic and by ignoring the scientific literature addressing it. The IPCC authors do not appear to be interested in exploring solar amplifiers or alternative solar forcing proposals. Connolly et al. (2021)[174] propose how a properly critical analysis of this debate should be conducted. The views of the IPCC must be revised, it is just a matter of time.

---

168  Matthes, K., Funke, B., Andersson, M. E., Barnard, L., Beer, J., Charbonneau, P., Clilverd, M. A., Dudok de Wit, T., Haberreiter, M., Hendry, A., Jackman, C. H., Kretzschmar, M., Kruschke, T., Kunze, M., Langematz, U., Marsh, D. R., Maycock, A. C., Misios, S., Rodger, C. J., Scaife, A. A., Seppälä, A., Shangguan, M., Sinnhuber, M., Tourpali, K., Usoskin, I., van de Kamp, M., Verronen, P. T., and Versick, S.: Solar forcing for CMIP6 (v3.2), Geosci. Model Dev., 10, 2247–2302.

169  Egorova, T.; Schmutz,W.; Rozanov, E.; Shapiro, A.I.; Usoskin, I.; Beer, J.; Tagirov, R.V.; Pete, T. Revised historical solar irradiance forcing Revised historical solar irradiance forcing. *A&A* 2018, 615, A85.

170  Hoyt, D.V.; Schatten, K.H. A discussion of plausible solar irradiance variations, 1700–1992. *J. Geophys. Res.* 1993, 98, 895–906.

171  Scafetta, N.; Willson, R.C. ACRIM total solar irradiance satellite composite validation versus TSI proxy models. *Astrophys. Space Sci.* 2014, 350, 421–442.

172  Scafetta, N.; Willson, R.C.; Lee, J.N.; Wu, D.L. Modeling Quiet Solar Luminosity Variability from TSI Satellite Measurements and Proxy Models during 1980–2018. *Remote Sens.* 2019, 11, 2569.

173  Ibid 1

174  Ibid 1

The public should not ignore the fact that the IPCC is a politically controlled organisation in which the IPCC bureau handpicks authors and review editors. Scientists supporting a stronger role of natural climate change in modern climate are typically excluded from the authorship of IPCC reports. But in the long run, facts and observations will prevail and the current uncertainties will be properly solved and the numerous empirical findings supporting a significant solar (or otherwise astronomical) effect on the climate will be confirmed.

Many topics in the climate sciences are poorly understood. The science is far from settled. Nevertheless, evidence for large natural forcings firms up with every year of additional research. It will therefore get easier and easier to fully acknowledge the significant influence that the Sun, ocean cycles ("modes of variability") and other natural drivers have on the complex climate system.

# 7. Misty Climate Sensitivity

**BY CLINTEL TEAM**

**How sensitive the climate is for greenhouse gases like $CO_2$ remains one of the most important issues of our time. In most reports the IPCC claimed a doubling of $CO_2$ will give 3°C of warming with an uncertainty range of 1.5°C to 4.5°C. In AR6, based on one highly influential paper, the best estimate for climate sensitivity remains 3°C but they narrowed the likely range considerably to 2.5°C to 4°C. This strongly suggests that lower values of climate sensitivity, which have also been published in the literature, are now rejected by the IPCC. But is this justified?**

T he climate sensitivity to $CO_2$ and other greenhouse gases (GHGs) is arguably one of the most important numbers in the climate change debate. Put very simply, if the climate is very sensitive to greenhouse gases and therefore climate sensitivity is high, then we can expect substantial warming in the coming century if greenhouse gas emissions are not severely reduced. If climate sensitivity is low, then future warming will be substantially lower, as will the rise in sea level.

Climate sensitivity is defined as the amount of global surface warming we get when the concentration of $CO_2$ in the atmosphere doubles. The term generally refers to the rise in temperature once the climate system has fully warmed up, a process taking over a thousand years due to the enormous heat capacity of the world ocean. This so-called 'equilibrium climate sensitivity' (ECS), is the traditional and still most widely used measure. In practice what is more commonly estimated[1] is 'effective climate sensitivity', a close approximation to ECS that is more practical to work with.

A shorter-term measure of sensitivity, transient climate response (TCR), represents the extent of global warming over a 70-year timeframe during which $CO_2$ concentrations double.[2] TCR can be estimated more easily than ECS, and is more relevant to projections of warming – although less for sea level rise – over the rest of this century.[3]

Historically, since the Charney report in 1979[4], the best estimate for climate sensitivity has been remarkably stable (see table 1). The best estimate in the IPCC sixth assessment report (AR6) is similar to the one in the Charney report, although the latter was based on very limited information.

Van der Sluijs (1998)[5] considered the reasons why the range for climate sensitivity has changed so little over a period in which the science has evolved enormously. He concluded that the range was only partly determined by the science itself and that many other factors played a role. One of these was 'a need to create and maintain a robust scientific basis' for policy action. We believe this observation by Van der Sluijs in 1998 is still valid today.

---

1   Including for global climate models. Note that by convention equilibrium climate sensitivity excludes adjustment by slow components of the climate system (e.g. ice sheets, vegetation).

2   The increase in $CO_2$ is specified to occur at a constant compound rate over the period, but modest fluctuations in the rate are unimportant. Estimation of TCR is unaffected by the actual rate of increase provided that the increase in global temperature is scaled appropriately, and TCR is little affected by moderate variations in the ramp period: between 60–80 years, at least.

3   Although TCR is easier to estimate, unlike ECS it does not have a useful interpretation in terms of the physics of the climate system. TCR is lower than ECS because heat going into the ocean contributes to the value of ECS but not to TCR.

4   Charney JG, Coauthors (1979) Carbon dioxide and climate: a scientific assessment. National Academy of Science, Washington, DC, 22 pp

5   J.P. van der Sluijs et al. (1998).

**Table 1:** Evolution of equilibrium climate sensitivity estimates in the last 42 years and the range for transient climate response since 2007

| | ECS RANGE (°C) | ECS BEST ESTIMATE (°C) | TCR RANGE (°C) |
|---|---|---|---|
| Charney Report 1979 | 1.5–4.5 | 3 | |
| NAS Report 1983 | 1.5–4.5 | 3 | |
| Villach Conference 1985 | 1.5–4.5 | 3 | |
| IPCC FAR 1990 | 1.5–4.5 | 2.5 | |
| IPCC SAR 1995 | 1.5–4.5 | 2.5 | |
| IPCC TAR 2001 | 1.5–4.5 | 3 | |
| IPCC AR4 2007 | 2–4.5[6] | 3 | 1–3[7] |
| IPCC AR5 2013 | 1.5–4.5[8] | None given | 1–2.5[9] |
| IPCC AR6 2021 | 2.5-4 | 3 | 1.4-2.2[10] |

Table 1 shows the evolution of both the range and the best estimate of ECS over the last 42 years. As one can see, the best estimate has not changed much. However, in AR6, the IPCC narrowed its likely range considerably claiming values below 2.5°C are now less likely.

## A Sensitive Matter

After the publication of the IPCC AR5 report, the British independent scientist Nic Lewis and the Dutch independent science writer Marcel Crok wrote an extensive report[11] – titled *A Sensitive Matter* – in which they explained how the IPCC "hid good news about global warming". In their report they detailed that during the production process of AR5 several papers had been published, based on observations in the 'historical period' (the period since 1850), that indicated a considerably lower estimate of ECS than those estimates based on General Circulation Models (GCMs). The CMIP5 models, on average, had a climate sensitivity of more than 3°C while observations indicated ECS values between 1.5 and 2°C.

In their report Crok and Lewis observed that the IPCC was confronted with a dilemma:

> In our view, the IPCC WGI scientists were saddled with a dilemma. How should they deal with the discrepancy between climate sensitivity estimates based on models and sound observational estimates that are consistent with the new evidence about aerosol cooling? In conjunction with governments – who have the last say on the wording of the SPM – they appear to have decided to resolve this dilemma in the following way. First, they changed the 'likely' range for climate sensitivity slightly. It was 2–4.5°C in AR4 in 2007. They have now reduced the lower bound to 1.5°C, making the range 1.5–4.5°C. By doing this they went some way to reflect the new, lower estimates that have been published recently in the literature.

They also decided not to give a best estimate for climate sensitivity. The tradition of giving a best estimate for climate sensitivity goes all the way back to the Charney report in 1979, and all subsequent IPCC reports (except the third assessment report in 2001) gave one as well. In AR4 the best estimate was 3°C. At the time of approval of the SPM by governments in September 2013, the decision not to give a best estimate for climate sensitivity was mentioned only in a footnote in the SPM, citing 'a lack of agreement on values across assessed lines of evidence and studies'. Only in the final report, published in January 2014, was a paragraph added in the Technical Summary giving slightly more explanation.

---

6     Likely (17–83%) range. Prior to AR4, ranges were not clearly defined in probabilistic terms.

7     10–90% range.

8     Likely range.

9     Likely range.

10    The IPCC now also gave a best estimate for TCR of 1.8°C (page 927)

11    A Sensitive Matter, Nicholas Lewis and Marcel Crok, GWPF (2014)

So, to deal with the new estimates for ECS that were based mostly on observations since 1850 (instead of on models) the IPCC *lowered* the lower bound of their likely range back to 1.5°C, where it was most of the time since the Charney report. Furthermore, due to a lack of agreement between different lines of evidence (i.e., mainly between observational estimates and models) they gave no best estimate, which was quite remarkable because the same report claimed the IPCC was more certain than ever that humans were the main cause of global warming.

The AR5 report was published in 2013 and the AR6 Working Group 1 report was published in 2021. So, the IPCC community has had eight years of time to figure out how they had to deal with these 'different lines of evidence'.

Now before we go into the details of how the climate community 'solved' this problem, let's dwell on the importance of a 'high' or 'low' climate sensitivity for a moment. A high climate sensitivity possibly makes the climate problem 'urgent' as we can expect a lot of warming with continuing $CO_2$ emissions. A high climate sensitivity means the climate models – in which the community has invested a lot, both in terms of money and their credibility – are 'right'. A 'low' climate sensitivity is good news for all but makes the case for urgent climate mitigation measures much weaker. A 'low' climate sensitivity would also mean that well-known climate sceptics – like Richard Lindzen or Roy Spencer – who have claimed for many years that sensitivity is 'low', were right after all. Stephen Schneider, a well-known climate scientist who passed away in 2010, wrote the book *Science as a Contact Sport*. Of course, it is. Science is about careers, citation indexes, funding, fame etc. There is a lot at stake here, a 'low' climate sensitivity would not only mean that the climate community was 'wrong' for a long time, it could also mean that in the future less money will flow to climate science departments and institutions.

## Models versus observations

Suppose for a brief moment that climate change was not a highly polarized and politicized issue. Scientifically speaking we have a normal situation that happens all the time in science, a discrepancy between theory and observations. Climate models (theory) have a climate sensitivity of more than 3°C, mainly as a result of positive feedbacks to greenhouse gas warming, i.e., water vapour and cloud feedbacks amplify the initial warming that is caused by $CO_2$ and other greenhouse gases. However, estimates based on post-1850 observations suggest climate sensitivity is much lower, perhaps between 1.5 and 2°C, or even lower.[12] In 'normal' science, scientists would give the observational estimates the benefit of the doubt. They would conclude that something must be wrong with the models. However, given all the interests that are at stake – i.e., to keep climate change an urgent issue or as Van der Sluijs wrote in 1998 to 'maintain a robust scientific basis' for policy action – this was not really an option for the climate community and the IPCC. So, their 'favoured' outcome was to prove that the historical estimates since 1850 were 'wrong'. And they 'succeeded' in this. It's a fascinating story.

First the outcome. In AR6 the IPCC is back with a best estimate for ECS of 3°C. Looking at table 1 this isn't spectacular, as this estimate is like the estimate in the 1979 Charney report. But in fact, it is quite spectacular because observations over the period since 1850 indicate a much lower ECS, between 1.5 and 2°C. So apparently, the IPCC has found a way to discard those estimates in favour of higher values.

But even more 'spectacular' is the new likely range of 2.5 to 4°C. Remember, in AR4 (2007) they raised the lower bound of ECS to 2°C, but then had to lower it back to the 'normal' 1.5°C in AR5 (2013) due to the new observational estimates for ECS. Now they raised it to 2.5°C and narrowed

---

12 These lower estimates of 1.5 to 2 are based on the energy budget method, in which scientists like Lewis assume – just like the IPCC – that all the warming since 1850 is due to anthropogenic greenhouse gases. Even lower estimates have been published in the literature by Lindzen and Choi (0.7) and Christy (1.1) but such estimates are ignored by the IPCC. Lindzen, R., & Choi, Y.-S. (2011, August 28). On the Observational Determination of Climate Sensitivity and Implications. *Asia-Pacific Journal of Atmospheric Sciences, 47*(377). Christy, J., & McNider, R. (2017). Satellite Bulk Tropospheric Temperatures as a Metric for Climate Sensitivity. Asia-Pac. J. Atmos. Sci., 53(4).

the likely range to 2.5-4°C. The likely range has never been smaller (narrower) than this. The message is clear: the IPCC is now more certain about ECS than ever and finds values below 2.5°C and above 4°C unlikely. Or in their words:

> Based on multiple lines of evidence the best estimate of ECS is 3°C, the likely range is 2.5°C to 4°C, and the very likely range is 2°C to 5°C. It is virtually certain that ECS is larger than 1.5°C. [page 926]

It is now tempting to think that the IPCC 'chose' the ECS from climate models and discarded the observational estimates. But this is not the case. In fact, the IPCC made it clear that climate models (i.e. GCM's) were not used to directly estimate ECS. From AR6 (our bold):

> All four lines of evidence rely, to some extent, on climate models, and interpreting the evidence often benefits from model diversity and spread in modelled climate sensitivity. Furthermore, high-sensitivity models can provide important insights into futures that have a low likelihood of occurring but that could result in large impacts. But, **unlike in previous assessments, climate models are not considered a line of evidence in their own right** in the IPCC Sixth Assessment Report. [page 1024]

Models (AR6 calls them ESMs) are used as one of the several sources of information to constrain possible values of ECS and TCR, as AR6 explains on page 993. However, the final ECS and TCR values presented were not computed from models as was done in previous reports.

## Ringberg Castle

In 2015 over thirty experts attended a week-long workshop in Ringberg Castle to assess gaps in understanding of Earth's climate sensitivities. The workshop was organised under the auspices of the World Climate Research Programme (WCRP) Grand Science Challenge on Clouds, Circulation and Climate Sensitivity. Nic Lewis attended this workshop and gave a talk.

This WCRP-initiated assessment process culminated in the publication in 2020 of a 92-page review paper by Steven Sherwood and 24 co-authors titled "An Assessment of Earth's Climate Sensitivity Using Multiple Lines of Evidence".[13] The paper has been extremely influential, including in informing the assessment of ECS in AR6 WG1; it was cited over twenty times in the relevant AR6 chapter 7. Lewis was not asked to be a co-author of that paper.

Since the Ringberg workshop was held, Lewis published papers concerning how to combine multiple lines of evidence regarding climate sensitivity using an Objective Bayesian statistical approach.[14] Disappointingly, for Lewis, Sherwood et al. instead used the common Subjective Bayesian method that, while simpler, his research had showed may result in unrealistic estimates and uncertainty ranges. Lewis therefore decided to replicate the Sherwood et al. paper, to implement an Objective Bayesian approach, and to review the paper's choice of probabilistic estimates for the input assumptions used. In doing so Lewis discovered another, more fundamental and potentially more serious, statistical problem in the Sherwood paper, as well as important conceptual errors and inconsistencies. He also found that after fixing these problems and substituting values derived from more recent sources of evidence, including AR6, for certain of the data-variable estimates used, the resulting estimate of climate sensitivity fell substantially.

Here is the full abstract of the paper that Lewis published in 2022, so a year after the IPCC report was published (our bold):[15]

> Recent assessments of climate sensitivity per doubling of atmospheric $CO_2$ concentration have combined likelihoods derived from multiple lines of evidence. These assessments were very

---

13  Sherwood, S.C. et al., 2020: An Assessment of Earth's Climate Sensitivity Using Multiple Lines of Evidence. Reviews of Geophysics, 58(4), e2019RG000678, doi:10.1029/2019rg000678

14  All his peer reviewed papers can be found here: https://nicholaslewis.org/peer-reviewed-publications/

15  Lewis (2022) Objectively combining climate sensitivity evidence. Climate Dynamics, doi.org/10.1007/s00382-022-06468-x

influential in the Intergovernmental Panel on Climate Change Sixth Assessment Report (AR6) assessment of equilibrium climate sensitivity, the likely range lower limit of which was raised to 2.5°C (from 1.5°C previously). This study evaluates the methodology of and results from a particularly influential assessment of climate sensitivity that combined multiple lines of evidence, Sherwood et al. (2020). That assessment used a subjective Bayesian statistical method, with an investigator-selected prior distribution. This study estimates climate sensitivity using an Objective Bayesian method with computed, mathematical priors, since subjective Bayesian methods may produce uncertainty ranges that poorly match confidence intervals. Identical model equations and, initially, identical input values to those in Sherwood et al. are used. This study corrects Sherwood et al.'s likelihood estimation, producing estimates from three methods that agree closely with each other, but differ from those that they derived. Finally, the selection of input values is revisited, where appropriate adopting values based on more recent evidence or that otherwise appear better justified. **The resulting estimates of long-term climate sensitivity are much lower and better constrained (median 2.16°C, 17–83% range 1.75–2.7°C, 5–95% range 1.55–3.2°C) than in Sherwood et al. and in AR6 (central value 3°C, very likely range 2.0–5.0°C).** This sensitivity to the assumptions employed implies that climate sensitivity remains difficult to ascertain, and that **values between 1.5°C and 2°C are quite plausible**.

This is quite spectacular. After correcting the Sherwood et al. methods and revising key input data to reflect more recent evidence, the central estimate for climate sensitivity comes down from 3.1°C per doubling of $CO_2$ concentration in the original study to 2.16°C in the new paper. The results of Lewis' analysis determined a likely range of 1.75 to 2.7°C for climate sensitivity, even narrower than the new range used by the IPCC (2.5-4°C), but much lower.

The central estimate from Lewis' analysis – 2.16°C – is well below the IPCC AR6 likely range of 2.5-4°C. This large reduction relative to Sherwood et al. shows how sensitive climate sensitivity estimates are to input assumptions. Lewis' analysis implies that climate sensitivity is more likely to be below 2°C than it is to be above 2.5°C. One wonders what would have happened if the Sherwood group (consisting of the who's who in the climate sensitivity field) had invited Nic Lewis to join their effort. Now the estimates in the Sherwood paper have instantly become the new 'golden standard' in the IPCC report and this will remain so until at least the next IPCC report will be published, but that will be only six or seven years from now. So, in the coming years policy makers will all assume that a 'low' climate sensitivity is more unlikely than ever since the Charney report in 1979 was published. Although their own method, if used correctly, shows the opposite, that ECS is likely on the low side.

## Different lines of evidence

So what did Sherwood et al. use exactly as their evidence? They combined evidence based on several different lines of evidence: process understanding (feedback analysis), the historical period (instrumental) record, and paleoclimate data from both warm and cold periods. The cold paleoclimate evidence concerned changes between the last glacial maximum (LGM) and preindustrial periods. Sherwood et al. analysed paleoclimate data from two warm periods, the mid-Pliocene warm period (mPWP) and the more distant Paleocene-Eocene Thermal Maximum (PETM) but did not use PETM data in their main results. Thus, Sherwood et al. used three main lines of evidence (Process, Historical and Paleoclimate), with LGM and mPWP evidence being combined to represent Paleoclimate evidence.

Lewis agrees that "this is a strong scientific approach, in that it utilizes a broad base of evidence and avoids direct dependence on GCM climate sensitivities. Such an approach should be able to provide more precise and reliable estimation of climate sensitivity than that in previous IPCC assessment reports."[16]

---

16    Lewis published a detailed commentary about his peer reviewed paper, available here: https://nicholaslewis.org/wp-content/uploads/2022/09/Lewis_Objectively-combining-climate-sensitivity-evidence_2022-Clim-Dyn-Detailed-Summary.pdf

In a detailed explanatory article[17] about the paper Lewis showed how the changes in input data and statistical methods led to different outcomes for climate sensitivity compared to the original paper by Sherwood et al:

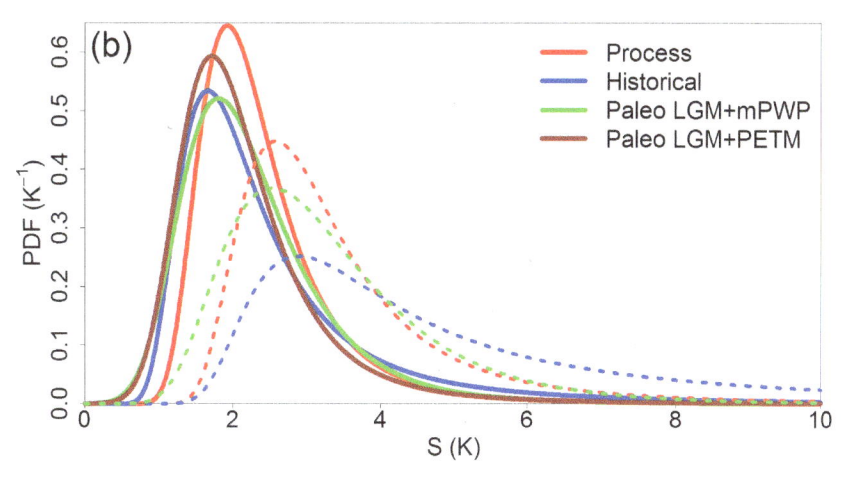

**Figure 1:** Posterior Probability Density Functions for S (= climate sensitivity) based on the revised (solid lines) and original data variable assumptions (dotted lines).

The probability density functions (representing the probable values of the estimated parameter for climate sensitivity) in the Lewis paper are much more constrained than in the original Sherwood paper and the different lines of evidence also show similar outcomes.

## Historical estimates

The 2014 Lewis/Crok report stated that at that moment observational estimates for climate sensitivity, based on the historical (instrumental) period, were 'superior':

> So, to conclude, we think that of the three main approaches for estimating ECS available today (instrumental observation based, palaeoclimate proxy observation based, and GCM simulation/feedback analysis based), instrumental estimates – in particular, those based on warming over a substantial period extending to the twenty-first century – **are superior by far**. Observationally based estimates give the best indication of how our current climate has actually been reacting to the increase in greenhouse gases. [A Sensitive Matter, page 37]

Now of course scientists involved in the IPCC report are free to disagree with that view. In the years since AR5 Lewis together with well-known climate scientist Judith Curry has published two peer reviewed papers[18], based on the historical period since 1850, to estimate both ECS and TCR. The most recent paper was published in 2018 and is referenced several times in the AR6 report. Here is the abstract (our bold):[19]

> Energy budget estimates of equilibrium climate sensitivity (ECS) and transient climate response (TCR) are derived based on the best estimates and uncertainty ranges for forcing provided in the IPCC Fifth Assessment Report (AR5). Recent revisions to greenhouse gas forcing and post-1990 ozone and aerosol forcing estimates are incorporated and the forcing data extended from 2011 to 2016. Reflecting recent evidence against strong aerosol forcing, its AR5 uncertainty lower bound is increased slightly. Using an 1869–82 base period and a 2007–16 final period, which are well matched for volcanic activity and influence from internal variability,

17  https://nicholaslewis.org/wp-content/uploads/2022/09/Lewis_Objectively-combining-climate-sensitivity-evidence_2022-Clim-Dyn-Detailed-Summary.pdf

18  Lewis, N. and J.A. Curry, 2015: The implications for climate sensitivity of AR5 forcing and heat uptake estimates. Climate Dynamics, 45(3–4), 1009–1023, doi:10.1007/s00382-014-2342-y; Lewis, N. and J. Curry, 2018: The Impact of Recent Forcing and Ocean Heat Uptake Data on Estimates of Climate Sensitivity. Journal of Climate, 31(15), 6051–6071, doi:10.1175/jcli-d-17-0667.1.

19  Lewis, N. and J. Curry, 2018: The Impact of Recent Forcing and Ocean Heat Uptake Data on Estimates of Climate Sensitivity. Journal of Climate, 31(15), 6051–6071, doi:10.1175/jcli-d-17-0667.1.

**medians are derived for ECS of 1.50 K** (5%–95% range: 1.05–2.45 K) and for TCR of 1.20 K (5%–95% range: 0.9–1.7 K). These estimates both have much lower upper bounds than those from a predecessor study using AR5 data ending in 2011. Using infilled, globally complete temperature data give slightly higher estimates: **a median of 1.66 K for ECS** (5%–95% range: 1.15–2.7 K) and 1.33 K for TCR (5%–95% range: 1.0–1.9 K). **These ECS estimates reflect climate feedbacks over the historical period, assumed to be time invariant. Allowing for possible time-varying climate feedbacks increases the median ECS estimate to 1.76 K (5%–95% range: 1.2–3.1 K), using infilled temperature data.** Possible biases from non-unit forcing efficacy, temperature estimation issues, and variability in sea surface temperature change patterns are examined and found to be minor when using globally complete temperature data. **These results imply that high ECS and TCR values derived from a majority of CMIP5 climate models are inconsistent with observed warming during the historical period**.

Here is the table with their key results:

**Table 2:** Best estimates (medians) and uncertainty ranges for ECS and TCR using the base and final periods indicated. Values in roman type compute $\Delta T$ using the HadCRUT4v5 dataset; values in italics compute $\Delta T$ using the infilled, globally complete Had4_krig_v2 dataset. The preferred estimates are shown in boldface. LC15 refers to the earlier paper by Lewis and Curry in 2015. Source: table 3 in Lewis and Curry 2018.

| BASE PERIOD | FINAL PERIOD | ECS BEST ESTIMATE (K) | ECS 17%-83% RANGE (K) | ECS 5%-95% RANGE (K) | TCR BEST ESTIMATE (K) | TCR 17%-83% RANGE (K) | TCR 5%-95% RANGE (K) |
|---|---|---|---|---|---|---|---|
| 1869-82 | 2007-16 | 1.50 | 1.2-1.95 | 1.05-2.45 | 1.20 | 1.0-1.45 | 0.9-1.7 |
| | | *1.66* | *1.35-2.15* | *1.15-2.7* | *1.33* | *1.1-1.6* | *1.0-1.9* |
| 1869-82 | 1995-2016 | 1.56 | 1.2-2.1 | 1.05-2.75 | 1.22 | 1.0-1.5 | 0.85-1.85 |
| | | *1.69* | *1.35-2.25* | *1.15-3.0* | *1.32* | *1.1-1.65* | *0.95-2.0* |
| 1850-1900 | 1980-2016 | 1.54 | 1.2-2.15 | 1.0-2.95 | 1.23 | 1.0-1.6 | 0.85-1.95 |
| | | *1.67* | *1.3-2.3* | *1.1-3.2* | *1.33* | *1.05-1.7* | *0.9-2.15* |
| 1930-50 | 2007-16 | 1.56 | 1.2-2.15 | 1.0-3.0 | 1.20 | 0.95-1.5 | 0.85-1.85 |
| | | *1.65* | *1.25-2.3* | *1.05-3.15* | *1.27* | *1.05-1.6* | *0.9-1.95* |
| LC15 results for comparison | | | | | | | |
| 1859-82 | 1995-2011 | 1.64 | 1.25-2.45 | 1.05-4.05 | 1.33 | 1.05-1.8 | 0.9-2.5 |
| 1850-1900 | 1987-2011 | 1.67 | 1.25-2.6 | 1.0-4.75 | 1.31 | 1.0-1.8 | 0.85-2.55 |

These results are quite spectacular. Depending on the dataset used, their 'preferred' estimate of ECS is either 1.5°C or 1.66°C. Two years later, the IPCC, in AR6, decided that the best estimate for ECS is 3°C, so (almost) double the value found by Lewis and Curry and far outside the likely range of 1.35-2.15°C of Lewis and Curry. So apparently, the IPCC has reasons to dismiss the results of Lewis and Curry. What are those reasons. Now it becomes really fascinating. The IPCC invoked a so-called 'pattern effect' to dismiss the 'low' estimates based on the historical period. Here is what they write in the executive summary of chapter 7 in AR6 (our bold):

Radiative feedbacks, particularly from clouds, **are expected to become less negative (more amplifying)** on multi-decadal time scales **as the spatial pattern of surface warming evolves, leading to an ECS that is higher than was inferred in AR5 based on warming over the instrumental record.** This new understanding, along with updated estimates of historical temperature change, ERF, and Earth's energy imbalance, reconciles previously disparate ECS estimates (high confidence). [page 926]

## The pattern effect

As we will explain, this is quite a Houdini act by the IPCC community! Most readers will probably have never heard about a 'pattern effect', which is meant by 'the spatial pattern of surface warming' in the AR6 paragraph above. And it's rather complicated, so let's discuss it step by step. ECS is the long-term warming after a doubling of the $CO_2$ concentration. Since 1850 the $CO_2$ concentration has been increasing but not yet doubled. A high climate sensitivity of 3°C or higher implies that climate feedbacks are positive, i.e., they amplify the initial radiative warming effect of $CO_2$ and other greenhouse gases. The idea that the climate science community now has launched is that initially the climate feedbacks don't operate in full power yet (so they don't amplify to their full potential yet), but eventually they will. However, when you estimate ECS based on the historical period you are misguided because the implied feedbacks over this period are not representative for the full period over which ECS has to be determined (theoretically until a new equilibrium in the climate system has been established, which can take over a thousand years). Smart readers will now ask: how do we know the feedbacks in the system will eventually get stronger? Good question! The short and simple answer is: because models say so!

The Carbon Brief website had a good article[20] explaining the evolving thinking of the climate community with respect to observational estimates over the historical period. Here is a lengthy excerpt from that article (our bold):

> One important insight is that the strength of climate feedbacks is expected to change over time, with **stronger feedbacks taking longer to emerge**.
> A 2017 paper by Dr Cristian Proistosescu and Prof Peter Huybers at Harvard University found that **amplifying feedbacks** that play a large role in ECS in climate models **have not fully kicked in for current climate conditions**. A similar paper by Prof Kyle Armour of the University of Washington suggests feedbacks will increase by about 25% from today's transient warming as the Earth moves towards equilibrium.
> This means that **sensitivity estimates based on instrumental warming to date would be on the low side, as they would not capture the larger role of feedbacks in future warming**. The authors suggest that "**accounting for these…brings historical records into agreement with model-derived ECS estimates**".
> This is in part because feedbacks **depend strongly on the spatial pattern of warming**. Prof Armour elaborates in a discussion on the Climate Lab Book website:
>
> > "**Nearly all GCMs** [global climate models] show global radiative feedbacks changing over time under forcing, with effective climate sensitivity increasing as equilibrium is approached. As a result, **climate sensitivity estimated from transient warming appears smaller than the true value of ECS**…
> > As far as we can tell, the physical reason for this effect is that the **global feedback depends on the spatial pattern of surface warming, which changes over time**…One nice example is the sea-ice albedo feedback in the Southern Ocean: because warming has yet to emerge there, that positive (destabilising) feedback has yet to be activated.
> > This means that **even perfect knowledge of global quantities** (surface warming, radiative forcing, heat uptake) **is insufficient to accurately estimate ECS**; you also have to predict how radiative feedbacks will change in the future."
>
> Prof Andrew Dessler agrees, telling Carbon Brief that **an understanding of how the pattern of surface warming influences sensitivity is one of the major advances in our understanding of climate sensitivity in recent years**. He suggest that it "allows us to **resolve the discrepancy** between the 20th century [instrumental] estimates and other estimates that give higher values".

You have to admit, from their perspective – cherishing high estimates of climate sensitivity including those based on the GCMs – this is a brilliant way out. They dismiss the 'low' climate sensitivity estimates based on the historical period (the only period in climate history for which we have at least a reasonable amount of measurement data), because feedbacks **will get** stronger in the future. How do they know that? Well, again, the models say so. But how do we know the models are realistic? Answer: we don't know.

---

20  https://www.carbonbrief.org/explainer-how-scientists-estimate-climate-sensitivity/

Not surprisingly, Nic Lewis has also looked into this 'pattern effect' as it is used against his and other's 'low' estimates of climate sensitivity based on the historical period. So, in Lewis and Curry 2018 they responded to these concerns of the climate community concluding:

> We have also shown that various concerns that have been raised about the accuracy of historical period energy budget climate sensitivity estimation are misplaced. We assess **nil bias** from either non–unit forcing efficacy or **varying SST warming patterns**, and that any downward estimation bias when using blended infilled surface temperature data is trivial.

In 2020 Lewis published a paper with Thorsten Mauritsen, who was also lead author of the IPCC chapter 7, dealing specifically with the pattern effect. Its title was "**Negligible unforced historical pattern effect** on climate feedback strength found in HadISST-based AMIP simulations".[21] They concluded (our bold):

> In this study **we have found no evidence for a substantial unforced pattern effect over the historical period**, arising from internal variability, in the available sea surface temperature datasets, save for when the AMIPII and ERSSTv5 datasets are used. Our results imply that the evidence suggesting existing constraints on EffCS from historical period energy budget considerations are biased low due to unusual internal variability in SST warming patterns is too weak to support such conclusion, and suggest that any such bias is likely to be small and of uncertain sign.'

Their paper is mainly a reply to a 2018 paper by Andrews et al.[22] which claims that the pattern effect leads to an underestimation of ECS based on the historical period of 40%. Lewis and Mauritsen showed in their paper that climate feedback estimates are far from robust. They rely on the choice of historical SST dataset used, and that when the widely used HadISST1 dataset is used in place of the AMIPII SST dataset, no unforced historical pattern effect is found with the models they used. We also investigated the unforced historical pattern effect using five other SST datasets, finding a significant estimated effect only in one case.

Mauritsen was lead author of the relevant chapter 7 in the AR6 report and their paper is mentioned in the chapter. This is what they had to say about it (our bold):

> Using alternative SST datasets, Andrews et al. (2018) found little change in the value of α' within two models (HadGEM3 and HadAM3), while Lewis and Mauritsen (2021) found a smaller value of α' [climate feedback] within two other models (ECHAM6.3 and CAM5). The sensitivity of results to the choice of dataset represents a major source of uncertainty in the quantification of the historical pattern effect using atmosphere-only ESMs that has yet to be systematically explored, but the preliminary findings of Lewis and Mauritsen (2021) and Fueglistaler and Silvers (2021) suggest that α' could be smaller than the values reported in Andrews et al. (2018).

So, the IPCC admitted there was evidence against a large 'pattern effect' as claimed by Andrews et al. 2018. Note that the findings of Mauritsen and Lewis are called 'preliminary' while such a word is not used for the Andrews paper. Nevertheless, the IPCC concluded with 'high confidence' that the pattern effect 'reconciles previously disparate ECS estimates', meaning that there is no longer a disagreement between the 'low' estimates over the historical period and the 'high' estimates based on the models. So, this is how the IPCC 'solved' the 'dilemma' that Lewis and Crok discussed in their 2014 report *A Sensitive Matter*.

This of course is very unconvincing. The 'pattern effect' seems to be a nice example of 'adding an epicycle', a rather ad hoc hypothesis to save an overall scientific theory.[23] Here it was introduced to save the overall case for a high climate sensitivity. The case for a rather 'low' climate sensitiv-

21   Lewis, N. and Mauritsen, T., 2020: Negligible unforced historical pattern effect on climate feedback strength found in HadISST-based AMIP simulations. *Journal of Climate*, 1-52, https://doi.org/10.1175/JCLI-D-19-0941.1

22   Andrews T. et al., 2018 Accounting for changing temperature patterns increases historical estimates of climate sensitivity. Geophys. Res. Lett. https://doi.org/10.1029/2018GL078887

23   https://en.wikipedia.org/wiki/Deferent_and_epicycle#Bad_science

ity (around 2°C) is now even stronger than it was around the time of the AR5 report. The IPCC, instead of clearing up the smoke, drew up a new smoke curtain by claiming a so-called 'pattern effect' could explain the discrepancy between model and observationally based estimates for climate sensitivity.

Ironically, by doing that, the IPCC had to claim that internal variability of the climate over the historical period 'masks' the much higher 'real' climate sensitivity. This is ironic because climate sceptics have often been accused of being misleading by claiming that the recent changes in the climate were mostly natural. Now, the IPCC itself, needs a 'large effect' from 'internal variability' to save their case for a high climate sensitivity.

## Appendix: Different views within the Clintel Team

This chapter was very difficult to write. The observational data available to estimate climate sensitivity to greenhouse gases is generally of poor quality, limited in extent, and affected by natural variability. Further, our theoretical understanding of many elements of the highly complex climate system – especially, and this is admitted by the IPCC, cloud feedback – is still inadequate, and our understanding of natural climate change and natural climate variability is also poor.

Moreover, the various definitions of climate sensitivity, including multiple definitions of "ECS" and "TCR", confuse matters, and it is hard to nail down a measure of long-term climate sensitivity that is scientifically testable in the real world.

Some in the Clintel team believe, in line with the standard IPCC view, that $CO_2$ and other greenhouse gases are responsible for most of the warming since the end of the Little Ice Age in the 19th century. The estimates by Lewis and Curry assume, like the IPCC, that all the warming since 1850 is due to greenhouse gases. In that sense their estimates could be regarded as an upper bound. Even then, their estimates are much lower than those of the IPCC as we have explained in the chapter. Others, though, think that natural forces like the sun and internal variability are responsible for at least half the observed one-degree of warming since the 19th century. If that is true, then of course, climate sensitivity will be considerably lower than the Lewis and Curry estimates. We all agree the IPCC estimates of climate sensitivity to $CO_2$ (both ECS and TCR) and other greenhouse gases are too high.[24]

### CLIMATE MODELS VS. OBSERVATIONS

We had different views on the following quote from AR6 (page 1024):

> *"All four lines of evidence rely, to some extent, on climate models"*

It is clear from the following paragraph in AR6 that the reference to "climate models" here is to GCMs.[25] Some of us believed the quote was accurate and should be read as climate models were needed, to some extent, to compute the climate sensitivity for all lines of evidence, including from observations. Others thought it was misleading and inappropriately conflated model-based estimates of ECS with observation-based estimates.

It is appropriate to quote Sherwood et al here:

---

24  A four-part series on the climate blog of Andy May documents alternative views on this issue and also pays more attention to the important issue of cloud feedbacks than we have done in this chapter.
https://andymaypetrophysicist.com/2023/04/24/the-mysterious-ar6-ecs-part-1/;
https://andymaypetrophysicist.com/2023/04/25/the-mysterious-ar6-ecs-part-2-the-impact-of-clouds/;
https://andymaypetrophysicist.com/2023/04/26/the-mysterious-ar6-ecs-part-3-what-is-climate-sensitivity/;
https://andymaypetrophysicist.com/2023/04/27/the-mysterious-ar6-ecs-part-4-converting-observations-to-ecs/

25  No distinction is drawn between GCMs and ESMs (the term generally used in AR6), since in simulations relevant to estimating climate sensitivity ESM carbon cycle representations are inactivated and the extent to which ESMs simulate other processes not usually represented in GCMs varies.

"Importantly, each term in Equation 10 [the Bayesian probability model] is computed using a model (cf. section 2.2) and involves judgments about structural uncertainty including limitations of the model."

In this case the "model" referred to is a physical model, typically involving a relatively simple set of equations intended to represent relationships between variables that represent aspects of the climate system, such as equations (1) to (8) in Sherwood's section 2.2.

One of us wrote: "AR6, unlike previous Assessment Reports, doesn't directly use models in its assessment of climate sensitivity, although they are used indirectly." Here, "models" is intended to mean GCMs.

So, use of the word "models" is often confusing, and it is unclear to what extent estimates of climate sensitivity still rely on GCMs.

## ARE OBSERVATION-BASED ESTIMATES OF ECS INVALID BECAUSE THEY ASSUME THAT FEEDBACKS ARE CONSTANT?

The team members agree that climate feedback variation with average temperature is of little relevance to estimates of ECS based on observed warming, since there is very little evidence that feedbacks vary between the change in average temperature over the historical period, and the change when $CO_2$ concentration is doubled from its preindustrial level. However, any climate feedback variation with time since $CO_2$ concentration increases, whether due to evolving temperature change patterns or solar variations, does affect ECS estimates from instrumental observations. Some members have a problem with AR6 de-emphasizing lower observation-based estimates of ECS because these studies assume radiative feedbacks will remain constant over time as increasing $CO_2$ causes the atmosphere to warm, at least with respect to ECS. Following is from AR6 (page 996):

"Thus, studies employing this framework [Lewis and Curry's equations to compute ECS from observations] (Otto et al., 2013; Lewis and Curry, 2015, 2018; Forster, 2016) implicitly assume that the net radiative feedback has a constant magnitude, producing an estimate of the effective ECS (defined as the value of ECS that would occur if [the feedback] does not change from its current value) rather than of the true ECS. As summarized in Section 7.4.4.3, there are now multiple lines of evidence providing high confidence that the net radiative feedback will become less negative as the warming pattern evolves in the future (the pattern effect) ... implying that the true ECS will be larger than the effective ECS inferred from historical warming."

All team members agree that AR6 has overstated the evidence regarding the "pattern effect." It suggests net feedback becomes less negative as the warming pattern evolves over time. Moreover, the above quote misrepresents Lewis and Curry (2018). Contrary to what AR6 claims, in that paper Lewis and Curry did provide an ECS estimate that allowed for the pattern effect, by adjusting the main energy budget estimate to reflect the fact that net feedback is observed to become less negative over time in most climate model simulations.

Some members of the team recognize that the feedbacks may change with time (the pattern effect), but object to the AR6 argument that the feedbacks change as a function of surface temperature ("feedbacks on feedbacks"). They see evidence that the climate state changes, due to or synchronously with, long-term ocean oscillations (AMO, PDO, etc.) and changes in solar activity.[26] Thus, in their view, the changing climate state causes the changes in feedback and the resulting temperature changes. Changing temperatures are the result of natural climate state changes, at least in part, and not the cause of the change in climate state.

26   Vinos, 2022, *Climate of the Past, Present and Future*, p 182-187

Other team members, while not disputing the existence of longer-term natural climate changes, accept that feedbacks may change with (average) surface temperature as well as with time-varying temperature patterns, but consider changes in feedback with average surface temperature to be largely irrelevant for estimating ECS (except when estimating ECS from warming at much greater than a doubling of $CO_2$ from preindustrial, and possibly from glacial periods). Those team members also consider natural changes (except for volcanism, ENSO and the AMO) to be largely irrelevant for estimating ECS from historical warming, save that they increase uncertainty.

# 8

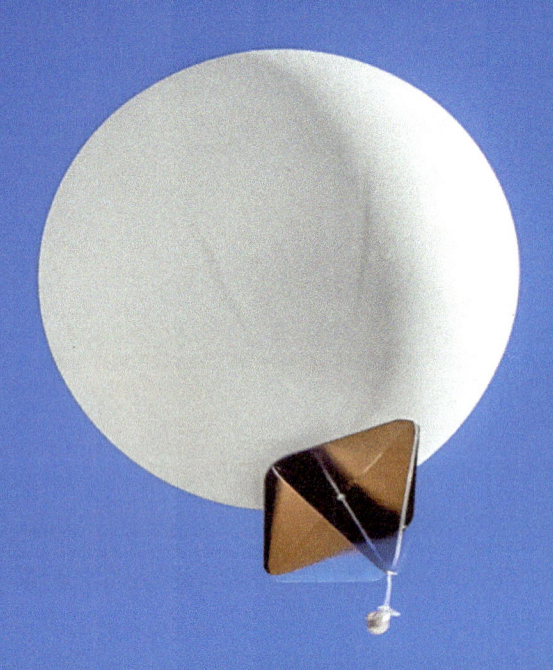

# AR6:
# More confidence that
# models are unreliable

**ROSS MCKITRICK**

8 AR6: MORE CONFIDENCE THAT MODELS ARE UNRELIABLE

**All climate models simulate amplified warming high up in the tropical troposphere. This area is therefore called the tropical "hot spot". The tropical hot spot provides a unique test for the models as they are not tuned to match observations there. Observations by weather balloons and satellites don't confirm the modelled hot spot. AR6 IPCC acknowledges there is a problem with the hot spot and therefore with the models. However, it does so in such veiled language that no one will notice.**

T he term "tropical hot spot" refers to the longstanding prediction from climate models that atmospheric warming in the tropics due to external forcing, including rising greenhouse gas (GHG) levels, should be amplified with altitude and should reach a maximum for the global atmosphere in the tropical mid-troposphere. Figure 1 shows the hot spot as generated by the Canadian climate model in a historical simulation of climate change over 1979-2017 in response to observed changes in external climate drivers, including GHG emissions. While the pattern of amplified warming aloft would arise in a model in response to any external positive forcing, the IPCC singles out GHGs as the only one that increased enough over the 20th and 21st centuries to have resulted in substantial atmospheric warming. The red coloration in the center of Figure 1 represents a hindcast of about 0.6 °C/decade of warming, while the model hindcasts around 0.2 to 0.3 °C/decade at the surface.

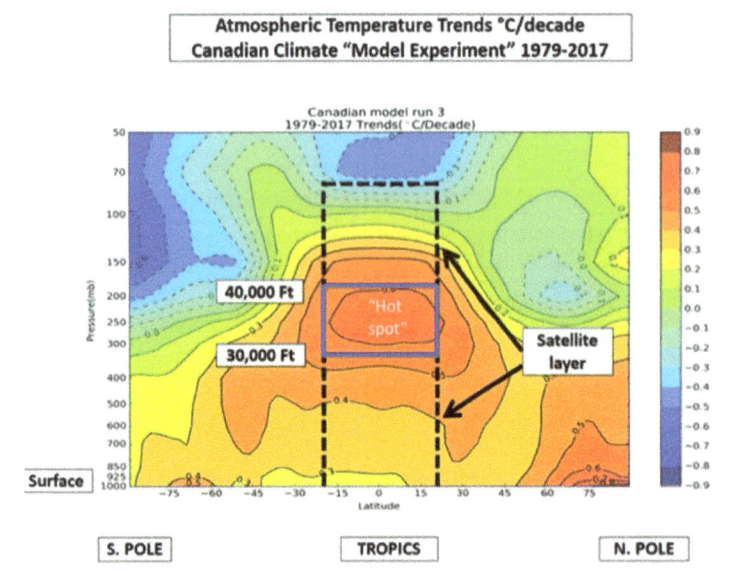

**Figure 1:** The tropical modeled "hot spot". The model used to generate this figure is the Canadian climate model. Source: (Christy, 2019).[1]

Model-generated warming in both the tropical surface and tropospheric layers exceed observed trends over the period. This can be seen in the atmospheric profile shown in Figure 2, which is from the 2013 IPCC Fifth Assessment Report (IPCC, 2013) or "AR5."[2]

---

1   Christy, J. (2019, June 18). *Putting Climate Change Claims to the Test*. Retrieved from Global Warming Policy Forum: https://www.thegwpf.com/putting-climate-change-claims-to-the-test/

2   IPCC. (2013). In T. Stocker, D. Qin, G.-K. Plattner, M. Tignor, S. Allen, J. Boschung, . . . P. Midgley, *Climate Change 2013: The Physical Science Basis. Contribution of Working Group I to the Fifth Assessment Report of the Intergovernmental Panel on Climate Change*. Cambridge: Cambridge University Press. Retrieved from https://www.ipcc.ch/pdf/assessment-report/ar5/wg1/WG1AR5_SPM_FINAL.pdf" https://www.ipcc.ch/pdf/assessment-report/ar5/wg1/WG1AR5_SPM_FINAL.pdf

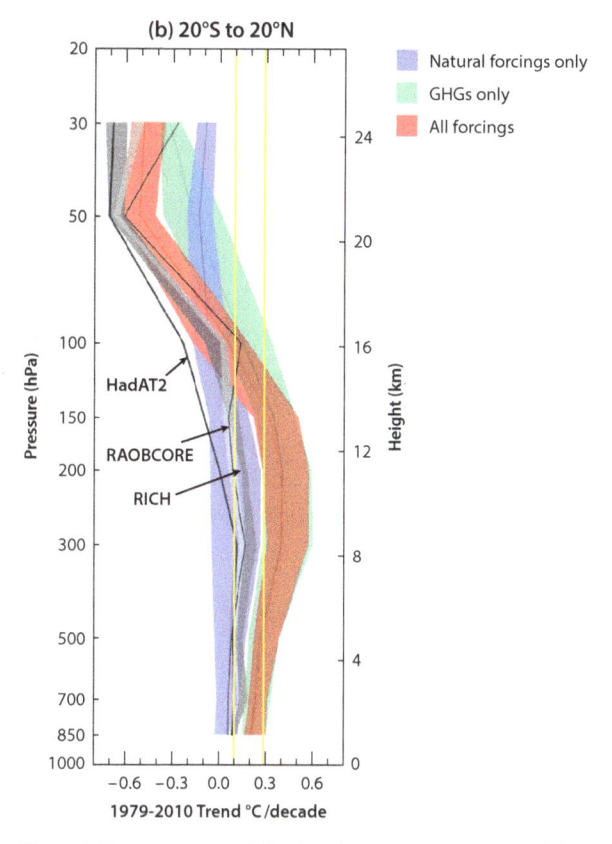

**Figure 2:** Temperature trends by altitude over 1979 to 2010: models compared to observations. Source: IPCC 2013 Figure 10.SM.1

The warming rates in three weather balloon-derived data sets (HadAT2, RAOBCORE and RICH) are shown.[3] The range of results from models run without greenhouse gas forcing is shown in the blue shaded area. Model results under greenhouse gas forcing only are shown in green and model results combining all climate forcings are shown in red. The "hot spot" region of the tropical troposphere is from around 300 hPa to 200 hPa (8-12 km altitude).

Here the observed warming trends are uniformly below all model runs and, interestingly, coincide best with the "natural only" model runs. In other words, the model runs that best correspond to observations are those in which GHG forcing is omitted altogether. When GHG forcing is included the models exhibit too much warming from the surface up to nearly the top of the troposphere (14 km or 150 hPa). This suggests that climate models are significantly overestimating the impact of GHGs on the climate system.

Note: this figure was published in a supplementary document, not directly in the IPCC AR5 report itself and was therefore seen by very few people. A similar figure is found in AR6, Chapter 3, page 444. The CMIP6 models used in AR6 have moved farther from the observations, not closer. See figure 3.

## Is Warming Amplified Higher in the Atmosphere?

There are two aspects to the hot spot prediction that need to be distinguished: whether the observed historical trend rate of warming matches model simulations and whether the warming aloft is amplified relative to the surface. At various points in the debate about whether the hot spot exists, some defenders of the models have argued that even if there is very little observed warming, the hot spot still exists in the sense that the warming aloft is greater than at the surface: there just happens to be very little surface warming. But there is good evidence that models exaggerate both the amplification rate and the resulting mid-troposphere warming rate.

---

3    HadAT2 is the Hadley Centre Atmospheric Temperature weather reanalysis dataset RAOBCORE is the Radiosonde Observation Correction using Reanalysis, a weather reanalysis dataset (Radanovics, 2010). RICH is the Radiosonde Innovation Composite Homogenization dataset, a weather balloon radiosonde dataset.

Climate observations are discussed in AR6 Chapter 2 (which covers the changing state of the climate system) and in Chapter 3 (which examines the human influence on climate). Chapter 2 Section 2.3.1.2.2 briefly surveys evidence showing that the troposphere has warmed. They point out that evidence from weather balloons goes back to the late 1950s and demonstrates a warming trend. They do not mention the paper by McKitrick and Vogelsang (2014)[4] that came out after AR5 which showed that the warming of the tropical troposphere from 1958 to 2012 was all attributable to a single step-change in the late 1970s, with no significant trend before or after. The IPCC reports with *medium confidence* that the tropical upper troposphere has warmed faster than the surface since 2001 but only *low confidence* for the interval prior to that.

There is no discussion in AR6 of the work of Klotzbach et al. (2009)[5] who showed that models project greater amplification with altitude than is observed. That paper gave rise to an online debate over whether the amplification rates were being correctly calculated, and whether the discrepancy between models and observations was statistically significant. The topic involves very advanced statistical theory and was dealt with by two experts in time series analysis, Vogelsang and Nawaz (2017).[6] They found that Klotzbach's conclusions were valid, and that observed amplification is significantly smaller than the model-projected rates. AR6 makes no mention of this work.

## Is the Stratosphere still Cooling?

However, AR6 does refer to evidence that stratospheric cooling (another greenhouse "fingerprint" the IPCC likes to highlight when it is observed) appears to have stopped. They note that the lower stratosphere has not cooled over the past 20 years (with one study even reporting slight warming) and that while the mid- and upper troposphere may have cooled over that interval there is low confidence in the trend magnitude.

In Chapter 3 Section 3.3.1.2 the IPCC picks up on the question of whether the observed tropical tropospheric warming rate is lower than in the models. They say that the AR5 assessed that it was, but only with *low confidence*. Based on evidence published since then they have upgraded their assessment to *medium confidence*. One wonders how much more evidence they need to claim high confidence, since in other areas they jump to that level with much less to go on.

From AR6:

> "Several studies since AR5 have continued to demonstrate an inconsistency between simulated and observed temperature trends in the tropical troposphere, with models simulating more warming than observations (Mitchell et al., 2013, 2020, Santer et al., 2017a, 2017b; McKitrick and Christy, 2018; Po-Chedley et al., 2 2021). … Over the 1979-2014 period, models are more consistent with observations in the lower troposphere, and least consistent in the upper troposphere around 200 hPa, where biases exceed 0.1°C per decade. Several studies using CMIP6 models suggest that differences in climate sensitivity may be an important factor contributing to the discrepancy between the simulated and observed tropospheric temperature trends (McKitrick and Christy, 2020; Po-Chedley et al., 2021), though it is difficult to deconvolve the influence of climate sensitivity, changes in aerosol forcing and internal variability in contributing to tropospheric warming biases (Po-Chedley et al., 2021). Another study found that the absence of a hypothesized negative tropical cloud feedback could explain half of the upper troposphere warming bias in one model (Mauritsen and Stevens, 2015)." (AR6, Chapter 3, p 443)

---

4    McKitrick, Ross R. and Timothy Vogelsang (2014) "HAC-Robust Trend Comparisons Among Climate Series with Possible Level Shifts" *Environmetrics* DOI: 10.1002/env.2294.

5    Klotzbach PJ, Pielke Sr RA, Pielke Jr RA, Christy JR, McNider RT. 2009. An alternative explanation for differential temperature trends at the surface and in the lower troposphere. *Journal of Geophysical Research* **114**: D21102.

6    Vogelsang, TJ, Nawaz, N 2017 Estimation and inference of linear trend slope ratios with an application to global temperature data. *J Time Series Analysis* 38: 640-667 DOI: 10.1111/jtsa.12209

Within this paragraph are some important admissions, although as is usual in IPCC work, they are minimized in the text. As pointed out in McKitrick and Christy (2018)[7] – though not mentioned in the IPCC's summary thereof – the tropical mid-troposphere is a uniquely important region for testing climate models. The modelers do not tune the models against observations there (as opposed to surface trends) so it is a genuine test of model performance. Also, all models make the same prediction, so it is a test that encompasses the overarching theoretical framework. And if the warming is present, it can only be explained by greenhouse gases since none of the IPCC's natural forcing estimates could account for it (see Figure 2). So, the model-observational discrepancy in the tropical mid-troposphere points to some genuine errors in climate model structure. The IPCC hints at this by mentioning that "differences in climate sensitivity" may be the cause – in other words models have excessively high climate sensitivity to greenhouse gases.

The final sentence refers to the absence of the Lindzen 'Iris Effect' – by which atmospheric warming reduces tropical cloud formation slightly and thus increases infrared radiation to space. When Mauritsen and Stevens (2015)[8] included an iris effect in a model, about half of the discrepancy disappeared.

The IPCC text goes on to point out that "Mitchell et al. (2013) and Mitchell et al. (2020)[9] found a smaller discrepancy in tropical tropospheric temperature trends in models forced with observed [Sea Surface Temperatures or SSTs]." What they mean by this is that perhaps the amplification rate is ok, but the SST trend is too high, so if they constrain the models to have the correct SST trend it will get the mid-troposphere trend correct as well. There are several problems with this line of argument. First, it papers over the problem of excess warming in the mid-troposphere by asking the reader to ignore the problem of excess warming at the sea surface. But the excess warming is present there as well. Second it ignores the studies mentioned above that found evidence of significantly exaggerated amplification. Third, and by implication, even when models are forced to reproduce observed SSTs they generate more warming aloft than observed. This is shown in Figure 3 which is taken from the AR6 Chapter 3 (p 444). The red symbols show temperature trend ranges from models that generate SST trends internally and the blue symbols show the tempera-

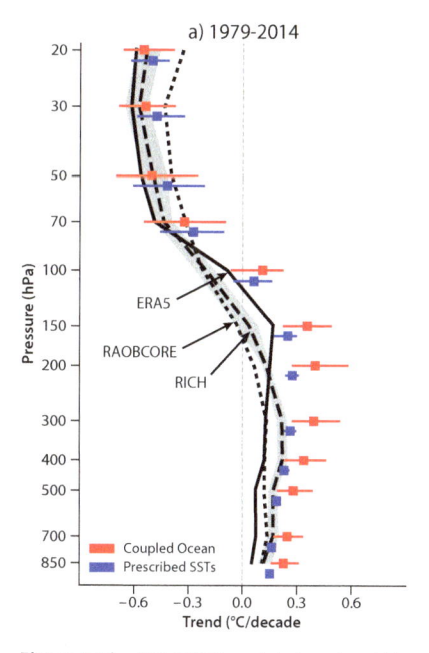

**Figure 3:** The AR6 CMIP6 models, in red and blue compared to observations in black. The red whisker plots use modeled SSTs and the blue symbols use observed SSTs.

7   McKitrick, Ross R and John Christy (2018) A Test of the Tropical 200-300mb Warming Rate in Climate Models. *Earth and Space Science* doi: 10.1029/2018EA000401.

8   Mauritsen, T. and B. Stevens, 2015: Missing iris effect as a possible cause of muted hydrological change and high climate sensitivity in models. *Nature Geoscience*, 8(5), 346–351, doi:10.1038/ngeo241

9   Mitchell, Dann M., Y. T. Eunice Lo, William J. M. Seviour, Leopold Haimberger, en Lorenzo M. Polvani. 'The vertical profile of recent tropical temperature trends: Persistent model biases in the context of internal variability'. *Environmental Research Letters* 15, nr. 10 (oktober 2020): 1040b4. https://doi.org/10.1088/1748-9326/ab9af7.

ture trend ranges from models that are forced to match the observed SST trends. As shown even the latter group exceed the indicated observations for the period 1979-2014 in the critical region 300 hPa to 150 hPa. (IPCC, 2021, pp. 444 ).

## AR6/CMIP6 Models are too Warm Globally

In McKitrick and Christy (2020)[10] we showed that not only do all models overstate warming in the tropical troposphere, but they now overstate it globally as well.

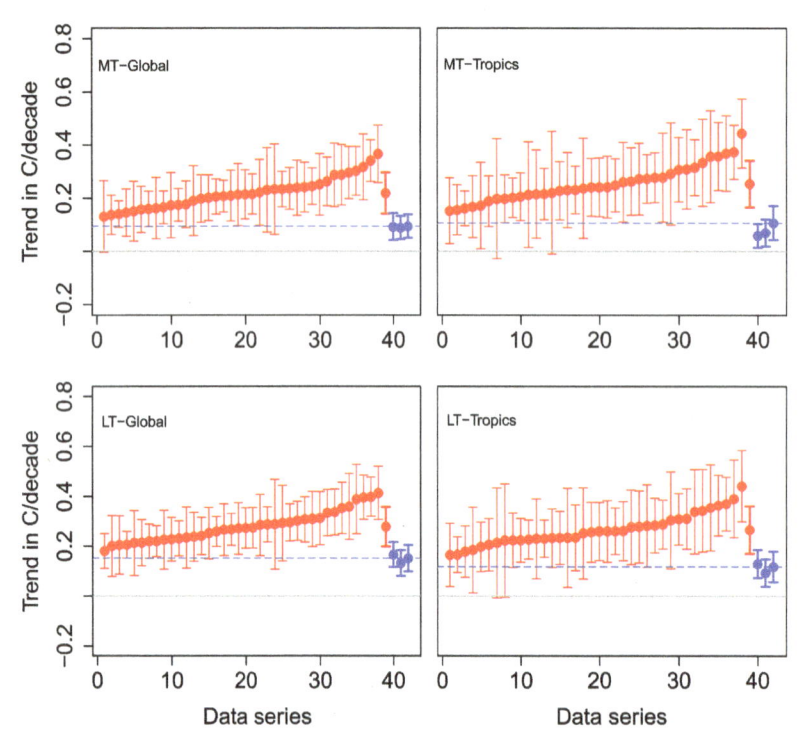

Trends and 95% CI's for individual models (red dots and thin bars), CMIP6 mean (red dot and thick bar), and observational series (blue). Horizontal dashed line shows mean satellite trend.

**Figure 4:** Models overshoot observed warming in the lower and mid-troposphere in the period 1979-2014, both in the tropics and globally. Source: McKitrick (2020)[11]

We examined the first 38 models in the CMIP6 ensemble, the results are shown in Figure 4. Here are the 1979-2014 warming trend coefficients (vertical axis, degrees per decade) and 95% error bars comparing models (red) to observations (blue). LT=lower troposphere, MT=mid-troposphere. Every model overshoots the observed trend (horizontal dashed blue line) in every sample.

This is to be expected if, as we argued in our 2018 paper, there is a problem with a critical core mechanism in the models. The problem is not simply natural variability or a transient temperature overshoot.

Most of the differences are significant at <5%, and the model mean (thick red) versus observed mean difference is very significant, meaning it's not just noise or randomness. The models as a group warm too much throughout the global atmosphere, even over an interval where modelers can observe both forcings and temperatures.

We found that models with higher Equilibrium Climate Sensitivity (>3.4K) warm faster (not surprisingly), but even the low-ECS group (<3.4K) exhibits warming bias. In the low group the mean ECS is 2.7K, the combined LT/MT model warming trend average is 0.21K/decade and the observed counterpart is 0.15K/decade. This figure (green circle added; see below) shows a more detailed comparison.

10   McKitrick, Ross and John Christy (2020) Pervasive Warm Bias in CMIP Tropospheric Layers. *Earth and Space Science* Vol 7(9) September 2020.

11   McKitrick, Ross and John Christy (2020) Pervasive Warm Bias in CMIP Tropospheric Layers. *Earth and Space Science* Vol 7(9) September 2020.

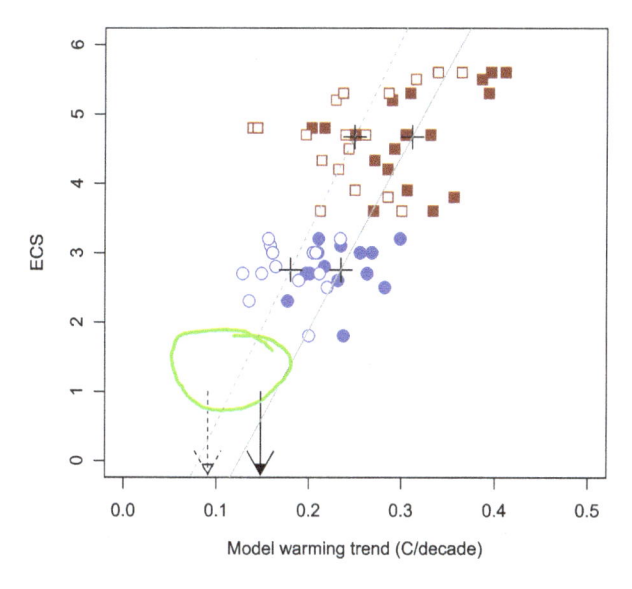

Model ECS values plotted against model warming trends. Red squares = high ECS group, blue circles = low-ECS group. Open shape = MT trend, closed shape = LT trend. Inverted triangles = mean observed LT trend (solid), mean observed MT trend (open).

**Figure 5:** even models with a 'low' climate sensitivity (blue) overestimate the observed warming in the lower and mid-troposphere. Source: McKitrick (2020)

The horizontal axis in figure 5 shows the model warming trend and the vertical axis shows the corresponding model ECS. The red squares are in the high ECS group and the blue circles are in the low ECS group. Filled shapes are from the LT layer and open shapes are from the MT layer. The crosses indicate the means of the four groups and the lines connect LT (solid) and MT (dashed) layers. The arrows point to the mean observed MT (open arrow, 0.09C/decade) and LT (closed arrow, 0.15 C/decade) trends.

While the models in the blue cluster (low ECS) do a better job, they still have warming rates in excess of observations. If we were to picture a third cluster of models with mean global tropospheric warming rates overlapping observations it would have to be positioned roughly in the area I've outlined in green. The associated ECS would be between 1.0 and 2.0K.

## Conclusions

The AR6 discussion has raised its confidence that models overstate tropical tropospheric warming from *low* to *medium*. For everything else, confidence that models give reliable forecasts should therefore move in the opposite direction.

I get it that modeling the climate is incredibly difficult, and no one faults the scientific community for finding it a tough problem to solve. But we are all living with the consequences of climate modelers stubbornly using generation after generation of models that exhibit too much surface and tropospheric warming, in addition to running grossly exaggerated forcing scenarios (e.g. RCP8.5). Back in 2005 in the first report of the then-new US Climate Change Science Program, Karl et al. pointed to the exaggerated warming in the tropical troposphere as a "potentially serious inconsistency." But rather than fixing it since then, modelers have made it worse.

If the discrepancies in the troposphere were evenly split across models between excess warming and cooling we could chalk it up to noise and uncertainty. But that is not the case: it's all excess warming. CMIP5 models warmed too much over the sea surface and too much in the tropical troposphere. Now the CMIP6 models warm too much throughout the global lower- and mid-troposphere. That's bias, not uncertainty, and until the modeling community finds a way to fix it, the economics and policy making communities are justified in assuming future warming projections are overstated, potentially by a great deal depending on the model.

# C

# Climate Change Scenarios

# 9 Extreme scenarios

**BY MARCEL CROK**

**The biggest news in the AR6 report is arguably that high-end scenarios like SSP5-8.5 and SSP3-7.0 are now believed to have low likelihood. That is extremely good news as it means that higher rates of warming in 2100 are thus viewed to be less likely than they were only a few years ago. Unfortunately, this news is deeply hidden in the report and few policy makers will see it. Worse, large parts of the report still emphasize these high-end scenarios. How did this happen?**

PCC reports are meant to be "policy relevant" and "policy neutral".[1] Policy makers deal—by definition—with an uncertain future. No one can predict with any certainty what the climate is going to do 50 or 100 years from now. However, climate scientists have tools to explore what the climate might look like in the future. These tools are called scenarios and since the first IPCC report in 1990 scenarios have played an important role in climate policy.

In AR6 we find a table and a figure showing how global temperatures might develop under the five scenarios that were selected for the report. Here is figure SPM.8:

## (a) Global surface temperature change relative to 1850–1900

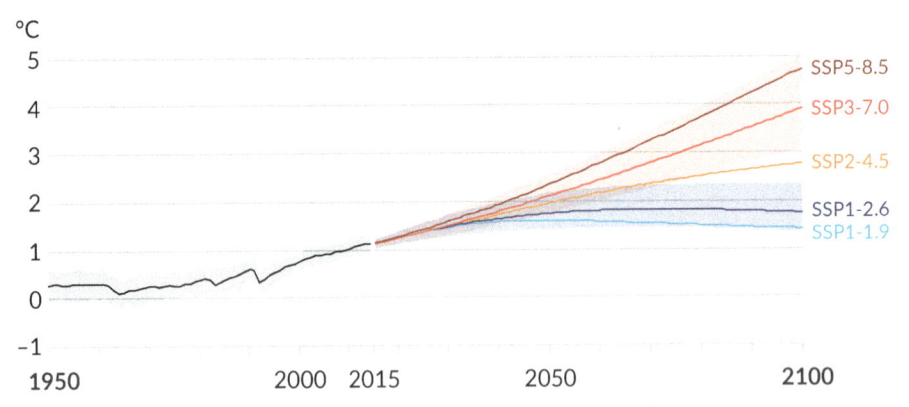

**Figure 1:** Global surface temperature change relative to 1850-1900 based on different scenarios. Very likely ranges are shown for SSP1-2.6 and SSP3-7.0. Source: AR6 figure SPM.8a.

Here we see that the two higher scenarios SSP3-7.0 and SSP5-8.5 reach 4 to 5°C of warming in 2100. That's quite dramatic. Remember, the IPCC estimates that the world warmed around 1°C since 1850, in a period of 170 years. These scenarios suggest we will get another 3 to 4°C of warming in just the next 80 years. This is a nightmare scenario for those who take the Paris agreement that we should stay below 2°C, or preferably below 1.5°C, seriously.

Therefore, a really important question for policy makers is: how plausible are each of the scenarios that underlie the projections of future climate? Well, this might be a big surprise for you, but the IPCC doesn't address this highly policy relevant question. In chapter 1 (page 238) it says: "In general, no likelihood is attached to the scenarios assessed in this Report."

This is quite an admission and one that actually should have been put in the *Summary for Policy Makers* with a big disclaimer such as: "Note, this report makes extensive use of scenarios. However, the likelihood of these scenarios itself hasn't been assessed!"

It becomes even stranger though. In the next paragraph on the same page there is this claim: "However, the likelihood of high emission scenarios such as RCP8.5 or SSP5-8.5 is considered low

---

1   "IPCC reports are neutral, policy-relevant but not policy-prescriptive." Source: www.ipcc.ch

in light of recent developments in the energy sector (Hausfather and Peters, 2020a[2], 2020b[3])." The IPCC has just said the likelihood of its scenarios is not assessed in the report and now it says the likelihood of RCP8.5 and SSP5-8.5 is "low". These statements are contradictory. How can you not assess the likelihood of the scenarios and then conclude that at least one scenario is low likelihood?

As we will see in this chapter the "low likelihood" of the IPCC extreme scenarios is quite an understatement. The RCP8.5 and the closely related SSP5-8.5 scenarios are – to use terminology of the IPCC itself – extremely implausible and it is more than correct for the IPCC to point this out. Again, this should have been pointed out prominently in the *Summary for Policy Makers* (SPM) with a disclaimer such as: "Note, the likelihood of the high emission scenarios RCP8.5 and SSP5-8.5 is regarded as low." However, no such disclaimers were shown in the SPM and most of the policy makers who took the effort to read the SPM will not know the highest scenario has a "low likelihood" of coming true.

## Baseline Scenario

Let's take a step back and describe what scenarios are and see how the IPCC used them in the past. The process starts with generating ideas about socioeconomic developments: future population growth, economic growth, technological changes, land use changes. Scientists use so-called Integrated Assessment Models (IAMs) to integrate all these inputs and assumptions. These models can then be used to project greenhouse gas emissions over the course of the century. The output of these models are used by the climate modelling community to project climate changes.

In its first report in 1990 the IPCC used scenarios in the same way as Shell and other energy companies use them. In general, you will have a business-as-usual, or baseline, or reference scenario. That scenario is supposed to show what is likely to happen without climate policies. The other scenarios will have some assumed greenhouse gas emissions reduction. The difference between the baseline and policy scenarios will give an impression of the potential effect of policy changes on global temperature.

In 2000 it was time for the IPCC to update its scenarios. After long discussions it was decided that the new scenarios would be presented without any consideration of their likelihood. This is a spectacular change as it means that each scenario is presented as likely (or unlikely) as the other scenarios. There was no longer a baseline scenario. The advantage was scientists don't have to go through the difficult process of determining how likely the scenarios are. The disadvantage though is that policy makers who are against strict climate regulations could use lower scenarios to claim things are not so bad and conclude no severe policies are necessary. Environmental NGOs were afraid this attitude would hamper active climate action.

So, in 2005 the process of making new scenarios started all over again. In hindsight this turned out to be a crucial moment. Scenario makers generally need a lot of time to generate new socioeconomic scenarios. However, the climate model community was very impatient and wanted to have the new scenarios as soon as possible. It decided, based on the extensive literature, that four so-called Representative Concentrations Pathways (RCPs) would be selected: one high scenario, one low and two in the middle. Two were put in the middle to prevent people from thinking the middle one was the most likely. These RCPs provide the greenhouse gas concentrations from 2005 until 2100. The climate model community could simply start using these new scenarios which were supposed to be 'representative' for different "families" of societal and energy system assumptions, and therefore used to project a small set of different climate futures.

---

2   Z. Hausfather, G.P. Peters, Emissions – the 'business as usual' story is misleading, Nature 577 (7792) (2020) 618–620.

3   Hausfather, Z. and G.P. Peters, 2020b: RCP8.5 is a problematic scenario for near-term emissions. Proceedings of the National Academy of Sciences, 117(45), 27791–27792, doi:10.1073/pnas.2017124117

Meanwhile in parallel the scenario community would start working on the so-called Shared Socio-economic Pathways (SSP), i.e., how could the global population and economy develop to reach the levels of radiative forcing in the four RCPs? However, this process took ten years. So, in 2013, when the fifth IPCC report (AR5) came out, the four RCPs were used without knowing what the fictitious worlds behind the RCP's would look like, or if they were even plausible futures. Nevertheless, the IPCC decided to use the highest of its four scenarios, RCP8.5, as the reference or business-as-usual scenario. As it turned out, this was misleading and unfortunately this error continues today.

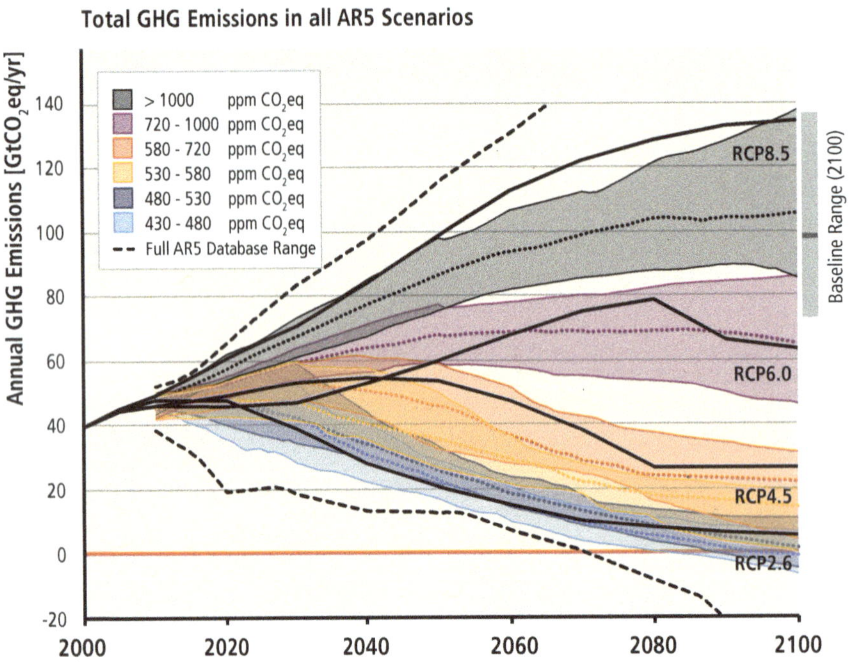

**Total GHG Emissions in all AR5 Scenarios**

**Figure 2:** Annual greenhouse gas emissions in the recent past and projected for the future based on the four RCP scenarios. Note how in the top right RCP8.5 was called the baseline range. Source: WG3, AR5, p. 52.

RCP8.5 would quickly become the favourite scenario of the climate model community because it generates such a clear signal-to-noise ratio compared to the background of natural climate variability. In plain English: climate models produce spectacular (or if you like dramatic) results if you feed them with the RCP8.5 scenario. The 8.5 by the way doesn't refer to temperature[4], but to the amount of climate forcing in 2100, i.e., 8.5 $W/m^2$. This is a huge amount of forcing[5], AR6 estimates the total increase in forcing since preindustrial to be 2.72 $W/m^2$. This increase took place over the period 1750 to 2020.

It all sounds rather technical, so why should ordinary citizens be bothered with this? Well, hardly a day or week passes without a new scientific paper based on RCP8.5 reaching you through the media. Such papers often have a message of doom and gloom. If you read in your newspaper that something terrible is going to happen with the climate in 2100, it is a pretty safe bet that the underlying research is based on the implausible RCP8.5.

A famous example is how the 2018 National Climate Assessment (NCA) in the US was communicated to its citizens. Here is the CNN headline: "Climate change will shrink US economy and kill thousands, government report warns."[6] The article said: "A new US government report delivers a dire warning about climate change and its devastating impacts, saying the economy could lose hundreds of billions of dollars – or, in the worst-case scenario, more than 10% of its GDP – by the end of the century." At least RCP8.5 is presented as a worst-case scenario—which it was not,

---

4    Sometimes people incorrectly think the 8.5 means 8.5°C of warming in 2100.

5    Doubling the $CO_2$-concentration gives a theoretical forcing of around 3.7 $W/m^2$. So 8.5 $W/m^2$ is the equivalent of more than two doublings of the $CO_2$-concentration in the atmosphere. Since preindustrial the $CO_2$-concentration in the atmosphere has increased from 280 ppm to 415 ppm.

6    https://edition.cnn.com/2018/11/23/health/climate-change-report-bn/index.html

as a worst case scenario also must be plausible—but in this case it was even worse: for the 10% estimate they used is an extreme upper limit of the already extreme RCP8.5 scenario. In that case Earth would warm a whopping 8°C in 2100. But even for RCP8.5 warming of 8°C is extreme. Normal warming rates for RCP8.5 are 4 or 5°C.

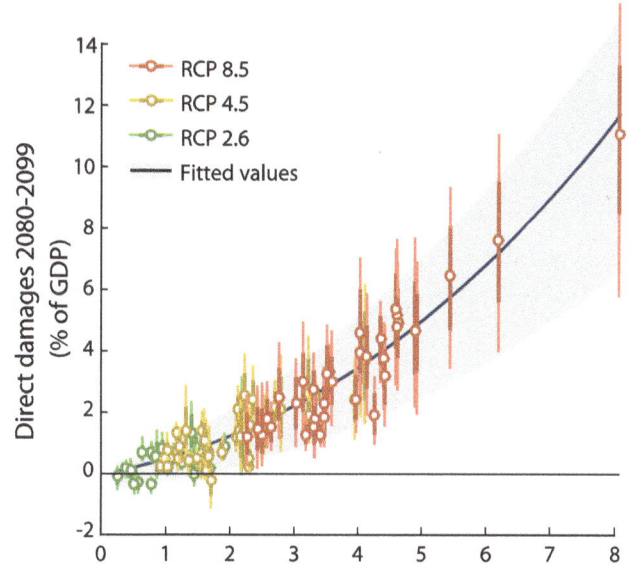

**Figure 3:** US GDP loss at the end of the century related to global warming rates.[7,8]

The NCA is being disingenuous, the underlying study they used only showed GDP losses of 3 to 4% (see figure 3).

In The Netherlands something similar happened after the publication of AR6. The Dutch KNMI published a report (in Dutch[9]) in which it showed some relevant conclusions from AR6 for Dutch policy makers. The relevant headline at the national public news broadcaster NOS read: "KNMI adjusts expected sea level rise upward".[10] It combined SSP5-8.5 with a very uncertain ice cap instability scenario to claim sea levels along the Dutch coast could rise by 1.2 meters in 2100 or even 2 meters. It was 18 centimetres in the past century with no sign of acceleration. Again, few news consumers (including Dutch policy makers) will realise what kind of assumptions are behind such grotesque predictions.

## How Plausible are the Extreme Scenarios?

So how extreme is RCP8.5 and its more recent version SSP5-8.5? Well, just to give you an idea, to get there the world would need to start using six times more coal per capita than we use now. Or to translate it into coal power stations: currently there are around 6000 coal power stations in the world. RCP8.5 (and SSP5-8.5) implies humanity will add another 33,000 between now and 2100. What about the next scenario SSP3-7.0? That still implies the building of 17,000 new coal power plants. Again, highly implausible.

Countries like China and India are still building coal power stations, but western countries are closing them and replacing them with natural gas-powered stations. Globally coal consumption seems to be at a plateau for a decade or so.

7   Hsiang S, Kopp R, Jina A, Rising J, Delgado M, Mohan S, Rasmussen DJ, Muir-Wood R, Wilson P, Oppenheimer M, Larsen K, Houser T. Estimating economic damage from climate change in the United States. Science. 2017 Jun 30;356(6345):1362-1369. doi: 10.1126/science.aal4369. PMID: 28663496.

8   https://fabiusmaximus.com/2018/11/29/scary-but-fake-news-about-the-national-climate-assessment/

9   https://cdn.knmi.nl/knmi/asc/klimaatsignaal21/KNMI_Klimaatsignaal21.pdf

10   KNMI adjusts expected sea level rise upwards

RCP-scenarios start in the year 2005 so there are now 15 years of real-world data to evaluate them. Such an evaluation is clearly something you might expect from the IPCC. After all it is highly policy relevant how their scenarios track with reality in order to know where we are going. However, apart from a short sentence about the likelihood, the IPCC said very little about the plausibility of its scenarios. It only referred to Hausfather and Peters 2020a and 2020b. These are indeed relevant pieces. One is a comment in *Nature*, the other is a reply to another paper in *PNAS*. They are not original peer reviewed works.

Several peer reviewed papers are available in the literature that deal with this issue. However, these papers were all ignored by the IPCC. A good starting point for this discussion is the 2017 paper "Why do climate change scenarios return to coal?" by Justin Ritchie.[11] The paper was very clear about RCP8.5 being an unlikely scenario because it assumes a return to coal. It said: "This paper argues SSP5-RCP8.5 is an exceptionally unlikely endpoint of future $CO_2$ forcing because it is biased by a return-to-coal hypothesis that distorts the future energy scenarios produced by IAMs [Integrated Assessment Models]." And elsewhere: "These four lines of evidence (i-iv) collectively indicate that RCP8.5 no longer offers a trajectory of 21st-century climate change with physically relevant information for continued emphasis in scientific studies or policy assessments."
This is a spicy remark, of course. Ritchie and his colleague specifically said RCP8.5 should no longer be used in policy assessments. That is, in IPCC reports. However, not only did IPCC ignore this paper, it also ignored the advice. Roger Pielke Jr, a well-known climate and policy scientist, in peer-reviewed papers, and summarized in his blog, documented how often RCP8.5 and SSP5-8.5 were mentioned in the AR6 report. The result is shown in the table below:

| SCENARIO | MENTIONS | PCT of MENTIONS |
|---|---|---|
| SSP5-8.5 & RCP8.5 | 1359 | 41.5% |
| SSP1-2.6 & RCP2.6 | 733 | 22.4% |
| SSP2-4.5 & RCP4.5 | 571 | 17.4% |
| SSP3-7.0 | 378 | 11.5% |
| SSP1-1.9 | 200 | 6.1% |
| RCP6.0 | 32 | 1.0% |

**Figure 4:** mentions of different scenarios in the AR6 report. Source: Roger Pielke Jr.

As you can see, of all the available scenarios, RCP8.5 and SSP5-8.5 are mentioned most. If you add the still extreme SSP3-7.0 scenario to it, then they are more than half of all scenario references in the report. Just to give some examples from the report:

- Under RCP2.6 and RCP8.5, respectively, glaciers are projected to lose 18% ± 13% and 36% ± 20% of their current mass over the 21st century (medium confidence). (77)
- Under RCP8.5/SSP5-8.5, it is likely that most land areas will experience further warming of at least 4°C compared to a 1995–2014 baseline by the end of the 21st century, and in some areas significantly more. (132)
- According to the SROCC, sea level rise in an extended RCP2.6 scenario would be limited to around 1 m in 2300 (low confidence) while multi-metre sea-level rise is projected under RCP8.5 by then (medium confidence). (188)

The reader gets the idea. All the scary messages from the report are based on RCP8.5 and SSP5-8.5. However, there is solid real-world evidence now, published in the peer reviewed literature that this scenario is not plausible. It is low likelihood according to IPCC, based on the implausible assumption of the explosive use of coal. It's a scenario that you simply should not use to inform policy makers. However, in AR6 it's the scenario that is used more than any other. How is this possible? Well, in a way it's quite understandable. IPCC is supposed to review all the available literature that was published in the period leading to the publication of the report.[12] Bloomberg news

---

11  J. Ritchie, H. Dowlatabadi, Why do climate change scenarios return to coal? Energy 140 (2017) 1276–1291.
12  The deadline for literature for AR6 was 31 January 2021.

did a google scholar search for the use of different scenarios in the literature. The figure below summarizes their results:

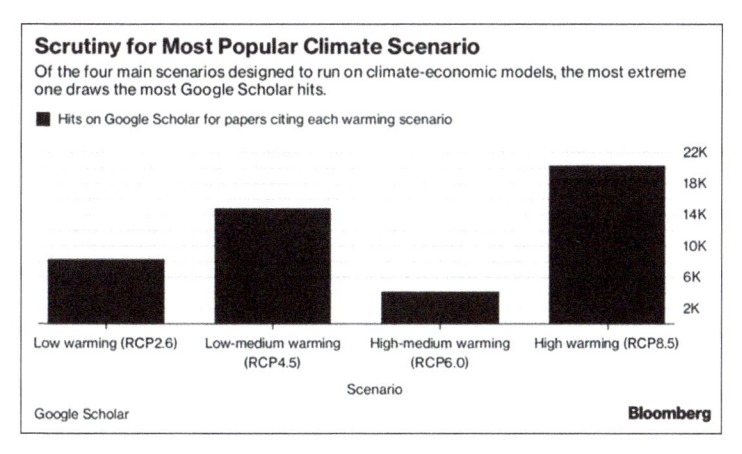

**Figure 5:** mentions in the literature of the different RCP scenarios.

RCP8.5 is not only the favourite scenario in AR6 but also in the literature. In this sense IPCC is simply doing its job, assessing and reviewing the literature. However, it's still highly problematic since RCP8.5 is such an unrealistic scenario.

## Scenario Reality Check

Another paper that was 'missed' by the IPCC was the 2020 paper "IPCC Baseline Scenarios Over-project $CO_2$ Emissions and Economic Growth" by amongst others Matthew Burgess, Justin Ritchie and Roger Pielke Jr. [13] Title sounds pretty relevant for an IPCC assessment, doesn't it? It showed this figure:

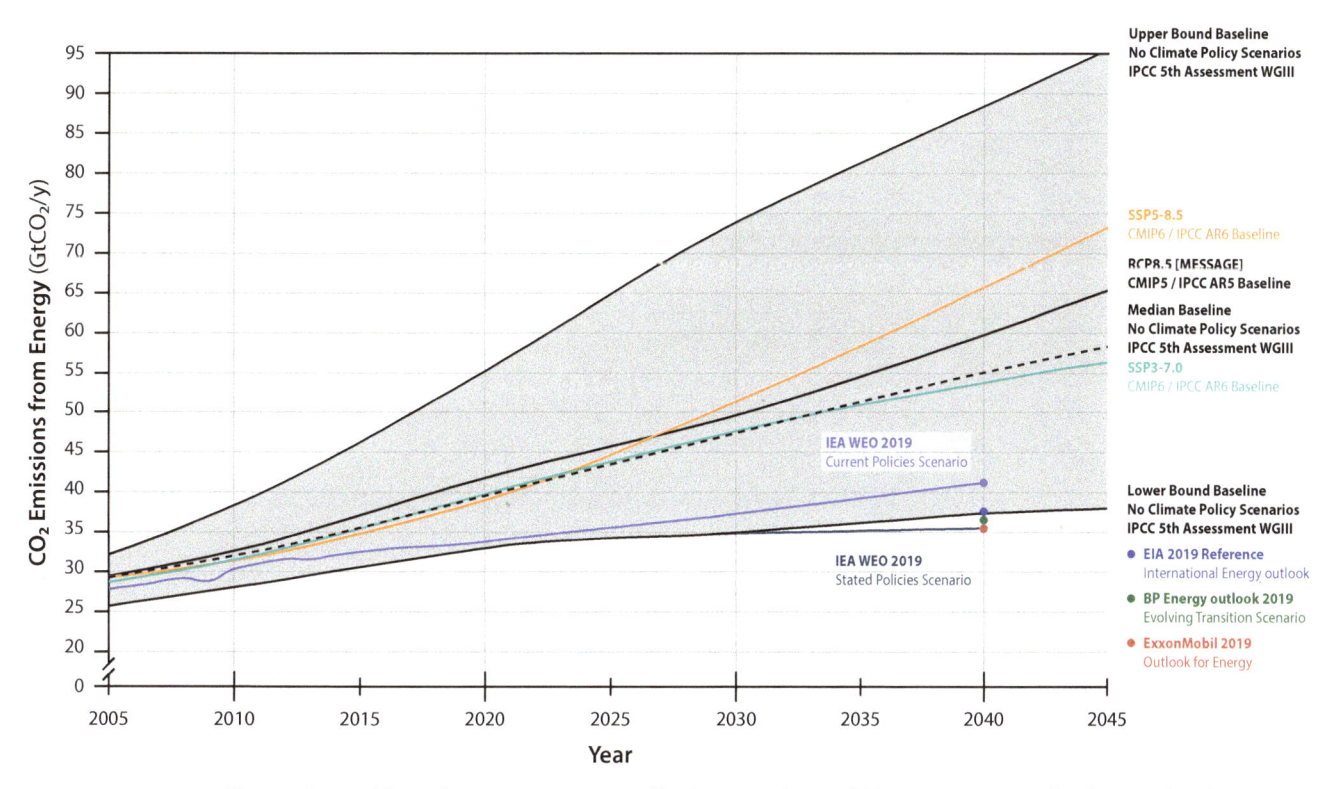

**Figure 6:** Past and future $CO_2$ emissions as projected by SSP3-7.0, RCP8.5 and SSP5-8.5 scenarios used by the IPCC. The coloured dots refer to several energy outlook scenarios of the International Energy Agency, the US Energy Information Administration, BP and ExxonMobil.

---

13    Matthew G. Burgess, Justin Ritchie, John Shapland, and Roger Pielke, Jr. IPCC Baseline Scenarios Over-project CO2 Emissions and Economic Growth. Environ. Res. Lett., 25 November, (2020), https://doi.org/10.1088/1748-9326/abcdd2.

The real-world emissions follow the lower boundary of the grey area closely and move farther and farther away from the SSP3-7.0, RCP8.5, and SSP5-8.5 scenarios. Notice the huge range for the extreme IPCC climate policy scenarios. According to the five SSP's, without climate policies, emissions in 2045 can be slightly higher than they were in 2020 (the lower bound baseline) or much higher. The upper bound is around 80 gigatonnes of $CO_2$/year in 2045. The SSP3-7.0, RCP8.5 and SSP5-8.5 scenarios all imply huge increases in $CO_2$ emissions between now and 2045. Increases that are not expected by the International Energy Agency, the US Energy Information Administration, BP, or ExxonMobil.

The Hausfather and Peters comment in *Nature* had a somewhat similar figure, combining emissions with expected temperature:

## POSSIBLE FUTURES

The Intergovernmental Panel on Climate Change (IPCC) uses scenarios called pathways to explore possible changes in future energy use, greenhouse-gas emissions and temperature. These depend on which policies are enacted, where and when. In the upcoming IPCC Sixth Assessment Report, the new pathways (SSPs) must not be misused as previous pathways (RCPs) were. Business-as-usual emissions are unlikely to result in the worst-case scenario. More-plausible trajectories make better baselines for the huge policy push needed to keep global temperature rise below 1.5 °C.

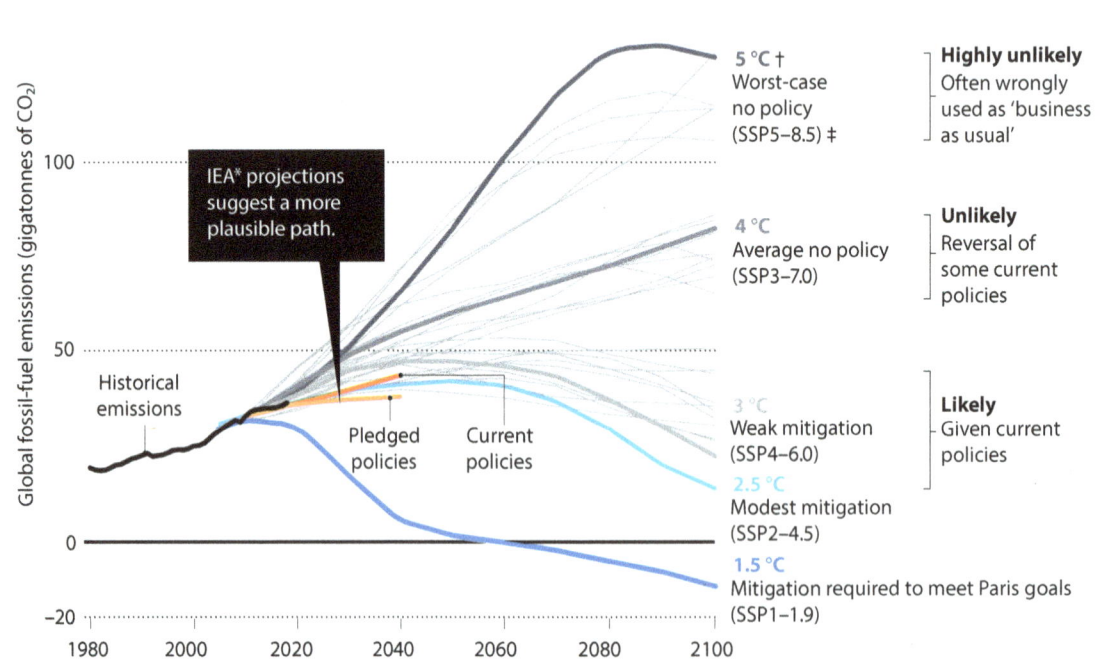

* The International Energy Agency (IEA) maps ou different energy-policy and investment choices. Estimated emissions are shown for its Current Policies Scenario and for its Stated Policies Scenario (includes countries' current policy pledges and targets). To be comparable with scenarios for the Shared Socioeconomic Pathways (SSPs), IEA scenarios were modified to include constant non-fossil-fuel emissions from industry in 2018.
† Approximate global mean temperature rise by 2100 relative to pre-industrial levels.
‡ SSP5–8.5 replaces Representative Concentration Pathway (RCP) 8.5.

**Figure 7:** Different scenarios and their potential relation with global temperature. Source Hausfather and Peters 2020.

The Hausfather and Peters' figure makes it clear that SSP5-8.5 is "very unlikely and often wrongly used as business as usual". SSP3-7.0 is "unlikely" as it requires a reversal of current policies, i.e., policies that are already in place independent of climate pledges.

It would have been helpful if a figure like this would have made it into the AR6 report. How else would policy makers have noticed this? There is no disclaimer or warning in the *Summary for Policy Makers* (SPM). There is only a short sentence in Chapter 1 stating that RCP8.5 and SSP5-8.5 have "low likelihood".

How could this have happened? How is it possible that such an extreme scenario became so dominant in the literature and in both the AR5 and AR6 report? The discussion about that has only recently started. A long essay with the revealing title "How Climate Scenarios Lost Touch with Reality" was published in the summer of 2021 by Justin Ritchie and Roger Pielke Jr, after the dead-

line for AR6.[14] It starts as follows: "A failure of self-correction in science has compromised climate science's ability to provide plausible views of our collective future."
One of the most striking sentences in the essay is this one:

> *"The continuing misuse of scenarios in climate research has become pervasive and consequential—so much so that we view it as one of the most significant failures of scientific integrity in the twenty-first century thus far. We need a course correction."*

This is a harsh conclusion. They talk about the "misuse of scenarios" and blame the climate science community for not yet correcting an error that has slipped into the literature. Therefore, they call it a "failure of scientific integrity". In a much longer peer reviewed paper Pielke and Ritchie dive even deeper into this issue.[15] This paper was available in 2020 although not yet officially published. Elsewhere in the report and in drafts the IPCC is not hesitant to use drafts of papers. But in this case they were not eager to fully discuss this issue in the report. The IPCC doesn't seem to be a big fan of the work of Roger Pielke Jr. Although Pielke Jr. has published authoritatively about scenarios, weather extremes, and about normalized damages due to disasters, AR6 only cited one of his papers, a rather old one from 2008. His more recent work is ignored. A recent report by the Global Warming Policy Foundation titled "The Hounding of Roger Pielke Jr" tries to explain where this attitude comes from.[16] In short: it has to do with politics.[17]

Several prominent climate scientists reacted to the essay by Pielke and Ritchie.[18] Chris Field (who has a long involvement with the IPCC) and Marcia McNutt (President of the National Academy of Sciences) rejected the criticism by Pielke and Ritchie. They wrote: "In particular, the high-emissions RCP8.5 scenario has long been described as a "business-as-usual" pathway with a continued emphasis on energy from fossil fuels with no climate policies in place. This remains 100% accurate, even if RCP8.5 does not appear to be the most likely high-emissions pathway."

They do admit that RCP8.5 is not the most likely pathway, but they still think it is right to call it a business-as-usual scenario.

Kate Marvel in her reply said: "I agree with Roger Pielke Jr. and Justin Ritchie's statement that we shouldn't call the high-emissions RCP8.5 scenario "business as usual," and they are right to call for the climate community to end this sloppy wording." However, she disagrees it is a matter of scientific integrity and emphasizes that AR6 doesn't call it that. "Neither the most recent Intergovernmental Panel of Climate Change report nor the National Climate Assessment claims RCP8.5 is "business as usual," but even an unrealistic scenario can yield interesting science if used appropriately."

Pielke and Ritch in their long peer reviewed article "Distorting the view of our climate future: The misuse and abuse of climate pathways and scenarios" show that scenarios such as RCP8.5 have become so endemic in the literature that it is hard to get rid of them. They agree with Marvel that there can be reasons of academic interest to study such 'extreme' scenarios, i.e., to study how the climate could react to such extreme increases in greenhouse gas concentrations. However, such studies should not be highlighted in scientific assessments as if they are plausible pictures of the future that are relevant for policy makers.

14    Jr., Roger Pielke, and Justin Ritchie. "How Climate Scenarios Lost Touch With Reality." Issues in Science and Technology 37, no. 4 (Summer 2021): 74–83 . https://issues.org/climate-change-scenarios-lost-touch-reality-pielke-ritchie/

15    Roger Pielke Jr. and Justin Ritchie, "Distorting the view of our climate future: The misuse and abuse of climate pathways and scenarios," Energy Research & Social Science 72 (2021): 101890.

16    https://www.thegwpf.org/content/uploads/2021/11/Laframboise-Pielke.pdf

17    More about this in chapter 12 about disasters.

18    https://issues.org/climate-scenarios-reality-pielke-jr-ritchie-forum/

At least climate scientists are beginning to openly acknowledge that RCP8.5 is not a realistic scenario. This raises the question, if RCP8.5 is not realistic which scenario is? Hausfather and Peters in their *Nature* comment (see figure 7) indicate that the weak to modest mitigation scenarios (SSP4-6.0 and RCP2-4.5) are currently in the likely range. This leads to warming of about 2.7°C in 2100, a number that is now frequently published as well.[19]

With a long and woolly sentence AR6 seems to agree with Hausfather and Peters:

> *"Studies that consider possible future emission trends in the absence of additional climate policies, such as the recent IEA 2020 World Energy Outlook 'stated policy' scenario (International Energy Agency, 2020), project approximately constant fossil and industrial $CO_2$ emissions out to 2070, approximately in line with the medium RCP4.5, RCP6.0 and SSP2-4.5 scenarios (Hausfather and Peters, 2020b) and the 2030 global emission levels that are pledged as part of the Nationally Determined Contributions (NDCs) under the Paris Agreement (Section 1.2.2; (Fawcett et al., 2015; Rogelj et al., 2016; UNFCCC, 2016; IPCC, 2018)."[20]*

Pielke, Ritchie and their colleague Matthew Burgess also looked into this issue: which of the scenarios is most likely and what would that imply for global temperatures?[21] In their paper they conclude that another SSP scenario, SSP3.4, fits best with the observed emissions. Note, this suggests that the world is on track for an even lower global forcing in 2100 than the SSP2-4.5 or the SSP4-6.0 that were used in the AR6 report. This SSP3.4 scenario isn't even mentioned in the AR6 report.

The median warming connected to this SSP3.4 scenario is 2.2°C of warming in 2100, close to the target of the Paris agreement. So according to them this would be the most likely warming in 2100. Again, this is very good news. Again and again we hear messages about the coming climate apocalypse in the media. We hear complaints that the world isn't doing enough to fight climate change. However, in reality, while emissions are still high, the world has moved away from the higher emissions doom and gloom world into a more moderate middle of the road scenario, where things don't look so bleak.

The IPCC had all the data and the literature available and should have highlighted this good news. However, for whatever reason, they didn't They make extensive use of a scenario that is completely out of touch with reality and highlight its results all over the report. No disclaimer was included in the *Summary for Policy Makers* warning policy makers of the situation. And week after week new publications appear using this extreme scenario to create screaming news headlines.

How to fix this unfortunate situation is not clear at the moment. If prominent leaders keep using this scenario and funding agencies keep funding research based on it, the use of this exaggerated scenario will continue for many years to come. Tighten your seatbelts.

---

19    "Earth will warm 2.7 degrees Celsius based on current pledges to cut emissions", https://www.sciencenews.org/article/climate-earth-warming-emissions-gap-pledges

20    AR6, p. 239

21    Pielke, R., Jr, Burgess, M. G., & Ritchie, J. (2021, March 23). Most plausible 2005-2040 emissions scenarios project less than 2.5 degrees C of warming by 2100. https://doi.org/10.31235/osf.io/m4fdu

# 10

# A miraculous sea level jump in 2020

BY OLE HUMLUM

**The IPCC launched a sea level projection tool which the public can use to 'make' different sea level scenarios for tide gauge stations around the world. Ole Humlum applied the tool to four Scandinavian capitals and shows the surprising results.**

Global, regional, and local sea levels always change. During the last glacial maximum about 20-25,000 years ago, global sea level was around 120 m below the modern sea level. Since the end of the Little Ice Age about 150 years ago, the global sea level has on average increased 1-2 mm/yr, according to tide gauges located at coasts. Observed data from sea level gauges worldwide can be accessed from the PSMSL Data Explorer.[1]

The issue of sea-level change, and in particular the identification of a hypothetical human contribution to that change, is a complex topic. Given the scientific and political controversy that surrounds the matter, people's high interest in this is entirely understandable. Nobody wants to be flooded.

Global (or eustatic) sea-level change is measured relative to an idealised reference level, the geoid, which is a mathematical model of the shape of the earth's surface and indicating a surface of equal gravity acceleration. The ocean surface will always try to adjust to this surface.

Sea-level is a function of the volume of the ocean basins and the volume of water that they contain, and global changes are brought about by three main mechanisms:
- changes in ocean basin volume caused by tectonic forces
- changes in seawater density caused by variations in ocean temperature and salinity
- changes in the volume of water caused by the reduction or growth of ice sheets, ice caps and smaller glaciers

Ocean basin volume changes occur too slowly to be significant over human lifetimes and it is therefore the other two mechanisms that drive contemporary concerns about sea-level rise. It is these mechanisms that IPCC are primarily concerned with in their modelling and discussion of this issue.

Higher temperature in itself is only a minor factor contributing to global sea-level rise, because seawater has a relatively small coefficient of expansion and because, over the timescales of interest, any warming is largely confined to the upper few hundred metres of the ocean surface.

The growth or decay of floating glaciers have no influence on sea level. However, the melting of land-based ice – including both mountain glaciers and the ice sheets of Greenland and Antarctica – is a more significant driver of global sea-level rise. For example, during the glacial–interglacial climatic cycling over the last half-million years, glacial sea-levels were about 120 m lower than the modern shoreline. Moreover, during the most recent interglacial, about 120,000 years ago, global temperature was higher than today, and significant extra parts of the Greenland ice sheet melted. As a consequence, global sea-level was several metres higher than today.

On a regional and local scale, however, factors relating to changes in air pressure, wind and geoid must also be considered. As an example, changes in the volume of the Greenlandic Ice Sheet will affect the geoid in the regions adjacent to Greenland. According to the climate models considered by IPCC, the Greenland Ice Sheet is expected to experience a significant loss of mass in the coming 100 years, caused by a modelled warming climate. In this case the overall mass in Greenland will diminish, the geoid surface will be displaced in direction of the planets centre, and sea level in the regions surrounding Greenland will drop. This will happen even though the overall volume of water in the global oceans will increase corresponding to the net loss of glacier ice.

---

1    Observed data; PSMSL Data Explorer. https://www.psmsl.org/data/obtaining/map.html

In northern Europe another factor must also be considered when estimating the future sea level. Norway, Sweden, Finland, and Denmark were totally or partly covered by the European Ice Sheet 20-25,000 years ago. Even today the effect of this ice load is clearly demonstrated by the fact that most of this region experiences an ongoing isostatic land rise of several millimetres per year. At many sites this more than compensates for the slow global sea level rise, so a net sea level drop in relation to land is recorded.

The relative movement of sea level in relation to land is what matters for coastal planning and is termed the relative sea level change. This is what is recorded by tide gauges.

## AR6 Sea Level Projection Tool

The most recent publication of the IPCC's work, the 6th Assessment Report from Working Group I, was released on August 9th, 2021. Modelled data for global and regional sea level projections 2020-2150 are available from the IPCC AR6 Sea Level Projection Tool.[2] The IPCC data considers the modelled future development of several factors, such as glacier mass change, vertical land movement, water temperature and -storage. Modelled sea level projections for different SSP scenarios are calculated relative to a baseline defined by observations 1995-2014.

It is instructive to compare the modelled data with observed sea level data, as illustrated below for the capitals of Norway, Sweden, Finland, and Denmark.

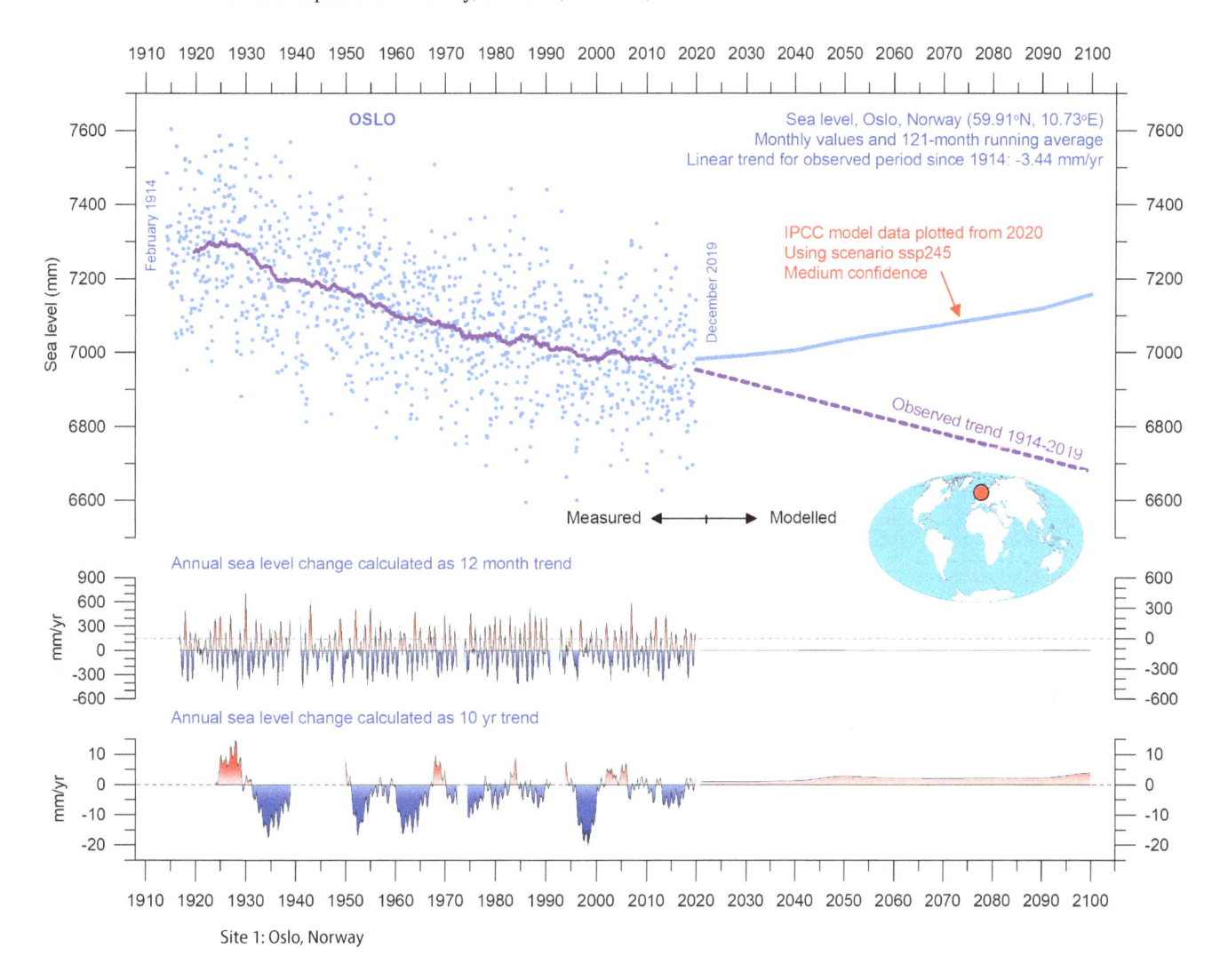

Site 1: Oslo, Norway

---

2    Modelled data; IPCC AR6 Sea Level Projection Tool: https://sealevel.nasa.gov/data_tools/17

Observed sea level at Oslo, Norway, February 1914 – December 2019 (blue dots and purple line). If the observed change rate continues (based on more than 100 years of observations), the relative sea level at Oslo (in relation to land) will have dropped by about 28 cm by year 2100, compared to now. In the diagram the blue line indicates the modelled sea level change 2020-2100 for Oslo, using the moderate SSP2-4.5 scenario. According to IPCC relative sea level (in relation to land) at Oslo will have increased about 17 cm by year 2100, returning to the Relative Sea Level seen at Oslo in 1945. Sea level increase is predicted to begin rather suddenly around 2020 at Oslo, in contrast to the previous sea level decrease of about -3.44 mm/yr recorded since 1914.

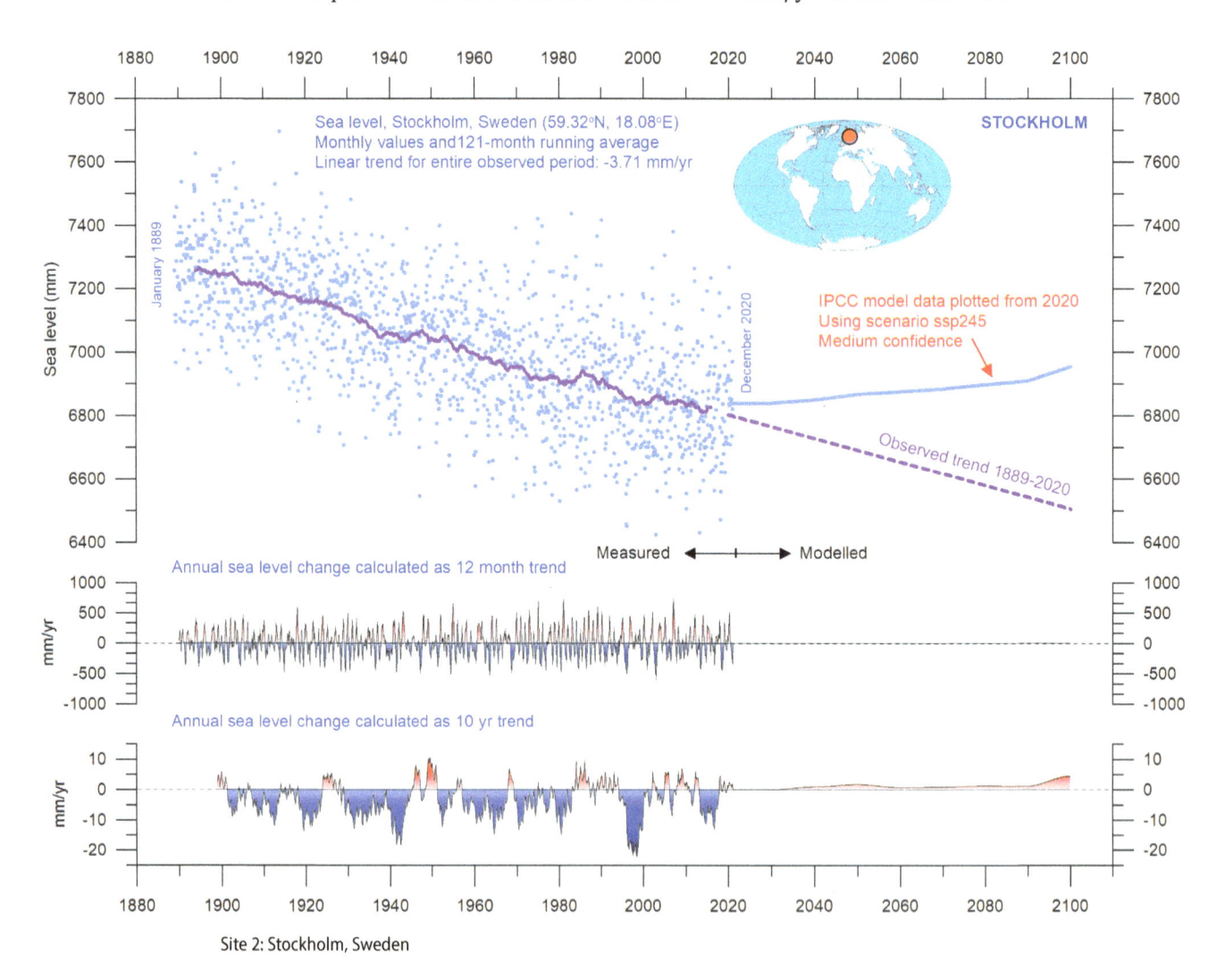

Site 2: Stockholm, Sweden

Observed relative sea level at Stockholm, Sweden, January 1889 – December 2020 (blue dots and purple line). If the observed change rate continues, the relative sea level at Stockholm (in relation to land) will have dropped about 30 cm by year 2100, compared to now. In the diagram the blue line indicates the modelled sea level change 2020-2100 for Stockholm, using the moderate SSP2-4.5 scenario. According to IPCC, the relative sea level (in relation to land) at Stockholm will have increased about 12 cm by year 2100, compared to now. A marked change from relative sea level decrease to -increase is predicted to begin around 2020, in contrast to the steady sea level decrease (about -3.71 mm/yr) recorded since 1889.

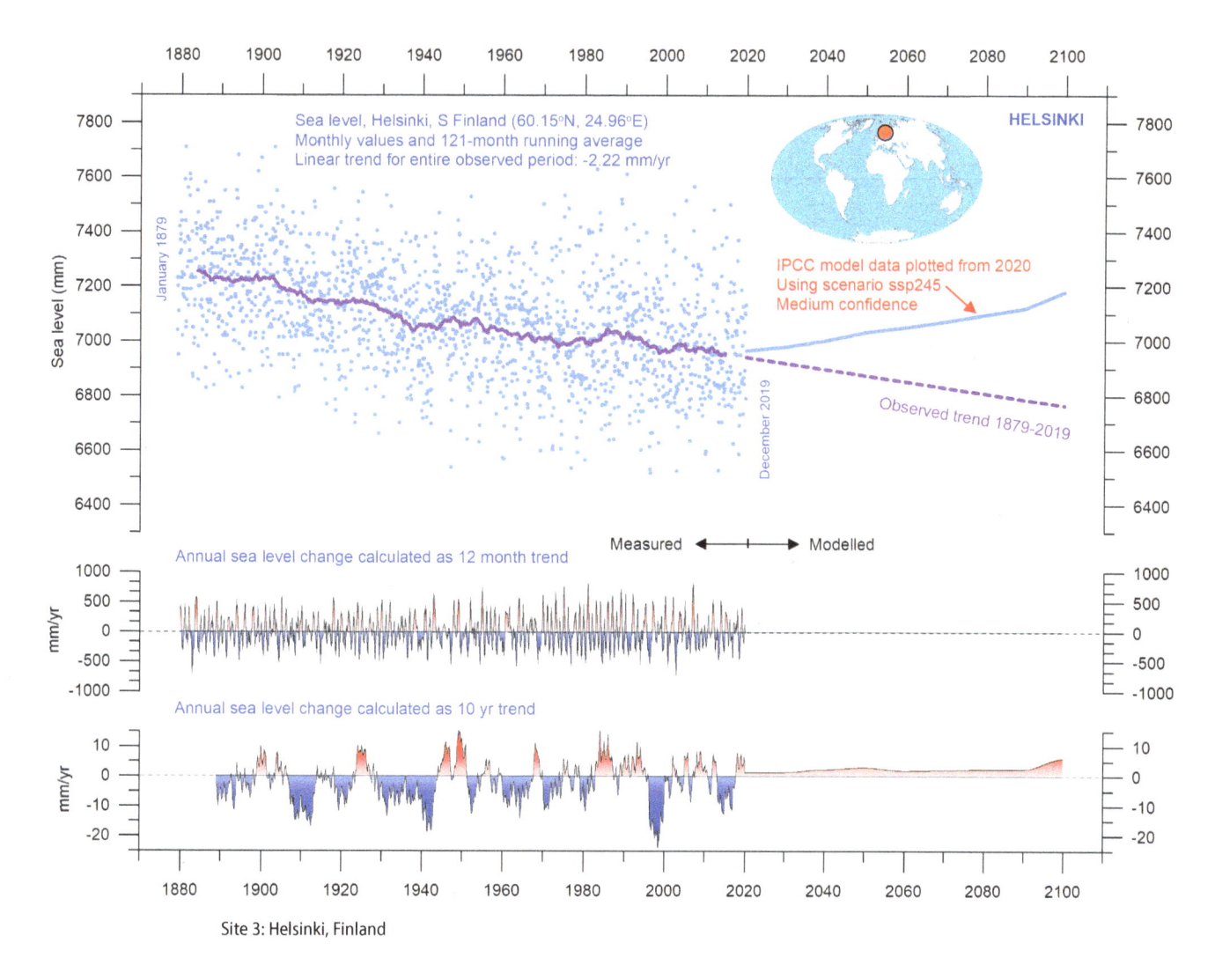

Site 3: Helsinki, Finland

Observed relative sea level at Helsinki, Finland, January 1879 – December 2019 (blue dots and purple line). If the observed change rate continues, the relative sea level at Helsinki (in relation to land) will have dropped about 18 cm by year 2100, compared to now. In the diagram the blue line indicates the modelled sea level change 2020-2100 for Helsinki, using the moderate SSP2-4.5 scenario. According to IPCC, the relative sea level (in relation to land) at Helsinki will have increased about 22 cm by year 2100, compared to now. A marked change from relative sea level decrease to increase predicted to begin around 2020, in contrast to the steady sea level decrease (about -2.22 mm/yr) recorded since 1879.

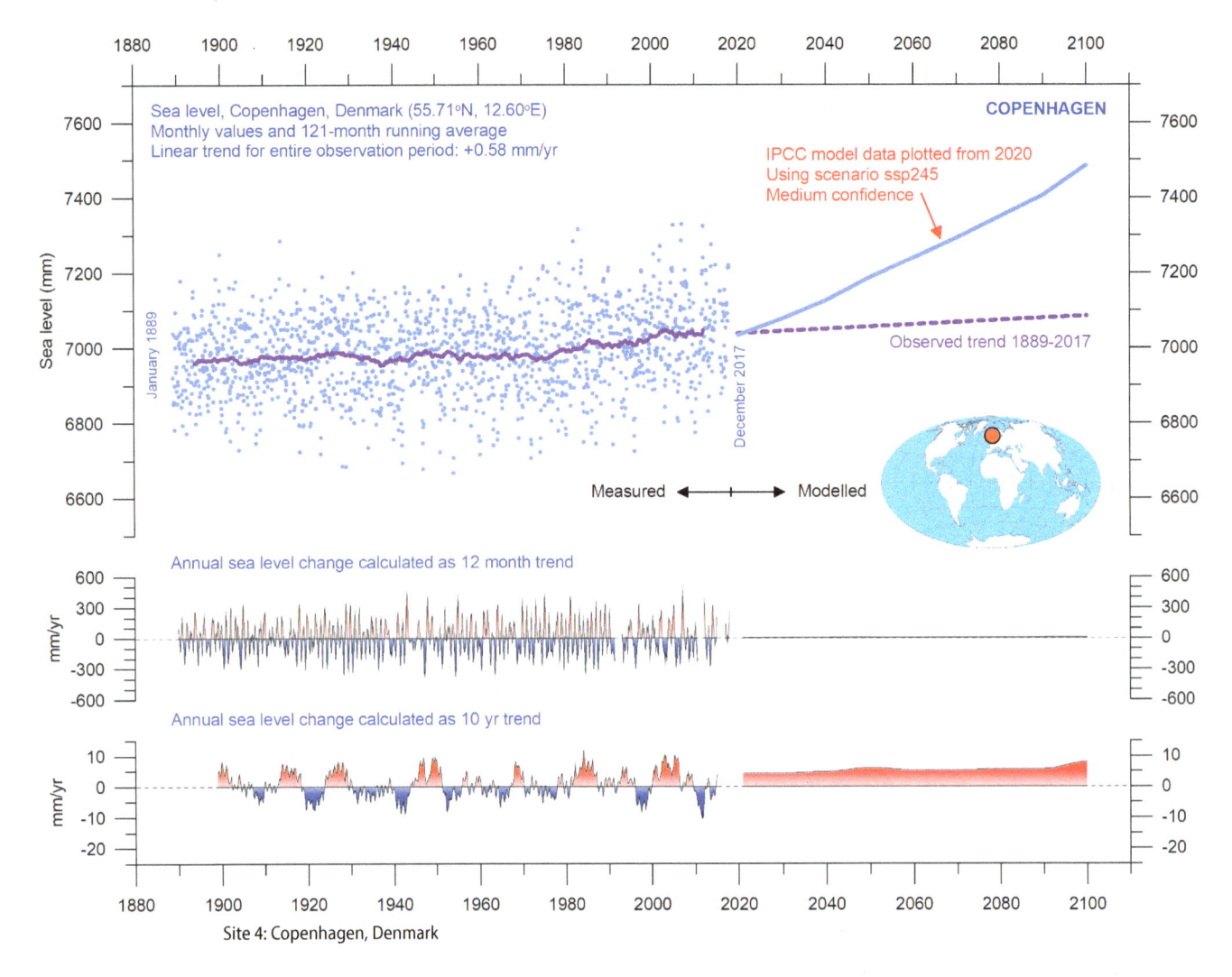

Site 4: Copenhagen, Denmark

Observed relative sea level at Copenhagen, Denmark, January 1889 – December 2017 (blue dots and purple line). Denmark was near the margin of the European Ice Sheet during the last glacial maximum, and the observed relative sea level change rate is therefore positive, although small. If the observed change rate continues, the relative sea level at Copenhagen (in relation to land) will have increased by about 4.6 cm by year 2100, compared to now. In the diagram the blue line indicates the modelled sea level change 2020-2100 for Copenhagen, using the moderate SSP2-4.5 scenario. According to IPCC, the relative sea level (in relation to land) at Copenhagen will have increased about 45 cm by year 2100, compared to now. A marked change in the relative sea level increase is predicted to begin around 2020, in contrast to the previous slow sea level increase recorded since 1889.

The step change in relative sea level for all four sites at year 2020 visually appear unrealistic and suggests that the modelled data was not tested appropriately against the measured relative sea level data before publication. This is surprising, as the modelled sea level projections for different SSP scenarios are calculated relative to a baseline defined by observations 1995-2014, for each station. The modellers must therefore have inspected the observed data.

According to the most recent (2021) publication of the IPCC's work, the 6th Assessment Report from Working Group I, human activities are estimated to have caused approximately 1.0°C of global warming above pre-industrial levels, with a likely range of 0.8°C to 1.2°C (*Summary for Policymakers*, A.1.3). Consequently, according to the IPCC, most or all of the warming experienced since about 1850 is due to man's activities. This IPCC finding is remarkable, as it downgrades the effect of natural climatic variations to nearly zero since 1850.

It is therefore extremely surprising that the modelled effect of this should first appear in 2020 as a rather marked step change in the relative sea level. Had the modellers instead modelled their sea level data from an earlier date, e.g., 1950, which would have been entirely possible, the conflict between measured and modelled data would immediately have become apparent. Usually, model improvements would then have been initiated as the next scientific step. It is highly disappointing that such a simple quality- or sanity check apparently was never requested or performed by the IPCC.

A study of several of the modelled sea level data demonstrates that major geoid changes in regions around Greenland are expected to take place in the future. This indicates a modelled substantial future reduction of the Greenland Ice Sheet. Summer temperature is a main control on the annual mass loss from glaciers, and winter precipitation is a main control on the annual accumulation of mass. The annual net mass balance is the numerical difference between these two numbers for any glacier and determines if the glacier is increasing or decreasing in volume. Erroneous or unrealistic input data for future air temperature and precipitation over Greenland therefore remain a main suspect for the unrealistic sea level modelling published recently by IPCC (the 6th Assessment Report from Working Group I).

Presumably, the startling recent IPCC finding that most or all of the warming experienced since about 1850 is due to man's activities, may have led to modelling of an unrealistic large future air temperature increase. Such erroneous input data may explain the above-described error in sea level modelling. The fundamental IPCC finding of no significant influence of natural variations since about 1850 should therefore be reconsidered.

For coastal planning, as usual, observations from traditional tide gauges remain the main data source to consider for planners and policymakers.

# D
# Human Impacts

# 11

# Hiding the good news on hurricanes and floods

BY MARCEL CROK

**Whenever an extreme weather event causes death and destruction, climate change becomes the culprit. The simple message always is "the climate is getting more extreme". But is that the case? The IPCC must answer such questions in a scientific and impartial way. Here we investigate whether the IPCC in their AR6 report succeeded in that task. The short answer is "no". Although deep inside the WG1 report the IPCC acknowledges some rather good news about extremes – i.e., that hurricanes and floods have not gotten worse – that good news is not communicated clearly to the policy makers and the media. In the WG2 report things got worse, and the IPCC even contradicts some of its own WG1 claims. The IPCC needs to do a much better job.**

Ever since hurricane Katrina hit New Orleans in 2005 and caused tremendous damage and deaths, climate change has been linked to extreme weather events. So, whenever a flood, drought, heatwave or hurricane occurs, scientists and the media quickly blame anthropogenic climate change for being the cause of it. Nowadays, there is even a sub discipline, called event attribution, that deals with the question whether a specific extreme event, like the terrible floods in Pakistan in 2022, have been caused by our emissions of $CO_2$. That is a dangerous question from a political and legal perspective, since countries that suffer loss and damage from an extreme event can consider claiming compensation from developed countries. Their idea is that rich countries have emitted most manmade $CO_2$ and are therefore to blame for the loss in more vulnerable developing nations. A huge fund for so-called "Loss and Damage" is now being negotiated at the yearly COP-meetings.[1]

So, given the importance of extreme events for the people who endured them, as well as for political, legal, and economic reasons, it is quite important for the IPCC to get the science about this 'right'. In this chapter we analyse what the IPCC has written about trends in extreme events. We compare what is written in the main WG1 and WG2 reports and how this is reflected in the *Summary for Policy Makers* (SPM).

## Pielke Jr.'s Assessment

Only days after the WG1 report was published in August 2021, the well-known US scientist Roger Pielke Jr summarised its finding with respect to extreme weather events in a long post on his personal website.[2] Pielke is very familiar with the literature about extreme events but was not involved in this (or any) IPCC report. He produced a table that is very revealing about what the IPCC had to say about all kinds of extreme weather, see table 1.

The IPCC uses 'detection' and 'attribution' as a framework to analyse trends in climate. Detection means that on climatic time scales a statistically significant change in some parameter has been 'detected'. The next step is to identify a 'cause' for that change, which in practice often means 'greenhouse gases', as these are the climate forcings assumed to dominate the total forcings by the IPCC.

1 https://unfccc.int/news/cop27-reaches-breakthrough-agreement-on-new-loss-and-damage-fund-for-vulnerable-countries
2 https://rogerpielkejr.substack.com/p/how-to-understand-the-new-ipcc-report-1e3

**Table 1:** Summary by Roger Pielke Jr of the AR6 WG1 report detection and attribution findings for different extreme weather phenomena.

|  | DETECTION | ATTRIBUTION |
|---|---|---|
| heat waves | yes | yes |
| heavy precipitation | yes | yes |
| flooding | no | no |
| meteorological drought | no | no |
| hydrological drought | no | no |
| ecological drought | yes | yes |
| agricultural drought | yes | yes |
| tropical cyclones | no | no |
| winter storms | no | no |
| thunderstorms | no | no |
| tornadoes | no | no |
| hail | no | no |
| lightning | no | no |
| extreme winds | no | no |
| fire weather | yes | yes |

As shown in figure 1, according to the IPCC, greenhouse gases have contributed most to an increase in radiative forcing since 1750. Changes in the sun have contributed close to nothing (for a different perspective about that see our chapter 6). The IPCC then attributes the detected trend to these anthropogenic forcings.

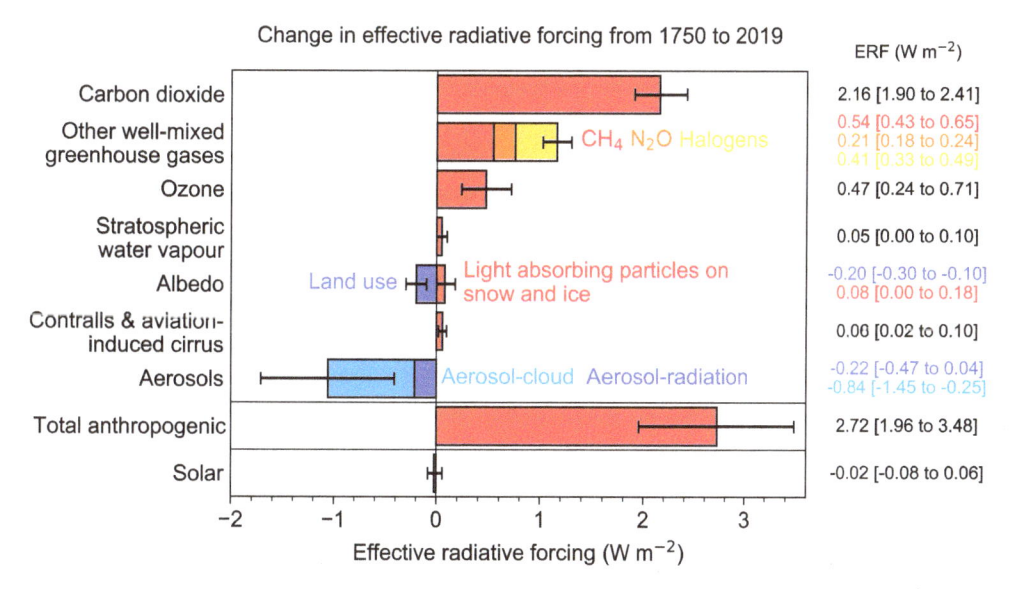

**Figure 1:** Reproduction of figure 7.6 from the WG1 report showing the change in radiative forcing since 1750.[3]

So, it's no surprise to see that, in the table provided by Pielke, the detected trends are also attributed by the IPCC to greenhouse forcing. However, what is most remarkable, and goes against most of the media coverage of extreme weather, is that for most extreme weather phenomena, no trend is detected. This is true for flooding, drought (meteorological or hydrological), tropical cyclones (in the Atlantic called hurricanes), winter storms, thunderstorms, tornadoes, hail, lightning, or extreme winds (so, storms of any type).

---

3    https://www.ipcc.ch/report/ar6/wg1/figures/chapter-7/figure-7-6

## Damage Trends

Globally, most damage by far (around 90%) from extreme weather is due to floods and tropical cyclones. So, Pielke's table, based on the WG1 report, is truly good news. The most damaging extremes, hurricanes, floods and (weather-related) droughts have not changed on climatic time scales. The earth has warmed by slightly more than one degree Celsius, the $CO_2$ concentration has gone up, but the most dramatic extreme weather events have not (yet) changed.

The IPCC did not provide a handy table like Pielke did in his blog post. They provided written evidence of the lack of trends, in chapter 11 of the WG1 report. We are not going to discuss all of them, but here are some examples from the chapter.

They claim an attributable trend in extreme precipitation but not in flooding. Here are the relevant sections (our bold):

> The **frequency and intensity of heavy precipitation events have increased** over a majority of land regions with good observational coverage since 1950 (high confidence, Box TS.6, Table TS.2). Human influence is likely the main driver of this change (Table TS.2). [TS page 84]
> However, **heavier rainfall does not always lead to greater flooding**. This is because flooding also depends upon the type of river basin, the surface landscape, the extent and duration of the rainfall, and how wet the ground is before the rainfall event (FAQ 8.2, Figure 1). [Page 1155]
> There is low confidence about peak flow trends over past decades on the global scale [Page 1568]
> In summary there is **low confidence in the human influence on the changes in high river flows on the global scale**. [Page 1569]

Citing these sentences Pielke commented on twitter: "So don't claim floods are increasing; Don't say they are "climate driven".""[4]

## Tropical cyclones

Next, we look at hurricanes (or tropical cyclones, TC):

> There is low confidence in most reported long-term (multi-decadal to centennial) trends in TC frequency- or intensity-based metrics due to changes in the technology used to collect the best-track data. [Page 1585]

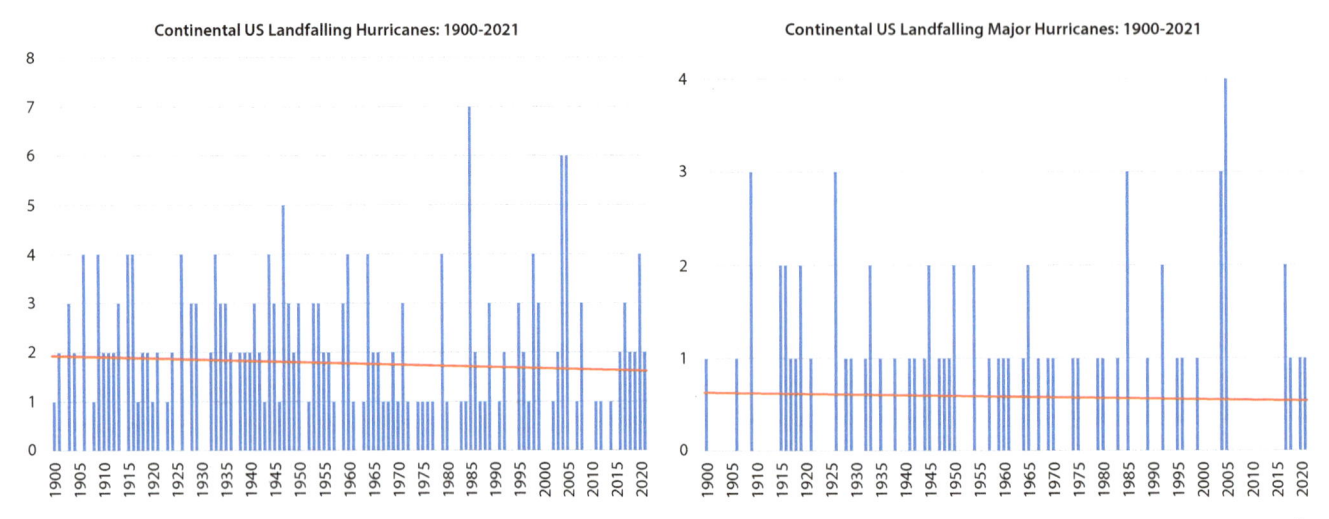

**Figure 2:** Number of US landfalling hurricanes and major hurricanes between 1900 and 2021. Updated graph from Klotzbach (2018)[5]

4   https://twitter.com/RogerPielkeJr/status/1424735415576104965
5   Klotzbach, Philip J., et al. "Continental US hurricane landfall frequency and associated damage: Observations and future risks." *Bulletin of the American Meteorological Society* 99.7 (2018): 1359-1376.

Pielke commented on the denigrating remark by the IPCC about the best-track data:

> The denigration of the TC "best track" dataset is bizarre. The dataset is the highest quality available on tropical cyclones around the world and widely used in research. It'd be a shame if the IPCC process were to have been used to promote certain work by denigrating the widely recognized best available data.

The IPCC decided not to show a graph in this section of the report, but here is a very relevant one, showing landfalling (major) hurricanes in the US. It shows that if anything there is a small decreasing trend. These graphs have been published in a peer reviewed paper by Phil Klotzbach in 2018 and are shown here in an updated version. The paper is not mentioned in the WG1 report.

This lack of trend in US landfalling hurricanes is important information, because they alone make up 60% of the global historical damage due to extreme weather events.[6]

Strangely, the IPCC decided to say nothing about trends in *global* tropical cyclone (TC) landfalls, although this 2012 paper, "Historical Global Tropical Cyclone Landfalls", by Weinkle et al. seems highly relevant.[7] That paper was co-authored by Roger Pielke Jr and Ryan Maue and concluded: "The analysis does not indicate significant long-period global or individual basin trends in the frequency or intensity of landfalling TCs of minor or major hurricane strength."

That paper showed this graph:

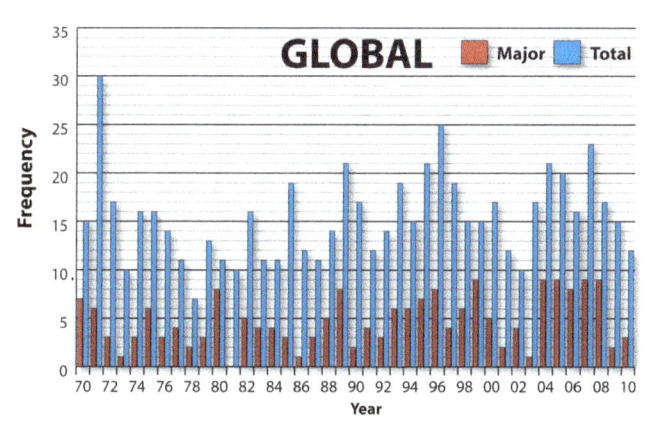

**Figure 3:** reproduction of figure 2 from Weinkle et al. (2012) showing global total and major hurricane landfalls.

Ryan Maue frequently updates this dataset on his website.[8] Here is the latest one:

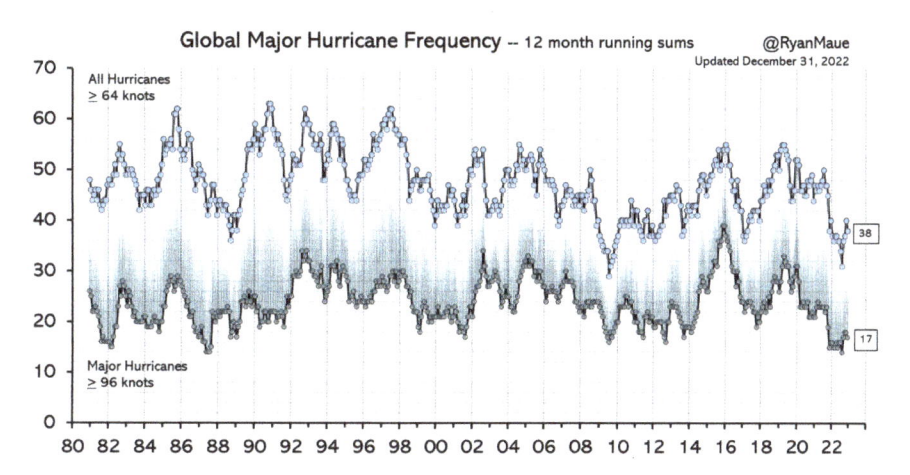

**Figure 4:** global hurricane frequency. On top all hurricanes, at the bottom major hurricanes. Source: Ryan Maue

6   Mohleji, S., & Pielke Jr, R. (2014). Reconciliation of trends in global and regional economic losses from weather events: 1980–2008. *Natural Hazards Review*, *15*(4), 04014009.

7   Weinkle, J., Maue, R., & Pielke, R. P., Jr (2012). Historical global tropical cyclone landfalls. Journal of Climate, 25(13), 4729–4735. https://doi. org/10.1175/jcli-d-11-00719.1

8   https://climatlas.com/tropical/

Clearly neither all nor major hurricanes show an up or down trend. There is large variability from year to year and from decade to decade. The calendar year with most hurricanes was 59 in 1992 and the least was 38 in 2009. The number of major hurricanes peaked in 2015 with 38 and the least occurred in 1981 with 15.

Now with these graphs the picture is quite clear that nothing unusual is going on with tropical cyclones.

Nevertheless, the IPCC manages to conclude this in their report (our bold):

> In summary, there is mounting evidence that **a variety of TC characteristics have changed over various time periods**. It is likely that the global proportion of Category 3–5 tropical cyclone instances and the frequency of rapid intensification events have increased globally over the past 40 years. [Page 1587]

That paragraph is confusing to say the least, especially without showing the graphs included herein. Pielke commented on twitter that using the latest forty years can also be misleading, as the 1970s and early 1980s were periods with relatively low tropical cyclone activity.

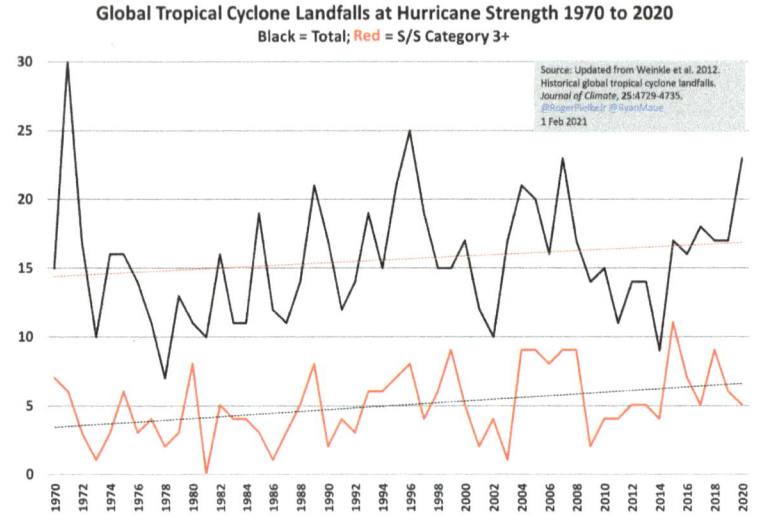

**Figure 5:** global tropical cyclone landfalls since 1970. Based on Weinkle et al. 2012. Source: Pielke Jr[9]

In figure 5 we see a trend up and it is tempting to think it is due to anthropogenic climate change. A truly global picture is missing before 1970, but there is good data for the North Atlantic and the Western Pacific, and those two areas account for about 70% of the global landfalls. The data for these two basins goes back to 1945:

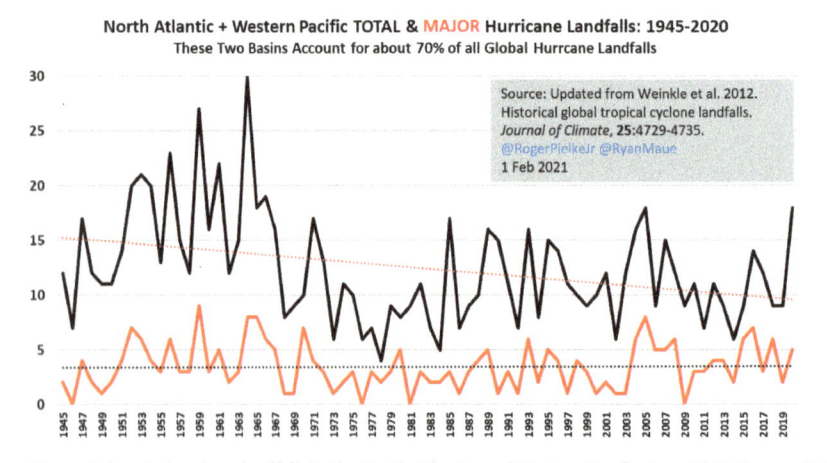

**Figure 6:** tropical cyclone landfalls in the North Atlantic and Western Pacific since 1945. Source: Pielke Jr

---

9    https://rogerpielkejr.substack.com/p/a-remarkable-decline-in-landfalling

Suddenly the upward trend that we saw from the 1970s is changed to a downward trend for all hurricanes and no trend for major hurricanes. It clearly shows one should be careful drawing conclusions from shorter periods of time.

## Drought

Next is drought. In AR6 the IPCC changed its definitions of drought (AR5 just talked about drought) and now distinguishes meteorological and hydrological drought (no trends) from ecological and agricultural droughts (trend detected).[10] Agricultural and ecological drought is related with abnormal soil moisture deficit (combination of precipitation deficit and excess evapotranspiration), meteorological drought with precipitation deficits and hydrological drought with streamflow deficit.

Here are some of the key conclusions:

On hydrological drought:

> There is still limited evidence and thus low confidence in assessing these trends at the scale of single regions, with few exceptions [Page 1578]

On meteorological drought:

> The regional evidence on attribution for single AR6 regions generally shows low confidence for a human contribution to observed trends in meteorological droughts at regional scale [Page 1579]

On agricultural and ecological drought:

> In summary, human influence has contributed to increases in agricultural and ecological droughts in the dry season in some regions due to increases in evapotranspiration (medium confidence).

So, based on the AR6 WG1 report you cannot simply state that drought in general is increasing.

## Extreme hot days and heatwaves

AR6 is most confident about trends in hot days and heatwaves (our bold):

> In summary, it is virtually certain that there has been an increase in the number of warm days and nights and a decrease in the number of cold days and nights on the global scale since 1950. Both the coldest extremes and hottest extremes display increasing temperatures. It is very likely that these changes have also occurred at the regional scale in Europe, Australasia, Asia, and North America. It is **virtually certain that there has been increases in the intensity and duration of heatwaves** and in the number of heatwave days at the global scale.

It is noteworthy though that they use 1950 as a reference year. It is well-known that at least in the US, the 1930s were the hottest. Here is a graph for the US:

---

10 Here is a footnote from the Technical Summary explaining the differences: "Agricultural and ecological drought (depending on the affected biome): a period with abnormal soil moisture deficit, which results from combined shortage of precipitation and excess evapotranspiration, and during the growing season impinges on crop production or ecosystem function in general (see Annex VII: Glossary). Observed changes in meteorological droughts (precipitation deficits) and hydrological droughts (streamflow deficits) are distinct from those in agricultural and ecological droughts and are addressed in the underlying AR6 material (Chapter 11)."

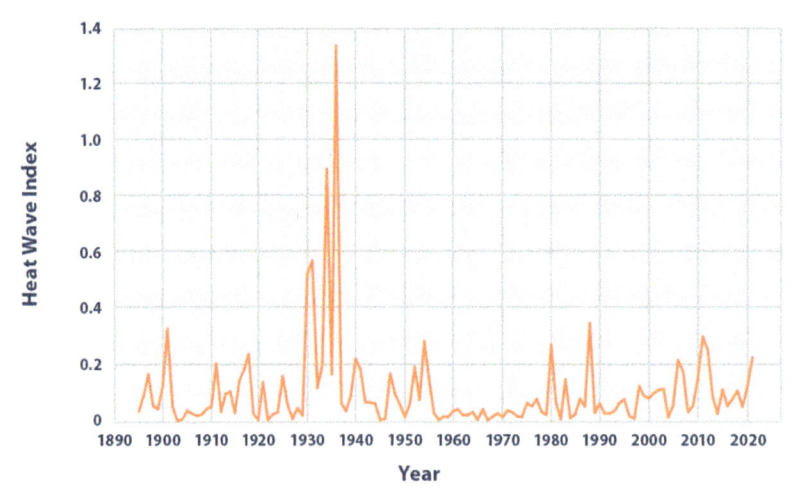

**Figure 7:** This figure shows the annual values of the U.S. Heat Wave Index from 1895 to 2021. These data cover the contiguous 48 states. An index value of 0.2 (for example) could mean that 20 percent of the country experienced one heat wave, 10 percent of the country experienced two heat waves, or some other combination of frequency and area resulted in this value. Source: EPA[11]

## AR6 WG1 Summary for Policy Makers

So, even if we take IPCC at face value and accept that some extremes (heatwaves, extreme precipitation, ecological and agricultural drought are increasing in frequency), the more impactful extremes (in terms of damage and deaths) such as flooding and tropical cyclones are not. This is good news. We are now going to see how the *Summary for Policy Makers*, arguably the most important part of the report, reflects these findings.

First let's look at tropical cyclones, as these, especially those landfalling in the US, dominate global disaster damages.

> Human-induced climate change is already affecting many weather and climate extremes in every region across the globe. **Evidence of observed changes** in extremes such as heatwaves, heavy precipitation, droughts, and **tropical cyclones**, and, **in particular, their attribution to human influence, has strengthened since AR5**. [AR6, SPM, A.3; Page 8]

Now this statement is highly misleading if not simply wrong. IPCC is simply hiding the fact that the frequency and intensity of tropical cyclones have not increased. It even claims the opposite, an observed 'change' in tropical cyclones, that can be attributed to human influence (i.e., the emission of greenhouse gases).

Point A.3.4 of the SPM goes into more detail (our bold):

> It is likely that the **global proportion of major** (Category 3–5) tropical cyclone **occurrence** has increased over the last four decades, and it is very likely that the latitude where tropical cyclones in the western North Pacific reach their peak intensity has shifted northward; these changes cannot be explained by internal variability alone (medium confidence). **There is low confidence in long-term (multi-decadal to centennial) trends in the frequency of all-category tropical cyclones**. Event attribution studies and physical understanding indicate that human-induced climate change increases heavy precipitation associated with tropical cyclones (high confidence), but data limitations inhibit clear detection of past trends on the global scale.

In three very detailed blog posts[12] Roger Pielke Jr showed how in different drafts IPCC changed the word "intensities" first in "instances" and then in "occurrence". The last change is not only flawed but happened outside the official review process. As Pielke observed: "This is not how assessments are supposed to work." Pielke also wrote that "a high-level participant in the IPCC

---

11 https://www.epa.gov/climate-indicators/climate-change-indicators-heat-waves#%20

12 https://rogerpielkejr.substack.com/p/a-tip-from-an-ipcc-insider; https://rogerpielkejr.substack.com/p/misinformation-in-the-ipcc; https://rogerpielkejr.substack.com/p/trends-in-the-proportion-of-major

(purposely vague to protect their identity) has confirmed to me that the major error on tropical cyclones that I recently identified was (a) indeed a major snafu and (b) a result of claims being inserted into the IPCC outside its review process." So, let's see if the IPCC will correct this error.

Ryan Maue published data on another metric, the so-called ACE, Accumulated Cyclone Energy. It is a measure of the total energy involved in tropical cyclones. If the proportion of major hurricanes increase, one would also expect an increase in the ACE. Here is the graph:

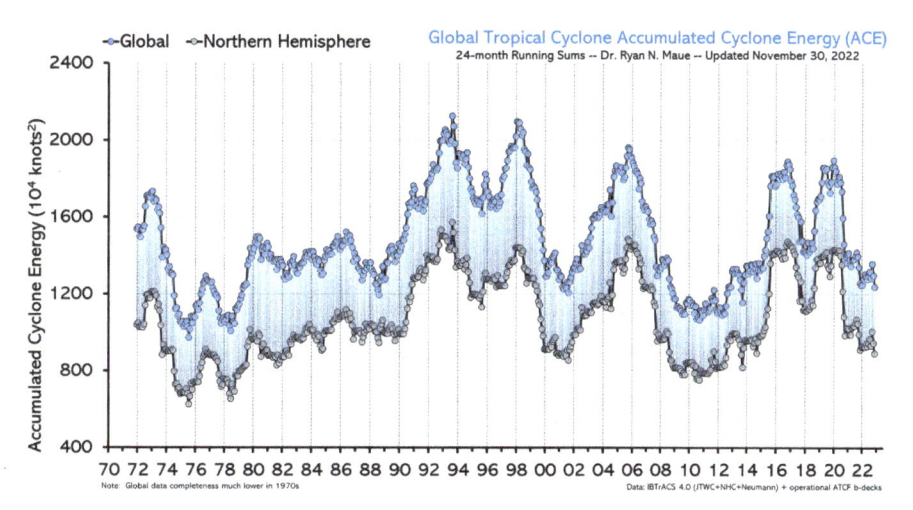

**Figure 8:** Last 50-years+ of Global and Northern Hemisphere Accumulated Cyclone Energy: 24 month running sums. Note that the year indicated represents the value of ACE through the previous 24-months for the Northern Hemisphere (bottom line/gray boxes) and the entire global (top line/blue boxes). The area in between represents the Southern Hemisphere total ACE. Source: Ryan Maue[13]

Again, we see large year-to-year and decade-to-decade variability but no clear trend. In summary, the IPCC is hiding good news about tropical cyclones.

## Floods

There is a statement about compound flooding in the SPM but not about the lack of trends in flooding in general. Remember, this is what the full report said: "In summary there is **low confidence in the human influence on the changes in high river flows on the global scale**." [Page 1569]

A statement like this is not highlighted in the SPM. It does mention this though (our bold):

> Human influence has likely increased the chance of compound extreme events[14] since the 1950s. This includes increases in the frequency of concurrent heatwaves and droughts on the global scale (high confidence), fire weather in some regions of all inhabited continents (medium confidence), and **compound flooding** in some locations (medium confidence).

We can therefore conclude that the two most important extreme events (from the perspective of damage) are not fairly covered in the SPM.

Now let's see if and how the IPCC treats heatwaves. They write:

> It is virtually certain that hot extremes (including heatwaves) have become more frequent and more intense across most land regions since the 1950s, while cold extremes (including cold waves) have become less frequent and less severe, with high confidence that human-induced climate change is the main driver of these changes. Some recent hot extremes observed over the past decade would have been extremely unlikely to occur without human influence on the climate system. [A.3.1]

---

13   https://climatlas.com/tropical/

14   Compound extreme events are the combination of multiple drivers and/or hazards that contribute to societal or environmental risk (Glossary). Examples are concurrent heatwaves and droughts, compound flooding (e.g., a storm surge in combination with extreme rainfall and/or river flow), compound fire weather conditions (i.e., a combination of hot, dry and windy conditions), or concurrent extremes at different locations.

On droughts:

> Human-induced climate change has contributed to increases in **agricultural and ecological droughts** in some regions due to increased land evapotranspiration (medium confidence). [A.3.2]

They mention an increase in agricultural and ecological drought, but not the lack of a trend in hydrological and meteorological droughts.

## WG1 Report

In general the WG1 report did a reasonably good job in describing trends in extreme weather events. However, the IPCC seems to be extremely focused on bad news and ignores good news. It tries hard to make the connection between climate change and more extreme weather. Deep inside the report it acknowledges (though grudgingly) that most extremes have not changed, such as flooding, drought (meteorological or hydrological), tropical cyclones, winter storms, thunderstorms, tornadoes, hail, lightning or extreme winds. So, there is a lot of good news available in the report, but one really has to look for it. The good news is not highlighted in the summary of the chapter, let alone in the *Summary for Policy Makers*. And did you ever hear an IPCC contributing scientist publicly acknowledge that there is no trend in tropical cyclones and flooding?

## WG2 report

The WG2 report was published nine months after the WG1 report. So, the authors of the WG2 report knew what was inside the WG1 report. WG2 covers the impacts of climate change so logically trends in extremes are also important in that part of the report. Let's focus on some of the most important extreme weather events, tropical cyclones, flooding and drought.

First, here is what WG2 has to say about tropical cyclones (our bold):

> Adverse impacts from **tropical cyclones, with related losses and damages, have increased due to sea level rise and the increase in heavy precipitation** (medium confidence). [SPM, page 9]

And

> Some extreme weather events are increasing in frequency and (or) severity as a result of climate change (Seneviratne et al., 2021) (high confidence). These include extreme rainfall events (Roxy et al., 2017; Myhre et al., 2019; Tabari, 2020); extreme and prolonged heat leading to catastrophic fires (Bowman et al., 2017; Krikken et al., 2019; van Oldenborgh et al., 2020); and **more frequent and stronger cyclones/hurricanes** and resulting extreme rainfall (Griego et al., 2020). These extreme events, coupled with high vulnerability and exposure in many parts of the world, turn into disasters and affect millions of people every year. [Page 588]

This is opposite of what the WG1 report said, namely "[t]here is low confidence in most reported long-term (multi-decadal to centennial) trends in TC frequency- or intensity-based metrics".

Instead of simply citing WG1 the WG2 claim of more frequent and intense hurricanes/cyclones goes to the paper Griego et al. (2020)[15], which has no analysis of hurricane/cyclone frequency or intensity.

WG2 is also claiming that floods are getting worse (our bold):

> Extreme weather events **causing highly impactful floods and droughts have become more likely and (or) more severe due to anthropogenic climate change** (high confidence). {4.2.4, 4.2.5, Cross-Chapter Box DISASTER in Chapter 4} [executive summary chapter 4, page 555]

---

15    Griego, A.L., A.B. Flores, T.W. Collins and S.E. Grineski, 2020: Social vulnerability, disaster assistance, and recovery: a population-based study of Hurricane Harvey in Greater Houston, Texas. Int. J. Disaster Risk Reduct., 51, 101766, doi:10.1016/j.ijdrr.2020.101766.

Remember what WG1 said, "there is low confidence about human influence on the changes in high river flows on the global scale." [page 1569]

Here is something about drought (our bold):

> **Anthropogenic climate change has contributed to the increased likelihood and severity of the impact of droughts (especially agricultural and hydrological droughts)** in many regions (high confidence). [executive summary chapter 4, page 555]

The WG1 report said human influence on agricultural and ecological drought but no trends in hydrological and meteorological drought. So again, there is a conflict between WG1 and WG2.

## Conclusions

If and to what extent extreme weather is changing is a very important question. This question has dominated political debates around climate change. It is therefore extremely important that the IPCC, which is, or should be, politically neutral, gets the science about this right. Here we have shown that in general the WG1 report did a reasonably fair job, except for the *Summary for Policy Makers*. However, the chapter about extremes (chapter 11) had a lot of good news to offer (no trends in hurricanes and flooding), but the IPCC failed to emphasize these results, both in the summary of the chapter and in the *Summary for Policy Makers*.

Policy makers therefore cannot be blamed for being unaware of the good news about recent changes in extreme weather, in particular, that the most impactful events (like hurricanes, floods, and hydrological and meteorological droughts) have not increased. We also show that global disaster losses normalised for GDP have not increased and that climate-related deaths have decreased in other chapters. These facts paint a far less bleak picture of climate change than the doom and gloom seen in the latest IPCC reports.

In WG2 things really get worse, the IPCC even contradicts many its own claims from the WG1 report. In 2010 several errors were discovered in the 2007 AR4 report. Those errors ultimately led to an investigation by the InterAcademy Council (IAC).[16] The IAC recommended many changes to improve the IPCC process. The bias and errors we have laid bare in this chapter and the chapters about disaster losses and climate-related deaths show that rather than improving, the IPCC, and especially the WG2 report, have deteriorated. It is more focused on advocacy than on a comprehensive, neutral science assessment.

---

16   Climate Change Assessments, Review of the Processes & Procedures of the IPCC (interacademies.org)

# 12 Extreme views on disasters

BY MARCEL CROK

**Economic losses caused by extreme weather are rising. Most of the scientific literature shows, however, that this increase is mostly due to increasing population and wealth. After normalising the data, evidence that anthropogenic climate change is contributing to the damage is non-existent. This is good news, but the IPCC completely ignored this literature for unknown reasons and instead comes to a cherry-picked alarmist result.**

Whenever a storm, flood or heat wave strikes, media reports quickly blame anthropogenic climate change for the disaster. So, an important question is whether extremes and economic losses due to those extremes are increasing. These questions of course have received a lot of attention in the scientific literature. In this chapter we address whether economic losses due to climate related disasters (hurricanes, floods, droughts, wildfires etc.) have increased.

During the past century the global population increased spectacularly from 2 billion to 8 billion people. So, we have a lot more people who can be hit by natural disasters. Potentially this is an explosive situation: if we have more extreme weather that is more severe, and it hits more people, we naturally expect more damage and more loss of life.

It is also quite easy to understand that if more people live in a hurricane prone area like Florida, then more people will suffer damage from hurricane events. So, each new hurricane will probably cause more damage since there are more houses and buildings and they are more expensive. Since 1998 scientists have developed methods to adjust for changes in population, economic growth (GDP), as well as adaption through more strict building codes etc.[1] The method is called 'normalisation' of disaster losses and can be used to check for trends in weather extremes. After all, if there is an increasing trend in, let's say hurricanes, the record of disaster losses, after correcting for societal developments, should also show an increase.

## Normalisation of damage

Since the first paper on normalisation in 1998 dozens of papers have been published in the scientific literature. One of the central scientists involved in this discipline, and the author of the first paper in 1998, is Roger Pielke Jr, professor at the University of Colorado in Boulder. In 2020 Pielke decided to review the 'normalisation' literature that has appeared between 1998 and 2020. The resulting paper was published in the journal *Environmental Hazards*.[2] The publication of this paper was rather timely. The IPCC was working on the second part of its sixth assessment report about "Impacts, Adaptation and Vulnerability". But before we delve into that very extensive report let's summarise what Pielke said in his review paper. Here is the full abstract with some sentences highlighted by us in bold:

**ABSTRACT**

Nowadays, following every weather disaster quickly follow estimates of economic loss. **Quick blame for those losses, or some part, often is placed on claims of more frequent or intense weather events. However, understanding what role changes in climate may have**

1   Pielke, R. A., & Landsea, C. W. (1998). Normalized hurricane damages in the United States: 1925–95. Weather and Forecasting, 13(3), 621–631. https://doi.org/10.1175/1520-0434(1998)0132.0.CO;2

2   Pielke Jr., R. (2021): Economic 'normalisation' of disaster losses 1998–2020: a literature review and assessment. Environmental Hazards, 20 (2). doi: 10.1080/17477891.2020.1800440

**Table 1:** Normalisation papers reviewed in Pielke 2020.

| Study (ordered by date of publication) | Phenomenon (region) | Detection claimed to be achieved? | Trend direction | Attribution claimed to be achieved? | Period (*italics* = <3 years) |
|---|---|---|---|---|---|
| **STUDIES FOCUSED ON SPECIFIC PHENOMENA** | | | | | |
| *Tropical cyclones* | | | | | |
| Martinez (2020) | United States | No | n/a | No | 1900–2018 |
| Grinsted et al. (2019) | United States | Yes | Increase | Yes | 1900–2018 |
| Chen et al. (2018) | China | No | n/a | No | 1983–2015 |
| Ye and Fang (2018) | China | Yes | Decrease | No | *1985–2010* |
| Weinkle et al. (2018) | United States | No | n/a | No | 1900–2017 |
| Klotzbach et al. (2018) | United States | No | n/a | No | 1900–2016 |
| Fischer et al. (2015) | China | No | n/a | No | *1984–2013* |
| Estrada et al. (2015) | United States | Yes | Increase | No | 1900–2005 |
| Bouwer and Wouter Botzen (2011) | United States | No | n/a | No | 1900–2005 |
| Nordhaus (2010) | United States | Yes | Increase | No | 1900–2005 |
| Zhang et al. (2009) | China | No | n/a | No | *1983–2006* |
| Schmidt et al. (2009) | United States | No | n/a | No | 1950–2005 |
| Pielke et al. (2008) | United States | No | n/a | No | 1900–2005 |
| Pielke et al. (2003) | Latin America and Caribbean | No | n/a | No | 1944–1999 |
| Raghavan and Rajesh (2003) | India | No | n/a | No | *1977–1998* |
| Collins and Lowe (2001) | United States | No | n/a | No | 1900–1999 |
| Pielke and Landsea(1998) | United States | No | n/a | No | 1926–1995 |
| *Floods* | | | | | |
| Du et al. (2019) | China | Yes | Decrease | No | 1990–2017 |
| Paprotny et al. (2018) | Europe | No | n/a | No | 1870–2016 |
| Wei et al. (2018) | China | Yes | Decrease | No | *2000–2015* |
| Fang et al. (2018) | China (Yangtze River) | Yes | Decrease | No | *1998–2014* |
| Perez-Morales et al. (2018) | Spain | No | n/a | No | 1975–2013 |
| Stevens et al. (2016) | United Kingdom | No | n/a | No | 1884–2013 |
| Barredo et al. (2012) | Spain | No | n/a | No | 1971–2008 |
| Hilker et al. (2009) | Switzerland | No | n/a | No | 1972–2007 |
| Chang et al. (2009) | Korea | No | Increase | No | 1971–2005 |
| Barredo (2009) | Europe | No | n/a | No | 1970–2006 |
| Downton et al. (2005) | United States | Yes | Decrease | No | 1926–2000 |
| Fengqing et al. (2005) | China | No | n/a | No | 1950–2001 |
| Pielke and Downton (2000) | United States | No | n/a | No | 1932–1997 |
| *Extratropical storms* | | | | | |
| Andres and Badoux (2019) | Switzerland | No | n/a | No | 1972–2016 |
| Stucki et al. (2014) | Switzerland | No | n/a | No | 1859–2011 |
| Barredo (2010) | Europe | No | n/a | No | 1970–2008 |
| *Tornadoes* | | | | | |
| Simmons et al. (2013) | United States | No | n/a | No | 1950–2011 |
| Brooks and Doswell (2001) | United States | No | n/a | No | 1890–1999 |
| Boruff et al. (2003) | United States | No | n/a | No | 1900–2000 |
| *Convective storms* | | | | | |
| Sander et al. (2013) | United States | Yes | Increase | No | 1970–2009 |
| *Wildfire* | | | | | |
| Crompton et al. (2010) | Australia | No | n/a | No | 1925–2009 |

| Study (ordered by date of publication) | Region (location & phenomena) | Detection claimed to be achieved? | Trend direction | Attribution claimed to be achieved | Period (*italics* = <3 years) |
|---|---|---|---|---|---|
| **STUDIES FOCUSED ON PARTICULAR REGIONS** | | | | | |
| *Region* | | | | | |
| Choi et al. (2019) | Korea (weather) | Yes | Decrease | No | 1965–2015 |
| Reyes and Elias (2019) | United States (crop loss) | Yes | Mixed | No | *2001–2016* |
| McAneney et al. (2019) | Australia (weather) | No | n/a | No | 1966–2017 |
| Paul and Sharif (2018) | Texas (hydrometeorological) | No | n/a | No | 1960–2016 |
| Bahinipati and Venktachalam (2016) | India (weather) | No | n/a | No | 1972–2009 |
| Zhou et al. (2013) | China(natural disasters) | No | n/a | No | *1990–2011* |
| Crompton and McAneney (2008) | Australia (weather) | No | n/a | No | 1967–2006 |
| Choi and Fisher (2003) | United States (weather) | No | n/a | No | 1951–1997 |
| *World* | | | | | |
| Pielke (2019) | All disasters & weather only | Yes | Decrease | No | *1990–2017* |
| Watts et al. (2019) | All disasters | No | n/a | No | 1990–2016 |
| Daniell et al. (2018) | Multi-hazard | Yes | Decrease | No | 1950–2015 |
| Mohleji and Pielke (2014) | All-weather related | No | n/a | No | *1980–2008* |
| Neumayer and Barthel (2011) | All-weather related | No | n/a | No | *1980–2008* |
| Visser et al. (2014) | All-weather related | No | n/a | No | 1980–2010 |
| Miller et al. (2008) | All-weather related | No | n/a | No | 1950–2005 |

**played in increasing weather-related disaster losses is challenging because, in addition to changes in climate, society also undergoes dramatic change.** Increasing development and wealth influence exposure and vulnerability to loss – typically increasing exposure while reducing vulnerability. In recent decades a scientific literature has emerged that seeks to adjust historical economic damage from extreme weather to remove the influences of societal change from economic loss time series to estimate what losses past extreme events would cause under present-day societal conditions. In regions with broad exposure to loss, an unbiased economic normalisation will exhibit trends consistent with corresponding climatological trends in related extreme events, providing an independent check on normalisation results. **This paper reviews 54 normalisation studies published 1998–2020 and finds little evidence to support claims that any part of the overall increase in global economic losses documented on climate time scales is attributable to human-caused changes in climate, reinforcing conclusions of recent assessments of the Intergovernmental Panel on Climate Change.**

The abstract is rather clear: there is an increase in economic losses, but after the data have been 'normalised' for societal factors there is little evidence that anthropogenic climate change is a factor in the increase. This is in line with the conclusions of earlier IPCC reports. The paper analyses 54 studies dealing with different extreme weather events in different regions of the world. The paper provides a table showing all the studies. See table 1.

The IPCC uses a two-step process to analyse climate change and its impacts. The first question is whether a statistical change in some climate phenomenon has been 'detected'. If so, then the next question is whether such a trend can be 'attributed' to anthropogenic climate change. If both questions are answered with 'yes', then the IPCC concludes that the human emissions of greenhouse gases have 'caused' the observed change. The table in Pielke 2020 notes several detected trends. The paper explains that eight of those trends are decreasing, five are increasing and one study finds mixed trends. However, of all 53 weather related studies (one deals with earthquakes) only one (Grinsted 2019) claims there is an increasing trend in disaster losses that is attributable to anthropogenic climate change.[3] That paper deals with hurricane losses in America since 1900, a topic that has been studied extensively in the literature. Two more papers (Estrada 2015 and Nordhaus 2010[4]) claim there is an increase in normalised losses due to landfalling hurricanes in the US although they don't attribute these trends to anthropogenic climate change. However, seven other papers described in the Pielke review paper conclude there is no trend in normalised US hurricane losses. So which conclusion is the right one or the more likely one?

## Landfalling hurricanes

In such cases it can be helpful to look at the climate records. For the US there is a good record of landfalling hurricanes since 1900. If there is an upward trend in hurricanes, it is more likely that there is also an upward trend in normalised losses.

Figure 1 shows there is no long-term increase in landfalling US hurricanes, either for all hurricanes or just for the major hurricanes. Both trends are slightly down. Surprisingly, these graphs, although they have been published in peer reviewed papers, have never been published in any of the 47 IPCC reports. This is surprising given the importance of US landfalling hurricanes for global disaster losses. Ninety percent of global disaster losses are due to global hurricanes (or cyclones) and floods. And 60% of global disaster losses are due to damage caused by US hurricanes. The reason for this is that so many prosperous people live in hurricane prone States like Florida.

3    Grinsted, A., Ditlevsen, P., & Christensen, J. H. (2019). Normalized US hurricane damage estimates using area of total destruction, 1900–2018. Proceedings of the National Academy of Sciences, 116(48), 23942–23946. https://doi.org/10.1073/pnas.1912277116

4    Estrada, F., Botzen, W. W., & Tol, R. S. (2015). Economic losses from US hurricanes consistent with an influence from climate change. Nature Geoscience, 8(11), 880–884. https://doi.org/10.1038/ ngeo2560 and Nordhaus, W. D. (2010). The economics of hurricanes and implications of global warming. Climate Change Economics, 1(01), 1–20. https://doi.org/10.1142/S2010007810000054

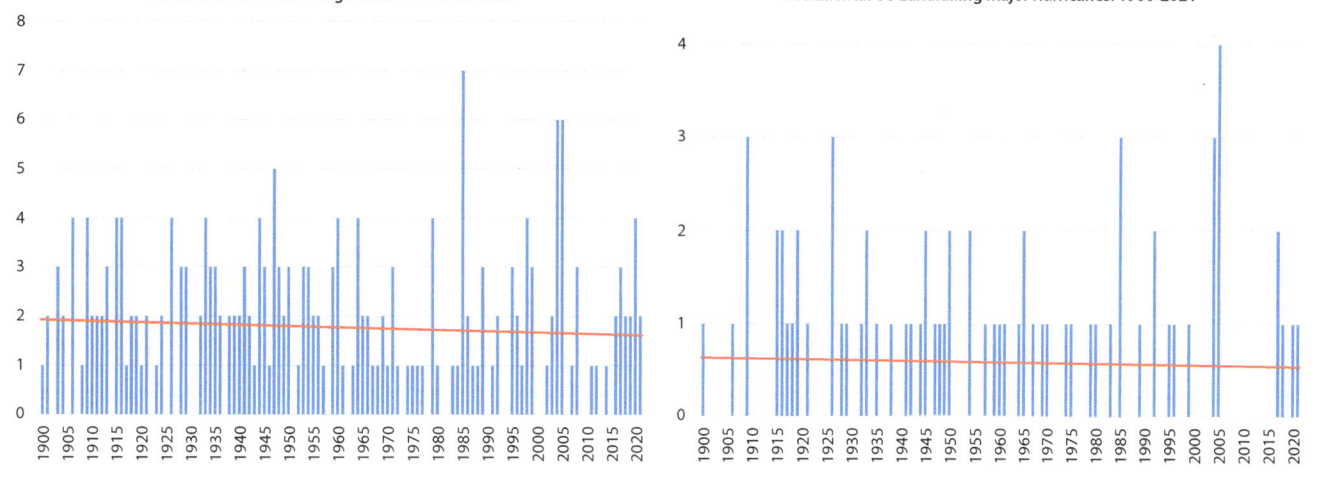

**Figure 1:** Number of US landfalling hurricanes and major hurricanes between 1900 and 2021. Updated graph from Klotzbach 2018[5]

So, the lack of a trend in hurricanes means that an unbiased normalisation of hurricane losses should also show no increase. This makes it more likely that the seven studies showing no trends in normalised US hurricane losses are correct. Pielke 2020 mentions another paper by Bouwer and Botzen (2011)[6] that concluded that Nordhaus 2010 did not sufficiently adjust for local increases in exposure (e.g., more people and therefore more houses that can be hit). Pielke (2020) also explains how the three papers by Nordhaus, Estrada and Grinsted severely underestimate the historic damage done by the 1926 Miami hurricane when compared to other well accepted estimates. So these studies underestimate historic losses, causing an artificial increase in damage over time, even though the frequency of hurricanes has not changed since 1900.

## Earlier IPCC reports

Pielke (2020) cites one of the conclusions of Bouwer and Botzen 2011: 'Our finding is important and indicates that climate change has not resulted in an increase in hurricane damage in the USA in the past.' Pielke then notes that his conclusions and those of Bouwer and Botzen are in line with several IPCC assessments reports. A key paragraph in Pielke (2020) is cited below (our bold):

> Taken together, the results of these studies reinforce and provide much stronger support for the 2014 conclusions of the IPCC that **'economic growth, including greater concentrations of people and wealth in periled areas and rising insurance penetration, is the most important driver of increasing losses'** and **'loss trends have not been conclusively attributed to anthropogenic climate change'** (IPCC, 2014).

> The 2014 IPCC assessment reinforced the conclusions of the IPCC (2012) special report on extreme events, providing even stronger evidence: 'There is medium evidence and high agreement that **long-term trends in normalised losses have not been attributed to natural or anthropogenic climate change'** and 'Increasing exposure of people and economic assets has been the major cause of long-term increases in economic losses from weather- and climate-related disasters** (high confidence)' (IPCC, 2014).

Pielke (2020) is further confirmation of conclusions already drawn in earlier papers and in several IPCC assessments reports. Yes, economic losses due to climate related disasters are increasing, but they do so because there are more people around who can suffer from damage. After you have 'normalised' the data for economic and social development there is no trend left to be attributed to anthropogenic climate change.

5    Klotzbach, Philip J., et al. „Continental US hurricane landfall frequency and associated damage: Observations and future risks." *Bulletin of the American Meteorological Society* 99.7 (2018): 1359-1376.

6    Bouwer, L. M., & Wouter Botzen, W. J. (2011). How sensitive are US hurricane damages to climate? Comment on a paper by WD Nordhaus. Climate Change Economics, 2(01), 1–7. https://doi.org/ 10.1142/S2010007811000188

## Global weather losses

Another paper published by Pielke (in 2019) presents a graph with global estimates of normalised disaster losses.[7] Such estimates are relevant in the context of the UN Sustainable Development Goals (SDG). One of the SDG's state the following sub goal:

> By 2030, significantly reduce the number of deaths and the number of people affected and substantially decrease the direct economic losses relative to global gross domestic product caused by disasters, including water-related disasters, with a focus on protecting the poor and people in vulnerable situations.

This is a reasonable goal. The economy is allowed to grow but we try to decrease disaster losses *relative* to this growth. The paper shows how we are doing so far in this respect. Below is an updated figure from that paper:

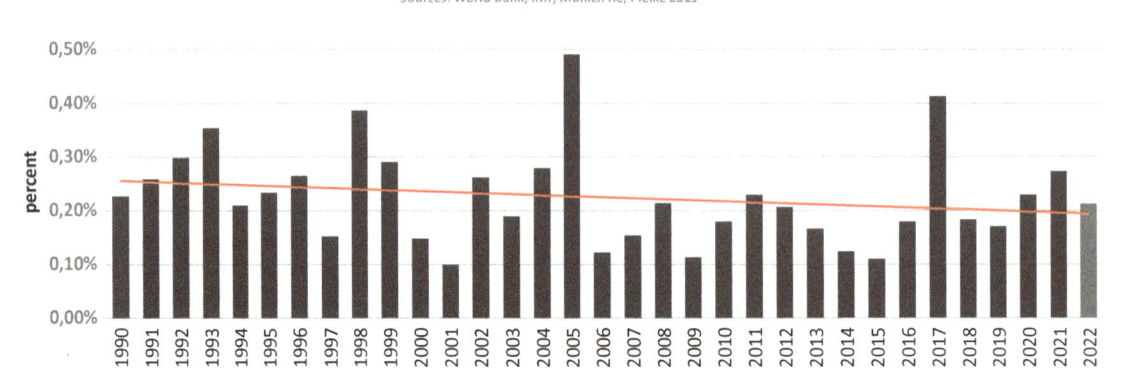

**Global Weather Losses as Percent of Global GDP: 1990-2022**
Sources: World Bank, IMF, Munich Re, Pielke 2019

**Figure 2:** Normalised global disaster losses as a percentage of global GDP. Source: Updated from Pielke (2019), from Pielke's website here.[8]

The trend since 1990 is also down, from about 0.25% of the global GDP in 1990 to now 0.20%. Note that the global economy in this period doubled. So, in absolute numbers the losses increased, but not as a percentage of the GDP. That is also good news.

## Working Group 2 report

Now it is time to have a look at the second part of the AR6 report (Working Group 2 report, WG2) titled *Climate Change 2022: Impacts, Adaptation and Vulnerability*.[9] Just like the Working Group 1 report it is a huge report, covering 3068 pages. It was published on 28 February 2022. The report deals with "impacts" and of course there is no bigger impact than a disaster caused by extreme weather. So, terms like "disasters" (598), "loss" (3504), "damage" (1464), "mortality" (1345) and "vulnerability" (3568) are used extensively throughout the report. But let's focus on the "disaster losses" that we have discussed so far in this chapter. That term is surprisingly used only four times.

Here is what the *Summary for Policy Makers* has to say about "losses" (our bold):

> **Human-induced climate change, including more frequent and intense extreme events, has caused widespread adverse impacts and related losses and damages to nature and people, beyond natural climate variability.** Some development and adaptation efforts have reduced vulnerability. Across sectors and regions, the most vulnerable people and systems are observed to be disproportionately affected. The rise in weather and climate extremes has

---

7   Pielke, R. (2019). Tracking progress on the economic costs of disasters under the indicators of the sustainable development goals. Environmental Hazards, 18(1), 1–6. https://doi.org/10.1080/ 17477891.2018.1540343

8   https://rogerpielkejr.substack.com/p/weather-and-climate-disaster-losses

9   IPCC AR6, WG2, 2022. https://www.ipcc.ch/report/ar6/wg2/

led to some **irreversible impacts as natural and human systems are pushed beyond their ability to adapt**. (high confidence)

This is the sort of language that policy makers apparently must understand. The first sentence in bold actually contains a pleonasm. The terms "human-induced climate change" and "beyond natural climate variability" have a similar meaning. The IPCC is saying here that the observed changes are not natural anymore, i.e. they are caused by greenhouse gases. Which changes? Well, at least more frequent and intense extreme events. These events cause "losses and damages", but be careful, they don't claim there is a global increase in "losses and damages".

In chapter 1 there is a section (1.4.4.2) titled "Emerging Importance of Loss and Damage". On page 171 it has this to say about losses (our bold):

> There is increasing evidence of economic and non-economic losses due to climate extremes and slow onset events under observed increases in global temperatures (Section 8.3.4; Coronese et al., 2019; **Grinsted et al., 2019**; Kahn et al., 2019)

A rather strange sentence if you read it carefully. And look at the references. Grinsted et al (2019) is the *only* paper out of the 54 studies on normalised losses that Pielke (2020) discussed, that claims there is an increase in losses that is attributable to greenhouse gases. What about the other papers? Kahn (2019)[10] is not about losses due to extremes but about future economic impacts of climate change in general, so it is irrelevant for the discussion about disaster losses. Coronese (2019)[11] is titled "Evidence for sharp increase in the economic damages of extreme natural disasters". As the title clearly indicates, the paper claims a rise in losses due to disasters. Pielke (personal communication) explains that the analysis in Coronese 'forgot' to correct for underreporting of disasters before 2000. Pielke: "It is a horrible paper. They take the EM-DAT database from 1960 without accounting for the fact that it is only complete from 2000."

The following graph was posted on twitter that clearly shows the underreporting before 2000:

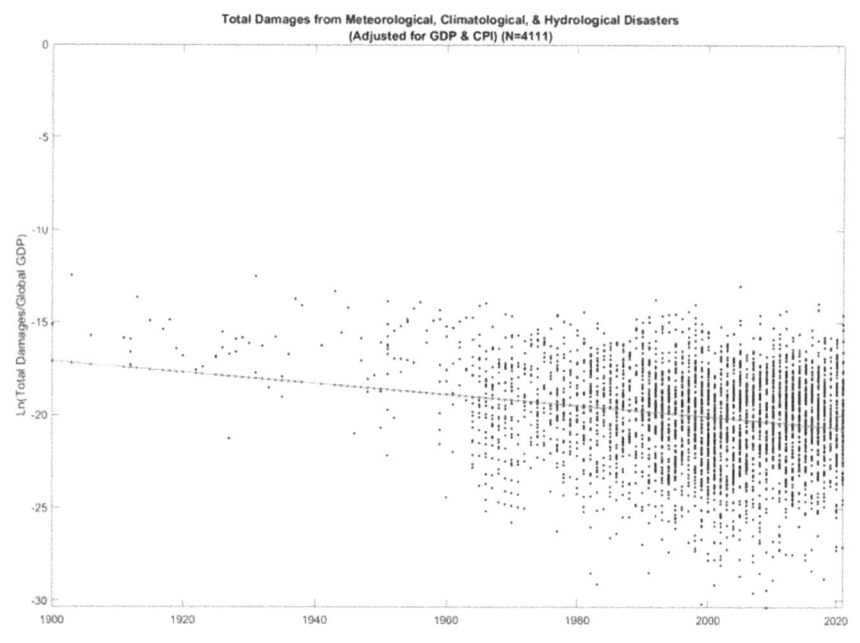

**Figure 3:** The missing data in the CRED EM-DAT database before 2000. Source: Pielke (twitter)[12]

So, this short paragraph in chapter 1 of the AR6 WG2 report looks like a severe form of cherry picking. It claims an increase in disaster losses without considering the full range of peer-reviewed literature on the subject.

---

10    Kahn, M., et al., 2019: Long-Term Macroeconomic Effects of Climate Change: A Cross-Country Analysis. doi:10.3386/w26167

11    Coronese, M., et al., 2019: Evidence for sharp increase in the economic damages of extreme natural disasters. Proc. Natl. Acad. Sci., 116(43), 21450, doi:10.1073/pnas.1907826116.

12    https://twitter.com/RogerPielkeJr/status/1591001971221475328

## AR6, Pielke Jr., and Normalisation

Let's simply look for the Pielke (2020) review paper in the full report. It is mentioned zero times. The term 'normalisation' is mentioned only once but not in the context of disaster losses. Maybe as a verb? So, we look for "normalised" and "normalized". Finally, a few relevant hits pop up. The most interesting paragraph is this one (our bold):

> **Over the last decades, losses due to natural disasters including those from events related to extreme weather have strongly increased** (Mechler and Bouwer, 2015). There is a need for better assessment of global adaptation costs, funding and investment (Micale et al., 2018). Potential synergies between international finance for disaster risk management (DRM) and adaptation have not yet been fully realised. **Research has almost exclusively focused on normalising losses for changes in exposure, but not for vulnerability, which is a major gap, given the dynamic nature of vulnerability** (Mechler and Bouwer, 2015[13]).

This is the final paragraph of Chapter 2 (page 318), a chapter that deals with "Terrestrial and Freshwater Ecosystems and Their Services". Here they claim that research has almost exclusively focused on "normalising losses for changes in exposure", but then the IPCC, whose task it is to assess the available literature, completely ignores all the published papers about "normalised losses".

Looking deeper and deeper in the report we found two more cases where (normalised) losses were mentioned. The chapter about North America has this to say on the page about economic losses due to hurricanes:

> Studies of US hurricanes since 1900 have found increasing economic losses that are consistent with an influence from climate change (Estrada et al., 2015; Grinsted et al., 2019), although another study found no increase (Weinkle et al., 2018).

Remember, these studies were all discussed in the Pielke (2020) review article. At least the IPCC acknowledges here that there is one other study (Weinkle (2018)) that found no increase. Roger Pielke Jr was co-author of that study. But this is all the IPCC had to say about it, also giving the impression that more studies claim an increase than no increase, while the review article by Pielke gives good reasons why Estrada (2015) and Grinsted (2019) should be dismissed.

The other short mention (page 1626) of a normalised losses study is that by McAneney (2019), a study that is also discussed in the Pielke review paper (our bold):

> **However, there is no trend in normalised losses** because the rising insurance costs are being driven by more people living in vulnerable locations with more to lose (McAneney et al., 2019).

So, it is fair to say that the whole literature about normalising economic losses is completely ignored by the IPCC. This raises a lot of questions of course. How could this happen? Who decided to do this? Why didn't reviewers protest? These questions are beyond the scope of this chapter, in which we merely document the failure of the IPCC to cite relevant literature about disaster losses.

## AR6 misrepresents Mechler and Bouwer

There are two names though that pop up frequently in the IPCC WG2 report with respect to this topic. Those are the names Reinhard Mechler and Laurens Bouwer. They both published extensively in the literature on the topic of loss and damage. Mechler, who works in Austria for the International Institute for Applied Systems Analysis (IIASA), was a lead author of chapter 17 (Decision-Making Options for Managing Risk) and a drafting contributing author of the *Summary*

---

13   Mechler, R. and L. M. Bouwer, 2015: Understanding trends and projections of disaster losses and climate change: is vulnerability the missing link? Climatic Change, 133(1), 23–35.

*for Policy Makers*. Laurens Bouwer (a Dutchman working in Hamburg) was a contributing author of chapter 17. They both co-edited a 2019 book titled *Loss and damage from climate change*.[14] The AR6 WG2 report frequently cites this book and both Mechler and Bouwer wrote a chapter for it.

The name "Mechler" is mentioned 113 times in the WG2 report and "Bouwer" more than 40 times. The chapter in the book *Loss and damage from climate change* by Bouwer is most relevant for our current discussion on disaster losses. Here is a fragment from its abstract (our bold):

> Studies into drivers of losses from extreme weather show that increasing exposure is the most important driver through increasing population and capital assets. **Residual losses (after risk reduction and adaptation) from extreme weather have not yet been attributed to anthropogenic climate change. For the Loss and Damage debate, this implies that overall it will remain difficult to attribute this type of losses to greenhouse gas emissions.**

This paragraph is fully in line with earlier work by Bouwer, it is also in line with the Pielke 2020 review paper (in which the work of Bouwer is also cited, see our table 1), and with earlier IPCC assessments. Bouwer mentions the work of Pielke extensively in his chapter in the book. However, this important and relevant conclusion of the book "Loss and damage from climate change" was not mentioned in the AR6 WG2 report.

Instead, the chapter by Bouwer was mentioned in the following ways:

> A further increase in the frequency and/or intensity of water-related extremes (Section 4.4) will also increase consequent risks and associated losses and damages (Section 4.5), primarily for exposed and vulnerable communities globally (**Bouwer, 2019**).
> (Chapter 4, page 652)

And

> Cascading and compounding risks arise from multiple climate hazards coinciding to produce impacts, for example, in mountainous regions, where the combination of glacier recession and extreme rainfall result in landslides (Martha et al., 2015). There is robust evidence that this effect has been observed around slow- and rapid-onset climate events related to drought (i.e., rising temperatures, heatwaves and rainfall scarcity), with devastating consequences for agriculture (Vogt et al., 2018; **Bouwer, 2019**).
> (Chapter 8, page 1178)

Bouwer was not involved in these chapters so it is very possible that he was not even aware of how his work was used in those chapters. These two paragraphs, by the way, are not representative of the full chapter by Bouwer. Bouwer also presents a table (his table 3.2) similar to our Table 1. His table was also ignored by the IPCC.

Our analysis matches with that of Steven Koonin in his bestseller book Unsettled.[15] On page 183, Koonin observes (our bold):

> It's clear that media, politicians, and often the assessment reports themselves blatantly misrepresent what the science says about climate and catastrophes. Those failures indict the scientists who write and too casually review the reports, the reporters who uncritically repeat them, the editors who allow that to happen, the activists and their organizations who fan the fires of alarm, **and the experts whose public silence endorses the deception.** The constant repetition of these and many other climate fallacies turns them into accepted "truths."

Here we have also documented clear shortcomings in the assessment and scientists who can know about this, like Mechler and Bouwer, have remained silent.

---

14  Mechler, R., et al., 2019: Loss and damage from climate change: concepts, methods and policy options. Springer Nature, Berlin Heidelberg

15  Unsettled: What Climate Science Tells Us, What It Doesn't, and Why It Matters, Steven E. Koonin, 2021

The IPCC works according to a set of principles.[16] The most important one is this (our bold):

> The role of the IPCC is to assess on a **comprehensive, objective, open and transparent basis** the scientific, technical and socio-economic information relevant to understanding the scientific basis of risk of human-induced climate change, its potential impacts and options for adaptation and mitigation. IPCC reports should be neutral with respect to policy, although they may need to deal objectively with scientific, technical and socio-economic factors relevant to the application of particular policies.

Here we have documented that, with respect to the literature on disaster losses, the latest AR6 WG2 report was neither comprehensive, open and transparent (it ignored most of the published literature on the topic), nor objective (it cherry picked the few studies that claimed an increase of losses due to greenhouse gases while the majority of the published studies show the opposite, no increasing trend after normalisation of the data). This is very poor performance by the IPCC.

In 2010 there was a lot of criticism of the IPCC after several errors were discovered, mainly in the fourth (2007) WG2 report. A well-known error at the time was the claim that the Himalayan glaciers would be gone in 2035, a claim which turned out to be based on a popular science article in *New Scientist*. The exposed errors led to an international investigation into the procedures of the IPCC by the InterAcademy Council.[17] The IAC Review stated upfront that the main conclusions of the IPCC were not questioned.

The most relevant part of the IAC Review is the following section on page 17-18 (our bold):

> **Handling the full range of views**
> An assessment is intended to arrive at a judgment of a topic, such as the best estimate of changes in average global surface temperature over a specified time frame and its impacts on the water cycle. Although all reasonable points of view should be considered, they need not be given equal weight or even described fully in an assessment report. Which alternative viewpoints warrant mention is a matter of professional judgment. Therefore, Coordinating Lead Authors and Lead Authors have considerable influence over which viewpoints will be discussed in the process. **Having author teams with diverse viewpoints is the first step toward ensuring that a full range of thoughtful views are considered.**
>
> **Equally important is combating confirmation bias—the tendency of authors to place too much weight on their own views relative to other views** (Jonas et al., 2001). As pointed out to the Committee by a presenter and some questionnaire respondents, alternative views are not always cited in a chapter if the Lead Authors do not agree with them. Getting the balance right is an ongoing struggle. However, concrete steps could also be taken. For example, chapters could include references to all papers that were considered by the authoring team and describe the authors' rationale for arriving at their conclusions.

It is evident that if a scientist such as Roger Pielke Jr would have been involved in the production of this WG2 report, the clear bias with regards to the normalisation of disaster losses would not have taken place. We asked Pielke Jr if he was ever asked to contribute to an IPCC report. He wrote back that a senior US IPCC contributor had once told him that "he would never be involved in IPCC".

If the IAC—12 years after their review—were asked to do another review and they took the treatment of "disaster losses" in the WG2 report as a test case, they would conclude that the IPCC didn't

---

implement their recommendations. They would also conclude that the more recent reports have made far more consequential errors than those that led to their 2010 review.

# 13

# Say goodbye to climate hell, welcome climate heaven

BY MARCEL CROK

**"We are on a highway to climate hell", said UN-boss Guterres recently. But an in depth look at mortality data shows that climate-related deaths are at an all-time low. Well-known economist Bjorn Lomborg published this excellent news in a 2020 peer-reviewed paper, but the IPCC, again, chose to ignore it.**

"We are on a highway to climate hell with our foot on the accelerator", said UN Secretary-General Antonio Guterres during his speech to delegates of the COP27 conference in Egypt.[1] "A climate hell", what would that mean? It can't be a good thing for sure. When global leaders talk about climate change, by which they mean anthropogenic climate change, they are using ever stronger and stronger language. A climate hell must mean death and destruction.

In the former chapter about disaster losses, we have documented how the IPCC failed to report honestly about disaster losses as it relates to greenhouse gases. The underlying literature is nearly unequivocal, after normalising the data for GDP there is no trend left in economic losses due to climate disasters. Apparently, the IPCC didn't want to bring this 'good news' to its readership.

## Bjorn Lomborg

In this short chapter we will have a look at climate-related deaths. Before we look into the IPCC AR6 WG2 report, we again start with an interesting scientific paper that covers this topic. It is by Bjorn Lomborg, a very well-known economist in the public and political arena, and the founder of the Copenhagen Consensus Center. In 2020 Lomborg published the paper "Welfare in the 21st century: Increasing development, reducing inequality, the impact of climate change" part of which discusses climate-related deaths.[2] The key graph is figure 17 in his paper, reproduced here as our figure 1:

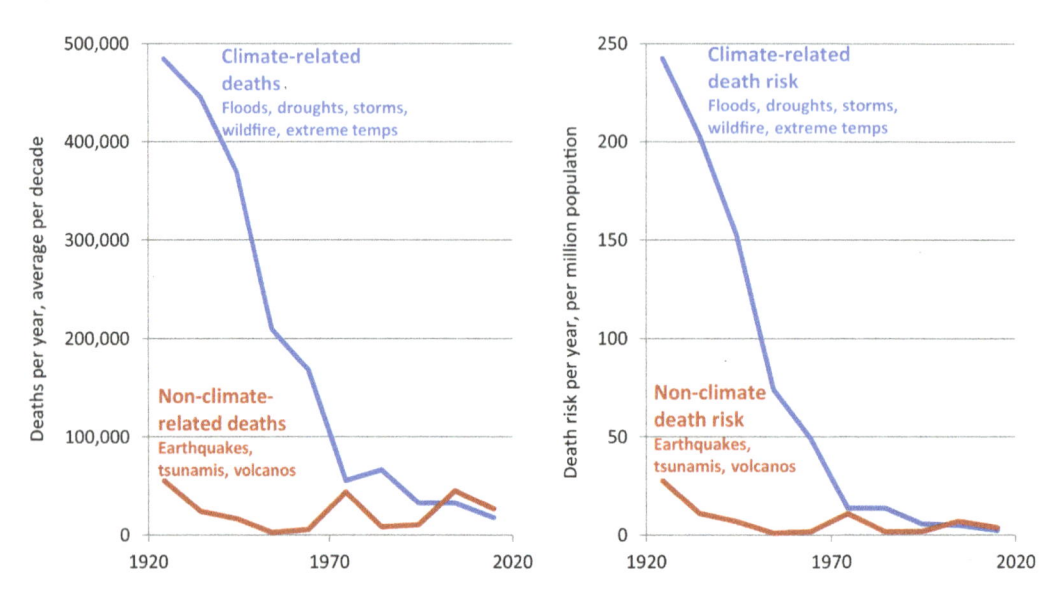

**Figure 1:** Climate and non-climate-related deaths and death risks from disasters 1920–2018, averaged over decades. Data comes from EM-DAT (2019), using floods, droughts, storms, wildfire, and extreme temperatures for climate-related deaths, and earthquakes, tsunamis, and volcanos for non-climate-related deaths. Source: Lomborg (2020).

---

1   https://www.youtube.com/watch?v=T4bpyjcFLVo

2   Bjorn Lomborg, Welfare in the 21st century: Increasing development, reducing inequality, the impact of climate change, and the cost of climate policies, Technological Forecasting and Social Change, Volume 156, 2020, 119981, ISSN 0040-1625, https://doi.org/10.1016/j.techfore.2020.119981.

On the left absolute numbers of deaths (averaged over decades) are shown, in blue for climate or weather-related disasters (such as floods, hurricanes and droughts) and in red for non-climate-related disasters such as earthquakes, tsunamis and volcanos. The steep decline of the blue line is remarkable (a 96% decrease), even more so if you realise that the world population increased from two to eight billion over the period. So even when extreme weather didn't increase, which the IPCC itself admits is true for hurricanes, floods and meteorological and hydrological droughts (see chapter 11) one would expect that more people would lead to more weather-related deaths. The opposite is the case. Around 1920 around *half a million* people died yearly due to extreme weather events. Droughts were especially deadly. In the past decade this declined to 20.000 on average and the past few years these numbers were even lower, around 6000 per year.[3] To put such numbers in perspective, yearly 1.2 million people die due to road injuries, more than 750,000 people die from suicide, more than 400,000 from homicide, and 237,000 people drown.[4] Extreme weather in that respect is a minor risk.

In relative terms the numbers are even more spectacular. Risk of death due to extreme weather has declined *99%* over the past century.

## Climate heaven

How is this possible? This is far from Guterres' climate hell, it is more a 'climate heaven'. Lomborg himself has this to say about these trends (our bold):

> It is to be expected that it is much harder to avoid death from non-climate-related disasters, since these are mostly earthquakes that are hard to predict. Hence, only better building standards can help. However, the large reduction in climate-related deaths from disasters shows a dramatic increase in climate resilience, **likely mostly brought about by higher living standards, a reduction in poverty, improvement in warning systems, and an increase in global trade, making especially droughts less likely to turn into widespread famines**.

Lomborg finishes that section of his paper with the following observation (our bold):

> Fig. 17 shows that **we are now much less vulnerable to climate impacts than at any time in the last 100 years**. It is possible that climate change has made impacts worse over the last century (although the discussion on floods, droughts, wildfire, and hurricanes suggests this is not the case), but **resiliency from higher living standards has entirely swamped any potential climate impact**.

These are important observations, based on data from a well-known database (EM-DAT) and published in a peer reviewed paper in 2020. Let's now move on to the WG2 report and see how the IPCC reported on climate-related deaths.

A first search for the term "climate-related death" gives zero results as does "Lomborg". So, his paper is not mentioned. The term "deaths" gives a lot of hits (298) as does "mortality" (1345). We can't discuss them all of course but let's look at some claims in the *Summary for Policy Makers* (our bold):

> Widespread, pervasive impacts to ecosystems, people, settlements, and infrastructure have resulted from observed increases in the frequency and intensity of climate and weather extremes, including hot extremes on land and in the ocean, heavy precipitation events, drought and fire weather (high confidence). Increasingly since AR5, these observed impacts have been attributed to human-induced climate change particularly through increased frequency and severity of extreme events. These include increased heat-related human **mortality** (medi-

---

3   Lomborg is frequently updating figure 1 on his social media accounts like Linkedin, see e.g.: https://www.linkedin.com/feed/update/urn:li:activity:7028351713191342082?updateEntityUrn=urn%3Ali%3Afs_feedUpdate%3A%28V2%2Curn%3Ali%3Aactivity%3A7028351713191342082%29

4   https://ourworldindata.org/causes-of-death

um confidence), warm-water coral bleaching and mortality (high confidence), and increased drought-related tree mortality (high confidence). [B.1.1]

And:

Climate change has adversely affected physical health of people globally (very high confidence) and mental health of people in the assessed regions (very high confidence). Climate change impacts on health are mediated through natural and human systems, including economic and social conditions and disruptions (high confidence). In all regions extreme heat events have resulted in **human mortality** and morbidity (very high confidence). [B.1.4]

And:

Between 2010–2020, **human mortality from floods, droughts and storms** was 15 times higher in highly vulnerable regions, compared to regions with very low vulnerability (high confidence). [B.2.4]

And:

Climate change and related extreme events will significantly increase ill health and **premature deaths** from the near- to long-term (high confidence). [B.4.4]

## EM-DAT database

So, there are all kinds of claims that climate change is leading to or will lead to more deaths. But what about the EM-DAT database and the fact that those data suggest a strong decrease in climate-related deaths? EM-DAT is mentioned only 7 times in AR6. The IPCC does show data from EM-DAT. Here is the figure:

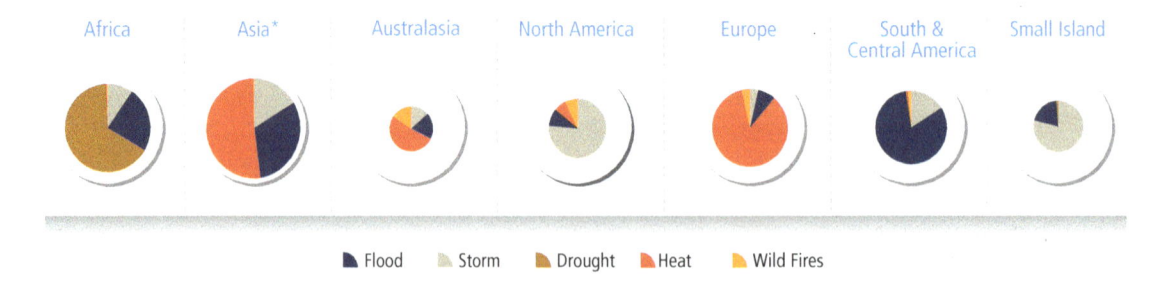

(c) Average mortality per hazard event per region between 2010 and 2020:

Africa    Asia*    Australasia    North America    Europe    South & Central America    Small Island

■ Flood    ■ Storm    ■ Drought    ■ Heat    ■ Wild Fires

Average mortality per hazard event is indicated by size of pie charts. The slice of pie chart shows absolute number of deaths from a particular hazard

\* The large size of the pie chart and the strong representation of heat waves is caused by the significant number of deaths from a single event in a single country. This single extreme outlier affected the overall average mortality per event in Asia.

**Figure 2:** reproduction of figure TS.7(C) of the WG2 AR6 report (p 77), showing relative mortality per hazard on different continents for the period 2010-2020.

They only show the average mortality per hazard for the period 2010-2020, and the pie charts are quite odd. Heat-related deaths dominate in Europe. However, in Africa, South America and the Small Island states, there are no heat-related deaths at all. How (un)likely is that? Their message is fair: less developed continents are more vulnerable to extreme events than richer continents. However, the IPCC fails to mention or reproduce the very important peer-reviewed figure from Bjorn Lomborg based on the same EM-DAT database. We learn more from what the IPCC leaves out, than from what it includes.

However, somehow the good news did slip into the report, on page 2435 (of 3068!) the IPCC mentions a paper by Formetta and Feijen which has the revealing title *Empirical evidence of declining global vulnerability to climate-related hazards* (our bold):[5]

> Formetta and Feyen (2019) **demonstrate declining global all-cause mortality and economic loss due to extreme weather events over the past four decades**, with the greatest reductions in low-income countries, and with reductions correlated with wealth.

The research for this paper was funded by the European Commission and therefore coming from an unsuspected source. Here are the highlights mentioned at the top of the paper (our bold):

> Highlights
> - We quantified the dynamics of socio-economic vulnerability to climate-related hazards.
> - **A decreasing trend in both human and economic vulnerability is evident**.
> - **Global average mortality and loss rates have dropped by 6.5 and nearly 5 times, respectively, from 1980 to 1989 to 2007–2016**.
> - Results also show a clear negative relation between vulnerability and wealth.

This is, as far as we know, the only place in the report where the IPCC reveals the good news about global mortality due to climate-related events. However, these important results didn't make it into the *Summary for Policy Makers* (SPM) or the Technical Summary (TS) of the WG2 report, let alone the press release or headline statements.

So again, the IPCC is ignoring a key paper (by Bjorn Lomborg in this case), that shows very good news about the decreasing impact on humans from extreme weather. It fails to bring this good news, either in the full report or in the *Summary for Policy Makers*. As a result, policy makers get an overly negative picture of climate change and its impacts on humans. There is no excuse for this. The IPCC is aware of the EM-DAT database and Bjorn Lomborg is one of the most visible public voices about climate. The IPCC deliberately chose to ignore this good news.

Looking at disaster losses and climate-related deaths the IPCC could have easily repeated its conclusions from the AR5 WG2 report (chapter 10, executive summary)(our bold):

> For most economic sectors, **the impact of climate change will be small relative to the impacts of other drivers** (medium evidence, high agreement). Changes in population, age, income, technology, relative prices, lifestyle, regulation, governance, and many other aspects of socioeconomic development will have an impact on the supply and demand of economic goods and services that is large relative to the impact of climate change. {10.10}

---

5    Giuseppe Formetta, Luc Feyen, Empirical evidence of declining global vulnerability to climate-related hazards, Global Environmental Change, Volume 57, 2019, 101920, ISSN 0959-3780, https://doi.org/10.1016/j.gloenvcha.2019.05.004.

# Epilogue

BY ANDY MAY

**It has been over two years since Marcel and I began working on this book. In that time, we learned a great deal from the IPCC and the writers and reviewers of the chapters herein. We find it notable that this is a review of the *sixth* major IPCC Report on climate change in the thirty-two years since the first IPCC report was published in March 1990. In total the IPCC has produced 47 reports[1] and either spent or caused many billions of dollars[2] to be spent since March 1990.**

All these reports attempted to convince the public, news media, and politicians that their "$CO_2$ control knob"[3] hypothesis is correct. This hypothesis concludes that observed climate change is caused by mankind and their emissions of key non-condensing greenhouse gases, mainly $CO_2$, which they claim regulate the "planetary temperature, with water vapor concentrations as a feedback."[4]

The current $CO_2$ control knob hypothesis has its origins in the 1960s and 1970s and culminated in the U.S. National Research Council's "*Charney Report*" in 1979.[5] Arguably, as you can see in this volume, the uncertainty regarding the effect of additional $CO_2$ and other non-condensing greenhouse gases is just as uncertain as it was in 1979. This lack of measurable progress after 43 years is a sign that the hypothesis is missing a major component and/or process. Have the IPCC developed "tunnel vision?" Are they so devoted to their hypothesis they are missing the obvious? Scientists sometimes suffer from confirmation bias and cannot see the weaknesses in their hypotheses.

The AR6 report reveals that they have ignored the very important multi-decadal ocean oscillations discovered in the 1990s and 2000s[6] long after the IPCC had focused exclusively on anthropogenic causes. These ocean oscillations, collectively, have a large effect on our climate, but are unrelated to "non-condensing greenhouse gases." AR6 states that "there has been negligible long-term influence from solar activity and volcanoes,"[7] and acknowledges no other natural influence on multidecadal climate change despite the recent discoveries, a true case of tunnel vision.

We were promised IPCC reports that would objectively report on the peer-reviewed scientific literature, yet we find numerous examples where important research was ignored. In Ross McKitrick's chapter on the "hot spot," he lists many important papers that are not even mentioned in AR6. Marcel gives examples where unreasonable emissions scenarios are used to frighten the public in his chapter on scenarios, and examples of bias and hiding good news in his chapters on extreme weather and snowfall. Nicola Scafetta and Fritz Vahrenholt document that over 100 papers showing solar activity correlates with climate change have been ignored by the IPCC. Numerous other examples are documented in other chapters. These deliberate omissions and distortions of the truth do not speak well for the IPCC, reform of the institution is desperately needed.

1   https://www.ipcc.ch/reports/
2   https://www.heritage.org/environment/commentary/follow-the-climate-change-money
3   (Lacis, et al. 2010 & 2013) and (AR6, page 179).
4   AR6, page 179.
5   (Charney, et al., 1979)
6   (Vinós, 2022) and (Wyatt & Curry, 2014)
7   AR6, page 67.

Perhaps this is why, after 47 reports and 32 years, they have yet to convince a majority of the people on Earth,[8] or in the United States,[9] that manmade climate change is our most important and serious societal problem. Other problems are always considered more important and urgent. In a 2018 Pew Research poll[10] climate change ranked 18th, of 19 issues in importance, in a similar 2014 poll,[11] climate change ranked 14th in a list of priorities. A 2022 poll by the Pew Research Center[12] also found climate change ranked 14th. In the UN *My World 2015 Report*, a poll of 10 million people around the world, climate change ranked last of 16 issues in importance. Minds are not being changed.

Are we at a fork in the road? Will the United Nations, the IPCC, and politicians finally realize that their 50-year-old hypothesis is out of date and incorporate the new natural warming forces discovered in the past thirty years into their work and projections? In the past the IPCC has fought off attempts to independently review their work.[13] It is unfortunate, but the IPCC has an opaque process for choosing their lead authors and contributing authors, the very people who choose what is included and what is ignored in each report. As one of our authors, Ross McKitrick has written:

> "The [IPCC] Bureau has, effectively, a free hand in picking Coordinating Lead Authors, Lead Authors and Contributing Authors of the report.
>
> Past Lead Author selections have been criticized by other Lead Authors as being overly dominated by political considerations.
>
> Coupled with the deficiencies in the peer review process, this opens up the possibility that the IPCC Bureau can pre-determine the conclusions of the report by its selection of Lead Authors."[14]

Any like-minded group, with inadequate infusions of new blood, runs the risk of becoming fossilized in their thinking. Independent, open, honest, and transparent peer-review is essential to good science. There are indications that this is not happening in the IPCC. Ray Bates, a long-time expert reviewer of major IPCC reports is particularly critical of the IPCC review process.[15] Bates points out that very eminent scientists, such as Prof. Aksel Wiin-Nielsen, have been excluded from IPCC leadership because they would not "toe the party line."

After every major IPCC report, the same complaints surface over and over again. The choice of lead authors and authors is "arbitrary,"[16] the IPCC's own procedures are often not followed.[17] Yet, time and again, nothing changes. Improper political interference during the second IPCC report was widely criticized when a past president of the United States National Academy of Sciences, Frederick Seitz, called the report a "Major Deception on Global Warming."[18] The third report included the deceptive and incorrect "Hockey Stick,"[19] a flaw *repeated* in AR6. The fourth report

8 https://news.gallup.com/opinion/gallup/321635/world-risk-poll-reveals-global-threat-climate-change.aspx, also see the UN My World 2015 report (http://about.myworld2030.org/my-world-2015/)

9 https://www.washingtonpost.com/politics/2022/10/10/half-voters-say-climate-change-is-important-midterms-poll-finds/

10 https://www.investors.com/politics/columnists/global-warming-polls-priorities/

11 https://news.gallup.com/poll/167843/climate-change-not-top-worry.aspx

12 https://www.pewresearch.org/politics/2022/02/16/publics-top-priority-for-2022-strengthening-the-nations-economy/

13 https://climateaudit.org/2011/06/18/ipcc-sabotages-an-interacademy-recommendation/

14 (McKitrick, 2011)

15 (Bates, 2020)

16 InterAcademy Council, 2010, Climate change assessments, Review of the processes and procedures of the IPCC, Link: http://intleval.cipa.cornell.edu/simulation/Climate%20Change%20Assessments,%20Review%20of%20the%20Processes%20&%20Procedures%20of%20the%20IPCC.pdf

17 McKitrick, Ross, 2010, "Submission to the Inter-Academy Council Independent Review of the Policies and Procedures of the Intergovernmental Panel on Climate Change." Link: http://www.rossmckitrick.com/uploads/4/8/0/8/4808045/iac.ross_mckitrick.pdf

18 May, Andy, 2020, *Politics and Climate Change: A History*, page 234, link: https://www.amazon.com/POLITICS-CLIMATE-CHANGE-ANDY-MAY-ebook/dp/B08LJSBVBC/ref=sr_1_3?crid=GJLUTQV2YMKQ&keywords=Politics+and+Climate+Change&qid=1674065419&sprefix=politics+and+climate+change%2Caps%2C87&sr=8-3 and in the *Wall Street Journal*, June 12, 1996.

19 Montford, Andrew, 2010, The Hockey Stick Illusion, link: https://www.amazon.com/Hockey-Stick-Illusion-W-Montford/dp/0957313527/ref=sr_1_1?crid=1CBSL88079DOV&keywords=The+Hockey+Stick+Illusion&qid=1674065645&sprefix=the+hockey+stick+illusion%2Caps%2C88&sr=8-1

resulted in the critical InterAcademy Council report,[20] and so on. AR6 repeats past flaws and is, in many ways, worse than the previous reports.

All the chapters in this volume have been independently peer-reviewed. All reviewer comments have been carefully considered and dealt with appropriately. This is not to say that all the authors and peer-reviewers agree on every point, disagreements among us remain in some cases, but we all had an opportunity to freely and openly debate our views. Consider this volume an independent assessment of the most important parts of AR6, an assessment that, unfortunately, was not done within the IPCC.

## Works Cited

Bates, R. (2020). Should Ireland Relinquish Authority over its Vital National Interests to the Intergovernmental Panel of Climate Change (IPCC)? *Irish Times*.

Charney, J., Arakawa, A., Baker, D., Bolin, B., Dickinson, R., Goody, R., ... Wunsch, C. (1979). *Carbon Dioxide and Climate: A Scientific Assessment.* National Research Council. Washington DC: National Academy of Sciences.
Retrieved from http://www.ecd.bnl.gov/steve/charney_report1979.pdf

McKitrick, R. (2011). *What is Wrong with the IPCC.* The Global Warming Policy Foundation.
Retrieved from https://www.thegwpf.org/images/stories/gwpf-reports/mckitrick-ipcc_reforms.pdf

Vinós, J. (2022). *Climate of the Past, Present and Future, A Scientific Debate.* Spain: Critical Science Press. Retrieved from https://www.amazon.com/Climate-Past-Present-Future-scientific-ebook/dp/B0BCF5BLQ5/ref=sr_1_1?crid=3DADACCQN7CX3&keywords=Climate+of+the+Past%2C+Present+and+Future%2C+A+Scientific+Debate&qid=1669221503&sprefix=climate+of+the+past%2C+present+and+future%2C+a

Wyatt, M., & Curry, J. (2014, May). Role for Eurasian Arctic shelf sea ice in a secularly varying hemispheric climate signal during the 20th century. *Climate Dynamics, 42*(9-10), 2763-2782. Retrieved from https://link.springer.com/article/10.1007/s00382-013-1950-2#page-1

---

20  InterAcademy Council, 2010, Climate change assessments, Review of the processes and procedures of the IPCC,
Link: http://intleval.cipa.cornell.edu/simulation/Climate%20Change%20Assessments,%20Review%20of%20the%20Processes%20&%20Procedures%20of%20the%20IPCC.pdf

Ingram Content Group UK Ltd.
Milton Keynes UK
UKHW050917170723
425264UK00005B/14

**Dr. Willie Soon** offers his congratulations to the Clintel Team for producing a "very excellent and readable report."

**David Siegel** writes: "Wow, the report you guys did is great! Gorgeous and very well done. The [Clintel] team of highly qualified scientists worked hard to expose the many biases in AR6. They show that the AR6 authors deliberately hide evidence that the so-called 'climate crisis' is not a crisis at all."

**Dr. Judith Curry**, Professor Emerita of the Georgia Institute of Technology and author of *Climate Uncertainty and Risk*: "Clintel's new Report provides an important critical evaluation of the exaggerated claims about climate change published by the IPCC in AR6. This much needed counterpoint highlights the inconvenient topics and publications that don't support the IPCC's narrative of dangerous human-caused climate change."

**Dr. William Happer**, Professor of Physics, Emeritus, Princeton University: "Clintel reveals how [AR6 uses] dodgy data … to support the narrative of impending climate doom. Much sounder data that implies the opposite is ignored. … The review is well illustrated and has numerous references …"

**Tom Nelson**, podcaster: "This report is a superb takedown of IPCC junk science."

**Andy May**
Andy May is a writer and lives in The Woodlands, Texas. He enjoys golf and traveling in his spare time. As a retired petrophysicist, he is a member of Clintel and the CO2 Coalition; and the author of four books and seven peer-reviewed papers on various scientific topics.

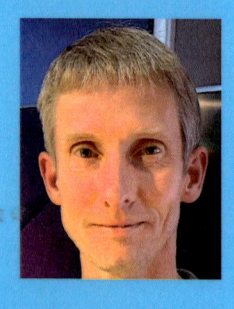

**Marcel Crok**
Marcel Crok (The Netherlands) studied physical chemistry before becoming a professional science writer. In 2005 he published an award-winning critical article on the notorious hockey stick graph. He decided to work full-time on the climate debate. In 2019 Crok and emeritus professor of geophysics Guus Berkhout created the Clintel Foundation.

clintel.org

ISBN 979-8-89074-862-1
53300
9 798890 748621

# Alderley Park Discovered

# Alderley Park Discovered

George B. Hill

*In* the Park and *from* the Park:
Stanley barons, ICI, AstraZeneca and medicines for the world

**Within the infant rind of this small flower**
**Poison hath residence, and medicine power**

William Shakespeare
*Romeo and Juliet*

For all who have worked in Alderley Park; and in memory of Dudley Thomas and Pete Smith and any others who departed from the laboratory of life far too soon.

20% of any royalties received by the author will be donated to medical charities.

## The author

George B. Hill was born in 1956. He joined ICI after a first class honours degree in chemistry at UMIST (Manchester), and worked as a synthetic organic research chemist in 'the Park' for 34 years, half of it on projects in oral health, arthritis and antibiotics and the other half in the cancer area; he was a team member on the **Faslodex**™ and **Tagrisso**™ projects. His name is on 12 scientific papers, three as lead author. Outside science, he wrote the texts for the five Alderley Park History & Nature Trail guides; and edited the annual Alderley Park & Radnor Mere Wildlife Report from 1978 to 2011. He left AstraZeneca in 2012.

George lives with his wife Christine in Cheshire where they are active at their local Baptist church. They have three wonderful grown-up children (one is a chemist). George now writes in several genres combining fiction, science, faith, history and poetry. His '*My Writing*' website at www.hillintheway.co.uk has details of his (mostly self-published) work.

Copyright © George Hill, 2016
First edition, 2016

Published by Palatine Books (an imprint of Carnegie Publishing Ltd)
with the support of AstraZeneca plc and Manchester Science Partnerships

Palatine Books/Carnegie Publishing
Carnegie House,
Chatsworth Road,
Lancaster, LA1 4SL
www.carnegiepublishing.com

ISBN 978-1-910837-04-7

Designed and typeset by Carnegie Book Production

Printed and bound in the United Kingdom by Cambrian Printers Ltd

# Contents

# Acknowledgements

I owe many debts to many people both inside and outside the company, including some sadly no longer with us. When I refer to 'the company', this may include any of ICI Pharmaceuticals Division, Zeneca plc or AstraZeneca plc, and infrequently MSP. When I refer to people, to avoid errors, many personal titles are omitted – see the section on Terminology.

**Definitions and Disclaimers Note:** *The Terminology section also defines all matters of corporate brand names and trademarks. Important disclaimers will also be found in this Acknowledgements section.*

On the estate's older history and form, I thank: Anglesey Antiquarian Society (with the late Lord Stanley of Alderley) for their anthology; Duncan Broomhead for information on the Alderley Mummers; former Site Manager George Driver for his anecdotes and account of the Stanleys of Alderley; Roger Foden for his study of the views of the Park; Victoria Frost for her *A Brief History of Alderley Park* (1996); Helen Houghton for her 1986 Manchester University dissertation on *The Changing Landscape of Alderley Park*; former Site Manager John Robinson for his notes on research into the building of the Radnor Mere dam; Ged Ryan for the anecdotes of Lady Victoria Stanley; Harold Storey of the Macclesfield Heraldry Society for his study of the Tenants' Hall; Craig Thornber for his history website; Ian Urquhart for his information on the Water Gardens and the wonderful horticultural efforts of the Matthews family; Mary Watts for her wartime memories of the Park; Dr John Wilkinson for his memories of the Park in the 1920s; and particularly Paul Woodcock, whose historical studies into the Park compare with my own, and led via his fascinating series on the estate's history in *Park Life* in 2001–2004 to the beautiful *Alderley Park Estate History* large booklet compiled by Clare Power with the help of Melanie Pennington, Alicia Edmonds-Smith *et al*.

The present Lord and Lady Stanley of Alderley have been both generous in their practical support and encouraging with their warm interest.

I also acknowledge and recommend two external sources. Firstly, Jenny Youatt, church archivist at St Mary's Church, Nether Alderley, kindly pointed me to the church's invaluable history CD (Volume 2, *The Stanleys in Alderley*) – this is still is available from www.stmarysalderley.com.

Secondly, the Alderley Edge Landscape Project (jointly with the National Trust) has its own book in press, *The Story of Alderley: Living with the Edge* (to be published February 2016) which is nearly complementary to this work in describing the whole Alderley district. Both its editor, Professor John Prag of the Manchester Museum, and historian Clare Pye, have been extremely helpful; I recommend that book's coverage of other aspects of the Stanleys' history (their support for the local community, relations with other landowners, and local maps and lands); also Matthew Hyde's architectural perspective on Alderley Old Hall, the church and (usefully) some of Alderley Park's buildings. Major stories peripheral to the Park's, including full accounts of Nether Alderley Mill and of the Edge's geology and mines, have been referred by me to there. Natural history content is also complementary; for example, bats, mammals, waterbirds and dragonflies are covered only here, while some other classes are well covered there. I reference it in the text as *The Story of Alderley*. I also acknowledge permission to quote from the Project's Oral Archive.

I am grateful to the writers of many in-house articles published down the years, particularly in *Fulshaw Times*, in *Scan* (especially Jane Powell's history articles of 1971; and she credited earlier sources, including Noel Worthington, one-time butler to the Stanleys).

On Alderley Park's wildlife, I thank everyone who has contributed over the last 35 years to the annual *Alderley Park and Radnor Mere Wildlife Report* (edited by me) produced for ClubAZ's former Natural History Section. Some observers have studied many species but specifically helpful were records from: for the bats, Clare Sefton and Ged Ryan of the Cheshire Bat Group; for the birds (with many others), John Barnes, John Bird, Mike Crawley, Mark Hamilton, Cheshire county bird recorder Hugh Pulsford, and John Rayner (the latter also dragonflies); for the fish, John Bowden and Paul Woodcock; for the flora, Roger Bowling, the Cheshire county flora recorder Graeme Kay, and Dave Tinston; for the fungi, Alistair Glen, Nigel Taylor and the North West Fungus Group; and for other mammals, Katie Stevens for her MMU small mammal survey results.

Also, I must acknowledge the remarkable and inspiring (despite its early finish) survey and recording the work of the AstraZeneca Global Biodiversity Scheme, coordinated by Ross Brown from AZ Brixham, supported by Andrew Broome, Hugh Pulsford and John Rayner, plus surveyors from the Cheshire Wildlife Trust. Alderley Park was the global pilot project for this visionary scheme, whose results remain a landmark.

On Alderley Park's estate management and life, I am grateful for much information from estate manager Andrew Broome and his predecessor Brian Owen. I must also mention the warm support of all the past Site Managers of Alderley Park; of those not already mentioned above, Chris Jump, Bill Rutter and Will Spinks particularly deserve note, as does Lorraine Crossland. A major source of this book was the work on Alderley Park's earlier and later History and Nature Trail series, in which the labours of Heather Green and Helen Hayes, respectively, were surpassing. Lesley Stafford and Dave Tinston helped me create the 1998 guide to the Water Garden and its plants.

The story of the social club (now ClubAZ Ltd) and Mulberry's was kindly corrected by Rachael Vipond; and that of the company's wider community involvement by published accounts of warm-hearted work by many people.

The earlier history of pharmaceuticals in ICI was written with help among many sources from Hilary Ford (née Curd), including the achievements and tragic death of her father Frank. Audrey Cooper, Julie James and Rachel Stewart helped me to mine the Mereside Library or company archives over the years (this amongst other finds unearthed Sam Ellingworth's remarkable 227-page unpublished history of ICI pharmaceuticals research up to 1957). Clare Wildman's research on the opening of Mereside was helpful. External works by academics given access to company archives, such as Dr Viviane Quirke, were informative. The site building development sequence included information on the site's growth and its architects from Andrew Shore and Neil Osborne.

Brief descriptions of many business functions were assisted by Graham Palmer, Paul Bowler and John Jones (on computing in the company); Lucy Padget (on patents and IP); Nigel Crossland (on quality assurance); Professor Joe Sweeney of Huddersfield University (on collaboration with academia); Roy Yates (on HR); Martin Haywood (on IT); Jeff Morris (on drug design); and Dave Rudge (on robotics). Roger Langley, *Scan*'s former Editor, kindly confirmed details of the company's various publications over the years; and others. Articles published by AZ staff also helped. Clearly very many more could have been asked, and some may seriously deserve an apology for being overlooked.

For my limited history of the actual research (mainly) and development sequence, I thank *all* those people who read incomplete texts I offered them or helped in one way or another. It is important to explain that these people have mostly corrected some of my errors and omissions, *not* endorsed the final text. This has been repeatedly edited (and in places simplified for a wider audience); differing accounts have been harmonised; and it must be understood that the entire responsibility for it, and for mistakes and misrepresentations in it (before publication), is mine.

*No statement should therefore be taken as the final view* of any of the following (or of the business, or of any other contributors). The key scientific patents and papers behind the great stories (some are listed in **Appendix 5**) must be considered the only authentic statements by their named inventors and authors and others who did the work. And no statements made in this book form affirmations, propositions, claims or statements, legal or otherwise, made by AstraZeneca plc (or any other company), even if I quote an existing member of staff. Comments may also be derived from unchecked public domain and media sources.

Only on these understandings do I sincerely thank, for contributions or comment on my stories of **'Inderal'**, Ralph Howe and Bernard McLoughlin; **'Tenormin'**, Roy Hull and Dave LeCount; **'Nolvadex'**, Mike Dukes, John Patterson and Sandy Todd; **'Meronem'** and **Invanz™** (ertapenem), Mike Betts and Mike Swain; **'Diprivan'**, Barry Arnold, Roger James and Hugh Pulsford, *et al.*; **'Zoladex'**, Mike Dukes, Mike Giles and Sandy Todd; **'Casodex'**, Mike Dukes and Howard Tucker; **'Faslodex'**, Sir Tom McKillop, Gary Nunn, Brian Tait and Robert Stevenson; **'Arimidex'**, Mike Dukes and Phil Edwards; **'Tomudex'**, Les Hughes and Trevor Stephens; **'Iressa'**, Andy Barker, Darren Cross, Keith Gibson and Maxwell Kirkby and Andrew Mortlock; **'Caprelsa'**, Andy Thomas; **'Tagrisso'**, Darren Cross, Ray Finlay and Andrew Mortlock; and several people for wider overviews. Neville Crossley and Paul Bowler commented on **'Equimate'** and **'Estrumate'**; I regret I could do no more. Also, those who held weighty (and often less visible) higher responsibility for the work have mostly not had opportunity to comment on matters that occurred on their watch. But those who know the real facts, whose opinions matter, will surely not be misled.

Works by external researchers, particularly Professor V. Craig Jordan's many publications on **tamoxifen** (and his historic images, very helpfully supplied when company sources no longer held them), gave important wider perspectives.

In the later company history, Sir Tom McKillop very kindly corrected some errors in my accounts of the demerger and merger in which he was pivotal; and John Patterson did the same for the pipeline record. Wendy Russell's history of Zeneca Pharmaceuticals was helpful.

The full story of the transfer from AstraZeneca to Manchester Science Partnerships was kindly elucidated by Graeme Bristow (Alderley Park Site Manager for AZ) and Chris Doherty (Alderley Park Site Director for MSP).

On a more general level, larger sections of the text were tested at one stage or another by Barry Arnold, Rob Bradbury, Graeme Bristow, Roger Foden, Les Hughes, Clare Pye and Paul Woodcock, to whom I am particularly grateful.

In formal terms, for the company, Lucy Padget with support from her colleagues provided an intellectual property viewpoint, but also added much keen detailed help and advice on several of the drug stories ('The draft has been on at least four transatlantic flights with me!').

Jim Lynch bravely reviewed the entire late stage text, for which he holds my great admiration.

A few parts of the text derive from nature or history articles previously commissioned by me as Editor of the *Alderley Park and Radnor Mere Wildlife Report* (1978–2011). I have already named all the authors concerned; but I have adapted their original texts which were of great help.

Most images not created by me were from historical or company sources; individual photographers for the latter are not usually identifiable (although Barry Smith certainly took some of them, and also gave me helpful advice) but are acknowledged gratefully. The wonderful nature photos of Paul Flackett deserve particular note. I thank Brenda Scott for her line drawings of Water Garden plants, and Gordon Yates for permission to use his lovely nature images. Other photographers have been gratefully credited *in situ*.

Lord Stanley of Alderley kindly gave permission to reproduce his family portraits, images, maps and heraldry both from his own collection and from the estate books and collections in the Chester Records Office. (Note: the heraldry shown comprises, or contains within it, many shields (coats of arms). Acknowledgements for the images of the *older* coats of arms by themselves name only the photographer, since in English law a coat of arms belongs only to that individual, not to the family – there is no such thing as a 'coat of arms for a surname'. But I gladly acknowledge Lord Stanley's courtesy regarding all his family's emblems.)

The seven Tunnicliffe drawings are reproduced by permission of the Estate of C.F. Tunnicliffe, RA. The dovecote sketch is by the late Barbara Freer with agreement from Jean Hansell's book. Karen Lewis pointed me to the sketches of William Wykeham Keyworth.

Every effort has been made to trace the creators of other images. Some photographers are difficult to identify or unknown even by those pictured.

Tracking down contemporary images of people was a particular problem, which explains some glaring omissions. Even more inaccessible were product images from before AstraZeneca's era (who keeps really old medicine packets??). Hilary Ford and Mike Dukes also helped greatly by providing historic photos of earlier ICI researchers. James Parker helped with sourcing recent company images and computer generated graphics.

Finally, this book itself, such as it is, would never have come into existence without the enthusiasm of a gallery of other people, including Colin Bath and Annabelle Hendrick of AZ and Michelle Hill of MSP. Jessica Bignell of AZ worked wonders in the final phase.

And, of course, Anna Goddard at Carnegie Publishing and her staff, especially Lucy Frontani, have endured much from me.

More embarrassingly, I must also add my grateful thanks to all of my company supervisors over the years! Brian, Bill, Graham, Neil, Nicola, Andrew, Kevin, Sam, Chris, Thorsten and Ray all inspired me in the amazing research efforts in which they led me, while permitting me to learn about the Park – and to field the numerous odd queries that floated my way due to my extracurricular knowledge of the site!

Then there are those who I have certainly and unforgivably forgotten; and others who I have not forgotten, but who are no longer with us (but taught me so much of what I know).

Without the financial support of AstraZeneca plc and MSP this book would exist, if at all, as a shadow of what it is.

And finally, there is the one person without whom this story would, for me, be both empty and unwritten: my lovely and long-suffering wife, Christine.

# Terminology

1. A few places have alternative names, including Tenants' Hall (the name I use in the older history) which became the Sir James Black Conference Centre after its renaming in 1989 (and is now expected to revert to the old name). Others include the Old Drive or Chestnut Avenue (slowly becoming the Lime Avenue), and the Dovecote or Columbarium. Coach Pasture Pond is also known as Coach Pool. Where even company records differ, I use Water Garden rather than Watergarden; and Mereview Restaurant rather than Mereside Restaurant (which also avoids confusion with the earlier building). After 1950, 'the Park', 'the estate' and 'the site' mean the same in this book.

2. Imperial rather than metric distances are used throughout, as the reverse would be strange in the historical text.

3. Personal titles were always complex – and identifying them correctly was even more so! For brevity and informality *only*, except in the Forewords I have omitted academic qualifications (e.g., 'Dr') for company staff (it is in any case difficult to confirm these in the older record); and job roles are often uncapitalised and truncated. For historical precision, titles awarded later may be omitted in earlier life, e.g., to Winston Churchill before he became 'Sir' Winston.

4. The Stanley family often repeated names – and even *they* do not entirely agree on their early genealogy! At a later date, the sixth baronet Sir John Thomas Stanley's son bore the same name as his father but I have followed family practice and called the son Sir John Stanley until he became John Thomas, the first baron.

5. For names of drugs or other commercial products, '**Name**', when quoted thus, is a trademark, the property of AstraZeneca plc; while **Name™** means the same in *Appendix 5* but elsewhere indicates a non-AstraZeneca trademark (wherever known). Each drug's '**Brand Name**' is capitalized while its **generic name** is not. I generally use the former until final patent expiry. Only first uses in a section are in **bold**.

6. Medicines often have different brand names in different territories; only the UK ones are given.

7. Drugs are often marketed as salts (hydrochloride, phosphate, etc.). For simplicity, the salt names have often been omitted (but are essential to the marketed products).

8. UK rather than US spelling is used for both scientific and non-scientific terms (e.g., 'oestrogen' rather than 'estrogen'); although 'sulfur' is now recommended global usage rather than 'sulphur' and has also been used for sulfonamides, etc., rather than the historical spellings.

9. Double terms are mostly hyphenated for reading ease, e.g., 'anti-oestrogen' and 'anti-malarial', which may not always be standard, (but not antibiotics). Drug numbers have commas added (eg. ICI 46,474) for the same reason.

10. Some scientific terms have been approximated for brevity and to help non-expert readers. This means that some words may not be strictly accurate, e.g., if 'heart' replaces 'cardiovascular', or 'lung' for 'pulmonary'. Thus, no *definitive* scientific meaning should be placed on *any* statements made.
11. Terms with unavoidably technical meanings are explained *in situ* rather than in an inconvenient glossary; and as few acronyms as possible have been used!

I hope that scientific as well as non-scientific readers will understand why these approaches have been necessary.

Readers of many levels are thus invited to embark on the same voyage of historical and medical discovery that others made, often dramatically, in the past …

# Guest forewords: 'voices for the park'

It was impossible to decide on one person who could speak for all of Alderley Park's affairs, from great to small, from ancient to state-of-the-art, from personal to global.

Each of the following has therefore, at my invitation, generously and modestly offered his or her voice in an eminent congress of appreciation and hope for the Park and its future. Between them they have witnessed all of Alderley Park's human history, neighbours, natural history, estate life and staff community, as well as its scientific spheres, world-class discoveries and service to the world. Only a fraction of all their roles and distinctions is listed. I commend their views first of all.

*** 

The Park has seen many changes and this process continues. The constant, though, has been the dedication and commitment of the people. There have been many exceptional scientists who worked there and I have been lucky enough to interact and become friends with many of them. Entering the main atrium and seeing a statue of a Nobel Prize winner says it all. The contribution this site and group of people have made to medicine cannot be underestimated.

*Dr Andy Barker, medicinal chemist who discovered anilinoquinazolines as building blocks for kinase inhibitors in cancer treatment leading to Dr Keith Gibson's chemistry for the lung cancer treatment **'Iressa'**.*

Alderley Park is simply an astonishing place where brilliant people have come together and pushed the boundaries of science to deliver life-changing medicines. There is something quite magical about the place, and I am delighted that George, who is uniquely placed, has captured the story so far. I also look forward to a tremendously bright future for the Park and the stories that are yet to unfold.

*Graeme Bristow, current AstraZeneca Site Manager of Alderley Park.*

Alderley Park is undoubtedly a unique environment, with a fascinating history harking back to the days of Lords, Ladies and tenant farmers, to its more recent setting as the home of ICI and AstraZeneca. This succession of ownership had allowed the wider estate to remain relatively undisturbed, enabling much of its character to be retained. Careful and sensitive estate management over the years has helped maintain and improve the woodlands and parkland, and the estate today provides rich habitats for a huge variety of flora and fauna. It has been a privilege to help contribute to this, and I look forward to continuing this important work as the site moves into new ownership.

*Andrew Broome, current Estate Manager of the Alderley Park farm and estate.*

This book stands at a key moment in the long history of Alderley Park. The future plan for the site is as a national centre for bioscience and this remains the focus for the new owners, who were tasked with bringing 100+ new companies to Alderley Park and providing the ecosystem for science to flourish. The site will be a superb environment in which to work, live, play and to stay and thus securing the enduring legacy of the site for the region and the country.

*Dr Chris Doherty, Alderley Park Site Director for Manchester Science Partnerships.*

On starting at Mereside in 1967 I was utterly overawed, not just by the superb setting and facilities, but by the amazing, youthful enthusiasm of Mereside's scientists (especially the senior ones) for their research and open exchange of ideas, experience and techniques. Whilst the personal joys of drug discovery were greatest for those closest to the research, the collegiate culture meant that all could take pride in all the drugs that Mereside produced. I look with pride on our collective achievements and with gratitude for the marvellous camaraderie. This amazing site still has a great future ahead. I hope the new enterprises within Mereside can recapture the spirit of earlier years.

*Dr Mike Dukes, bioscientist behind **'Arimidex'** and the patent research for **'Nolvadex'**, the two world-leading breast cancer drugs. With his colleagues Professor Barry Furr and Dr Alan Wakeling, he helped build on Dr Arthur Walpole's historic foundation of cancer drug discovery.*

As a chemist in charge of the Beta-blocker team, it was a pleasure to work with Jim Black and his team of biologists. When I first saw the test result card on ICI 45,520, the compound made by Bert and Les which became 'Inderal', I wrote on the top of the card, 'This is it!' Indeed it was, and where it led medically, scientifically and commercially is history.

*Dr Ralph Howe, medicinal chemistry team member on the project leading to the heart drug **'Inderal'**; and the 'Chemist in Charge' of test results for it.*

Alderley Park has a special place in my heart having been privileged to work there for so much of my career engaged in the discovery and development of many important medicines. Fine scientists and great characters built a stimulating environment which nurtured ICI's infant pharmaceutical business to full maturity leading to the demerger which created Zeneca and then the merger with Astra to form AstraZeneca. Exciting times indeed! Hopefully, this glorious past will help inspire and lead to success the many new businesses which occupy such a splendid location.

*Sir Tom McKillop, FRS, first chief executive of Zeneca Pharmaceuticals and first Chief Executive Officer of AstraZeneca plc.*

Bruntwood and Manchester Science Partnerships are extremely proud to have acquired Alderley Park; both to continue its impressive legacy and help to shape its future. As an iconic place, deeply rooted in the science community, Alderley Park will evolve and be transformed into a desirable location to work, live and visit.

*Chris Oglesby, Chief Executive and founder of property company Bruntwood, the major partner in MSP, Alderley Park's new owner.*

From the Domesday Book to world-leading pharmaceutical research, Alderley Park has enjoyed a rich heritage of innovation. Today, the park finds itself at the centre of Britain's thriving life sciences' industry and provides an essential platform for businesses to deliver greater economic security to the North West. I have been privileged to have been able to contribute to the latest chapter of this incredible story.

*The Rt Hon George Osborne MP, Member of Parliament for Tatton (the constituency containing Alderley Park), Chancellor of the Exchequer.*

ICI gave birth to its Pharmaceuticals Division following from its WWII penicillin fermentation projects. It was supported through many loss-making years to what we will look back on as the golden era of the organic chemistry based pharmaceutical industry, which by the late 80s had led this international business to become half of ICI's profit base. The Alderley Park roll of honour is made up of a long list of medicines that have and continue to change the health of the world. I know of no other single site in the world that has such a proud record of innovation and medicines development. No one person 'creates' a medicine, it requires many different skills, dedication and a bit of luck. I hope that the new occupants of Alderley Park can use the BioHub to create as many world beating medicines as their predecessors.

*Dr John Patterson, CBE, led clinical development of **'Nolvadex' (tamoxifen)**, and was later Executive Director for Development.*

Alderley Park has benefited from dedicated naturalists documenting, recording and protecting the wildlife on the site for over five decades and providing that data to both local and national natural history organisations. This has revealed the diverse flora and fauna that occupies this area of the Cheshire landscape and highlights the need to continue to preserve the various habitats on the site for the future.

*Hugh Pulsford, current Cheshire county bird recorder, and a former global portfolio manager for **'Diprivan'**.*

As the 9th Baron Stanley of Alderley I am proud that my family's long association with Alderley Park continues to be remembered. Perhaps the recent changes can be seen in the historic context of leading intellects and dynamic individuals like Thomas Huxley, Matthew Arnold, H. H. Asquith, Bertrand Russell, Gertrude Bell, and Winston Churchill, who found time to think and relax with the Stanleys at Alderley Park, through to the more recent past where the scientists of ICI and AstraZeneca produced major medical advances. I hope I can look forward to the next chapter in Alderley Park's contribution to science and human endeavour with confidence.

*Richard Stanley of Alderley, Lord Stanley's family owned the Park for five centuries. He succeeded to the baronial title in 2013 on the death of his warmly remembered father, 'Tom'.*

As archivist of St Mary's Church, Nether Alderley and a keen local historian, I have much appreciated AstraZeneca's varied support within our local community. I have also been impressed, on my visits to their site, how AstraZeneca has maintained so well the historical features, such as the Tenants' Hall, the Water Garden and the Dovecote.

*Jenny Youatt, church archivist of the parish church of Nether Alderley, adjoining the Park.*

A notable omission from the above is Professor Barry Furr, OBE, who very sadly passed away as this book was being written. His contribution is much missed.

The scientific giants of Mereside past, who included Sir James Black OM FRS, Jean Bowler, Dr Dora Richardson, Dr Frank Rose FRS and Dr Arthur Walpole, still speak, of course, by their achievements.

## Academic voices

The above include none of the constellations of leading academics who have known the Park only briefly or intermittently over the last six decades. But the reflections on Park life of Professor V. Craig Jordan, OBE, FMedSc, international researcher on and advocate for **tamoxifen**, who began his career as a summer student in the Park in 1967, are also worth noting, as also are those of Professor Joe Sweeney, Head of Chemical Sciences at Huddersfield University, a regular consultant to AstraZeneca chemists (and a past scientific joint author with me). They may be found on pages 263–4.

# Author's preface

I was standing on a flower-sprinkled heath near Cape St Vincent, the south-westernmost point of mainland Europe, when a text arrived on my mobile phone. My wife Christine and I were in Portugal on a bird-watching holiday with A Rocha, the Christian conservation charity. It was noon on a glorious day of blue skies, windless despite the creamy breakers pounding the sheer cliffs. Minutes earlier, a magnificent short-toed eagle had flown past, followed iridescently by the first bee-eater of the summer. I was surprised and irritated to find I even *had* a phone signal there. Then I read the text.

Nine months earlier I had left, after a fulfilling thirty-four years, my job as a synthetic chemist at Alderley Park in Cheshire, UK. The majority of my career there was spent on cancer drug research at (latterly) AstraZeneca plc, once the fifth biggest pharmaceutical company in the world and the source of world-class medicines for the world. At its peak in 2007 the Park employed over 6,000 people; I had left behind numerous friends and colleagues, many of them more clever and gifted than me. Incidentally to my career there, I had acquired knowledge of the wildlife and history of the ancient estate of Alderley Park; this too seemed a thing of the past. Yet the text on my phone screen brought it all joltingly back. It was from a family friend with a link to the City; and it read:

Presumably you know that AZ is announcing permanent closure at Alderley Park at 1pm …?

Its author was wrong; I had not seen *this* change coming – even though his claim was exaggerated (the company's research was to relocate but 700 non-science staff would remain). I was stunned, though much less so than thousands of current employees that afternoon, which was a chill one back in England. There was slight relief that the terms and timescale of the upheaval might have been worse. Yet I could only reflect this was yet *another* great trauma – for it was not the first – in the long history of the Park. Until recently, our company of AstraZeneca plc had considered the Park its 'jewel in the crown'. But jewels have dramatic histories and gemstones are often re-cut; Alderley Park had proved again to be no exception.

The decision to close one of the leading scientific sites of our day is not one on which I am placed to comment, despite the pain of many old friends. And we all understand the huge economic forces, the scientific challenges, and the still unmet medical needs that drive change. Yet the hole is a large one. Thankfully, the challenge of turning Mereside into a thriving science park is making promising progress. And the Park itself will surely survive, in some form. Yet its entire amazing story should be told, while it still can be.

## Previous accounts

Many have published brief accounts of Alderley Park from an external viewpoint; and others have created unpublished in-house records.

A few pages on the Park's work appeared in Carol Kennedy's diamond jubilee history of ICI of 1986, and a few on the Division's origin in W.J. Reader's 1975 formal ICI company history (to 1952). A history of CTL appeared in 2008. But no full book on all of Alderley Park by someone from within the life of the Park itself has appeared since Nancy Mitford's collections of Stanley family letters in the 1930s (the famous Mitford sisters were descendants of the Stanleys of Alderley).

And no-one has previously, to my knowledge, even attempted to tell Alderley Park's whole tale, except for one person – who could have done it far more justice than I have. That was Dr Dora Richardson, the medicinal chemist who at the Park in 1962 made the first ever sample of **'Nolvadex' (tamoxifen)**, later the world's most prescribed hormonal cancer drug. She was one of the Park's outstanding (although in her case, sometimes overlooked) scientists. Many years ago, I had the privilege of being invited, by her over the phone, to help her in such a project using my knowledge of the Park. I gladly agreed; but heard no more before her death in 1998. But perhaps some of her vision might be recaptured here.

## The story

Why have I set out to recall this story before some details of it fade? Firstly, for *that* reason alone; we should recall what Alderley Park discovered and what has been discovered about Alderley Park (the book's title can read both ways). The Park's *good* stories have often been overlooked, for more than eighty years; they should be told. As novelist Thomas Berger put it: 'Why do writers write? Because it isn't there.'

Secondly, I have a close personal connection with the Park. Over thirty years ago, I held in my hand the very first sample of a (different) new class of potential breast cancer drug. That particular research led nowhere useful, otherwise I would never have mentioned it, for in research all success is team success. Yet just a decade later, I had cause to reflect deeply on that memory when, on the afternoon of our fifteenth wedding anniversary, my wife underwent surgery for breast cancer. At the time, we had three children below teen age. After surgery she was treated with Dora's drug for five years: a drug discovered by a team I later joined, in a lab I worked in myself, on a project in whose successor I took personal part. Last year, Christine and I celebrated a healthy thirty-second wedding anniversary, for which we thanked God and then science. Very many people work to defeat the terrible scourge of cancer; few have the privilege to do so directly. I, for one, will not forget Alderley Park.

In writing about it, nevertheless, I offer only an outline of its huge story: not least because I am, sadly, what my father used to call 'a jack of all trades and master of none'. Worse still, I am more a beachcomber than a historian, one who merely displays attractive flotsam and evidence fallen from great enterprises floated by the great. Obviously, my limited view needs higher assistance; I have acknowledged help from many people. Yet there will surely be errors; I can only hope they are not irritating ones.

## Informal

In Appendix 3, I explain my general approach. But this book is not a formal study of the science or its results, or of medical possibilities. Nor does it plumb the depths of the huge ethical dilemmas of pharmaceutical research and drug testing, except in personal testimony. This is not because any of us can overlook them; but I expect all readers will agree I could only trivialise them by attempting any comment here. That belongs in wiser writings.

I have celebrated everything that has seemed worthwhile. But I have described little of the Herculean labours that have so far proved barren; although my focus on the successes should inspire rather than dismay those of us whose work (as yet!) has borne little fruit.

I also explain properly in Appendix 3 why I have named *some* individuals at *some* points. It is essential to say that I have chosen to do this only as the least *bad* option (and as the only readable one) for telling the story of all. But I have also tried to illumine the whole human Park and the teamwork across its community and history – to include *everyone* who worked in the Park – by offering brief generic snapshots of each business area; I hope these will be seen as respectful recognition, rather than as insultingly inadequate.

And I have included the whole company record (from the first creation of ICI's pharmaceutical branch in 1936, long before Alderley Park was bought); for, without it, the towering characters of Mereside's opening years might appear colourless and their programmes rootless. In this 90th anniversary year of the start of ICI, and the 80th anniversary of 'Pharms', the great stories should be told in full. (Sadly, I could not cover the work of AstraZeneca's Macclesfield Works or other later company sites away from the Park, or the stories of discoveries from these.)

Unhappily, my account is unbalanced by containing rather too much of my own experiences – and perhaps of my own science of chemistry. I apologise for the latter flaw, to aggrieved colleagues from other sciences. But all stories can be typed by only one set of fingers.

Finally, I have tried to make this detailed record readable, by inserting many notes and anecdotes of what I have generally called 'Life in the Park'. A few of these may merely reflect my own erratic sense of humour or pathos; and one or two may irritate; but I hope that they will enrich rather than belittle.

For, Alderley Park was always its people, first of all.

The new 'Patent Wall' at Radnor Reception, Mereside

PHOTO ASTRAZENECA

# Introduction

The ancient, elegant and surprisingly little-known estate of Alderley Park in Cheshire, UK, eighteen miles south of Manchester, lies below the westernmost foot of the English Peak District. Since the fourteenth century its parkland, woodland, water, gardens and buildings have increased to cover some 400 acres (1.5 square kilometres), eighty per cent of which is still green. The Park's colourful history has involved kings, lords, prime ministers and at least one Nobel prize-winner, pharmacologist Sir James Black. The scientific research done within its bounds has provided medicines for millions and touched perhaps every country in the world. Its entire story has never been told.

In Tudor times, the Park became the home estate of the Stanleys of Alderley. This was the younger arm of a prestigious family. Their elder siblings lit up the pages of history and Shakespeare when one of them lifted Richard III's fallen crown of England (from below a Bosworth battlefield bush) and his brother placed it on the head of Henry Tudor. Yet the quieter Alderley Stanleys also acquired a substantial amount of England. When they found the Park, they created a domain of green beauty that drew renowned visitors as well as generations of naturalists and artists.

The Stanleys of Alderley became barons and their descendants illuminated an era; one of these condescended to marry Winston Churchill, who as a guest planted a tree in the Park soon after the engagement, a sweet chestnut that still grows, oddly small but determined – like its famous plantsman.

Those who followed the Stanleys in the Park preserved much of its beauty. But nobility and power were replaced by striking medical adventure. The finest scientific minds were intrigued by what Sir James Black later described as 'this fairytale place that ICI were building at Alderley Park'.

The Mereside laboratories grew from a remarkable ICI Pharmaceuticals Division project into the heart of demerged Zeneca, then of the research colossus of AstraZeneca plc. Mereside saw the invention of some of the world's bestselling heart and cancer drugs, amongst a gallery of other important discoveries. Many thousands of clever people researched and worked in Alderley Park, over more than five decades; and the whole world's health benefited.

Yet, time and chance happen to all. The Park's guard has changed dramatically more than once; and now has again. Yet the story of the Park, now refracting into a rainbow of parts, still continues – we hope, for much longer.

Any of the parts of Alderley Park's story could fill volumes: but this is not one of them. Only a rich gleam from each strand of Alderley's tapestry of time can be captured here. The Park's ancient history, living nature, aristocratic elegance, scientific adventure, medical discovery and human individuality are all described – but only in broad detail. That is, this account tells many stories badly and few well. You, interested reader, may well be able to add more as you read.

In any case, you are invited to share in that spirit of invention for which all who have lived, worked and researched here are remembered. For, in the great and changing estate of Alderley Park, discovery has always been the order of the day.

# Brief timeline

| | |
|---|---|
| Before 1066 | Held by two Saxon overlords, Godwin and Brun. |
| From 1087 | Held by two Norman overlords, Bigot and William. |
| Until the 1200s | Held by the Lords of Aldford & the Barons of Halton then Montalt. |
| Until the 1400s | Held by the Ardernes and other families, plus a local monastery. |
| Until the 1600s | Owned by various Stanleys plus the Fitton and Bentley families. |
| Until the 1800s | Owned by six Stanley baronets and the Deane family. |
| 1779 | Alderley Hall (outside the Park) burnt and the Stanley family moves into the Park. |
| 1931 & 1933 | Alderley House mansion also damaged by fire, then demolished. |
| Until 1938 | Owned by six Lords Stanleys of Alderley. |
| 1938–1950 | Left in the hands of Hambling, Crundall & Company. |
| **25 October 1949** | Outline planning permission request filed by ICI. |
| 1950–1993 | ICI Pharmaceuticals Division. |
| **22 June 1950** | Purchase of Alderley Park by ICI announced. |
| **12 October 1950** | Outline permission officially granted for Mereside. |
| **February 1955** | Construction of Mereside started. Former Deer House demolished. |
| **1 October 1957** | Mereside officially opened by Lord Waverley. |
| **December 1961** | Permission granted for building of Alderley House by Radnor Mere. |
| **1962** | Planners opt for different Alderley House site in the south of the Park. |
| **1963** | Industrial Hygiene Laboratory (later CTL) settles in Alderley Park. |
| **1964** | Alderley House completed. |
| **1973** | Block 19 extension and new Mereside Library opened. |
| **1976** | ICI reaches 150,000 trees planted in Alderley Park. |
| **1977** | Former Tenants' Hall converted to a Conference Centre. |
| **Early 1980** | Block 19D completed. CPU block built on legs over Radnor Mere. |
| **1985** | Alderley Park farm changes back to a normal sheep farm. |
| **September 1988** | Block 11 (Clinical Data Centre) opened. |
| **17 October 1988** | Nobel Prize awarded to Sir James Black for discovery of beta-blockers. |
| **October 1988** | First site History and Nature Trail opened. |
| **1989** | Radnor Mere dam renovated to EU standards. |
| **7 March 1989** | Block 24 (Drug Kinetics) opened. |
| **9 June 1989** | Former Tenants' Hall renamed the Sir James Black Conference Centre. |
| **9 July 1990** | Mereview Restaurant opened. Block 26 opened. |
| 1993–1999 | Zeneca plc. |
| **1 June 1993** | Ownership of Alderley Park passes from ICI to new company Zeneca plc. |
| **1994** | Woodland regeneration programme begun. Mulberry's opened. |
| **1998** | Alderley House Stage IV completed. Park enlarges to 400 acres. |
| 1999–2014 | AstraZeneca plc. |

| | |
|---|---|
| **1 June 1999** | Ownership of Alderley Park passes to AstraZeneca plc. |
| **2001** | First tiered car park opened. Lower courtyard buildings refurbished. |
| **2002** | Block 5 completed. Block 23 Safety Assessment building completed. |
| **2003** | Parklands opened. New energy centre opened. |
| **2004** | Blocks 50–52 opened. |
| **2005** | Logistics centre Block 109 opened. |
| **2006** | RA1 (Blocks 30, 33 & 35) and new Radnor Reception opened. |
| **2007** | 50th anniversary of Mereside. Staff peak at 6000+. CTL closed. |
| **2008** | Alderley Park Conference opened. Alderley Edge bypass begun. |
| **21 March 2013** | Closure of R&D in Alderley Park announced, with move to Cambridge. |
| **19 November 2013** | Death of Tom Stanley, 8th Baron Stanley of Alderley (1927–2013). |
| *2014–* | Manchester Science Parks (now Partnerships) |
| **1 April 2014** | Sale of Alderley Park to Manchester Science Parks is completed. |
| **Oct 2014** | Manchester Science Parks changes to Manchester Science Partnerships. |
| **c.2018** | Likely end of AstraZeneca's R&D in the Park. |

Nether Alderley

Mill

Church

Parkhead Pond

Hocker Lane

Beech Wood

Coach Pasture Pond

Radnor Mere

Church Lodge entrance

Sold in 2014

Alderley Park Farm

Conference Centre

Car Park

Mereview Restaurant

Larchwood

former CTL site

Bypass

Mereside

Old Drive or Chestnut Avenue

Car Park

Heatherley Wood

A34

Bentley Brook

Parklands

Alderley House site

Tenants Hall

Boys Walks

Bollington Lodge entrance

Water Garden

Ice House

Arboretum

Serpentine

Mulberry's

former Matthews site

Eagle Lodge entrance

ALDERLEY PARK

in 2014

Map © AstraZeneca plc & George B. Hill, 2016

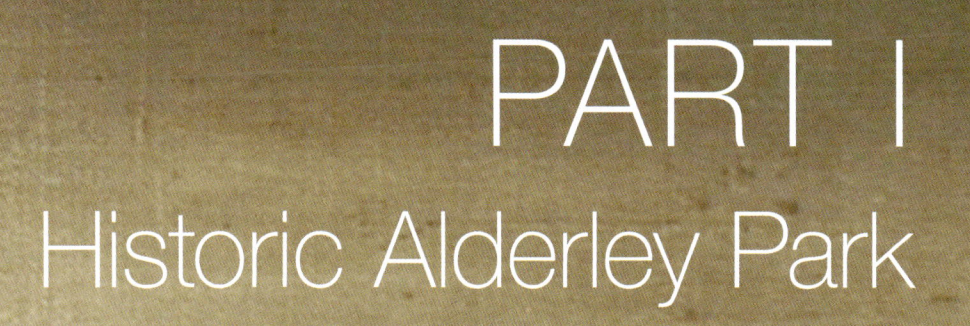

# PART I
# Historic Alderley Park

'If I want to get anything out of an antique hunk of masonry lying in a field with a chain around it, I have to shut my eyes, and tell myself, very slowly and emphatically … "This – is – a very, very old piece of stone, and it's terribly fascinating and marvellous."'
*J. Basil Boothroyd*

# The early park & land

## When was Alderley Park named?

The fascinating history of Alderley Park begins in 1390. In that year King Richard II was ruling England, while taking ruthless revenge on the old guard who had controlled him until then. Geoffrey Chaucer was vainly trying to complete his *Canterbury Tales* before he died.

Europe's population was 50 million, one fifteenth of that now. Queen Margaret of Denmark was about to unite the countries of Scandinavia, with Swedish support. But great swathes of mainland Europe were being plundered by armies of mercenaries. Many Jews were being expelled from France. The Turkish armies had crushed the Byzantine Empire and were threatening Constantinople. In Egypt, the Mamluks had recaptured Cairo; whilst their enemy Tamerlane, heir of the Mongols, had conquered Persia, where he had piled up 30,000 heads of the citizens of one obstinate city outside its walls, and would shortly invade India and sack Delhi.

In China, the Ming dynasty was creating the finest porcelain in history. The Ashikaga Shogunate was dominating Japan. In New Zealand, the Maoris were pushing the flightless moa to extinction; and Australia was still a land outside record. In North America the Apache and Navajo peoples were migrating south west across the Great Plains. In Mexico the Aztecs were building their greatest temple in Tenochtitlan. In South America, the Incas were founding the most glittering empire of the age. In southern Africa, the huge stone city of Great Zimbabwe was at the height of its power.

In England, an age of fine cooking had dawned! The dish of the moment was blancmange, made from chicken and rice boiled in almond milk and flavoured with aniseed.

Only forty years had elapsed since the greatest medical emergency of the millennium, the Black Death. Probably some 40% of the population died in England (where there are still abandoned villages that have never been rebuilt); in Spain twice that proportion succumbed. The cause was the bacterium *Yersinia pestis*, carried by the fleas of black rats, which caused several sorts of plague including bubonic plague in the lymph nodes, septicaemic plague in blood vessels and pneumonic plague in the lungs. Interestingly, plague is highly responsive to antibiotic therapy – commonly with streptomycin or chloramphenicol – and antibacterial research was carried out for many years in Alderley Park.

And, in that same year of 1390, 'the park of the Lords of Alderley' was first mentioned, in a historical deed created by one of its near neighbours, Emma, the widow of Adam de Acton of Over Alderley.

### Three parts

To begin with, 'Alderley Park' applied to one of three parts of the estate as it was later known. The second was added in 1602; and the third, the Alderley House site (whose earlier history is unclear), around 1750.

The Park's later history centres on the famous Stanley family. Yet to understand its early history (up to 1602), we must start with that of the two parishes of Alderley within which it stands: Nether and Over Alderley. Nether and Over mean 'lower' (or 'inferior') and 'upper', respectively. Most of the present Park is within Nether Alderley parish.

## Nether Alderley

The beautiful village of Old or Nether Alderley, which adjoins Alderley Park, has been the site of settlements since the Stone Age. Later, a Roman military road carried salt from the Cheshire mines through Alderley to stations in the Derbyshire Peak. The road's original surface or *pavimento*, now buried, has a width of 5' 3", the usual breadth for three soldiers marching abreast, which was the custom in the legions.

An important crossroads existed where this road crossed the main north–south highway, an old trade route of Saxon origin. The roads met at the foot of Artist's Lane, where the stub of the old Market Cross still survives.

Alderley Park in 2005 from the SSE with Nether Alderley and Alderley Edge beyond
PHOTO ASTRAZENECA

A Saxon settlement was founded around 500, probably beside the mill stream. Its focus was probably a precursor of the beautiful medieval Nether Alderley Mill still visible beside the former A34. Other ancient settlements have disappeared, although an aerial survey of Alderley Park in the 1990s observed faint ridges running east–west across the field in front of Alderley House, suggesting medieval occupation there.

## Alderley in the Domesday Book

The name Alderley appears more than once in the Domesday Book of 1086, in the form of *'Aldredelie'* (Aldred's meadow or clearing) or *'Aldredslic'*.

We do not know who Aldred was, though her name shows she was English. The following translations are thought to relate to Nether and Over Alderley respectively:

… the same Bigot holds Aldredelie. Godwin held it as a free man … There is one acre of meadow, a wood one-and-a-half leagues long and one league broad and two hedged enclosures. In the time of King Edward (the Confessor) it was worth 20 shillings, now ten shillings. (The Earl) Found it waste.

William son of Nigel holds of the Earl Aldredslic. Brun held it and was a free man … It was and is waste. There is a wood two leagues long and two leagues broad. In the time of King Edward it was worth twenty shillings.

By the time the Domesday surveyors arrived, Nether Alderley was held by Bigot, a Norman who for his services to William during his Conquest of 1066 had evidently been rewarded with a fiefdom. Prior to him, Godwin – and his countryman and neighbour Brun – had held their lands as Saxon freeholders until dispossessed by the Norman invaders. It seems likely they fought for their freedoms and lost their holdings and perhaps their lives – the first recorded transfer of the occupancy of Alderley Park, and doubtless the most brutal. The lands of Bigot and William Fitz-Nigel can be traced through numerous holders, but remained largely distinct (in the Park's story) until the seventeenth century.

## The manor of Nether Alderley

The land held by Bigot passed to his son Hugh, then to their successors, the lords of Aldford. The inheritance included not only the manor of Nether Alderley but also its church.

About 1220 it passed by marriage to Sir John de Arderne. In 1337 Sir John de Arderne was referred to as the Lord of Nether Alderley.

In 1340 Sir John granted the lands to his brother Peter; and it was a female descendant of his (Matilda) who brought Alderley lands for the first time (and only temporarily) into the hands of the Stanley family when she married Thomas de Stanley, the third son of John I Stanley, King of Mann.

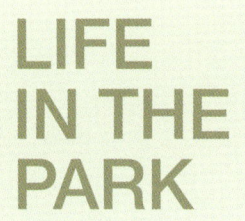

### Deer address

**The first identifiable occupants of Alderley Park had antlers! It may have been the Ardernes who created a medieval deer park that was in existence by 1423 at Nether Alderley, probably in the northern part of the present Park (although no map shows it at any date before 1777).**

LIFE IN THE PARK

## Stanleys, Eagle and Child

Who were the famous Stanleys?

It seems clear they were not quite what they suggested! They claimed to be descended from Charlemagne and to have come over with William the Conqueror. The Alderley Stanleys can certainly later claim descent from Charlemagne (through the wife of the first baron) and the family is related to several English royal lines. But an oft-quoted 1774 history of the Stanley family makes them Norman and gives their original descent from Adam de Aldithley and his sons Lydulph (or Liulf) and Adam, from a place called Aldithley in Normandy.

In fact Aldithley (now Audley) appears to have been a village in Staffordshire! Lydulph (born around 1115) and Adam were thanes (an early Saxon form of knight). The family presumably later invented a fictitious (or at least more prestigious and perhaps safer) Norman origin story. Some believe they were Normans of lower rank, rather than Saxons. But Staffordshire had been laid waste in a revolt and attracted few Normans, although Lydulph certainly acquired his property in return for military duties (called 'socage') to Bertram de Verdun, an important baron under the Conqueror.

### Family name

Around 1190 the latter's grandson gifted Lydulph's son Adam another Staffordshire manor, Stanleigh (now Stanley, five miles SW of Leek). The name means a craggy or stony meadow. Later, Adam exchanged his new Manor of Stanleigh with his cousin William, son of Lydulph's brother. Before that time, surnames were rare and it was William (1170–1236) who adopted the new manor's name, partly in honour of his wife, an heiress of ancient Saxon descent. By 1203, now known as 'William de Stanleigh', he had become the first proper bearer of the famous Stanley name.

Thereafter, many of William's descendants became soldiers, fighting in numerous wars including the invasion of Brittany with Henry III in 1230, for the Stanleys were a forceful clan.

Peter Stanley, in his book *The House of Stanley from the 12<sup>th</sup> Century* comments that:

> Much of the family's rise to power and influence can be attributed to their military reputation which led to successive Kings seeking their support and endeavouring to retain the family's loyalty by granting them generous gifts of honours and lands.

In 1284, on the death of his wife's father, William de Stanley acquired his father's office and his extensive estate between the Rivers Dee and Mersey. William and Joan then moved from Stoneleigh to Stourton. Their son John died from wounds received in the battle of Crecy in 1346.

# LIFE OUTSIDE THE PARK

## Elopement

A typically bold Stanley story comes from 1282, when William's (probably) grandson, another William, fell in love with Joan de Baumville, whose father Sir Philip held the hereditary office of Master Forester of the Wirral. But a marriage had been arranged for Joan with her stepmother's son, for which she had no liking! While Sir William and her stepmother were at a banquet at Stourton on the Wirral, Joan slipped out to where William was waiting with saddled horses. Together they galloped secretly to Astbury, where they were wed in the church by William's uncle, John de Stanley.

## Sir John Stanley KG, and the Kings of the Isle of Man

This John's grandson, another and more famous John (c.1350–1414), just escaped the Black Death that had swept the Wirral in 1349, and married in 1385 Isabel Lathom, heiress to the estates of Lathom and Knowsley in Lancashire.

John Stanley travelled all over Europe, famed for his skill in single combat and in tournaments. But he and his brother had violent reputations; when he was accused of killing a neighbour he needed the King's pardon (given because of his military service) to escape judgment. In 1385, Richard II made John Lord Lieutenant of Ireland and appointed him Steward and Master Forester of Macclesfield and Governor of Chester.

In 1399 Henry IV granted the Isle of Man as a fiefdom to Henry Percy, Earl of Northumberland, who (improperly) styled himself 'King of Mann'. When Percy rebelled against the King, Percy's 'kingdom' was confiscated and the tenure was granted in 1405 to John Stanley as a reward for his services to the King at the Battle of Shrewsbury, in return for two falcons at each Coronation. He thus became the Stanley family's first king – John I of Mann (John I Stanley). He was also made a Knight of the Garter, the highest English order of chivalry for a commoner, in the same year.

The Isle of Man was ruled by more Stanleys (not of the Alderley branch) for another 331 years. Thus the Stanleys acquired their first and only kingdom, even if it was a long fought over – and then relatively worthless – island in the Irish Sea. The stained glass windows in the Sir James Black Conference Centre (Tenants' Hall) still bear the symbol of the Legs of Man.

The Legs of Man window in the Tenants' Hall
PHOTO GEORGE B. HILL

## The Eagle and Child

The marriage of Sir John Stanley KG had brought into the family the Eagle and Child, the ancient crest of the Lathoms of Northenden. The legend of this arises from a positive tangle of possible tales!

The usual story is that a child was heard crying by a childless member of the Lathom family, a Sir Thomas Lathom, while walking in Tarlescough Woods with his wife. Servants found the boy found lying in grass below an eagle's nest – or, in another version, in the nest; or, in yet another, the child was seized from his cradle in a courtyard then dropped back into the yard unhurt. That sounds unlikely, although children have been dropped by eagles, as in a case in Italy in 1959 when a three year old child was carried away and found after 36 hours. Legends (perhaps related ones) date back to at least Alfred the Great. Sir Thomas called the boy Oskatel (his wife's surname) and brought him up as his own.

A less creditable story suggests that the child was Sir Thomas's own from an illicit liaison with a gentlewoman on his estate, and that because he had no son he arranged the whole thing, to persuade his wife to accept the child! In this version he made the boy his heir (and Oskatel de Lathom certainly existed) ahead of his legal daughter. Sir Thomas planned to leave his estate to Oskatel, but is said to have confessed the scheme on his deathbed. Or it was under pressure from the Stanleys who threatened to cancel the marriage to Isabel, his only legal heir.

But this story also blurs into the historical fog, for Isabel was not the only legal heir of John Stanley's father-in-law. If there is any truth behind the legend, perhaps a more likely candidate for the Sir Thomas in question is Isabel's brother, to whom she was also eventually heir (he died in 1383) although only after the death of his posthumous daughter, Ellen in 1390. A tangled tale, indeed!

After Sir Thomas's death, it appears from legal records that Sir John Stanley may have tried (characteristically) to get his way illegally, and 'muscled in' on the Lathom property. But Ellen had been made a ward of John of Gaunt, who would have been a daunting antagonist to Sir John, as one of the richest men who have ever lived. Parliament ruled against Sir John, doubtless with his opponent's backing; and Sir John was forced to litigate around 1386.

Even after Ellen's death, when Isabel should have expected to inherit by right, there was more delay (perhaps due to Oskatel) and the result was a sixteen-year intermission until 1406 before the estates of Lathom and Knowsley were finally granted to Sir John in his wife's right. (Oskatel took two other manors; he and his crest appear in a stained glass window of St Wilfred's Church at Northenden).

The Stanley family are said then to have adopted the Eagle and Child as a badge and constant reminder of how easily the Lathom lands might have been lost to them. The original Lathom eagle had been looking backward; the Stanley one looked down on the child as if about to devour him, to emphasise the triumph of the true heiress.

Stanley of Alderley crest
COURTESY LORD STANLEY OF ALDERLEY

A Stanley Eagle and Child beside Eagle Lodge

# LIFE IN THE PARK

**Another Eagle!**

The above histories should not be confused with another similar adventure of a juvenile Stanley in Italy in 1701, when a baby of the family was snatched and then released by a hungry eagle. This event is commemorated by a tall sandstone obelisk, erected by Sir Edward Stanley in 1750 near the head of Radnor Mere (outside the present park). It bears the Stanleys' spread eagle – but no child. The only link between the different tales is that more than one family baby appeared tasty to a bird of prey!

The *Eagle and Child* was also the name of a former inn in Nether Alderley village, known locally as the 'Brid and Babby'. (The sign that used to hang outside the pub was found in an outbuilding in Alderley Park in the 1950s). Stone eagles still appear in various places around the Park itself and are described later.

## John II of Mann

Sir John Stanley KG had never visited the Isle of Man, but his eldest son Sir John (1390–1437), John II of Mann, visited it several times. On two occasions (in 1417 and 1422) he put down rebellions there. He took a great interest in the island with beneficial results. Amongst other things, he introduced written laws there for the first time.

It was his second brother Thomas, whose wife was Matilda de Arderne, who was probably the first member of the family to hold lands at Nether Alderley when she inherited them. Nevertheless, they did not belong to his line or remain Stanley property for long. Thomas's son Sir John Stanley fought and was created a knight at the Battle of Tewkesbury in 1471; then his son (yet another Sir John) sold the Nether Alderley estate between 1474 and 1495 to Thomas's great-nephew, the celebrated Sir William Stanley of Holt, the richest man in the kingdom.

## Sir Thomas Stanley, Controller, and his three sons

Sir William was the grandson of John II and the second son of Sir Thomas Stanley who was Controller of the King's Household (Chancellor to Henry VI) and a baron.

This Sir Thomas (1405–1459) had three sons, and despite the cascade of Stanleys up to this point they are worth paying attention to, as all three – Thomas, William and John – are key figures in the story of the Stanleys and Alderley Park. Their story will be told shortly. In the meantime, we shall merely note that Sir William was in the end the unluckiest of the three, despite his wealth. Despite earlier loyalty, he offended King Henry VII by seeming to support the rebellion of Perkin Warbeck in 1495. Although the charge was very doubtful (it was said the king wanted his wealth), it led to the loss of his head that year. This was an exception to the family's usual efficacious attempts to stay on the right side. In general, as Peter Stanley explains,

> The Stanleys managed to steer a successful course during those troubled times, always ending up with the winning side …

## More owners

Forfeited to the Crown after Sir William's beheading, the title to the Nether Alderley manor was granted successively to a William Brereton (who also mislaid his head), to Edmund Peckham, the King's treasurer; and to Margaret Moreton, one of the Gentlewomen of the Palace.

Finally, in 1557 it was granted to Sir Edward Fitton of Gawsworth. It was probably his son of the same name who sold it for the sum of £2,000 on the twentieth of June 1602 to Sir Thomas Stanley (1576–1606) of Over Alderley and Weever, of whom more later. With this purchase the manors of Nether Alderley and Over Alderley were thus first united under the Stanleys.

## The manor of Over Alderley

But how had they acquired the Manor of Over Alderley?

The most likely version is that the lands of William Fitz-Nigel, mentioned in the Domesday Book of 1087, passed first to his successors, the Barons of Halton. Before 1294 they were in the possession of fellow barons, the Montalts, then at this date they came into the hands of the self-styled 'Lord of Alderley', Roger Throstle. In 1297 they passed to the Downe family, then in 1337 they were acquired by the Ardernes, thus temporarily uniting both Nether and Over Alderley under the ownership of the same family. However, although as recorded above the Ardernes retained Nether Alderley until the fifteenth century, they soon relinquished control of Over Alderley and by 1360 the former Fitz-Nigel lands had passed by marriage to a family by the name of Weever.

WEEVER · ARDERNE
ORREBY · MONTALT

Arderne and Weever coat of arms in the Tenants' Hall windows
PHOTO GEORGE B. HILL et seq.

## First mention

It was during this period that Alderley Park received its first historical mention in deeds of 1390, when Emma, the widow of Adam de Acton, granted all her lands in Over Alderley 'lying near the park of the Lords of Alderley' to her son Nicholas.

The last of the Weever line was Elizabeth de Weever, an heiress who was a minor when her father died in 1445 and so became a ward of King Henry VI, who granted her hand in marriage between 1446 and 1461 to (yet another!) Sir John Stanley (who died in 1485). We shall follow the story of the descendants of Sir John pictorially, as shown on the wonderful heraldic shields (which bear the coats of arms) preserved on the oak wainscot paneling in the Tenants' Hall. On the shields, the small crescent distinguishes the Alderley Stanleys from other branches of the Stanley family. Each shows the arms combined through that marriage.

But this time we are able to tie all the loose ends together. For this Sir John was the *third* son of Sir Thomas, Controller of the King's Household named above for his three famous sons. And Sir John was the ancestor of the Stanleys of Alderley.

But if the hapless Sir William was the second of the three sons and Sir John was the third, then who was the first son and the most famous Stanley of all?

JOHN STANLEY
b. 1440 ? d. 1480 ?
ELIZABETH WEEVER

## Senior and junior Stanleys; Shakespeare

The best known of all the Stanleys was Sir Thomas Stanley KG (d.1504), who became the first Earl of Derby.

He was not of the Alderley branch of the family, which descended from his younger brother Sir John. But his story was celebrated by all of the family, so a little of it, and that of his descendants, is worth including here.

He was made the first Earl of Derby after the epochal Battle of Bosworth Field in 1485. This was fought between Richard III and Sir Thomas's own stepson Henry Tudor.

In the War of the Roses, Sir Thomas had no desire to get involved in the struggle between the rival royal houses of Lancaster and York. He walked a hazardous path for many years. Both Sir Thomas and his brother Sir William secretly swore fealty to both sides. At Bosworth they contrived to stay on the outskirts until the outcome was apparent, then they charged onto the battlefield. In *Richard III*, Shakespeare immortalised the dilemma Sir Thomas faced:

**King Richard**: What says Lord Stanley? Will he bring his power?

**Messenger**: My lord, he doth deny to come.

**King Richard**: Off with his son George's head!

**Norfolk**: My lord, the enemy is past the marsh. After the battle let George Stanley die.

After Richard fell in the battle, Sir William found King Richard's ornamental crown under a bush and gave it to Sir Thomas. (Legend has it that Richard died with the word 'traitor' on his lips.) Sir Thomas briefly knelt, then placed it on the head of his stepson, proclaiming him Henry VII.

In return, Henry made him Earl of Derby two months later. Was Sir Thomas's betrayal of King Richard treachery? Perhaps; but it is known that had Richard won the battle he planned to rid himself of the Stanleys once his throne was secure.

## Senior and junior Stanleys

The new Lord Derby was, as stated, elder to his brothers William and John. Later the Stanleys of Alderley, although descended from John, mistakenly claimed to be the senior branch of the family, a claim recorded by a descendant of the family, the novelist Nancy Mitford, eldest of the famous Mitford sisters. Her mistake probably arose from confusion with yet *more* Stanleys – the Stanleys of Hooton, Chester, who were senior to both of the brothers.

The Earls of Derby themselves went on to produce many famous historical figures. The second Earl – Shakespeare's George – lived peacefully until 1503, with his head intact. His brother Sir Edward later commanded the rear of the English army at another great battle, Flodden Field in 1513. His son was uncle to both Ann Boleyn and Catherine Howard.

The third Earl was a great survivor, who by judicious changes of loyalty managed to stay in favour successively with Cardinal Wolsey, Henry VIII, Queen Mary and Queen Elizabeth.

## Shakespeare himself

Most fascinating of all was the sixth Earl of Derby, William Stanley, known as the 'wandering earl' and famed as a soldier and traveller. Peter Stanley reports 'some very interesting speculations' that he had an even closer link with Shakespeare, in that he might have been the real writer of his plays! Or at least, one of those behind them.

Most scholars doubt the link; but many have puzzled over how Shakespeare, who never left England, was schooled only to age eight, then became a glover and butcher until he turned actor, could have written early plays with an excellent knowledge of France and Italy and also shown a deep one of Latin, Greek, the law, history and English. Tudor nobles at the time customarily published poetry or plays under an agent's name and William Stanley is known to have written plays in secret for his brother's group of actors whom Shakespeare joined. Could he have chosen William Shakespeare for their common initials? At least two early distinctively Shakespearean poems appeared under the initials of 'W.S.', at a time when Shakespeare himself was still a lad in his father's slaughterhouse. And Shakespeare himself suddenly ceased to publish anything in print after 1594, which was the year in which William Stanley's father died and William inherited the peerage. As the new 6th Earl, William was an important man, potentially in line to the throne; so publishing plays even under a pen-name would have become too risky.

Later, Shakespeare himself suddenly became rich and left London abruptly – had he attempted to blackmail someone and been bribed to keep silent? Oddly, when Shakespeare died, no-one outside his home town paid any notice. In 1599, just as Shakespeare's finest comedies were appearing, a letter comments that: 'The Earl of Derby is busy only in penning comedies for the common players.' And the mystery will remain, for all the contents of the Earl's personal library were destroyed during the Civil War.

## The Stanleys of Alderley at Alderley

John Stanley, ancestor of the Lord Stanleys of Alderley, became owner of the manor of Weever and of the Weever lands in (mainly) Over Alderley. After 1512 the lands in Over Alderley became known as the manor of Over Alderley.

We now move on three generations. The great-grandson of John Stanley was another Thomas Stanley. In his case, however, we start a new era. For it was he who brought the famous house of Stanley to Alderley for a prolonged stay.

THOMAS STANLEY
b. 1470? d. 1526.
DOUCE LEVERSEDGE

THOMAS STANLEY
b. 1507. d. 1556.
JOAN DAVENPORT

THOMAS STANLEY
b. 1530. d. 1591
URSULA CHOLMONDELEY

Coat of arms of Sir John Stanley's son Thomas

Coat of arms of Sir John Stanley's grandson Thomas

Coat of arms of Sir John Stanley's great-grandson Thomas

RANDLE STANLEY
b. 1556? d. 1595.
MARGARET MAISTERSON

SIR THOMAS STANLEY K.
b. 1577. d. 1606.
ELIZABETH WARBURTON

Coat of arms of Thomas Stanley's son Randle

Coat of arms of Sir Thomas Stanley who united the Park

Although the family was already prominent in local affairs, having been settled in Cheshire since the reign of Edward IV, it was this Thomas Stanley (c.1532–1591), High Sheriff of Chester in 1572, who was first styled 'of Alderley' and who made Alderley his main seat and Hall.

And it was *his* grandson Sir Thomas Stanley (knighted by King James I in 1603) who bought Nether Alderley from the Fittons in 1602, thus uniting its parts to create the core of the modern Alderley Park.

## The Alderley House site and later

One piece of the jigsaw still remained; the site originally known as Park House and later as Alderley House. This name was attached in turn both to the Stanleys' huge mansion, and then to the ICI Divisional Headquarters building. Its acquisition around 1750 is recounted later. Expansion of the estate continued after this date (in 1850 the Hall was surrounded by a park of 200 acres, increasing to 300 by 1892, 365 after the sale in 1938, and a final peak of over 400 acres). It was with the addition of the Alderley House site that the estate of Alderley Park, as we know it, became a single entity.

## The Park's situation and weather

Alderley's wood, park and water stand where the lowest foothills of the Pennines descend to the broad Cheshire Plain. The Congleton to Alderley Edge road marks its western boundary, with the new A34 bypass now taking much traffic away from the northern section of this. Along the Park's NE boundary runs the unsurfaced Hocker Lane. Elsewhere, the perimeter of the roughly oval estate is marked by fence, hedge and ancient drystone wall, sometimes where the woodland of the Park gives way to adjacent farmland.

### Topography

The estate forms a shallow bowl opening to the west, dropping some 160 feet to its lowest point, the road bridge over Bentley Brook. There are few steep inclines; natural slopes are due to stream erosion or quarrying. Extensive landscaping has taken place. During their

Alderley Park in 2013 from the SW
PHOTO ASTRAZENECA

Home for rabbits
and sheep

tenure from 1602 to 1938, the Stanley family undertook many projects including diversion of streams, building of several dams, the excavation of the Water Garden and the enclosure of this and other walled gardens.

Some of the work may have been directed by (or more likely done in the style of) the landscape gardener Humphry Repton (1752–1818), to make a 'contrived' natural-looking landscape. One large hillock in the parkland, which has gained the nickname of Rabbit Hilton, is believed to have been built for rabbits used for food. Extensive tree-planting and erection of many large and small buildings and various roads, tracks and paths created a home estate which was both secluded and relatively self-contained. After the purchase of the estate by ICI in 1950, the scale of operations increased further, with very considerable construction of modern buildings, roads and car parks, and tree-planting on an enormous scale.

## Climate

Cheshire lies on the wetter side of England's backbone Pennine range of hills. Most of the year is mild but there is rarely a shortage of rain!

Big fallen trees have blocked the Park's roads once or twice and there was major damage in the woods in the 'hurricane' of 1987, with many conifers uprooted. After an earlier great storm from the WNW on 21 January 1802, Maria Stanley wrote to her sister:

… bewailed our misfortunes in Thursday's storm. You can sympathize with the groans of the wood, and are ready to shed tears with our unhappy hamadryads. I went with great anxiety to see if your tree with the sunny seat was safe, and was right glad to see it unhurt … We rode a tour of the estate the next day, and saw nothing but mischiefs, which, however, all hide their diminished heads before the thirty [fallen] trees in the park …

Account of a storm
in 1802
CHESHIRE ARCHIVES
AND LOCAL STUDIES Ref.
DSA 5/10

THE EARLY PARK & LAND

Intense winter cold is unusual; Radnor Mere is rarely known to freeze completely and has done so only four times since the 1940s, the last time on 7 December 2010, although the Stanley family regularly collected ice from it in the nineteenth century.

Radnor Mere in winter

# Rocks and views of the Park

## Geology

Underneath the human Park lies the hidden one. The geology of the Alderley area is described thoroughly in *The Story of Alderley*. Suffice it to say that Alderley Park lies on coarse Triassic sandstones partly overlain by softer rocks covered in turn by a silty red clay or 'marl'. Cores from old boreholes drilled for agricultural purposes in the Park showed that the northern part of the Park is underlain by about 400 feet of the older sandstone (from which many local Alderley buildings have been constructed), while the southern half lies on younger mudstone.

Several geological faults lie under the Park. One of these is the Alderley Fault, which follows roughly the line of the A34. This is not a cause for concern – we are unlikely ever to see a major earthquake in the Park, although at least two minor tremors have disturbed staff in the past, including one that shook the bottles in the chemistry labs and left the liquid contents stirred if not shaken!

No bare rock is visible in the Park although two old quarries exist, one of which may have been used to build one or other of Radnor Mere's two dams (the earlier one is now submerged). The local bedrock is not visible nearer than near the mediaeval Mill; the Moat millpool above this spot was cut 'through the living rock'.

## Water underneath the Park

The clay under the Park's loamy, acid soil separates the Park's numerous ponds from the groundwater-holding strata beneath. In 1971 an old borehole in Alderley Park itself, near the current farm, was re-opened to test the possibility of water extraction. Sadly, faults in the surrounding rock prevented a reasonable flow. Also, on pumping a test sample the water soon became salty and the strata also appeared contaminated with material used in the past to infill the old mines on Alderley Edge.

Likely geology of the Park
COPYRIGHT ASTRAZENECA

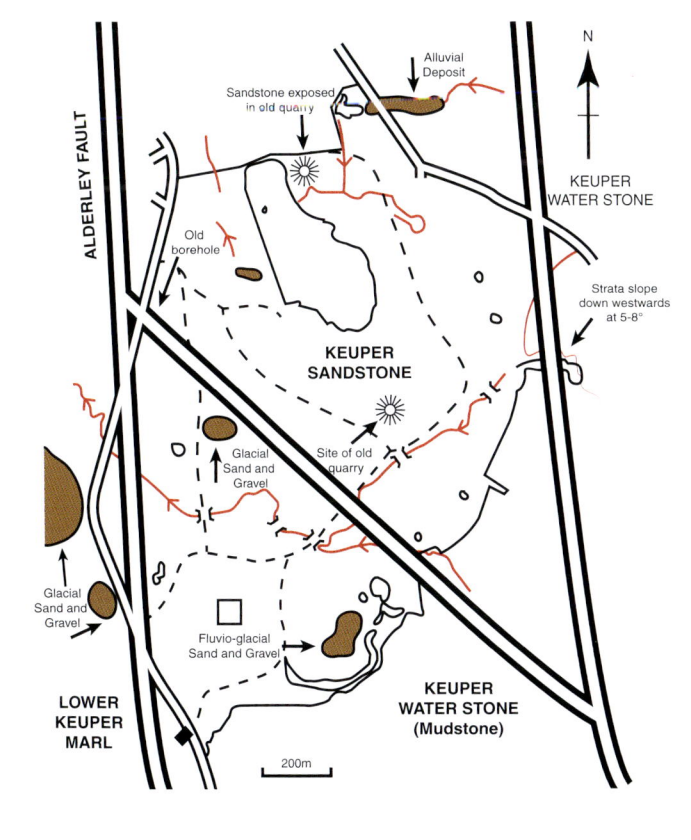

## Views of the Park

Views across the Park are attractive, including north from the site of Alderley House and south from Mereside. But views into the Park are remarkably scarce. The ornamental water of Radnor Mere has a reputation as a 'hidden water' and can scarcely be glimpsed from the Park's perimeter even in winter, or from any public footpath.

## Views from the Park

Distant views from ground level are few, although in 1836 Rev Edward Stanley, speaking almost certainly of Radnor Mere, recorded that:

> (the mere's) western margin is bounded by an artificial dam, which as the water is upon a much higher level, commands an extensive view over a flat, rich country, the horizon terminated by the faint outline of the first range of Welsh mountains.

Trees now mostly obscure this view in winter of Moel Fammau, just across the Welsh border. For most of the year, the only visible point is the square tower of Nether Alderley parish church.

From the buildings the view is a reduced one of that from the Edge. Clockwise, it extends from Shining Tor on the Cheshire border (just visible), through Shutlingsloe, Croker Hill, Bosley Cloud, Mow Cop, the Wedgwood Monument on Bignall Hill, then more remotely the south Shropshire hills, with Brown Clee Hill and The Wrekin, Caer Caradoc, Stiperstones and, just across the Welsh border Corndon Hill, at 59 miles the furthest visible. Nearer are Rail House in Crewe and the Jodrell Bank radio telescope. Beyond the Peckforton Hills and Beeston Castle are Moel Sych (57 miles) in the Berwyn range, the highest peak in NE Wales, then Moel Fammau in the Clywd hills.

Further north are the Stanlow complex by the Mersey, Overton Hill, Billinge Hill, Rivington Tower and Winter Hill. This brings the total number of counties visible to nine. But the Park is well wooded and its buildings are not obvious themselves from most distant points.

# Baronets to baron

## The Old Hall and its Baronets to 1779

### THE STANLEY BARONETS

| | | |
|---|---|---|
| Sir Thomas Stanley | (1597–1672) | created baronet in 1660 |
| Sir Peter Stanley | (1626–1683) | second baronet |
| Sir Thomas Stanley | (1652–1721) | third baronet |
| Sir James Stanley | (1680?–1746) | fourth baronet |
| Sir Edward Stanley | (1690?–1755) | fifth baronet, brother of James |
| Sir John Thomas Stanley | (1735–1807) | sixth baronet |
| Sir John (Thomas) Stanley | (1766–1850) | seventh baronet; created Baron in 1839 |

The history of Alderley Park pivots around the year 1779, in which Alderley Old Hall burnt down. The records available after this date would fill a bookshelf, let alone a chapter. But the fire that destroyed the old Jacobean mansion also destroyed most of the family's archives to that date (as did the later fire of 1931).

So this is most of the little we know.

### Location of the original Hall at Alderley

As stated, the ancient Alderley Old Hall did not stand within the bounds of today's Alderley Park, but in Nether Alderley village beside the mill. Its site and gardens were sold off separately from the Park at the great auction of 1938.

The Hall, in Nether Alderley village, dated from the 1570s and was apparently built on the site of a former house perhaps first occupied around 1420. It was described in the 1580s as the Stanleys' 'new mansion house', and is stated, on the tombstone of the Thomas Stanley who died in 1591, to have been rebuilt by him. A niece of the first Lord Stanley of Alderley, Louisa Stanley, wrote in 1843 that:

> the house stood in the village of (Nether) Alderley close to the mill. It was surrounded by a moat, spreading out into a large sheet of water on the east side, and on the west filling a channel cut out of the solid rock.

The whole mill pool is now called the Moat.

### Baronets

The first of the family to inherit the combined manors, at the age of seven, was the future Sir Thomas Stanley (1598–1672), the first baronet. Around 1630 Sir Thomas also greatly enlarged Alderley Old Hall, with extensions which virtually amounted to yet another rebuilding. The Hall now had a fine stone arched gateway and spacious

Coat of arms of Sir Thomas Stanley, the first baronet

SIR PETER STANLEY
2ND BART. b.1621. d.1683.
ELIZABETH LEIGH

SIR THOMAS STANLEY
3RD BART. b.1652. d.1721.
CHRISTIANA LENNARD

SIR JAMES STANLEY
4TH BART. b.1680? d.1747.
FRANCES BUTLER

Coat of arms of Sir Peter, the second baronet

Coat of arms of Sir Thomas, the third baronet

Coat of arms of Sir James, the fourth baronet

Coat of arms of Sir Edward, the fifth baronet

SIR EDWARD STANLEY
5TH BART. b.1690 d.1755.
MARY WARD

stables. Its owner was made High Sheriff of Cheshire in 1634.

At the outbreak of the Civil War the Stanleys backed both sides! Thomas's brother James backed Charles I and was beheaded by the Roundheads at Bolton in 1652 for his error. Sir Thomas himself was a staunch Puritan, who supported the Parliamentarians but with moderation and without taking any military action. He foresaw the end of Cromwell well before the Restoration of Charles II in 1660. Clare Pye records in *The Story of Alderley* the complex story of how after the Restoration Sir Thomas steered cautiously through to being made the first baronet in the county (a title between a knight and a peer), to the great annoyance of his neighbours, 'probably because the king recognized his long administrative experience in the county and fundamental acceptance of the realities of power.'

Little of note happened in the next century. The fourth baronet Sir James, who died in 1747, was: 'A fair man … and of remarkably placid character. He drove up to the Edge almost daily in his carriage drawn by four black long-tailed mares, always accompanied by a running footman of the name of Critchley.'

Among his papers when he died were found the following lines:

The grace of God and a quiet life,

A mind content and an honest wife,

A good report and a friend in store,

What need a man to wish for more?

Sir James and his father had been Roman Catholics but his brother Sir Edward, who succeeded him as the fifth baronet, had been brought up a Protestant. A more active man, he was unusual in having married away from the district, to a Miss Ward, the daughter and heiress of a London banker.

## Deer House

Sir Edward erected in 1750 the four-square sandstone Deer House beside Radnor Mere, after the design of one built by Lord Vernon at Sudbury. With corner towers as high as a two-storey house, it provided a feeding point and shelter for the herd of fallow deer that grazed the open parkland. From a distance it looked like a small castle, with its brick battlements and castellated walls ten feet high enclosing an open courtyard planted with elms. The deer could come and go through the many large gates that ran through the lower half of the mock turrets. In winter extra rations of grain and hay would have been laid within the walls. Once within, the deer could be inspected closely and caught for any veterinary treatment or individual marking which might be necessary.

The deer grazed the 'Deer Park' which originally extended from the banks of the mere across to the Drive. Later, a fenced five-acre park north west of the mansion was used. The fallow deer herd was once large and provided a source of venison for the Park House larder. However by 1908 it had declined to 15 'head' and all had gone by 1938. Featured in many sketches, the ivy-laden ruin was demolished in 1955 when Mereside was built on the same site.

Deer House
COURTESY MSP (IMAGE ASTRAZENECA)

Deer House plus Mereside as of 2005
COPYRIGHT ASTRAZENECA

Mereside when first completed
ASTRAZENECA ARCHIVES

Near to the Deer House was the Keeper's Meadow:

A house stood before called the keepers lodge in which Andrew Whittaker lived. He had a croft
or two near the house of which the fences are still traceable from a few old thorns which grew
in them. He had a meadow near the wood called the keepers meadow and part of which was
planted with oaks.

## Alderley Old Hall

Sir Edward also built far more, although he was baronet for only nine years. In the early
1750s Alderley Old Hall was, according to Daniel Lysons' account of 1810:

newly fronted and in great part rebuilt by Sir Edward Stanley; the hall of a former mansion built
(i.e. rebuilt) by Sir Thomas Stanley Esq. in the reign of Queen Elizabeth was suffered to remain;
it was fifty feet by forty.

The Old Hall was then of high architectural quality and according to Nancy Mitford, 'a
contemporary picture shows it to have been a beautiful Jacobean red brick house'.

The watercolour shown was drawn by the (later) first baron from memory around 1798.

A floor plan and detailed description of the Hall appears in Matthew Hyde's account in
*The Story of Alderley*. It was quadrangular, with a grand brick frontage to the road, and older
parts behind. In 1779 the house, as remembered to Louisa Stanley,

Deer House and
the Alderley Park
Volunteers drilling,
1803
COURTESY LORD
STANLEY OF ALDERLEY

consisted of three sides of … a mansion, a large hall of an older date occupying the other side, and offices behind the hall. A handsome stone bridge of two arches crossed the moat from the ground entrance and west side, to a stone terrace which commanded views of the Park, the Church, and the plain of Cheshire, and by a flight of steps led to a handsome stone arched gateway close to the road, built by Sir Thomas, the first Baronet.

The Old Hall was a necessary centre of village life; in its spacious courtyard a bear was kept and the local militia was drilled when village forces mustered in time of danger. The painting shows them drilling in 1803 – on the site where Mereside now stands.

Sadly Sir Edward did not enjoy it for long, as in 1755 he was suddenly seized with a fit of apoplexy and died in his carriage before he could be conveyed home.

## The Alderley Old Hall fire

Sir Edward's son, the sixth baronet Sir John Thomas the elder, is best known for the fire which in 1779 burnt his Hall down, the fire first being discovered by 'returners from the Alehouse at 10' and spreading, 'from some Wood work communicating with the Kitchen Chimney. The old Brew House, Laundry and Stewards office now remain and are now called the Hall.'

The Hall was partly a timber building plastered to look like brick and stone, which may account for its efficient destruction in the fire.

A more authoritative account appears in the personal *Praeterita* of Sir John Thomas Stanley's son, Sir John Thomas the younger (who is called just Sir John in this account):

In the summer of 1776 my father and mother took me with them to Alderley … I date from this time my accurate remembrance of the houses, rooms, moat, mill and gardens &c. at Alderley, the stone entrance gates, the stone terrace, the bridge, the hall, the parlour and its pictures, the dining room, the great stairs, and the chiming clock on them; the whole house as it surrounded its inner court with all outbuildings … Alderley Hall was burnt in the month of March 1779. I read the account of it in a newspaper while at my French lessons, and burst into tears. It was never known what caused the fire; it began in the kitchen chimney, and it was supposed a beam reached the flue of the fireplace in the library, where my father, some evenings before, had been burning papers. All the books and papers in the house were burnt … My father had gone to Chester, and was on his return the night of the fire; he was first told of it at the Allerton toll-bar.

Coat of arms of
Sir John, the sixth
baronet

The house stood on a rock (a breaking-out of the sandstone of Alderley Edge) surrounded by a moat, to complete which the rock was in part cut through, and from some of the windows a view of the extended plain of Cheshire was magnificent. I mourn over its destruction, and never pass by what were once its accompaniments, the mill, the Glastonbury thorn, the pillars, the stone walls of the terraces, without a regret that my father, instead of occupying the Park House, had not laid out the same sum he did there in rebuilding, and so keep what had been for many centuries the home of the family. The hall was nearly a square of between forty and fifty feet, with an immense fireplace … A full-length picture of my father, by Gainsborough, was too large to be thrown out of the window when the house was burning; the people did not think of cutting or tearing it out of the frame, so it perished.

A later chapter will describe how after the fire in 1779 the Stanleys then moved to their bailiff's house, an old mansion called the Park House, later called Alderley House.

Sir John Thomas the elder married Margaret Owen, a Welsh heiress who brought the estate of Penrhos in Anglesey into the family, and they had two sons and five daughters.

Later he was little at Alderley as his main home, being mostly in London where he was Gentleman of the Bedchamber to King George III, or at Hoylake where he was developing the resort town there; he was largely separated from his wife who spent most of her time at Penrhos. He spent nights at Alderley only en route; the Park House was thus available for his son to occupy when the latter and his wife settled there in 1797.

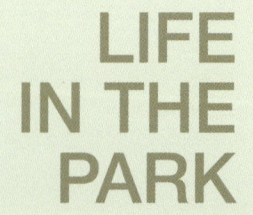

## LIFE IN THE PARK

### Priorities?

A different, possibly apocryphal account of the fire is recorded by Nancy Mitford:

'It is said that Sir John and Lady Stanley were at their town house in Chester at the time, and that the news was broken to them by a tearful housekeeper in the following terms:

'Yes, Sir John, all the furniture – yes, My Lady, all the silver – yes, Sir John, all the pictures, but (brightening considerably) your Ladyship will be very glad to hear that we managed to rescue all your Ladyship's hams and preserves."

In old age Sir John Thomas the elder was said to have been dirty, curmudgeonly and crippled with gout. Nancy Mitford perhaps mistakenly attaches this description to his son Sir John, but he seems a very different character.

# Beeches and avenues of the Park

To the casual observer, Alderley Park seems full of buildings and roads; and its farmland holds what at one time became the largest head of sheep in Cheshire. It comes as a surprise, therefore, to find that over one third of the historical Park is woodland. Large areas of the estate have been continuously wooded for nearly four centuries.

## Ancient woodland

In early times, lowland oak forest filled most of Cheshire. By the eleventh century most of this had been burned, either by one of the various invaders or to smoke out outlaws. The Domesday Book of 1087 mentions only two large woods near Alderley, one being three miles across. The majority of the district is described as 'waste', probably meaning heather moor and uncultivated scrub; and so it long remained.

## The beeches of Beech Wood

Sir Thomas Stanley (1598–1672), the first baronet, is said to have brought beeches to the Park at the time of his marriage to Elizabeth Pytts at Kyre in Worcestershire on 17 January 1621. They were planted for the sake of his young homesick bride using beech mast from her father Sir James's estate there, where the tree was common (and native). Presumably first raised in a nursery at Alderley, they are said to have been the first beeches in Cheshire. There is a tradition that they were planted about the time of the sailing of the *Mayflower* (6 September 1620).

The beechwood originally stretched right down to the Moat pool behind the medieval mill, but many trees were felled before 1755 by Sir Edward Stanley, leaving only the well-known Beech Wood above Radnor Mere. His grandson Lord Stanley, commenting that 'thanks will not be paid to him', followed the family tradition on his marriage and sowed more with seed from Sussex in 1798. But the cleared area above the millpool became the Hall's new gardens (now vanished) and the eagle statue mentioned earlier was erected here.

*An ancient beech*

Although some thirty were blown down in the storm of 1802, the beeches were in 1850 said to be unequalled in the county, and their noble position across the Mere was then a noted feature of the estate. The Stanleys had a log house in the wood. Beside Radnor Mere, the last of the mighty beeches were until recently still the most obvious trees in the wood. One of the biggest beeches in Cheshire was felled there in 1959; its circumference was about 22 feet and its height 85 to 90 feet. Another opposite Mereside was thought to be even taller and had 265 annual rings – a considerable age for a beech. It was actually observed falling, during a fierce gale in 1980.

Most were felled for valuable timber between or during the two World Wars (in the Great War reputedly for use as trench supports). The survivors are all near the end of their life (and rich with interesting fungi). The rot-cored giants provide homes for many hole-dwelling creatures, from grey squirrels and bats to stock doves and tawny owls.

Artist's fungus on one of the old beeches

A Park Beech in winter

Beeches by Radnor Mere in summer

The Stanleys' log house in Beech Wood
COURTESY LORD STANLEY OF ALDERLEY

## Old Drive or Chestnut Avenue

A fine (but diminishing) avenue of old trees lines the Old Drive or Chestnut Avenue that bisects the parkland. This was once part of the north–south highway dating back to Saxon, or possibly even Roman, times and was once one of the busiest parts of Alderley Park, which the first Lord Stanley knew in his childhood as: 'the Road which was the only one to Manchester or to London'.

The A34 was then merely a meandering cart-track. The Drive was turnpiked in 1775 by 'Blind Jack of Knaresborough', but still carried all traffic past the Stanleys' new residence until they took exception and moved the road west to its present line by 1818. The Drive then became a mere carriage drive, giving the family privacy.

## Churchillian guests

The Drive must have seen many famous guests who visited the Stanleys. It would also have carried the cars and carriages of the 2,000 guests who attended the wedding of Blanche Stanley to Major Eric Serocold in July 1912.

Among them was the young Winston Churchill, then First Lord of the Admiralty, who brought a wedding gift of oriental rugs. Winston had himself recently married Clementine Hozier, a great-granddaughter of the second Lord Stanley. The wedding was a huge one at Nether Alderley Church, with seven bridesmaids. It was intensely hot and many ladies fainted; some men and the press resorted to the school roof for a view. The guest list included two viscounts, three viscountesses, a count, two countesses, eight lords, one earl and one marchioness.

A reception followed in Alderley Park, with music by the Cheshire Yeomanry. The presents were arranged in the billiard room. 350 tenants of the estate were entertained in the gardens in a large marquee.

In 1949 it was initially proposed by ICI that Chestnut Avenue be re-opened for traffic; happily, another solution was found. Another longer avenue planted by the fourth baron in 1913, now gone, ran NE from the site of Alderley House for half a mile.

# The Alderley House mansion

After the burning of Alderley Old Hall in 1779, the Stanleys moved to the property called the Park House.

## The Park House

This stood in the south of Alderley Park, on land which bordered the common (enclosed in 1778) known as Monks Heath. The Park House was thought by Louisa Stanley to have been:

> … formerly a part of the Estates which were held in Alderley by the Monastery of Dieulacres in Staffordshire [near Leek] … After the dissolution of the monastery, the property which had belonged to it in Alderley was granted to two brothers of the name Sheldon. They sold it to a family of the name Greene, who then owned Fernhill, the estate bordering on the land granted to the Sheldons, and under which family of Greene, Ralph Bentley, and Thomas Deane de Park, who died in 1629, successively held the land so sold to them by the Sheldons.

**History of the Park House / Alderley House site (originally part of the monastic lands):**

| | |
|---|---|
| part of the estates of Dieulacres Monastery | before dissolution |
| two brothers Sheldon | after dissolution |
| Greene family | ? |
| Ralph Bentley | ? |
| Thomas Bentley/Deane de Park | until 1629 |
| Thomas Deane de Park (son) | 1629–1695 |
| *(the latter built the Park House, 1668, on the site of an older Park Tenement)* | |
| William Stanley | from 1695 |
| William Stanley de Park | |
| Thomas Stanley de Park | until 1748 |
| purchased by the main Stanley family | 1748–1779 |
| became the principal Stanley residence | 1779–1930s |
| *(the Park House was largely demolished in 1818 and a new mansion built on the same site)* | |
| *(the Alderley Park mansion was damaged by fire in 1931 and demolished in 1932)* | |
| Hambling, Crundall & Co. Ltd | 1938–1950 |
| Imperial Chemical Industries | 1950–1993 |
| *(the present Alderley House was built on the same site in 1963/64)* | |
| Zeneca plc | 1993–1999 |
| AstraZeneca plc | 1999–2014 |
| Manchester Science Partnerships | 2014– |

History of the Park House site

Thomas Deane and his son of the same name were styled Deane de Park, and the former:

> occupied what was called the Park Tenement. The House called the Park House was built on the site of the old one, by Thomas Deane, the son. The house (was) first built by him in 1668 … After Thomas Deane de Park's death in 1695, the Park House was inhabited successively by William Stanley of Astle (a distant relative of the baronets' line) … William Stanley, called de

Park, son of the said William Stanley, and his son, Thomas Stanley, also called de Park, who in 1748 left the place to live at Astonhurst in Staffordshire.

The house was evidently bought by the main family, the Stanleys of Alderley, who had already bought the lands pertaining to the then Dieulacres (God's Acres) property between 1558 and 1567. A spinster,

Mary Stanley, the daughter of Sir Thomas Stanley, Bart. then occupied it, the house having been let to her by life by her brother Sir James Stanley, and on her death, in 1766, Mr Peter Stockton (who had been the steward successively of Sir James, Sir Edward, and Sir John) had a lease of it granted to him for life by Sir John Stanley. He died in the same year that it became a refuge to Sir John and his family after the burning of Alderley Hall … and, adding some rooms, made it their permanent residence …

The deceased steward's wife and daughter cooked for Sir John and his wife. At the time of their arrival the Park House was described as 'a respectable gable end house with a handsome hall and oak panelled Parlour' and a 'fair specimen of the dwellings of rich yeomen or small squires of Cheshire.'

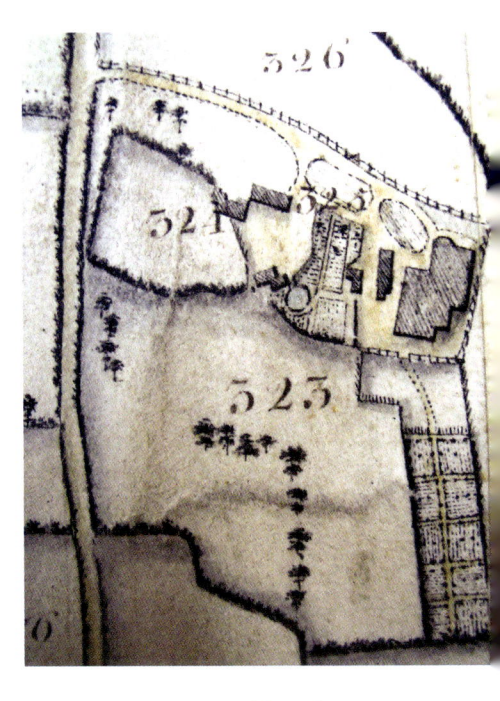

## Park House discovery

No image of the earlier Park House (which was largely or all replaced by the new mansion) was known to us when this account was first written. But while I was leafing with Lord Stanley through one of his family sketchbooks, we found a small handwritten card bearing the date 1807, then realised by chance that it had a sketch of the Park House on the back. On the right of this appears 'Sir John's Justice Room'. Later, in an old ICI review, a reproduction of a second, nearly similar, sketch turned up, this one (mislabeled as the later mansion) drawn in 1819 by Isabella Louisa Stanley, aged 18. (She married the Arctic explorer Sir William Parry.) The location of Isabella's original is not currently known.

Comparison of the two sketches with the two maps in this section shows that the Park House site was under the east and rear of the later mansion (and of ICI's Alderley House). The mansion was then apparently built with the same bow window design shown in the Park House sketches.

## Building into a mansion

According to Nancy Mitford, the family's original intention was to rebuild the burnt Old Hall: 'By degrees, however, they found themselves so comfortably installed that they gave up the project and built on to the Park House instead.'

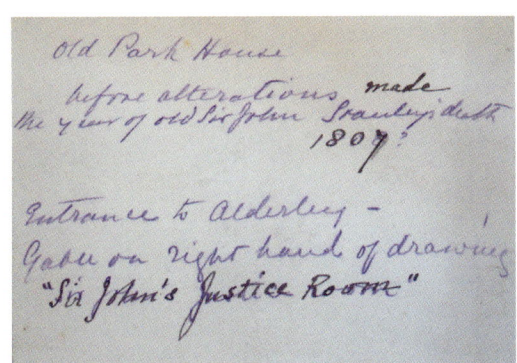

Old Park House before alterations, made the year of old Sir John Stanley's death 1807.

Entrance to Alderley – Gate on right hand of drawing "Sir John's Justice Room"

# LIFE IN THE PARK

### Regrets

Sir John was himself unhappy with the result of the family's work, commenting in about 1840:

'I have made the house a Mansion, but in my ignorance of what was wanting for Comfort, for hospitality, and even for use if many were concerned, the Place … comes back to my Remembrance as one spoilt and not improved by all that has since been done to it.'

1872 Ordance Survey map of Alderley House site

Alderley House or Hall mansion by c.1925
COURTESY MSP (IMAGE ASTRAZENECA)

The east face of the former mansion
COURTESY MSP (IMAGE ASTRAZENECA)

The building was extended, then mostly demolished in 1818–19. A much larger house was built on the site by the first baron. The Tenants' Hall (later the Sir James Black Conference Centre) was only one of six great entertaining rooms in the final design, as the family 'continued enlarging it to suit the size of their families and requirements, and in the end the little Park House had become a north-facing mansion of some sixty bedrooms and of very small architectural merit.' It was built of brick with a stone frontage and a loggia on the NE side.

By 1842, when the first local tithe maps were printed, several acres of gardens and pleasure grounds had been created around the new house, field boundaries had changed and plantations had been established throughout the estate.

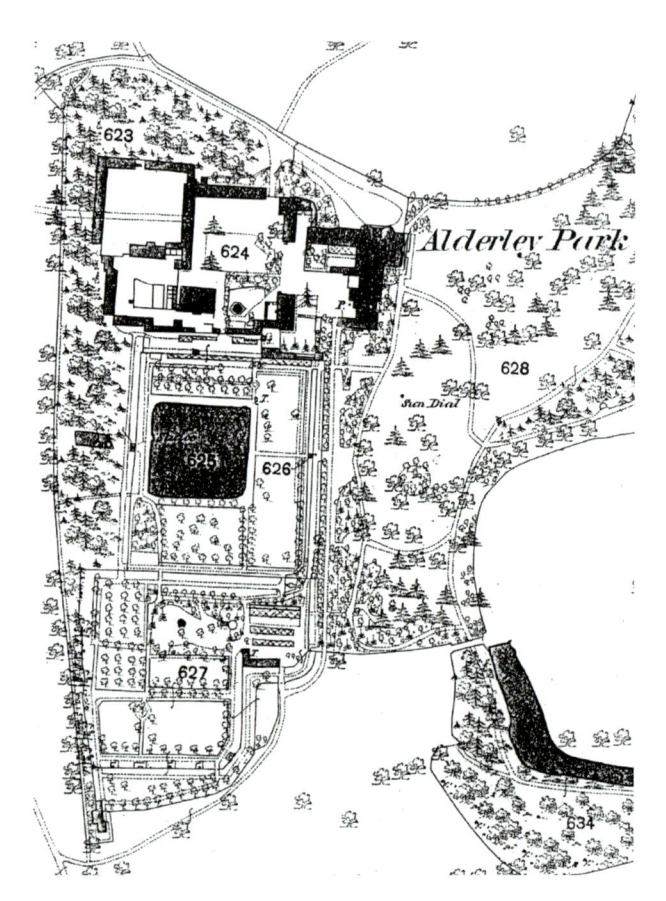

## The big house

At Christmas, grand events were held in the 'big house' while the estate workers were given their own celebrations. Presents were more modest than today, with Lady Stanley receiving on one occasion a tortoiseshell comb and a novel, while she gave her husband a silver-handled whip. Winters were then long, with snow lying for weeks; the Stanley children always built a large snowman on the drive in front of the mansion doors, which the coaches-and-four, ferrying guests to and from the house, had to avoid so as not to upset the children.

## Asquith and Venetia

The new mansion was the home seat of the first five Lords Stanley of Alderley. Under the fourth Lord Stanley, the vast building that had obliterated the original old house became a meeting place of people with power, especially Prime Minister Herbert Asquith, who stayed there every Christmas.

Mr Asquith became a close friend of his host's daughter Venetia Stanley, and wrote to her often, even during Cabinet meetings and even asking her opinion of questions of military strategy during World War I. His visits to Alderley, although accompanied by junior colleagues including Winston Churchill (as described later), were clearly more than just ones of political friendship. Eventually, Venetia found his interest too strong and wrote to tell him that she had decided to marry elsewhere, although her friendship with him was restored years later.

## The first auction

In 1931 the Alderley House (or Hall) mansion was badly damaged by fire. Either before or after this the mansion was closed and the household staff was dispersed. The extent of the fire is not clear but was only partial, for the Hall was stripped but in part still intact in the following year by the time a big furniture auction was held over four days from 27 September 1932 at the direction of the executors of the 5th baron.

Two days prior to this, nearly 5,000 mature trees had been sold:

> The first indication of material change in the fortunes of the estate came with the sale by auction of thousands of trees. Before the rhythmical blows of the feller's axe had ceased to echo through the woodlands, treasures from the extensive library were despatched to London as a preliminary to the cataloguing of family treasures, heirlooms and furniture. The demolition of the greater part of the hall followed, but the spacious Tenants' Hall was converted into a drawing room to be used in conjunction with other apartments which did not come within the range of the demolition scheme.

Over 80 merchants were there to bid for the wood, and the auctioneer began with a lament: 'You can imagine that it cannot be the wish of anyone to cut this fine timber, particularly so close to the house …' Three lots were withdrawn to save 511 trees on Alderley Edge, but 171 mature beeches in Beech Wood went for £4 each.

1932 auction notice
CHESHIRE ARCHIVES
AND LOCAL STUDIES Ref.
SC/2/43

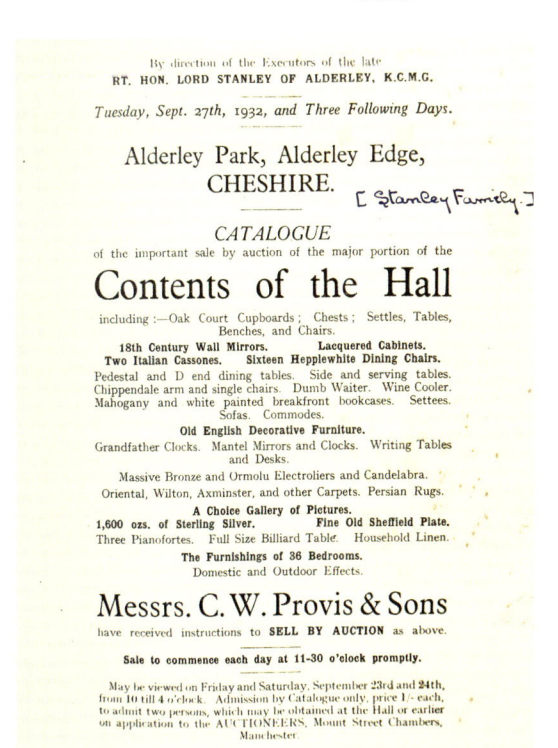

By direction of the Executors of the late
RT. HON. LORD STANLEY OF ALDERLEY, K.C.M.G.

*Tuesday, Sept. 27th, 1932, and Three Following Days.*

Alderley Park, Alderley Edge,
CHESHIRE. [ Stanley Family. ]

*CATALOGUE*
of the important sale by auction of the major portion of the

# Contents of the Hall

including :—Oak Court Cupboards ; Chests ; Settles, Tables,
Benches, and Chairs.

**18th Century Wall Mirrors.**      **Lacquered Cabinets.**
**Two Italian Cassones.**      **Sixteen Hepplewhite Dining Chairs.**
Pedestal and D end dining tables.  Side and serving tables.
Chippendale arm and single chairs.  Dumb Waiter.  Wine Cooler.
Mahogany and white painted breakfront bookcases.  Settees.
Sofas.  Commodes.
**Old English Decorative Furniture.**
Grandfather Clocks.  Mantel Mirrors and Clocks.  Writing Tables
and Desks.
Massive Bronze and Ormolu Electroliers and Candelabra.
Oriental, Wilton, Axminster, and other Carpets.  Persian Rugs.
**A Choice Gallery of Pictures.**
**1,600 ozs. of Sterling Silver.**      **Fine Old Sheffield Plate.**
Three Pianofortes.  Full Size Billiard Table.  Household Linen.
**The Furnishings of 36 Bedrooms.**
Domestic and Outdoor Effects.

# Messrs. C. W. Provis & Sons

have received instructions to **SELL BY AUCTION** as above.

**Sale to commence each day at 11-30 o'clock promptly.**

May be viewed on Friday and Saturday, September 23rd and 24th,
from 10 till 4 o'clock.  Admission by Catalogue only, price 1/- each,
to admit two persons, which may be obtained at the Hall or earlier
on application to the AUCTIONEERS, Mount Street Chambers,
Manchester.

The sale of over 1,000 lots from the mansion by Provis & Sons in one of the biggest household sales in Cheshire for many years included the 'major portion of the Contents of the Hall'. The list was remarkable: Hepplewhite, Chippendale, Sheraton, Chesterfield, Collard, Jacobean, Louis, George III, Derby, Dresden and Spode lots all jostled for attention. The furnishings of 36 bedrooms were on offer. Curiously, a full concert grand piano was on offer in the servants' hall. 1,600 ounces of sterling silver goods, some fine old Sheffield plate, 60 engravings by Hogarth and the mansion's billiard table were to go as well. A huge number of local people attended, with both courtyards and the nearby parkland all packed with cars.

Demolition followed that winter. After the demolition all that remained were the Green Room and Tenants' Hall, along with the estate farmhouse, described later. Most of the site of the former mansion was itself bare until ICI Pharmaceuticals Division decided to build its headquarters building on the site, beginning in 1963.

# John Thomas, 1st Lord Stanley of Alderley

The seventh baronet put Alderley Park on the map when he was elevated from being Sir John to John Thomas, first Baron Stanley of Alderley, in 1839.

In youth, Sir John had a fiery spirit, but his was also a dreamy and introspective nature. He was described as 'dignified, high-minded and intellectual' and, on becoming fluent in German while staying with his mother in Germany, he became one of Goethe's earliest translators.

Unusually he went not to Oxford but to Edinburgh, where he met Adam Smith. He was certainly strongly adventurous; when young, for no particular reason, 'in 1789 he sailed with a few companions for Iceland in his own ship, the 'John', an enterprise then almost unprecedented. His Journals and his drawings remain to tell the story of his adventures in the North.' He reached the summit of Mount Hekla on his exhausting scientific expedition and investigated the volcano and its lava flows; his report on Iceland remains an important historical record of life there at the time. He remained interested in science all his life and was elected a Fellow of the Royal Society the next year. Later, he entered Parliament, spoke four languages and mixed in the court circles of three European capitals.

A sketch of him in the National Portrait Gallery has the same features and a bush of light-coloured hair:

Sir John suffered a personal tragedy when his first fiancée, Maria Jones, who came from Wicklow, died just before their wedding in June 1792.

## Marriage to Maria

In 1796 he married Maria Josepha, eldest daughter of John Holroyd, the first Earl of Sheffield, whom he had met while serving (he was a major) with the Royal Cheshire Militia in Sussex. His second love, Maria loved John devotedly all her life and bore him nine children, twin sons Edward and William and seven daughters. She

Hon. Maria Josepha Holroyd. idge. A.R.A. in 1793.

Sir John as a young man
CHESHIRE ARCHIVES AND LOCAL STUDIES Ref DSA 180

The first Lord Stanley of Alderley
NANCY MITFORD, 'THE LADIES OF ALDERLEY' 1938 (NATIONAL PORTRAIT GALLERY)

John's first fiancée died just before their wedding
CHESHIRE ARCHIVES AND LOCAL STUDIES Ref DSA 180

Maria Josepha Stanley before her marriage
CHESHIRE ARCHIVES AND LOCAL STUDIES Ref DSA 180

Coats of arms brought by Maria

Coat of arms of John and Maria

married him in Sussex and she delighted in her new home of Alderley Park. She was surprised to find the Park House so comfortable, and her pleasure enchanted her new husband:

> Maria is happy and placid, and pleased with everything about her. To live at Alderley with such a being to love, has ever been the wish of my heart.

## Politics and science

Maria's youth had been as unusual as her husband's. In 1792, at the age of nineteen she had been a personal witness of the French Revolution, hearing many tales of horror after she and her parents arrived in Paris on the same day that the king and Marie Antoinette were brought back under arrest from their escape attempt. During the same trip she became a friend of the historian Edward Gibbon, famous author of *The Decline and Fall of the Roman Empire* with whom she later corresponded after she and her father visited him in Lausanne. After his death in 1795, they arranged his papers. Gibbon had a high opinion of her intellect and Lord Sheffield reported that 'Gibbon used to lament that Maria Josepha was not a boy, saying she would maintain a contest well with Charles James Fox.'

One notable visitor, who came to Alderley Park, in 1859 when Lady Stanley was very old, was the celebrated writer Thomas Carlyle, whose great book on the French Revolution she must have discussed with him with interest. Carlyle was a man of forthright views who told the Stanleys that '… we did not want liberty & it was all nonsense …'!

**LIFE IN THE PARK**

### Early science?

Maria had a strongly enquiring mind. In 1803 a letter to Louisa showed this extended to an interest in science:

'I wish I knew how to make a battery and try some experiments on the ox's and sheep's heads, and I wish I could hear what the wise folk say about it. Mr. Holland does not seem disposed to think it different from the electric fluid, and he is the only person I have seen who knows the difference between Galvanism and Calvinism.'

## In Alderley Park

The impact of Sir John and Maria on Alderley Park, where they lived for over half a century, was great. Sir John was a hard worker who undertook all sorts of projects, including the sowing of hundreds of bushels of tree seeds around Alderley. He contrasted sharply with his father, whom he described in his *Praeterita*:

> The year after the burning of the Hall all the land round the Park House was let by my father to one Stephen Penncroft. It was a favourite expression of his that he hated the occupation of a field, even to having a cow. To be without a care of any kind was his elysium, and he has said in my hearing often, that if he could have his choice of life, it would be living in a hotel.

It would clearly not have been Sir John's own choice, however, since Maria Josepha could write of her husband during 1797 that 'The dear old Man is as busy as fifty bees all day long, and plenty of employment there is, for it would vex anybody to see some of the most beautiful ground possible in a state of wilderness beyond anything you can conceive'.

> We are as busy here as possible. He is making me seats and walks and bosquets, intermixed with more necessary operations of planting, fencing and draining, &c. In the few months we have been here much has been accomplished, and yet the very first moment it can be done, we must build quite a new mansion. The dear old rogue … delights in his idle, busy, lounging, active life. He is never indoors at meals, and does not come in to dinner till six o'clock.

The marriage of the first Baron and his wife was a turning point for the Stanleys. The fusion of what Maria Josepha called 'the sluggish blood of the Stanleys' successively with hers and that of her daughter-in-law Henrietta Maria – 'the Ladies of Alderley' – brought a radical change. It created a family whose descendants, from these two intelligent and masterful women, were remarkable and sometimes eccentric. Their history is inextricably entwined with that of Alderley Park itself.

Later chapters will describe the creation by Sir John of the special projects of the Water Garden, the Arboretum, the woodland Boys Walk and the new (current) Radnor Mere dam. Most of the fine old buildings of the Park date from his time, built from bricks made on the estate, and bear his initials.

He kept meticulous records of the estate farm and its output in his estate books. When ICI bought the Alderley Park estate in 1950, several thousand letters and many estate records and ledgers, some now preserved in the county archives at Chester Records Office, were found in the lofts of the disused former coach-houses and stables.

## Man of the people

Sir John Stanley stood out from his neighbours, the Cheshire squires of the time. Unlike them, he was politically reform-minded, being a strong Liberal. Moreover, fox-hunting, their chief occupation and topic of interest, held no attraction for him. Said to have been a very gentle and kind person, he was not a hunter but an observer of nature, and an enthusiast for it from his childhood, recording that:

> How I gained it I know not, but a love of Nature had grown up with me to be a passion, and I cannot but think that a sense of the beautiful in the natural is by some inexplicable law united in the human mind with a sense of the beautiful in the moral world.

## Community care and medicine

Sir John deeply regretted the loss of the family's records in the 1779 fire of Alderley Hall (which included those of their business interests). Clare Pye describes in *The Story of Alderley* the ways in which he 'did his best to piece together the history of Alderley and his own family's involvement in the area.' He interviewed many of the old men of the parish to discover what life was like before the fire and the Enclosure (the first fencing of open land as fields) of Alderley Edge.

He and his wife were heavily involved in helping their own people and local community as Maria wrote to her aunt Serena Holroyd in letters around 1800:

> We have just succeeded in getting a schoolmistress established in the village; by giving house, rent and £6 a year for teaching fourteen children, and I have no doubt she will have a very large school.
>
> We are buried in snow, and the trees are looking beautiful with their powdered wigs … The necessaries of life are so dear, I do not know what would become of the poor without assistance. There has been a subscription raised in the two Alderleys which amounts to above seventy guineas. This is allotted to selling meal or other food at a reduced price to the families who are thought proper object by the committee … I have tried, but find a soup shop would not answer here, where the population is scattered.

## Maria gives advice on innoculation

In 1802, Maria wrote to her sister Louisa Clinton:

> You and Lady S are two ninnies to insist on Baby Lou's being twice cowpocked. I have had thirty-six children of all ages inoculated [against smallpox] in Alderley within the last month and all doing well.

The latter was, however, probably the confidence of an experienced mother, since by the end of 1802 Maria had six children. Two years before she had not been so sure:

> Since I wrote my mind has been a good deal engaged and good deal anxious about my poor little baby. She has been inoculated a fortnight ago, and the place seemed to die away, so Mr. Holland thought it advisable to try the other arm. On the eleventh day after the first inoculation, fever and inflammation had increased so considerably that Mr. Holland applied five leeches to the poor little arm, which in a very few hours gave great relief. Last night she had fever, and the other arm is very much inflamed in the usual way.

This nevertheless shows Maria's courage and innovation, since Edward Jenner had only carried out the first ever inoculation with cowpox (to protect against the deadly smallpox) just four years earlier, in 1796.

It is interesting to reflect that two centuries later in Alderley Park it required at least double that period to move from the discovery of a new drug to its launch into widespread clinical use.

After a long and famous life, Lord Stanley died aged eighty-four, in 1850. He lies at ancient Nether Alderley church, within sight of his beloved Park.

# Alderley Edge and its railway; the Cottontots

The kindness of Lord Stanley and his wife toward their own folk did not extend to all outsiders. In particular they were extremely unfriendly toward the so-called 'Cottontots'. This was the name they gave to the cotton millionaires of Manchester who took to escaping the grime of the city by means of the new railway line that passed through the newly built village of Alderley Edge. The railway company popularised day trips and cheap excursions to the village, to the grief of the Stanleys. Mitford comments that the Stanleys 'whose Liberal principles seem to have rather deserted them over this affair, struggled to keep the public away, but gradually were obliged to give in.'

## Impudence

In September 1843 Henrietta Stanley wrote to her mother-in-law, Lady Stanley:

> As we were at luncheon yesterday we saw a man walk up whom his Lordship feared was a Parson. I feared a Rail Roadian and so the case proved, Mr. Waddington, Deputy Chairman B & M. [It was actually the Manchester and Birmingham Railway.] … Now what do you think was the impudent man's proposal, that Lord Stanley should allow the managing clerk of the R. R. to give orders on one of our private days!
>
> The Cottontots forsooth prefer the days without a crowd … I said Lord Stanley was already extremely liberal in granting 3 days and that we must have the other 3 for ourselves. 'Oh, but these would be respectable people who would show you respect.'
>
> I answered I did not want respect but privacy and that it was perfectly immaterial to us who the people were …

> The man was pertinacious. 'Would we not try it?' 'Impossible … 'Oh, Lord Stanley lived in the heart of the people' and something about the public. I said we had nothing to do with the public and owed them nothing and after all this man is a Tory.

## Great dogs

Lady Stanley wrote in agreement to her husband: 'The Manchester gentry are much more annoying to one's comfort and enjoyment, than operatives, as one can neither hand cuff nor great dog them if they are intrusive or offensive.' A year later Henrietta was still complaining, this time to her husband Edward:

> I had a talk with Simpson [the Stanleys' gamekeeper on the Edge] yesterday, he had been to Swain with a message from Lord Stanley to say if he did not keep Manchester people off on Private Days he should be removed, & he said he could any day but Sunday when the people sit all along the path from the Hough to Stormy Point.

# Heraldry of the Tenants' Hall

One thing preserved by ICI in the former Tenants' Hall (more recently known as the Sir James Black Conference Centre) was its heraldry. It richly reveals the Park's earlier aristocratic history.

## Shields of the Hall

A full survey of the Hall was carried out by Harold Storey of the Macclesfield Heraldry Society in 1996. There are three displays of heraldry in the Hall. Over the fireplace are the complete armorial bearings of the 4th Baron Stanley of Alderley, who built the hall, and his wife. (These combine the crest, coronet, supporters (stag and lion), shield bearing the coat of arms, and the family motto, *Sans changer*).

More importantly, the eighteen painted heraldic shields (coats of arms) on the oak panelling were, when examined, in very good condition (much more so than some others of the family in Nether Alderley Parish Church). Such a group of shields of all the family ancestors with their wives is rare.

Several coats of arms appear in the glass of the stone mullioned windows. They might be the work of the artist who designed the windows in Manchester Town Hall and Central

Library. They include the Legs of Man (recalling the Stanleys' kingship of that island), four shields with the arms of the four married daughters of the 4th Lord Stanley of Alderley, plus eight shields showing other coats of arms which the Stanleys are entitled to quarter as a result of marriages to heiresses.

## Titles held by the Stanleys of Alderley

Several titles are held by the family, the first being baronet, a title first created by King James I in 1611 to raise money. In 1660, Thomas Stanley (1597–1672) was created a baronet. The present peer, the 9th Baron Stanley of Alderley is the 15th Baronet.

The next oldest title is Baron Sheffield. This one is complicated. To start with, it is a peerage of the Irish House of Lords which no longer exists. In 1781 John Baker-Holroyd was created Baron Sheffield and was later promoted to Earl of Sheffield. But he could not pass this title onto his two daughters. So he arranged to be given a *second* title of Baron Sheffield in 1783 which *could* descend to the male heirs of his daughters. In the event, he married again and had a son; but his line died out in 1909. The 1783 baronry then reverted back to the eldest descendant of Maria Josepha, who as described earlier married the 1st Baron Stanley of Alderley. In 1909 the 4th Baron Stanley of Alderley became also the 4th Baron Sheffield.

Coats of arms below the Minstrels' Gallery
PHOTO GEORGE B. HILL

## Stanley of Alderley

The Barony of Stanley of Alderley is much simpler. It was created in 1839 for the then Sir John Stanley, 7th Baronet. (1766–1850). As a United Kingdom barony (unlike Sheffield), the holder held a seat in the House of Lords at Westminster until the House of Lords was reformed.

The Barony of Eddisbury was created in 1848 for Edward John Stanley (1802–1869). He was the elder son of the 1st Baron Stanley and had been the MP for North Cheshire. The government wanted him in the House of Lords (while his father still lived), so made him a baron also. Two years later his father died and he succeeded him as 2nd Baron Stanley of Alderley.

Of the three baronies above, Sheffield is first in order of precedence and that title has been used by some of the barons, but mostly the Stanleys have chosen to be known by the Stanley of Alderley title.

# 3 Elegance and water

## The Water Garden and grounds

The carefully designed sun-trap of the Water or Pond Garden (sometimes called Watergarden, though this account uses the original name) was part of a complex of walled gardens begun in about 1798 to the south west of the Park House, which became after 1779 the main residence of the Stanleys.

### Lady Margaret's design

The Garden is said to have been designed by Lady Margaret Stanley (1742–1816), who gained the idea while on a fashionable Grand Tour of Europe with her husband Sir John, the sixth baronet. Nevertheless, Matthew Hyde comments that the rather old-fashioned Garden 'seems to be a deliberate echo of the one they had left behind at the Old Hall'. The Garden is not shown on the estate map of 1798; and the work was doubtless not carried out by Lady Margaret's husband, who was not an active man, but by her son.

### Lady Margaret's Walk

The avenue of mature blue-green Lawson Cypresses descending from the Adam Gates toward the pool, framing a view of its fountain in one direction and the gates in the other,

bears the name Lady Margaret's Walk and presents perhaps the most beautiful vista in Alderley Park. Contrary to what most people think, the curious shape of the cypresses is not due to pruning; they have never been trimmed.

## Water Garden construction

The Water Garden takes the form of a terraced garden around a large square pond with a fountain, all surrounded by high walls of mellowed handmade bricks, with the main original entrances along the east wall.

The Garden was the focus of about 10 acres of formal garden, and its excavation required the removal of large quantities of earth by shovel, horse and cart from what was then called the Hodge Field. Terraces and conifer-lined walks were laid out in the style of a sixteenth-century Italian garden.

## The Water Garden pool

The pool has more than one water supply. The original inflow was from the nearby Serpentine. A more continuous supply was later obtained via a pipe from a spring on Hocker Lane nearly a mile away to the north east. There is little outflow of water, due to evaporation.

In 1974 the pond was drained by removing an oaken post which was the drain plug. The bed of the pond was found to be of peat covered by up to thirteen feet of mud and a mass of huge water lily stems. In the mud were found two very old model wooden boats which had been hand-carved. The larger – nearly a metre long – was constructed from a solid piece of pitch pine, with a lead keel fastened by brass rivets. It perhaps belonged to young Owen Stanley, son of the first baron's brother Edward, who was born at Alderley Rectory in 1811. A close friend of Lord Stanley's ten children, Owen sailed toy boats with them on the pond and had an early passion for ships and the sea. He grew up to achieve considerable success as a naval surveyor and explorer, charting the Great Barrier Reef in Australia and taking famous biologist Thomas Huxley, an Alderley Park visitor, with him on one voyage.

## The Gardens and their visitors

The Stanleys preserved their gardens rigorously for many decades. A gardener's bell once hung on the wall of the Garden, which would be rung whenever a member of the family wanted to speak to the head gardener. The century-old bell was found during excavation in the 1970s and identified by site manager and historian George Driver.

The summer house was added to the Garden in 1907. The initials over it are those of Mary Katharine, the wife of the fourth baron. A millstone set in its floor may have come from the mediaeval mill.

After the First World War, visitors were allowed into the estate. On Saturday afternoons from July to September, charabancs packed the stable yard and a brass band played for dancing on the well-kept lawns, as crowds arrived to enjoy the Park's gardens.

On 11 September 1931, the gardens were a fine show on the occasion of a local fete:

> In the park just beyond the garden boundary deer grazed peacefully, quite undisturbed by the band which was playing in front of the hall. It was one of those scenes which one likes to regard as typical of the English countryside … On the southern side of the lawn there are magnificent specimens of many species of trees, including the cedar of Lebanon and a long avenue of cypress parallel with the high wall of the beautiful pond garden.
>
> The pond garden at the rear of the hall is protected by high walls, and on passing through the gateway one's attention is arrested by a totally different aspect of natural beauty. This garden is planned on more symmetrical lines, with succeeding terraces down to the level of a large square lily pond in the centre, but the lay-out of the beds is such that one does not mind the severity of formal outline. In fact it possesses a distinctive charm; and the perfect orderliness of the beds, the neatly trimmed lawn terraces and the absolute freedom from weeds bear testimony to the skill and to the industry of Mr T. A. Summerfield (the head gardener) and his staff …

In 1936 the plants of the Water Garden were described in *Cheshire Life* magazine, including the unusual *Garrya elliptica*, then a rarity with its light jade catkins said to resemble those depicted in some Adam architecture.

## Neglect

The Garden fell into disrepair during the last war, with brambles growing to the water's edge and self-seeded oak trees growing out of the wall of the pool itself. Its restoration by Fred Matthews is described later.

## Plants of the Water Garden

A few plants survive from the Stanley period. Very many were later introduced, and one survey in ICI's time identified over 250 different species. The pool holds several colours of lilies, not just the native white and yellow ones.

Some plants are striking. Even the briefest examination during the late summer cannot miss the huge **prickly rhubarb** (*Gunnera manicata)* which grows at the north-west corner of the pool. This native of Brazil is the largest-leaved plant hardy in the British Isles. Each year its foliage reaches up to ten feet in height, with leaves sometimes up to six feet across completely shading the brush-like flowers. **Royal fern** grows nearby.

In spring the path along the west side is lined with beautiful magnolia trees. Nearby is a **New Zealand flax**, a species brought to the Water Garden by Fred Matthews from the famous gardens at Inverewe; and the strange **umbrella plant** (*peltiphyllum*) flowers by the water in spring.

A magnificent **copper beech** stands at the south-west corner and cherry trees along the south wall. In the south-east corner is a pink-flowered *camellia* that may be a century old and formerly grew in the middle of the walled garden. In the north-east corner is a Doric arch beside which grows a lovely *magnolia grandiflora*; while to its left, reputedly ninety years old, is Lord Stanley's **bay tree**.

In 1998, a guide to the plants of the Garden, *The Water Garden Through the Year*, was created by me with the help of Lesley Stafford and professional knowledge of Ian Urquhart, Fred Matthews's son-in-law. Illustrated with original line drawings by Brenda Scott, it is now out of date and only a collector's item.

The ancient *camellia*
PHOTO GEORGE B. HILL

Doric arch

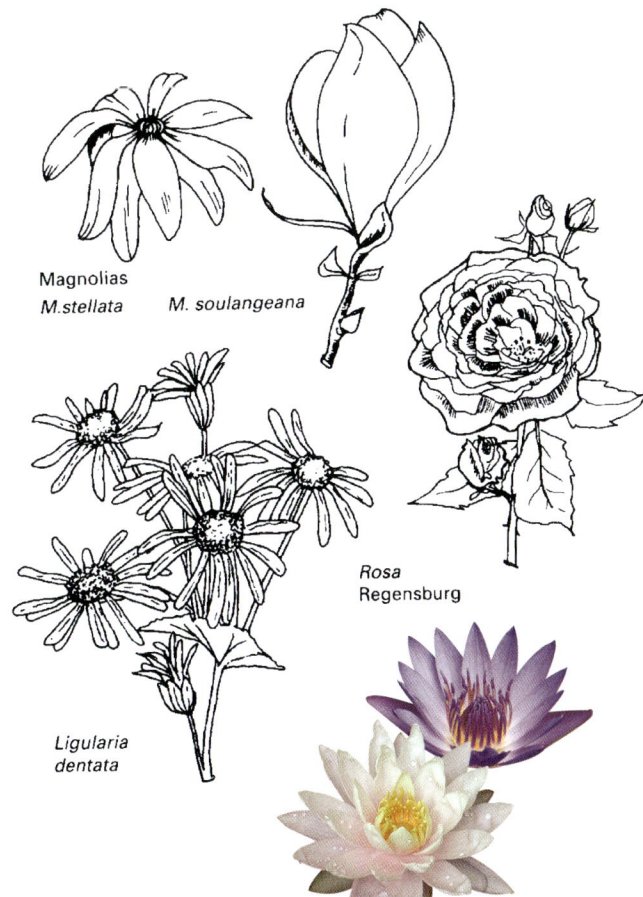

Magnolias
M.stellata    M. soulangeana

Rosa
Regensburg

Ligularia
dentata

Water Garden 1998
guide

Magnolias
DRAWN BY BRENDA
SCOTT (KIND
PERMISSION))

*Ligularia dentata*
DRAWN BY BRENDA
SCOTT (KIND
PERMISSION)

*Rosa regensburg*
DRAWN BY BRENDA
SCOTT (KIND
PERMISSION)

The Water Garden
holds several
coloured water lilies

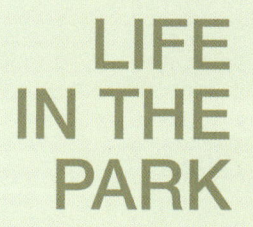

**Wrong trees!**

Some of the mulberries are said to have been sent to Alderley Park by King James I when he was endeavouring to encourage the silk industry in England. The last was felled in about 1964. King James's attempt failed, apparently because he sent the wrong trees: it was not then known that silkworms grow on white mulberries, not on the black mulberries he sent. It is related that when his idea did not work King James later sent in a bill for the trees!

### Walled garden

Adjacent to the Water Garden is the enclosed walled garden, later used as a cricket pitch with tennis courts. The walled garden would have provided fruit and vegetables all year round and also acted as a sheltered area in which to stroll. An underground network of flues, heated by coal or wood fires in regularly spaced hearths built into the hollow walls, was used to keep the soil warm. Peaches, figs, melons, apricots and vines flourished, although the alcoves where they grew no longer remain. In the rest of the area more common vegetables and flowers were grown in cultivated beds. The garden would have been protected by large lockable gates, making it inaccessible to the casual visitor. In 1931:

> The kitchen garden is a notable feature at Alderley Park for here we find a great variety of fruit and vegetables which are so essential in every home … Bulbs, of course are amongst the most prolific producers of spring flowers, and no less than 15,000 tulip, hyacinth, and narcissi bulbs have been or will be planted in the course of a few weeks. Already between five and six thousand leeks are planted out, and neat rows of spring cabbage numbering between three and four thousand will be doubled before the autumn is far advanced. Lettuce and onions are produced by the thousand, and next year the quantities will be considerably increased; for I was told that many trees and hedges which have occupied a considerable portion of the kitchen garden will be removed and the land put under cultivation.

The walled area was later again the site of a productive market garden during and at first after the Second World War.

In the centre of the garden there was formerly a walled cemetery where the dogs of the Stanley household were buried. The tradition dated back to 1700 and each dog had its own headstone. In the cemetery there grew, until the early 1960s, the camellia mentioned above together with several very old mulberry trees.

## The Adam gates

The main original entrances to the Water Garden were along its east wall. In the centre of this, looking down the fine view towards the central pool, there stand the impressive neo-classical wrought-iron Robert Adam style gates.

The gates, by Adam's acolyte Samuel Wyatt were brought to Alderley Park from Winnington Hall near Northwich, where they once guarded the poultry house. Winnington,

also later purchased by ICI, was obtained in 1808 by the Stanleys as a winter residence, being much more comfortable than the small house in Alderley Park. But Maria Josepha, the first Baron's wife, had no doubt about which home she preferred, writing often to her relatives with such comments as:

> To be sure, I do like Alderley a thousand times better as a residence. There is nothing in favour of Winnington but a better house … Winnington is remarkably dry and warm, well adapted to an autumn or winter residence, and during the summer we shall frequently come <u>home</u> for a few days.

The beautiful filigree work with its lead emblems holding the wire lacework is typical of this most popular architect of the 1700s. Lichen-encrusted eagle and child statues top both pillars.

## The day when an eagle flew!

The stone eagles on the Adam gates have surveyed their realm for many years. On one occasion, however, one of them disappeared! For reasons never explained, it was discovered sitting on the desk of chemistry department manager Bernard Langley! But it evidently flew back home afterward.

# LIFE IN THE PARK

ELEGANCE AND WATER **49**

# Alderley House Arboretum

The Arboretum is the tree collection on the lawned area behind the site occupied (in turn) by both Alderley Houses.

The most spectacular tree is the **giant redwood** or **Wellingtonia** from California, which stands on its own knoll and has an aromatic foliage and a deep, spongy, fibrous bark – which can be punched without hurting one's fist! Small birds like treecreepers roost in hollows in the bark. Two more were planted in recent years to replace it in due course. Along with a selection of attractive small deciduous trees there are also fine mature specimens of **Lawson cypress**, **western hemlock**, **American red oak**, **holm oak** and **yew**.

A curious feature of the Arboretum is a small gravestone in the grass marked 'Lena 1856–62'. Henrietta Stanley had more than one dog of that name, but why one was not buried with the rest of the family's dogs in the walled garden is unknown.

## Old mansion surrounds

The finest single tree in Alderley Park is the classic **cedar of Lebanon** that stood first at the rear of the mansion, at one end of the Arboretum, and then in the heart of ICI's Alderley House complex, which was built around it as a stunning feature. This tree marks the final resting place of the Stanley family's pet birds! Lady Victoria Stanley, who visited in 2003, remembered a series of parrots and similar birds being interred there.

At the front of where the old mansion stood there remain three fine trees. One, the famous **sweet chestnut** known as the Churchill Tree, is mentioned later. Beside another chestnut is a fine young **blue Atlantic cedar**, planted by ICI.

## Boys Walks

South east of the Arboretum, paths either side of the Serpentine lead to informal wooded parts of the pleasure grounds created by the first Lord Stanley.

The main path along the pool leads to the Boys Walks. This maze of informal paths was laid out through woods where the mixed plantation around the far end of the Serpentine now grows. The only remnant of the first baron's original planting is the mature deciduous woodland near the Alderley House site. The nature trails through the woods are described later.

Cedar of Lebanon and AstraZeneca's Alderley House in 2014

# Reverend Edward Stanley & his birds

Reverend Edward Stanley (later Bishop of Norwich), the younger brother of the first Lord Stanley, was another figure who put Alderley Park on the map.

Born in 1779, the year Alderley Hall caught fire, he inherited from his Welsh mother her brilliant colouring, distinctive eyebrows and flashing dark eyes that gave force as well as beauty to her face. From her, too, came the romantic Celtic imagination and fiery energy which inspired him to find interests everywhere.

## Liberal reformer

As a young man, Reverend Stanley saw France still smarting under the effects of the Reign of Terror, an experience that made him a Liberal, like the rest of his family and unlike the Stanleys' county neighbours. He then returned home at his brother's request and took command as a captain of the Alderley Volunteers – a corps of defence raised by him on the family estate in expectation of a French invasion and approved by King George in September 1803. In 1804 Prince William Frederick visited the Park to inspect as the men 'went through the various manoeuvres of a Batallion [sic] of Light Infantry, and were much praised by His royal Highness for their Discipline and appearance.'

# LIFE IN THE PARK

Edward also became an amateur playwright and poet as well as an accomplished writer. Legends of the countryside, domestic tragedies and comedies were turned into verse, including the fall of Sir John Stanley and his spectacles into Radnor Mere!

Edward would rather have gone to sea than entered the ministry, but when he did his father made him Rector of St Mary's. He was a successful minister; before he arrived the clerk used to go to the churchyard stile to see whether there were any more coming to church, for there were seldom enough to make a congregation. But before Edward left 33 years later his parish was one of the best organised of the day. He set up schemes of education throughout the county as well as at Alderley, and was a leader in local reform. He also had great personal courage. On one occasion he heard of a brutal prize fight nearby and rode rapidly alone into the centre of the huge crowd to stop it, before calling the fighters to him next day to give them a *Bible* each.

## Church and Park

Edward's family of five, with his wife Catherine at the Rectory and that of Sir John at the Park, created a lively social circle, and there was constant to-and-fro between the Park House and the squat tower of ancient Nether Alderley Church. (Edward's third son was Arthur Penrhyn Stanley, the famous Dean of Westminster who was tutor to King Edward VII).

## Bird book

Reverend Edward Stanley was a well-known ornithologist who never lost a chance of watching wild bird life, and including accounts of them in his ministry, as when he described the crowds of Starlings that fed on the lawn near his 'ivy-mantled parish church, with its massy grey tower' as they 'sociably twitter away their chattering song, as the vane creaks slowly round with each change of wind.' His major work outside his church activities was the writing of his *Familiar History of Birds* (1836, in two volumes), which is considered second only to Reverend Gilbert White's celebrated *A Natural History of Selborne* as a classic of its kind. He was highly respected and became President of the Linnaean Society and, like his brother, a Fellow of the Royal Society.

Edward's book is particularly interesting on Alderley Park, since in it he recorded many anecdotes that obviously refer to Radnor Mere or the Park. Some significant bird records are quoted later but others among his writings show an odd mixture of science and myth. Thus, he dismissed as 'ignorant reasoning and credulity' the ancient belief that barnacle geese hatched from sea shells. Yet he still mentioned the theory that swallows hibernated in mud at the bottom of lakes during the winter! He was still able to talk of a flock of over two billion [*sic*] of the now extinct passenger pigeon. His book is full of wonderful anecdotes, including those of a one-legged flamingo that used its neck as a second leg; and of a gentleman who was attacked by a shot bittern he was carrying in his (large!) pocket.

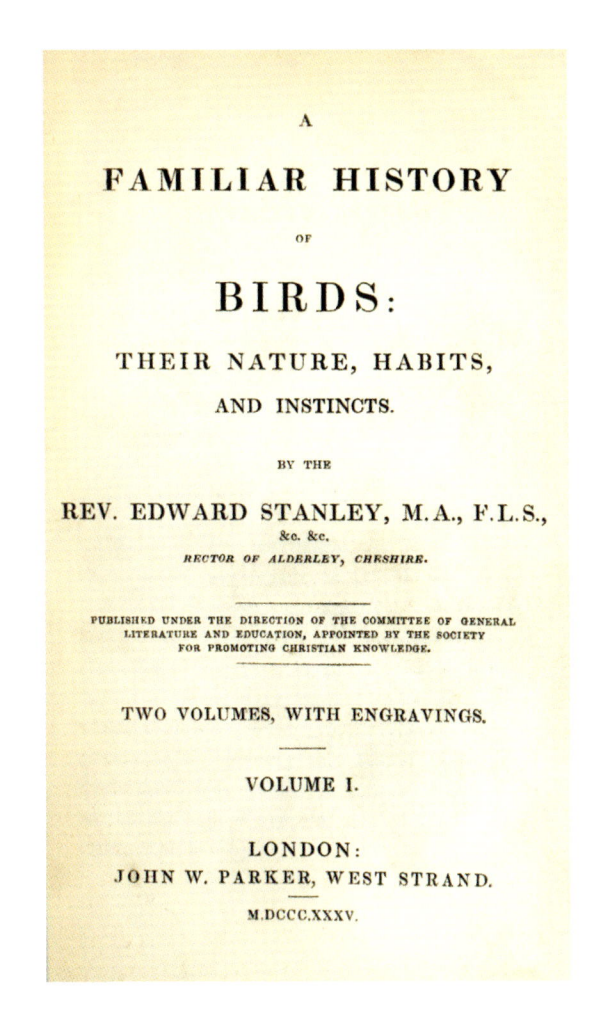

Hawk carrying off a Leveret.

He recorded a belief prevalent in the Faeroe Islands that herons had no intestines and that a heron catching a fish 'is under the necessity of placing its rump against a hillock or stone, to prevent the fish from getting out again', standing in that position until the fish was digested. Nonetheless his analytical mind rarely faltered, although he did take seriously a claim that swifts 'fly at nearly 180 miles per hour'!

## The History of Radnor Mere

The purplish brown of the wood rising above the softened reflection of it in the water; a few touches of brighter brown in the shrubs and ferns near the edge; the boat-house relieved by the dark wood behind it; a line of yellowish brown reeds breaking the reflection of it in the water; and another still brighter yellow-and-brown island coming immediately before it; the soft blue haze spread over the water and softening the reflected outlines of the wood without weakening the effect, contrasted here and there with the vivid and determinate outline of a few leaves or weeds lying on the surface of the water; the scene enlivened now and then by a wild duck darting from the reeds across the lake, making a flutter and foam before her, and leaving a line of clear light behind her on her path, her wild cry distinctly echoed from the wood, and sometimes both from the wood and deerhouse *together – such a simplicity, yet variety of tint, such a force of effect, and such a softness of shade and colour! Artists, one and all, hide your diminished heads!*

*17 December, 1810 Catherine Stanley*

Amongst the striking features of the natural landscape of Cheshire and Shropshire are the fifty or so picturesque lakes known locally as 'meres'. The origin of this Mere's name is unclear, although the name Radnor refers elsewhere in Cheshire to the local sandstone or 'red bank'.

## A mere?

The present Radnor Mere, a third of a mile long, is less a classical mere than a reservoir. Like most meres, nevertheless, it has only a small inflow and outflow for its size, and is partly fed by groundwater, which gives a very stable water level even during long dry seasons. It probably conceals no hidden spring. It has accumulated much silt and at its narrower end this has built up into a small swamp with ancient willows.

Even a glance shows that the present Mere lies on a gentle hillside. But a smaller and shallower mere existed before it, perhaps originating as a natural marshy pool or hollow, possibly formed by subsidence.

## Probable history

The Mere's form was dictated by the needs of the nearby Mill. The water supply to drive the mill wheels originally came from two brooks flowing down from Alderley Edge. It is unlikely that these were sufficient all year and so the mill was built with a large pool and a series of sluices or 'paddles' with which the miller could accumulate water overnight to work the mill the following day. At some point, two more brooks were diverted into a new Parkhead Pool and then down into a new or enlarged Radnor Mere, which was later enlarged a second time.

## First dam

The first dam may have been sixteenth-century and was certainly present by the 1620s when the beechwood was planted. By 1798 the Mere held extensive reedbeds which must have taken many decades to grow. The reeds were gathered each winter by estate workers and used for animal bedding and thatching.

This dam was built of local sandstone, possibly from the quarry near the northern end of the Mere, and sealed with a core of local marl (lime-rich clay). The digging of this might have created the current Mere overflow pool below Mereview Restaurant, which is clearly marked on the 1798 map and has been called Radnor Pool or the Stock-field Pool. From here the stream followed a new course to the mill pool via the route which still exists and is a diversion of the Mere's original outflow.

Once filled, this earlier lake occupied some 15 acres and provided substantial insurance against drought for the Nether Alderley Mill. This probably proved adequate for the needs of the mill for many years.

## High water!

On 17 August 1799, the original dam's time began to run out. There was torrential rain, leading to much flooding in the area. With the waters rising to dangerous levels, the dam was deliberately broken and the Mere emptied. Interestingly, 250lbs of eels were gathered from the residual mud. When the waters subsided, a more substantial sluice was constructed to replace the ancient one weakened by the floods.

But by 1826 either the old dam was beyond repair, or the volume of water in the Mere was inadequate for the needs of the Mill. A new, higher, straight dam was constructed between 40 and 70 yards behind the old dam, perhaps over the old overflow pool; the new dam was named Stockfield Dam. The estate records of 1826 show that Randle Clarke, Peter Winnington and others were subcontracted to do the construction. The building of the dam occupied about half a dozen men from July to December 1826. The same technique as before was used but the new dam was much longer and its height varied from six feet to a maximum of fourteen at the deepest point. The Mere's area increased by half and its depth by around three feet (though three feet of silt now covers its floor).

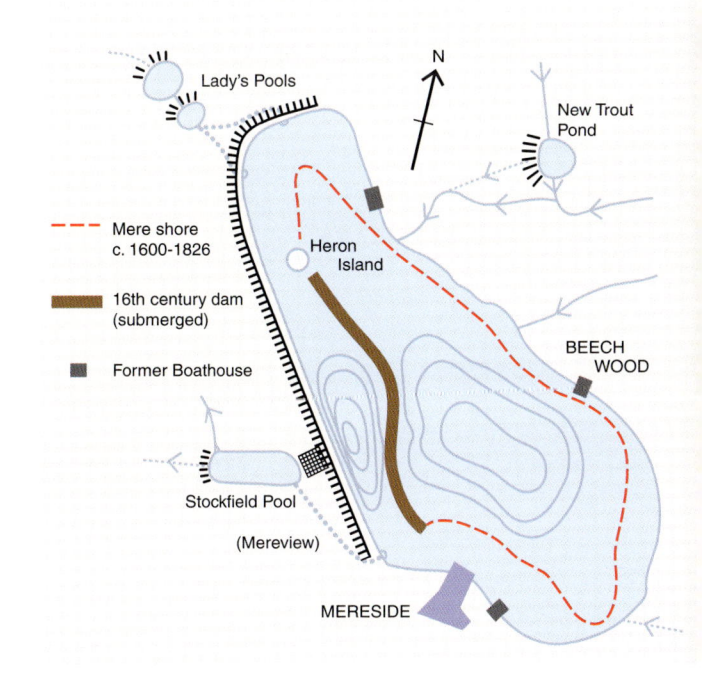

Map of Radnor Mere contours

Once the new dam was completed, the sluices on the old dam were removed and the Mere was allowed to fill to its new capacity, submerging the previous dam. The latter still remains largely intact below the surface; its sandstone capping stones and gravelled top have been kept free of silt by the scouring action of a current across the dam top, caused by it restricting circulation of the lake water during strong winds. Some of its line can be seen on aerial photographs, and an echo-sound map of the Mere's floor, made by the ClubAZ Angling Section, shows it clearly.

The new dam now has many alder trees growing along it, which shield it from view and whose roots probably do little good to the dam underneath. In 1989 an inspection found the dam did not meet current legal standards under the Reservoirs Act, so it was raised about a foot with limestone chippings. Prior to this, the water in the lake had nearly overtopped the old 1826 dam's crest in wet periods.

## Islands

The newly flooded Mere contained two islands, the larger being a deliberately raised, sandstone-reinforced, circular island near the north end. A small island at the southern end, shown on a map of 1872, has now merged with the bank. The island is the site of a thriving heronry, described later, which began around 1975. Previously to that a pair nested in the nineteenth century; but it is unlikely that the Stanley family would have tolerated herons which would have represented a threat to the rearing of food fish, particularly trout, in the estate's smaller pools.

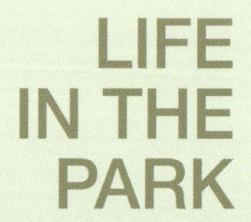

## LIFE IN THE PARK

### Ken's captives!

The recent herons in the Park may have descended from ones brought to the Mere in the early 1970s by Ken Haworth, who rescued them from Roman Lakes at Marple where a local fish farmer was intent on shooting them. Ken fed them until they could not fly, netted them then drove them to the Park. They settled because he kept feeding them for a while.

## Change of purpose

The Mill was not used commercially after 1939. And the new Mere became ornamental in purpose much earlier, as a feature and focal point in the Park. It became a major source of recreation; one sketch shows a sailing boat, and several of the Lords Stanley were keen fishermen. The estate records for 1827 refer to a fishing expedition on Radnor Mere. Several years later Lady Stanley recorded the excitement that the capture of a 10lb pike had caused. But in 1849 she declared that 'Fishing is the bane of this family', after Edward caught a heavy cold doing it.

The Mere always supported a healthy fish population; this would once have been an important source of food on the estate. Indigenous species such as pike, perch and eels would have been regularly cropped for food. In the early twentieth century, the estate permitted angling from the banks for a charge of one shilling, proceeds being distributed to local hospitals. During the Second World War, permits were available from the shop of Mrs Potts, opposite the Mill, and it was not uncommon to see a row of anglers along the bank.

## Boathouses and artists

The early boathouse referred to by Catherine Stanley was submerged when the Mere's depth was increased. This sketch from a Stanley family notebook probably shows it. The family kept a little ketch, the 'Ariel', on Radnor Mere.

The Ordnance Survey maps of 1870 show the presence of three later boathouses, two on the wooded shore in the north-east corner and east side, and one amongst the sedges

Early boathouse on Radnor Mere
COURTESY LORD STANLEY OF ALDERLEY

and reeds in the south-west corner. Radnor Mere is one of the meres featured in famous wildlife artist Charles Tunnicliffe's *Mereside Chronicle* (1948), a diary and sketch account of his local Cheshire meres. The book includes his sketch map of the Mere, drawn in 1946 and a drawing of the one boathouse then remaining on the wooded shore (the overgrown remains were visible in early ICI days but there is no visible sign now).

The boathouse on the south-west bank was still standing when ICI purchased the Park in 1950. It was renovated toward the end of the 1950s to house the 'Chairman's Punt' – the first Division Chairman Philip Smith being a keen angler! It was demolished in the early 1960s and the sunken punt removed, to allow rapid access for the fire brigade to the deepwater boat channel.

Map of Radnor Mere by Charles Tunnicliffe
FROM 'MERESIDE CHRONICLE', 1948. COPYRIGHT TUNNICLIFFE FAMILY

## Swimming

Radnor Mere was popular for swimming early in the twentieth century. Around the time of the Great War, some one hundred permits for people to swim in the Mere were issued annually. Regular camps of Territorial Army cadets and scout jamborees were attended on occasion, as in 1931, by the Chief Scout, Lord Baden Powell. In May 1919 the estate manager recorded that over 900 boys were camped in the Park, many of whom swam. Alderley Park was also used for bonfires on November 5th. These were organised by Mr Paton, headmaster of Manchester Grammar School. A great character renowned for his apparent ferocity, he was heard to say that: 'Any boy who falls into the mere will be put on the fire to dry; any boy who falls in the fire will be put in the mere to cool!'

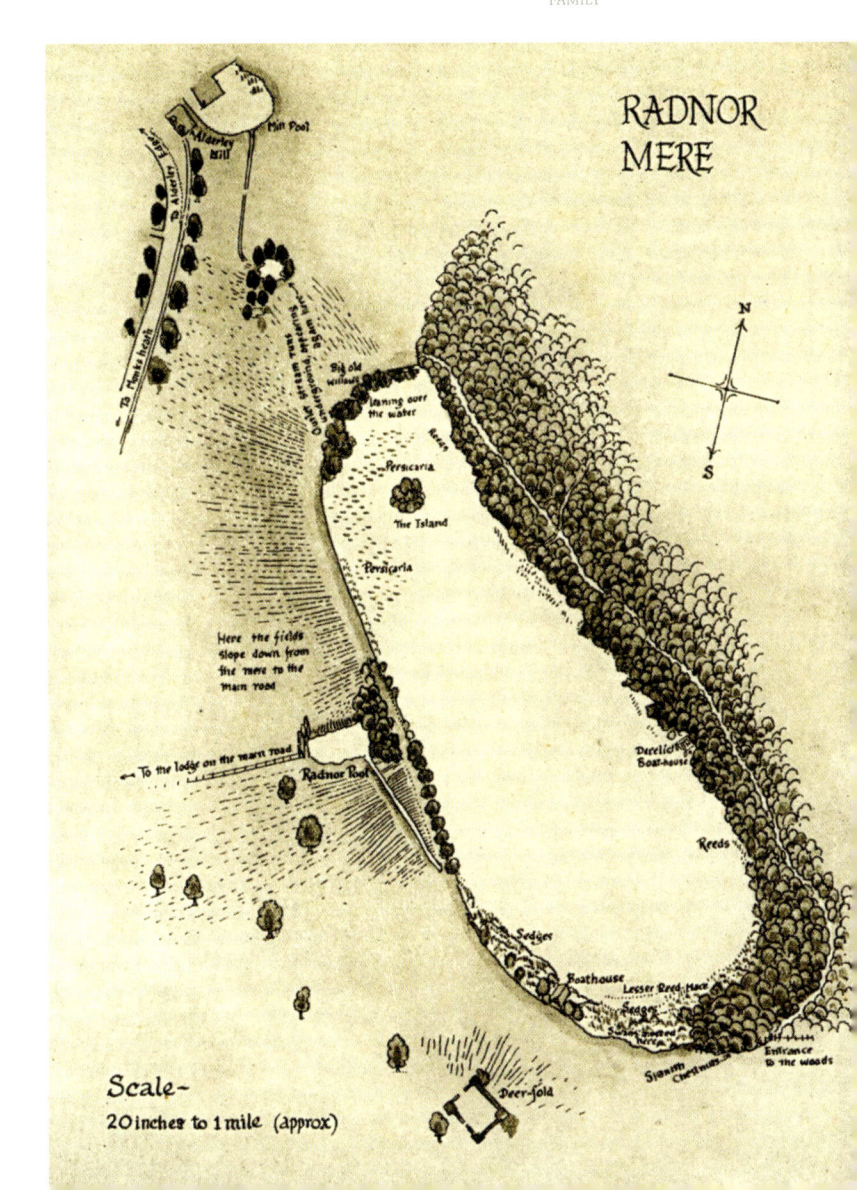

RADNOR MERE

Scale –
20 inches to 1 mile (approx)

## Outflows and management

Radnor Mere has four outflows, but only one normally operates and two are ancient. The main outflow is near the Mereside buildings. A large emergency slipway nearby is designed to allow water to escape rather than flood the restaurant, if the main outflow blocks. It has never operated thus far.

The Mere receives no particular maintenance. Projects to increase its wildlife value have included a sand martin nesting bank; unfortunately, its construction coincided with a national crash in the sand martin population (which it was designed to alleviate); its holes proved of more interest to other birds.

The Mere's reedbeds have been declining for many years. A serious loss occurred when some beds along the open shore below the buildings declined and collapsed after landscaping. Shadowing trees have been felled along part of the wooded shore and reeds planted there, which have only slowly taken root. But wildlife from birds to dragonflies uses the reeds, water lily beds and other greenery around the lake, and diving birds like tufted duck even explore the lake floor.

One other visible change in August 2007 was the completion of a new pump house at the south end of the Mere to provide an emergency supplementary water supply in case of a major fire.

Tufted duck
PHOTO GEORGE B. HILL

Brown hawker
dragonfly egg-laying
PHOTO PAUL FLACKETT

# Further Lords Stanley of Alderley

### THE LORDS STANLEY OF ALDERLEY

**John Thomas Stanley** (1766–1850)
*Seventh baronet. Created Baron Stanley of Alderley in 1839.*

**Edward John Stanley** (1802–1869)
*Second baron and first Lord Eddisbury.*

**Henry Edward John Stanley** (1827–1903)
*Third baron. Became a Muslim.*

**Edward Lyulph Stanley** (1839–1925)
*Fourth baron. Henry's brother. Inherited and used the title of Lord Sheffield.*

**Arthur Lyulph Stanley** (1875–1931)
*Fifth baron. His early death led to the sale of Alderley Park.*

**Edward John Stanley** (1907–1971)
*Sixth baron. Sold the Alderley estates.*

**Lyulph Henry Victor Owen Stanley** (1915–1971)
*Seventh baron, held the title only for three months. Brother of Edward.*

**Thomas Henry Oliver Stanley** (1927–2013)
*Eighth baron. Cousin of the previous two barons. Farmed in Anglesey.*

**Richard Oliver Stanley** (born 1956)
*Ninth baron.*

### The second baron: politics

The second Lord Stanley of Alderley, Edward John, the elder of twins, was already a peer in his own right when he became baron in 1850.

Handsome, and a friend of Prime Minister Lord Palmerston, he had a notable rather than famous political career. He was a Whig Member of Parliament from 1831. He rose to the Cabinet as Postmaster General in 1860 and established the Post Office Savings Bank in 1861. He served in five governments, and was Lord Palmerston's Chief Whip and President of the Board of Trade.

One of his grandsons was the famous mathematician Bertrand Russell, who described him as a man of caustic wit; indeed, he was known in London as 'Benjamin Backbite' for his sharp humour. Clementine Churchill and the Mitford sisters were all his great-grand-daughters. An inattentive husband and father, he apologised to his wife and children on his death bed for his great nastiness and neglect to them. Yet his bark was worse than his bite; his wife and children all adored him.

Unlike his father he was probably little involved in management at Alderley. On one occasion in 1853 when he was in the Park, he wrote to his wife:

> Yesterday I assisted at the washing of the sheep in the mere, and found to my amazement that I am possessed of 600 sheep, beside lambs. It was a very pretty sight, with the men in the water, the shepherd and his dog, and the young lambs calling with their mothers outside the pen; together with the beauty of the spot and the beech trees in better bright green foliage than I have generally seen them.

His wife Henrietta, an Irish Jacobite, spent her youth in Florence, where she was a friend of the widow of Bonnie Prince Charlie. She bore her husband a large and brilliant family. Nancy Mitford says that:

> She loved him so much in spite of everything that during his lifetime her personality was completely suppressed, and she appears in her letters to him a gentle complaining, creature.

After their first son and heir Henry, Edward and Henrietta had eight other children who grew to adulthood. The sons were John, who became a colonel in the Grenadier Guards; Edward Lyulph, who after the deaths in turn of John and Henry became the fourth baron; and Algernon, who became a Roman Catholic bishop. The daughters were Alice; Blanche, whose famous descendants are described below; Maud, who Nancy Mitford describes as both 'ugly' and 'goodness itself' and like her mother was a pioneer of girls' clubs; Kate, who was Bertrand Russell's mother and an early supporter of female suffrage who used as her family doctor Elizabeth Garrett Anderson (the first Englishwoman to qualify as a physician and surgeon in Britain); and Rosalind, who married George Howard, Earl of Carlisle, heir to Castle Howard.

Edward, the second baron
COURTESY LORD STANLEY OF ALDERLEY

Coat of arms of Edward

## Light Brigade

Lord Stanley's letters to Henrietta were full of current affairs, including details of reports sent to him from the Crimean War in 1854 when he was paymaster to the Armed Forces. The following, perhaps written home to Alderley, should need no explanation:

> Nothing could have been more brilliant they say than the cavalry affair of the 25th … from some misunderstanding of orders Ld. Lucan ordered Ld. Cardigan to charge, which he did under a protest as he said it was madness to attempt it … he galloped in front of them all, right over a battery of guns, sabring the gunners at their guns & over a Regt. of Hussars … he returned through them all & brought the remnant of his men back to their position without confusion … it was perhaps the most daring feat ever performed by cavalry. The French said that 700 English Light Cavalry charged the whole Russian army, which was the fact, & one of them said 'C'est magnifique mais ce ne'est pas le guerre'.

Henrietta's letters were more prosaic, dealing with small affairs of theirs and their neighbours, such as in 1859:

> Did I tell you there was a bad smell at Tatton – fancy my nerves when I found it was Lord Ellesmere. Lady Egerton was so alarmed she thought she was getting diphtheria till she also found out where the smell came from. His wife must be an angel to be so devoted.

### A changing world

Nancy Mitford, who published many of the letters between the two evidently found them remarkable, commenting herself that:

> It is quite hard to believe that a world so different from our own can have existed less than a hundred years ago [Mitford was writing in March 1938]. The very relationships between human beings seem to be changed. What wife, nowadays, would support year after year the neglectful gallivantings of an Edward, what husband would support the continual reproaches and complaints of a Henrietta?

Nevertheless, Henrietta remained subdued during her husband's lifetime, during which she was apparently distinguished only by the interesting letters that she exchanged with her mother-in-law and husband. But:

> When he died, however, she became the terror of all who knew her. She threw herself with unbounded energy into the championship of women's rights and higher education. She became a solid matron with beady eye and hooked nose. In 1869 she was one of the founding donors of Girton College, Cambridge (giving money to establish its first library in 1884) along with Queen's College, London and High Schools for girls. Bertrand Russell later described his formidable grandmother as 'a woman of vigorous but not subtle intelligence, with a great contempt for "nonsense"'.

The second baron was buried like his father at Nether Alderley parish church, in an underground crypt. The location of this was later forgotten and was only rediscovered during work on the church in 2008 when church archivist Jenny Youatt and an architect opened it to find the coffins of the first two barons and their wives, along with those of the first baron's youngest son Alfred, who died aged 3, Colonel John Stanley, son of the second baron, and 'Moomie', the family nurse.

### The third baron: a passion for the East

Henry Edward John, the third Lord Stanley, who was partly deaf, was one of the great Victorian eccentrics. Very different to his respected predecessors, he developed at an early age a passion for the East, asking for an Arabic grammar book as a birthday present at the age of twelve.

He and his three brothers were all strong-willed, with a habit of rudeness and quarrelling, great indifference to public opinion, the same thick legs and eyebrows and with widely different religious opinions. All of Henry's family, particularly his mother and oldest brother Johnny, united against him when he started to adopt what they then (incorrectly) called the Mohammedan religion.

At home Johnny, a violent child who mistreated his brothers, used to tease Henry 'by coming down early in the morning and dedicating the breakfast to a statuette of Buddha he had procured for the purpose. "Have you dedicated it all?" Henry would say, hungrily eying the delicious fare thus placed out of his reach. "All but the ham, dear boy."'

Henry was thrifty and noted for his meanness; a wit once described him as a 'man of rare gifts'. Nevertheless he was kind by nature. Once, while still young, he paid from his own money to educate an orphaned boy, the son of an Alderley servant who had died in a cholera outbreak on the estate. But his manner could be fierce. He quarrelled with most of the family, on one occasion much later getting his gamekeeper to turn off the estate his brother Lyulph, who eventually succeeded him to the title.

After Eton and a year at Cambridge, Henry wandered the Near East dressed as a Turk, generating regular odd reports in the *Times*, with 'sensational' headings such as 'Mr Stanley at Aden'. He became a friend of Sir Richard Burton, one of the discoverers of the source of the Nile.

During his wanderings, in 1862 Henry married a Spanish woman, Fabia, daughter of Santiago Frederica San Roman of Seville, delicately described by Peter Stanley in his book as a man of uncertain antecedents. The ceremony, in Constantinople, was conducted several times due to Henry's doubts, causing his brothers later to imagine he had a harem. At first the family was aware that Henry kept a woman in Geneva, but not yet that he considered himself married to her.

Seven years later, he told his mother of his marriage (though not, as legend later claimed, as her train was drawing away from Chelford Station after her husband's funeral), in the hope she would receive his wife. But old Lady Stanley was badly upset that Henry had concealed his marriage from his father and everyone he knew for so long. The family then lived in dread of him revealing a brood of children just as suddenly. According to his butler, Henry was sent abroad 'for treatment', after which the possibility of him producing an heir no longer existed. Later he married Fabia twice more; only the last marriages at first appeared valid, as her previous husband was still alive in Spain until 1870 and the earlier unions were bigamous. Henry's brother Lyulph, the lawyer in the family, then raised the strong possibility that Henry was not legally married to Fabia at all, and so precipitated a lasting quarrel. Fabia herself

Johnny Stanley
CHESHIRE ARCHIVES
AND LOCAL STUDIES
Ref. DSA 181

Henry, the third baron, Britain's first Muslim peer
COURTESY LORD STANLEY OF ALDERLEY

Coat of arms of Henry

was declared insane by the family doctor and kept housed with a nurse until she died. Her Spanish nephews then tried to claim the estate, but the family proved no registrar had been present at the later unions and defeated their claim.

Henry was frequently spotted wearing Turkish-style robes and he made himself unpopular when, as a Muslim, he closed down the public houses on the estate, including the Wizard on Alderley Edge, the Iron Gates at Monks Heath and the Eagle and Child in Nether Alderley village; only the first is still in business. He was a frequent visitor to Britain's first mosque in Liverpool, although he also restored various churches, a strange act by modern standards. According to Nancy Mitford, 'Bertrand Russell says he was the greatest bore he ever knew.' Russell himself, an atheist and Henry's nephew, called him 'definitely stupid.' But Mitford adds: 'He was deaf; but he was far from stupid.'

Henry was the first Muslim member of the House of Lords and was faithful in his regular prayers and greatly respected in the Muslim community. When he died, he was buried facing Mecca in unconsecrated ground outside the present Alderley Park. The chief mourner at his burial was Hamid Bey, First Secretary to the Ottoman Embassy in London, who he knew from his days as a diplomat.

After his death there was a long legal battle. His complicated life and the chaos he left seem sad, since had he not succeeded to the title he might have been left to follow his interests in peace, including one of fishing from a boat on Radnor Mere, where his keeper Mr Haydon rowed him for hours.

## The fourth baron: important guests

The fourth baron, Edward Lyulph Stanley, was always known as Lyulph. He discarded the title of Lord Stanley of Alderley when he came to the title in 1903, choosing later instead to be known by the alternative title of Lord Sheffield, a barony inherited through his grandmother Maria Josepha in 1909. As the third son of the second baron, Lyulph Stanley had a less privileged upbringing than his older brothers, as his father noted in a letter to Lady Stanley in 1862:

> Lyulph wants to buy a young horse and have it kept for him. Third sons must not expect to have hunters until they can make money for themselves.

A man of both high intellect and high principle, Lyulph Stanley had a brilliant career at Oxford, becoming a Fellow of Balliol, Oxford, and a barrister, before entering Parliament in 1880. However, his views were too radical and his tongue, like his father's, too sharp; he left politics. But Bertrand Russell records how Lord Sheffield took the chair for him at a pacifist meeting during the First World War, an action which in those days required the greatest moral courage.

As the cleverest of his family, Lord Sheffield's knowledge was encyclopaedic and his wit sharp. When Henry was being given his Muslim burial, a relative removed his hat in respect. Lyulph promptly rebuked him with the cry, 'Not your hat, you fool, your boots!'

### Christmas house parties

During the time of Lord Sheffield, many distinguished visitors came to Alderley Park, when the Christmas house parties became events of political significance. The Stanleys had always followed a Liberal tradition. As mentioned earlier, Prime Minister Herbert Asquith was a frequent visitor and, in the momentous years leading up to the First World War,

senior members of his government also became familiar figures at Alderley. At the festivities, according to Mr Summerfield, the Alderley Park head gardener of the time:

At Xmas time a tea party with Christmas tree was held in the Tenants' Hall for the school children and a ball was given for the staff and tenantry, together with a gentry's ball, when the Tenants' Hall was crowded with guests and local gentry. Lord Sheffield had 17 grandchildren and their parents all spent Xmas at Alderley; there were governesses, nannies, valets and ladies' maids galore. The Prime Minister and several Cabinet Ministers would be present. Sir Winston Churchill, Lord Asquith and one of the Maharajahs of Indore planted trees on the lawn.

The famous Arabian traveller Gertrude Bell was a visitor to the Park and in 1923 she reported that 1,000 Liberal supporters were attending a fête there.

Lord Sheffield's wife Mary Katherine was known as 'Aunt Maisie' to all. His and his wife's ashes lie in the mausoleum at Nether Alderley church.

### Bear cub

Another short and rounded figure to be seen at Alderley House at that time was ursine. A pet bear cub was a cute (and later alarming) member of the fourth baron's household! In 1909, Clementine wrote to her husband from Alderley Park: 'There are numberless dogs & a most beautiful little gray bear – He stands about 3 feet high & has a most lovely soft coat & makes nice grunty noises – He tried to get inside the beehives yesterday and was badly stung …'

### Churchill Tree

The Churchill Tree is the rather stunted (like its famous planter!) **sweet chestnut** at the lower east corner of the car park below the Sir James Black Conference Centre (Tenants' Hall). Its tubby shape certainly means it is in no danger of toppling and it currently appears healthy. The tree may have been planted at Christmas 1908, since local people believe it was planted to celebrate Winston's engagement to Clementine Hozier in August of that year. Winston was related to the Stanleys through his beloved wife 'Clemmie', who was a granddaughter of Blanche, Countess of Airlie, the second Baron's daughter.

Nancy later published collections of many Stanley family letters. The story of the Mitfords themselves is too well-known to repeat here and not relevant to Alderley Park, except perhaps that it was during the writing of this book that the last of them, Nancy's youngest sister Deborah, dowager Duchess of Devonshire and of Chatsworth House, finally passed away. She was the last of her generation, and one of the few, ever, to have known each of Sir Winston Churchill, John F. Kennedy and Hitler personally.

### Lady Victoria Stanley

In 2003, the fifth baron's daughter, Lady Victoria Stanley (now Woods) arrived at Alderley Park for a visit at the age of 86. Site staff member Ged Ryan was delighted to show her around the estate. Amongst other memories, she told him an anecdote about the gates beside Eagle Lodge, on the capstones of whose gateposts are the two spread eagle statues. She remembered the Stanley children being told that if, when the church bells struck noon they were still beside the gates, the eagles would turn round and frown at them. This was, of course, a way of ensuring the children were back at the house in good time for lunch!

Lady Victoria had left for Canada in 1938 when, as she described it, things got a bit topsy turvey for her family.

And one of the eagles has also moved! Originally the gap between the pillars, designed for a horse and carriage, was quite narrow. In recent times, one of the pillars was moved as the gateway was widened. However, the gates themselves were not stretched, so, to this day, the gates do not meet if closed.

ARTHUR LYULPH
5TH LORD STANLEY OF ALDERLEY
b. 1875. d. 1931
MARGARET EVANS GORDON

EDWARD JOHN
6TH LORD STANLEY OF ALDERLEY
b. 1907
AUDREY VICTORIA CHETWYND TALBOT

## The fifth baron

The fifth Lord Stanley, Arthur Lyulph, was a barrister and MP and a Captain in the Cheshire Yeomanry during the Boer War who became Governor of Victoria in Australia. His wife was Margaret Gordon. He held the Alderley title only for a short time and also like his father preferred the name of Lord Sheffield. He died in 1931 at only 56 of a bacterial infection, only six years after his father. This resulted in two sets of death duties very close together.

## The sixth baron: the last at Alderley

Arthur Stanley was succeeded by his second child Edward John, the unfortunate sixth baron. It was his spectacular and very expensive matrimonial career that accelerated the break-up of the estate, which was already under pressure from the two sets of death duties. Edward Stanley married successively: in 1932 Lady (Victoria) Audrey Chetwynd-Talbot; then in 1944 the society beauty Lady Ashley, who had formerly been Mrs Douglas Fairbanks and later became Mrs Clark Gable; then in 1951 Mable, the daughter of a French general; and finally, in 1961, Lady Crane.

In 1933 he had most of the fire-damaged Alderley House mansion pulled down, leaving a much more practicable remnant. Yet the high society theme continued amid all the difficulties. Another famous visitor to Alderley Park at this time was the emperor of Ethiopia, Haile Selassie, probably in July 1938.

The sale of the Alderley Park estate in 1938, partly as a consequence of the sixth baron's unhappy lifestyle, was a great local tragedy that caused much suffering among local people and the old estate residents, as described later. It was said the death duties were responsible but the real reason was its owner's extreme laziness and addiction to wine, women and, particularly, gambling.

At the outbreak of the Second World War, the sixth baron volunteered for all three services – to the Royal Navy as Lord Stanley, to the Army as Lord Sheffield and to the RAF as Lord Eddisbury. The Navy called for him first and he rose to Lieutenant Commander. He died in 1971.

# Lords after the Park

After the sad story of the sixth Lord Stanley of Alderley, the connection with Alderley Park ends.

## The seventh baron

The seventh baron, another Lyulph, held the title for only three months in 1971. He was described as being as dissolute as his brother: 'His death was the end of the direct line of the Stanleys of Alderley, when he died without a lineal heir.'

## The eighth and present (ninth) barons

The Barony of Alderley then passed to a cousin of the sixth and seventh barons in Wales (a grandson of the fourth baron through his third son), the eighth Lord Stanley, Thomas Henry.

'Tom' Stanley, the son of a war hero, rose himself to become a captain in the Coldstream Guards. He was a very different figure from his immediate predecessors. As the third son of a third son, he had not expected to inherit the title at all until his two elder brothers, also ex-combatants, were both tragically killed in a cliff collapse in Anglesey in 1947. He took a farm near Oxford in 1954, and also owned one in Anglesey.

He also had, until the changes in the House of Lords, a more respected political career than nearly all his forebears, despite their advantage as owners of the huge Alderley estates and the associated prestige. A poignant epitaph to his career appeared in the *Daily Telegraph* in 1999 shortly before the House of Lords was reconstituted, in an article that lamented the forthcoming loss of him from the House as 'that great Welshman, Lord Stanley of Alderley, one of the only people in Parliament who gets his hands dirty every day as a farmer.' He died in 2013, the same year that Alderley Park, sold by his forebear in 1938, was put up for sale again.

His son Richard Oliver, ninth Lord Stanley of Alderley, continues with the Oxford farm (which includes the excellent Rectory Farm Shop). He also holds the titles of 8th Baron Eddisbury, 9th Baron Sheffield and 15th Baronet. He and Lady Carla Stanley have been very interested in and helpful with this book, particularly with the images.

The late eighth baron, Thomas, painted by Sir Kyffin Williams
COURTESY LORD STANLEY OF ALDERLEY

Richard Lord Stanley of Alderley, the present (ninth) baron, on his farm
LADY CARLA STANLEY OF ALDERLEY

**Family Tree – Showing the male descent of the title Baron Stanley of Alderley to the present holder**

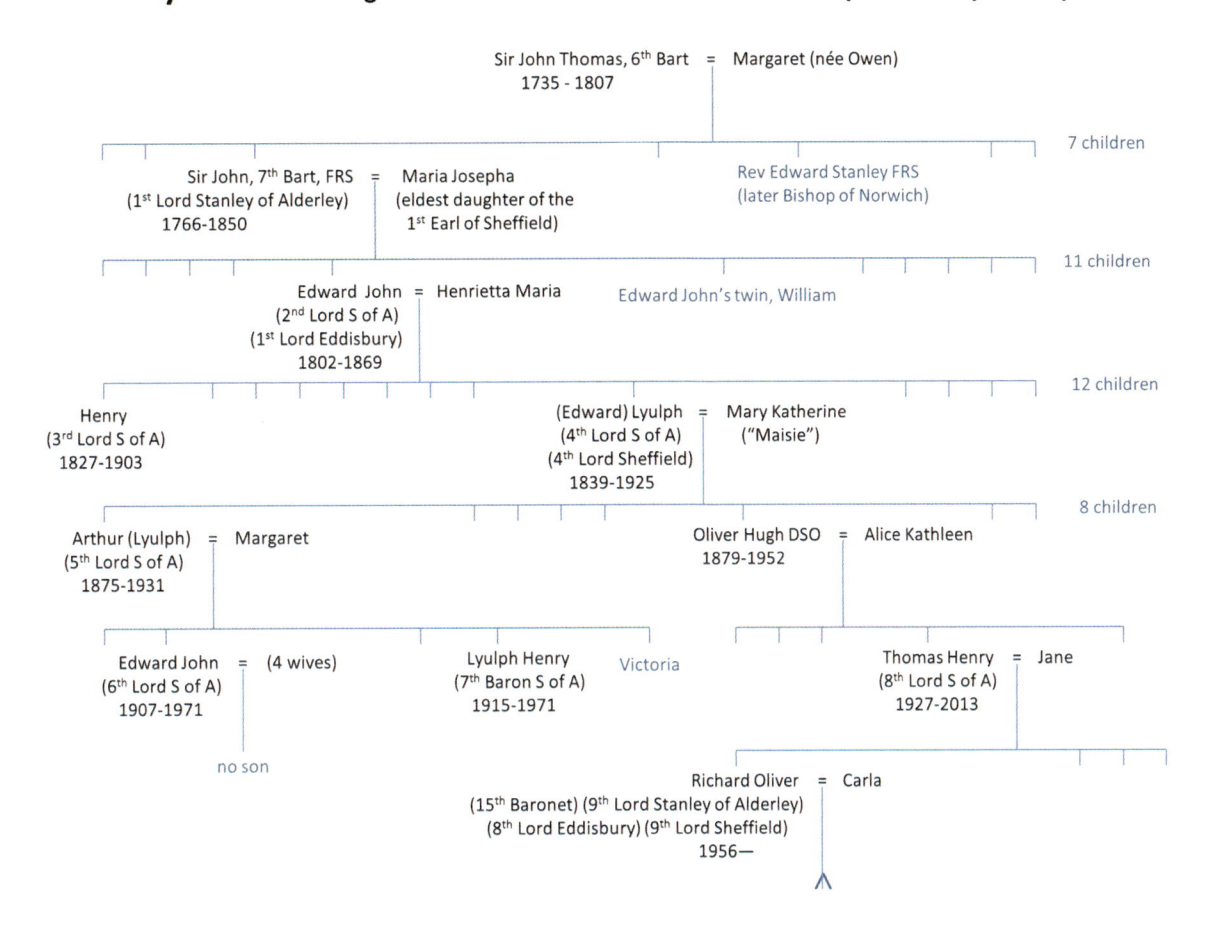

Family tree

# The smaller waters of Alderley Park

Alderley Park has lots of water. Over 30 acres of it is occupied by ditches, streams, ponds, pools and the gem in the tiara – Radnor Mere. Almost all were created to power machinery, provide alternative sources of food, and as recreational and landscape features. In early medieval times, when Alderley Park was mostly deciduous woodland, only streams would have existed. Over the years even these have been diverted, dammed and culverted for various purposes.

## Streams and ponds

All the streams in Alderley Park drain later into Pedley Brook, a tributary of the River Bollin which in turn runs into the River Mersey. The small brook that crosses the Park's centre is unnamed on modern maps but on the 1798 estate map it bears the name Bently [sic] Brook. Now called Bentley Brook, it leaves the Park near the A34 roundabout.

The present Park contains some fifteen smaller pools in addition to the Mere. Some are marl pits, dug out for the useful lime-rich marl, as the 1802 estate book notes: 'There is to be met with abt. 3 feet below the surface a large bed of Marle in Alderley Park which the farmers consider remarkably fine.'

## Serpentine

The Serpentine Pool in the south of the Park is the second largest area of water in Alderley Park after Radnor Mere. Its name is appropriate, for it forms a sinuous curve nearly a third of a mile long. It is narrow and two bridges cross it and it was created partly as a recreational boat channel, probably in the nineteenth century, by widening an existing ditch and several small marl ponds into a continous pool and lining it with clay behind a low earth embankment. It has no significant inflow apart from tiny springs, but three outflows.

The Serpentine originally drained underground to the Water Garden pool, where it drove a stone fountain long before the advent of electrically powered pumps. To increase the head of water, a paddle of wooden boards allowed extra water to build up. This outflow still exists but is now routed through a sandstone box structure originally constructed as a silt trap. The cleaning of this trap was doubtless a regular chore for gardeners when the estate was in its heyday. Probably, the original sluice was adjustable such that the level of the pool could be built up and then released to power the fountains high into the air for short periods during special events.

Punts were formerly used by the Stanley family on the Serpentine, and they stocked the lake with fish. In the 1938 sale brochure it was advertised as having excellent trout fishing.

In 1992 the Serpentine was drained and excavated down to the old clay lining. It is now of significant wildlife value.

## LIFE IN THE PARK

### Pulling the plug

The Serpentine also had two other outflows. One of these was via a drain point in the bottom of the pool. This was plugged with a large tapered oak drain plug some 12 inches in diameter and two feet long. During a working party of the site Angling Section in 1967, the plug was discovered in the pool bottom and pulled out. The result was disastrous, as Paul Woodcock recalled: 'We could see a whirlpool of water going down the hole but had no idea where it was coming out – until someone wandered up through Matthews to tell us that the water was running out across the A34!' The resultant flood stopped traffic on the main road for some time!

## Ponds in open areas

Most ponds in the parkland are surrounded by trees. Canada geese and occasional Mandarin ducks may nest. The Mere overflow pool and the Water Garden pool were described earlier.

A particularly rich pond is the Larchwood Field Pond, lying above the drystone wall at the top of Beech Wood. A feature on the estate Red Route nature trail, it is rich in dragonflies and also contains leeches and great-crested newts. It is the only pond in the Park that is not mostly tree-lined or in woodland. An old hornbeam beside it is nearly the last in the Park of a tree once common here.

Hatching four-spotted chaser at Darter Pond
PHOTO GEORGE B HILL

## Boys Walks and nearby ponds

A derelict woodland pond in Boys Walks near the end of the Serpentine was refurbished in 1997 when the surrounding plantation was felled, and given the name Darter Pond. Other ponds are more shaded.

## Beech Wood ponds

Beech Wood holds four ponds of which the two largest are used for angling. The Coach Pasture Pond, also known as the Coach Pool, is a long, narrow water formed by damming the course of an old ditch. It has a very low water flow and by the 1970s was almost totally silted up with sedge and willow. It was refurbished in the 1980s and holds a notable selection of wildlife. It has a few great-crested and smooth newts, and in midsummer large water lily beds appear, along with many dragonflies. Canada geese and tufted ducks have both nested.

Parkhead Pond is in the north of the estate. When first built, it extended beyond the present Hocker Lane, which cut it in two when the lane was re-routed to its present line in 1799. The upper half has long since disappeared.

The New Trout Pond has a particularly good flow relative to its size, which is doubtless why it has been used, both in the distant past, and in more recent times, as a trout water. It was undoubtedly originally excavated as a stew pond.

Parkhead Pond

New Trout Pond

The New Trout Pond has also held something larger than trout. In 1972 it was used by the BBC, to film a nude bathing scene for *Bridge Anstey*, a Granada TV drama!

# LIFE IN THE PARK

One quiet little pond at the top of the wood, Teal Pond, is an occasional refuge for waterbirds disturbed from elsewhere.

## Maintenance

Ponds slowly disappear. The 1798 map shows nearly twice as many waters as today's maps. Restoration work has been carried out since 1980 on many of the larger ponds which were close to disappearing. These have been major projects, involving draining the ponds and removing thousands of tons of silt. Over 800 tons of silt was removed from the 950 square yards of the Coach Pasture Pond alone.

Teal Pond

# 4 Remembering

## Tenants' Hall to Conference Centre

Undoubtedly the finest of the surviving buildings of the Stanley era in Alderley Park is the Tenants' Hall, which has commemorated the Park's most recognised scientific achievement. In 1989, it was renamed the Sir James Black Conference Centre, in a ceremony hosted by Principal Executive Officer David Friend, in which 'Jimmy' Black unveiled a plaque containing a replica of the Nobel Prize awarded to him the previous year. This account uses the old name, now to be restored, for describing the Hall's history.

The Hall was designed as a ballroom in 1819 and, together with the adjoining Green Room, is the last remnant of the former mansion. It is an elegant, originally oak-panelled room with square-paned, mullioned, lead-glazed windows and a roof with exposed tie beams. The fine carved oak fireplace is supported by classic figures. The Hall was rebuilt by Paul Phipps for Lord Sheffield (the fourth Lord Stanley) and a new date plaque added in 1904, when it was dedicated to Mary Katherine, Lady Sheffield.

An early photograph shows that the furnishings of the Tenants' Hall matched its appearance. A pre-Reformation oak refectory table, a Sheraton table of glass-like surface and beautiful Hepplewhite chairs were augmented by many fine family portraits. The oak

panelling bearing the family's coats of arms was at that time on the wall of the minstrel's gallery, where the arms are clearly visible.

When the Great War began, the paintings were evidently put elsewhere as Lady Mary supervised the Hall's use as a Red Cross hospital (along with the Granary and White Drawing Room) in which 824 soldiers were treated, many of whom died. Volunteer staff nursed others back to health on fruit and vegetables from the adjacent market gardens, as for three years the lawns around the great house were ploughed up and planted with potatoes.

After the War, the Hall was reconverted to an elegant drawing room, housing many of the Stanley hunting trophies. Later, these included the head of a rare two-horned African black rhinoceros, possibly shot by the eighth baron's father. Fastened to the wall by an eleven-inch bolt, it was taken down in 1992 and donated by ICI for study at the Manchester Museum.

Flags hung in the Hall included ones presented to Lord Stanley by King George IV for his services in raising a troop of infantry at the time of the Battle of Waterloo; and also the

Mary Katherine Stanley's plaque above the Hall entrance

Tenants' Hall as a hospital in the Great War
COURTESY MSP (IMAGE ASTRAZENECA)

THIS HALL AND THE ADJOINING ROOMS WERE USED AS A RED CROSS HOSPITAL DURING THE GREAT WAR, FROM 29 NOV. 1914 TO 28 FEB. 1919. AND 824 SOLDIERS WERE TREATED HERE DURING THAT PERIOD.

MARY KATHERINE LADY SHEFFIELD COMMANDANT.

Early photo of the Tenants' Hall richly furnished with paintings
COURTESY MSP (IMAGE ASTRAZENECA)

Hospital plaque

Tenants' Hall later, as a drawing room with trophies
COURTESY MSP (IMAGE ASTRAZENECA)

colours from the cruiser H.M.S. Southampton, which had carried the commodore's pennant of one of the family at the Battle of Jutland.

After the Second World War, when ICI took over, the Tenants' Hall was one of the few buildings still in good condition.

### Christmas Dinner

The Hall obtained its original name as the location of the annual dinner held by the family for the tenants of the estate, only the men being invited. A great meal for the traditional rent dinner was served with plenty of roast beef and afterwards everyone was given a church-warden clay pipe and tobacco on dinner plates. It was symbolical of the close link between landlord and tenant and the personal interest which each took in the other in earlier times.

The Hall was also the scene of many parties and balls. A mournful newspaper article in 1939 reminisced about:

> the ever-popular and picturesque tenants' ball, the assembly of yeomen with their wives and families, we hear again the hearty outbreak of applause as my Lord and Lady appeared through the doorway at the east end of the spacious hall and followed by members of the family and distinguished house party advanced and mingled with their guests preliminary to the start of the real festivities with a round ribbon dance, in which every person present was expected to join.

At one tenants' ball in January 1876, Lady Stanley had 300 guests. The local newspaper gave a flowery account:

Detail from Tenants' Hall fireplace

The Green Room

Former farmhouse, with Tenants' Hall

The Tenants Hall, in which the Terpsichorean recreation took place, was profusely ornamented with evergreens and suitably interspersed were flags, bannerettes, etc. … Lady Stanley came into the hall at about 10.30, when eight young females, previously stationed there for the purpose, sang the National Anthem … dancing (went on) until nearly 5 o'clock.

## Farmhouse

The attached corridor and Green Room were the link leading from the Tenants' Hall to the now demolished mansion. The former estate farmhouse was also separately attached to the Tenants' Hall. This was a capacious three-storey building, demolished in the late 1950s. The sixth baron moved into it after the 1931 fire, and the Tenants' Hall was then used as the main reception room. From 1977, the Tenants' Hall was used for conferences.

# The Alderley Mummers

A short traditional Mummers' play was for a time performed each year in the Tenants' Hall.

The play was a form of the well-known folk plays known all over Europe, a rumbustious romp handed down by word of mouth and perhaps the oldest play in England, at around a thousand years old – curiously consistent in every region yet everywhere salted with local and topical jokes.

In Cheshire, Mummers' plays were once performed in about forty locations, mostly on All Souls' Day (November 1st).The Cheshire plays are unusual in often having an extra character, in the form of a Wild Horse. The Alderley Park version differs further in being performed at Christmas – perhaps when audiences could be expected to give generously! Possibly it was moved to Christmas near the end of its active life, so that it could be performed at the annual tenants' ball.

The play was performed for nearly a century and a half until 1937 by members of the Barber family (tenants on the estate), originally in estate farmhouses but latterly in the Tenants' Hall from about 1900, when the play was revived after a lapse of twenty years and when its performance date may have been altered. A painting by Cyril Hodges shows the scene in the Tenants' Hall while the Stanley family, their guests and servants crowded the hall, foyer and balcony to watch as the 'Mummers' were re-enacting the slaying of the Dragon by St. George.

The Mummers slaying the Dragon by Cyril Hodges

## Last revival

A photograph of the performers from about 1920 also survives. After being broadcast on the radio in 1937 the play lapsed, when the estate was broken up and sold.

In 1976 it was revived by Duncan Broomhead with help from the Barber family, using costumes as close as possible to the originals, and performed annually around year end in local venues. In December 1995, a team from the Adlington Morris Men led by Duncan presented it in the Tenants' Hall itself to a full house, and received a great welcome.

The story, if such it can be called, opens with a prologue by the 'Enterer In', who makes his appearance and beats his sword on the floor. When performed in farm kitchens, this used to be an exciting affair with sparks flying from the stone floor and much noise:

Open the door and let us in,

We hope your favour we shall win,

We do our endeavour to please you all.

Now acting time is come and we do here appear,

The time of mirth and merriment to all spectators here.

A room, a room, a room to let us in.

We are not of the ragged sort, but some of the royal trim.

And if you don't believe me what I say,

Step in St George and clear the way.

This heralds the approach of St George, who swaggers up and down the Hall, swiping at imaginary dragons, Saracens and other phantom enemies, boasting of how he won the King of Egypt's daughter, and keeping the bucolic watchers at a respectful distance:

I am St George, that noble champion bold,

And with my glittering sword I won five crowns of gold.

'Twas I that fought the fiery dragon

And brought him to the slaughter,

And by fair means I won fair Sheba, the King of Egypt's daughter

But bring to me that man that there before me stand

And I'll cut him down with sword in hand.

Other characters include Prince Paradise and Colonel Slasher. The latter gives St George his most popular line:

**Slasher**

Thou mortal man that lives by bread,

What makes thy nose so long and red?

**St. George**

Thou silly fool, dost thou not know?

'Tis Lord Stanley's ale that is so stale

That keeps my nose from looking pale!'

Finally, there appears the Horse, a terrifying creature whose pranks bring screams of delight and fear from the audience, from whom he collects money as the play ends. After the play, a sumptuous meal of roast beef was washed down with copious amounts of Lord Stanley's own ale.

Mummers revival
in 1995
PHOTO ASTRAZENECA

# The art and photography of the Park

### Old Masters

Margaret Owen (later Stanley), the wife of the sixth baronet and mother of the first Lord Stanley, was painted three times before her marriage in 1763, by the famous Sir Joshua Reynolds.

A half-length version with pearls in her hair, a low-cut gown and arms folded on a cushion was later copied in watercolour by Blanche Airlie (died 1921), sister of the fourth Lord Stanley. The portrait of Lady Stanley's husband Sir John Thomas, father of the first baron, painted by Sir Joshua's rival Thomas Gainsborough, was lost in the fire of 1779.

### Alderley Park paintings

An unsigned painting of the old Hall itself before it was burned shows it standing beside the busy London to Manchester coaching road.

A painting of the Alderley Volunteers drilling in the Park has already been described, as has the 1798 estate book sketch of Alderley Hall.

### Great and lesser artists

A portrait not mentioned earlier is an attractive one of Lord John with his terrier, Crab, showing a strong-featured man with marked eyebrows and a determined expression. It is by John Hoppner and is reproduced elsewhere, in *The Story of Alderley*. The Tenants' Hall formerly contained various portraits by Hoppner, Angeli, de Laszlo and Gardner including works that 'would make a collector's eyes glitter'. A few remain with the family but many valuable works went in the auctions of 1932, 1938 and 1939.

Much more humble are the pleasant nineteenth-century drawings of William Wykeham Keyworth, of scenes including Alderley Old Hall. The famous artist and writer Edward Lear also visited Alderley Park and produced images of Alderley Edge and taught the family to draw and paint.

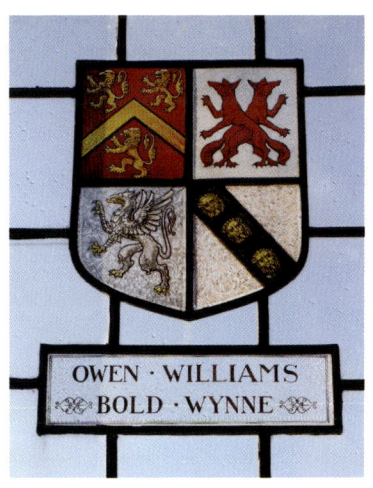

Margaret Owen, wife of the sixth baronet, by Sir Joshua Reynolds
COURTESY LORD STANLEY OF ALDERLEY

Welsh coats of arms acquired through marriage in 1763

Deer House from Radnor Mere by Edward Stanley
CHESHIRE ARCHIVES AND LOCAL STUDIES Ref DSA 192–4

Some of the Stanleys liked drawing; in the 1840s when the new railway brought tourists to the area, young 'Johnny Stanley' had to be careful of strangers while sketching alone in the woods. Around 1818, Edward Stanley filled several notebooks with sketches including the Deer House and its turret, the Mere and Church Lodge. An anonymous watercolour of 1860 is the only known interior image of the Alderley House mansion.

## Tunnicliffe

Charles Tunnicliffe's *Mereside Chronicle*, in addition to the Radnor Mere map mentioned earlier, holds some unique views of the Park as it was when he visited it in the 1940s. Tunnicliffe was a supreme book illustrator, contributing to over 250 books including, famously, *Tarka the Otter*. Born in 1901 in Langley, Macclesfield, he spent his early years living on a farm in the Macclesfield area, where he saw much wildlife, and later visited and recorded the wildlife of Radnor and other meres.

Tunnicliffe's Alderley Park work was not all natural history. In his early career, Charles earned a living through his etchings and from commercial work for manufacturers of cattle food, fertilisers and veterinary products such as ICI, Bibby's and Boots. One long-lost piece of Tunnicliffe art is a picture used as the cover for an ICI Pharmaceuticals Division Christmas card in the 1950s.

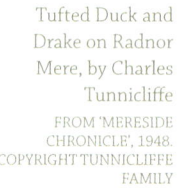

**TUFTED-DUCK AND DRAKE**

## Recent art

An early site manager of Alderley Park, Noel Cusa, trained as a chemist but in 1967 produced covers for several issues of the RSPB's *Birds* magazine, and also illustrated books by the naturalist R. M. Lockley.

Local artists such as Jackie Freeman have also painted the old buildings in the Park. The first estate Nature Trail guides held sketches by Tony Broome, a well-known Cheshire bird artist, as well as buildings sketches by ICI engineer A. W. Sywanyk. Brenda Scott's sketches of the Water Garden's plants were mentioned earlier. Graphic art is also a feature of the newer trail guides in the form of the attractive coloured maps showing the routes of the Trails.

## Sculpture and corporate art

One attractive corporate work is a painting of the newly completed Mereside buildings for ICI by the well-known artist Raymond Teague Cowern in 1957. Earlier, a signed drawing by Edward Adams (*c.*1955) shows that artist's impression of the planned Mereside buildings for Harry Fairhurst & Sons.

Other forms of corporate art such as the once world-famous ICI roundel (both in its original form and in the later, less choppy-waved version!) are hardly works of genius. But the present AstraZeneca plc logo combines the company's initials in an ingenious way faintly reminiscent of a Möbius strip.

## Historic photographs from the Park

The historic photographs shown earlier are from original glass slides mostly dating from between 1904 and 1930, or from official photographs for the 1938 auction. These or others were also sold as postcards, probably for sale to the public when they were allowed to visit the Park during the 1920s. Many copies doubtless exist in old postcard collections to this day; the slightly scratched one of Alderley House shows well the unremarkable mansion, so characterless by comparison with the lost Jacobean beauty of the older Hall.

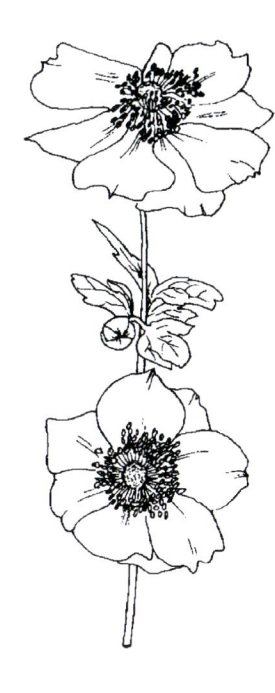

Japanese Anemones
DRAWN BY BRENDA
SCOTT (KIND
PERMISSION)

The new I.C.I. Pharmaceutical Research Laboratories

Reproduced from the original painting by R. T. COWERN, A.R.A.

The new 1957
Mereside Labs by R.
T. Cowern
COPYRIGHT
ASTRAZENECA

Old roundel
COPYRIGHT
ASTRAZENECA

New roundel
COPYRIGHT AKZO NOBEL
(BY PERMISSION)

One of the old Mereside
Restaurant murals
PHOTO ASTRAZENECA
(BARRY SMITH)

# LIFE IN THE PARK

## Eating in great company

Artworks remembered by very many were the murals on the interior walls of the original Mereside Restaurant (drawn by artists J. and M. Green). These showed several early staff members doing science – of a rather surrealistic kind!

## Aerial art

By 1973, pilot Bruce Martin had been taking images of the Park (latterly from a Beagle Husky flying at 60 knots) for over 20 years, having begun when there was a park but no modern buildings, and later having taken the first aerial shots of Mereside. More recent ones from drones include amazing views down the years.

1938 Sale brochure
COURTESY MSP (IMAGE
ASTRAZENECA)

### ALDERLEY EDGE — CHESHIRE

Illustrated Particulars with Plans and Conditions of Sale
of the

### ALDERLEY PARK ESTATES

MANCHESTER 15 miles; ALDERLEY EDGE 1 mile; MACCLESFIELD 5 miles;
extending to an area of about

#### 4,624 Acres

and comprising

#### "ALDERLEY PARK"

with nearly 400 Acres, Gardens and Parklands

77 FARMS, all in a good state of cultivation
166 HOUSES and COTTAGES   "THE OLD HALL"
"HEAWOOD HALL"   "BOLLINGTON GRANGE"
"CHORLEY HALL"   "SOSSMOSS HALL"
Freehold Ground Rents   Woodlands and Plantations

Many Valuable BUILDING SITES including the famous
ALDERLEY EDGE and WIZARD WOODS
with glorious views

The whole producing an actual income of

#### £9,699 Per Annum

### JOHN PRITCHARD & CO.

in conjunction with Messrs.

### GRANT STEVENSON & CO.

Will Offer for Sale by Auction
At the STANLEY HALL, MACCLESFIELD
On Tuesday, Wednesday, Thursday and Friday, 11th–14th October, 1938
commencing at 1.30 p.m. each day.

# The Great 1938 Sale

The disintegration of the Alderley estates came in 1938.

After the deaths of two Stanley heirs in 1925 and 1931, together with a fire in the latter year, it was known that the family fortunes were in decline. The end, however, was more sudden than most had expected, due to the costly matrimonial career of the sixth baron, Edward John.

His finances were hit by gambling losses as well as two expensive divorces and the death duties. In 1938 he sold the estate piecemeal. The immense auction of the Stanleys' entire landholding with little warning was a great local tragedy, many of the old residents being forced to leave the village and the only homes they had ever known.

## Newspaper accounts

The Press told the tragic story as sale day approached.

They are calling the lovely country district around here 'Heartbreak Pasture' tonight. Villagers for miles around are going to bed miserable and anxious with the knowledge that their homes and land are no longer their own.

For more than 300 years this green corner of Cheshire has … faced a changing world with the Stanley motto 'Sans Changer' (Without Changing). To-night, farmers and cottagers on the lovely estate of over 5,000 acres are looking at the motto and shaking their heads. At last, tradition has failed them.

On nearly every gatepost or house door hangs a 'death warrant'. It says, 'Alderley Edge Sale. Lot No …' For passers-by, the notice means merely that the property is 'on offer', just another of the 664 lots, including halls, farmsteads and cottages, which will come under the hammer at Macclesfield during the next four days. But for the inhabitants it spells doom. In the majority of cases it means the severing of family ties which have held strong for centuries. Alderley Edge estate has been in the hands of the Stanleys for more than 500 years. Now almost all of it, consisting of 4624 acres, is to be sold.

When the sale opens tomorrow few villagers will be present. 'What is the use … when so few of us can afford to bid?'

Early speculation suggested that the Park might become a housing estate, or a lido with an up-to-date hotel and swimming pool … or even a zoo! But the tenant farmers in the crowd of 400 crammed into the auction room at Stanley Hall in Macclesfield had no interest in such things. Their lives had been destroyed.

## The Sale

When begun, the auction went badly. Matthew Hyde records that: 'The sale itself … was an extraordinary fiasco … The papers were full of the Munich crisis and many were fearful for the future. The tenants turned up en masse, but not to bid. Instead they sat in rows in frozen silence.' Only 14 out of 148 lots sold on the first day. Many lots remained unsold for a long time.

… 77 farms, many black and white – 166 houses and cottages – 4 Halls – Alderley Edge & Wizard woods – Welsh Row (Artist's Lane) – Nether Alderley Mill …

Only 14 out of 148 lots sold on the first day. Many lots remained unsold for a long time.

Mrs Jane Bracegirdle, who has lived in a cottage on the Wizard for 60 years, is willing to spend her life savings to buy her home.

A corps of solicitors waited all the afternoon in an ante-room adjoining the hall to draw up sales contracts. There was complete divergence of opinion as to the value of scores of building plots, farm-lands, corner sites and woodlands, and these were monotonously withdrawn hour after hour till the auctioneer began to skip them.

The sale included not only the Park but also the home farm of the Stanley estates, which at the sale in 1938 extended to over seven square miles, the whole providing an income at that date of £9,699 per annum.

On the platform a long table contained microphones, catalogues, books containing details of the valuation, plans on every lot and maps showing the exact situation by number of every hall, farm, cottage and building plot. Behind this sat the veteran London estate broker, Mr A. E. Becheley Crundall, the new owner of the estate and the dominating figure in the proceedings, though he rarely raised his eyes from the documents before him and only by a nod, almost imperceptible, did he indicate to the auctioneer when a lot could be put on the market if the bidding had not quite reached the reserve.

… the greatest real estate sale ever held in Cheshire … out of 664 lots in the catalogue only 58 were actually sold under the hammer.

## Alderley Park itself

Most of Alderley Park comprised only the single Lot 126A.

The largest lot offered was Alderley Park, the recent home of the Stanleys, with its 340 acres of land, gardens, lodges, etc. (The auctioneer) endeavoured to put it in the bidding at £70,000;

but although he gradually brought the figure down to £30,000 he was unable to start the bidding …

Alderley Park itself was taken over by Mr Becheley Crundall himself. 'Earlier in the day Lord Stanley of Alderley, 31-year-old holder of the ancient title, had confirmed in London this morning's reports. He told the Daily Dispatch that he had received and accepted an offer for Alderley Park, his home and estate. "It is subject to contract, however,' he added, 'and I cannot make any further statement now. There are 200 separate holdings on the estate, and other people have various rights to be considered. It will be at least a month before I can say anything."'

> [Alderley Park] … was admirable for a country club and … an ideal golf course. Alternatively, it would provide an excellent area for development as a high class housing estate.

It was intended to convert it to a high-class housing estate but planning permission could not be obtained. At first this was due to the war, during which other needs became paramount; after the war, the green belt provisions of the Town and Country Planning Act 1947 frustrated development. (Twelve years later, ICI finally bought it from Hambling, Crundall & Co for £55,000.)

'Mr Whittaker, a Blackburn solicitor, acquired the Visitors' Book for £1. 12.0d. The book contains the names of many distinguished visitors to the park.' Its whereabouts are now unknown, sadly.

## Manor of Alderley

The sale included the manorial rights to the Manor or Lordship of Alderley. They were sold to the property developers in a conveyance dated 19 October 1938, and were presumably passed on to ICI when it bought the estate.

## Change but not restoration

From playing host to Winston Churchill, Alderley Park faced a grim new world. Ironically, one of the editions of the newspaper reporting the sale in 1938 held an editorial praising the then Prime Minister Chamberlain's agreement with 'Herr Hitler' and condemning a certain Mr Churchill as a 'Jeremiah'. Sadly for the Alderley tenants, their fortunes would never be restored as those of Churchill were shortly to be.

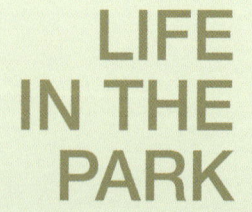

## LIFE IN THE PARK

### Down to earth

Even Lord Stanley gained little from the sale. Although he had sold his entire estate including Alderley Edge for a reputed £500,000 (of which the Alderley and Birtles estates comprised £255,300), he then sold his own home for £1,250. The newspaper article noted that, despite the fortune that had passed through his hands:

'… Lord Stanley's dustbin is still left outside his door along with the boot repairer's and the grocer's. The only difference between them is that the letters "S of A" are roughly emblazoned on the zinc bin.'

## Last contents

The final death knell came the following year, on 10 November, 1939, when an auction managed by W. H. Sutton and Sons disposed of the last remaining effects of the family and ancestral paintings still in the Tenants' Hall. The sale took place at short notice due to the requisitioning of the premises by the wartime military authorities. It was the last changing of the Stanley guard.

# Surviving old buildings of the Park

In addition to the Tenants' Hall (Sir James Black Conference Centre), other fine buildings remain in the Park. (An architectural study of many buildings across the Stanley estates is presented by Matthew Hyde in *The Story of Alderley*.)

Soon after ICI bought the Park, restoration began of the surviving structures of the former mansion's home farm. Some buildings deserve special mention.

## Upper courtyard buildings

In its heyday, Alderley Park was a substantial establishment, which included a brewery, laundry, mill-house and home-farm within a vast double courtyard. In effect, it was practically self-supporting. Around the once cobbled and grassed upper courtyard much has been demolished. But the former stables and coach houses, built in 1813 of handmade brick and with handsome oak beams and archways headed by pediments bearing the date, have been converted and preserved. One part of them became a training centre and is now known as The Cottages, with The Coach House on the other side of the arch. The upper courtyard provided a base for the administration of the Stanleys' estate and was probably maintained in a slightly more formal manner than the lower courtyard.

Next to these is an old water trough completely subsumed by the trunk of a large oak tree, which now appears to grow out of it.

## The Stanley Arms

One surviving building, known to thousands of ICI staff from 1964 as the Stanley Arms (its interior is pictured later), has always been connected with refreshment. Its first use is thought to have been as the granary annexe of Lord Stanley's brewery. By the middle of the nineteenth century, Lord Stanley's ale was a feature of local festive events.

When no longer used for the storage of grain, the Stanley Arms building took on other roles. During the Great War it was put to use, like the Tenants' Hall, as a hospital ward for casualties. After the War, it seems to have been a canteen for the gardeners and other estate workers, but on Saturday afternoons from July to September it was polished up and became the catering centre from which ladies of the estate sold cakes and teas to parties of visitors.

Upper courtyard buildings c.1925
COURTESY MSP (IMAGE ASTRAZENECA)

Former brewery (later the Stanley Arms) in 1925
COURTESY MSP (IMAGE ASTRAZENECA)

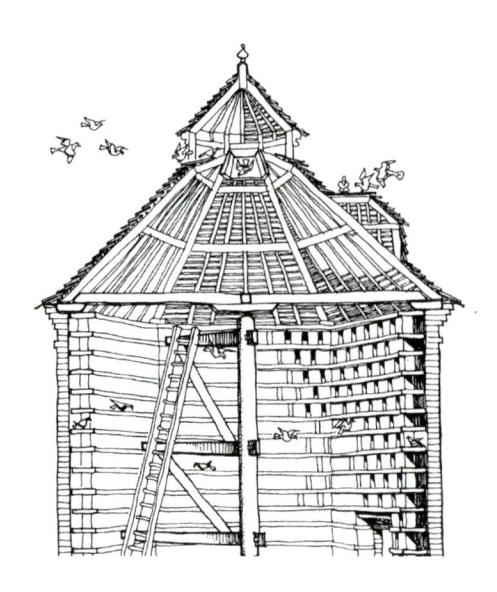

## Dovecote or Columbarium

The jewel of the upper courtyard, the fine Grade II listed six-sided Dovecote or Columbarium, is a beautiful example of its type and one which is rare in its hexagonal shape. A listed building of classical style, the Columbarium derives its name and design from a Roman building used to store cremated ashes. The 600 nesting places that line its inner walls show that the building was once useful as well as decorative. In former times, a Dovecote was an essential part of a gentleman's house. The squabs (tender unfledged pigeons) proved a very welcome addition – most commonly in a delicacy called 'Squab Pie' – to a diet of salted beef and venison that was bound to be monotonous, particularly when fish was not plentiful.

The Dovecote is in an excellent state of preservation. However, it sadly lacks its 'potence'! This mysterious-sounding equipment was actually the machinery used for collection. The potence (see sketch) was the rotating ladder system, supported on a central ash post revolving on an iron pin, which allowed a man on the ladder easily to move round and collect eggs and fat pigeon squabs from all of the nest holes. Never oiled, a well-made potence would have moved at a touch even after centuries of use. But Alderley Park's has gone. The sloping roof is still surmounted by a wooden bell cote and weather-vane, but the circular openings for access are sealed.

The Dovecote was probably still in use in the 1920s when the estate blacksmith was said to be able to get as many pigeons as he wanted from Lord Stanley for pigeon pie. When ICI arrived, no use was found for the building and it was rarely entered.

## Lower courtyard – Waterloo Barn

The cobbled lower courtyard was probably a working area given over to storage of grain, hay and straw, maintenance and parking of coaches, and a home to the working horses of the estate. The barns were open-fronted on the ground floor with hay lofts on the second floor. The outlines of the old loading holes and perforations in the brick walls can still be seen. Meadow Cottage was occupied until the 1980s.

Probably the most interesting building is the Waterloo Barn, the large stone-tiled corn barn on the left side, more recently used as a studio and office accommodation. The Barn was reputedly the place of a celebration in 1815 of victory in the battle of that name (whose bicentenary, as it happens, falls this year). Another, larger barn once stood nearby.

A detailed account survives of another great fête held in January 1803, in this or another nearby barn, to celebrate the birth of Sir John Stanley's twin sons. Before the latter fête, a rehearsal in the barn was nearly disastrous when the proud father fell through a hole in the stage floor while 'trying to perform harlequin to show the step to a very stupid clown'. The fête itself was equally exciting, as the twin's mother Maria Josepha recorded vividly:

> All is happily over, and we are alive and the barn is standing. Had you seen the crowd on Tuesday, you would hear with no little surprise that no damage whatever has been done … No description can give you any idea of the general effect, which was beautiful; and at night, when the lanthorns were

lighted, which were disposed with great taste among the beams and green boughs, it was the prettiest sight imaginable … There was a hundredweight of plum pudding, ten pounds worth of mincemeat for the pies, the greater part of a bullock and two sheep: nothing but bones left. The toasts were each announced by a discharge of artillery placed on the Hoblington, the green knoll near the gate [that is, the one now in the field near the Tenants' Hall] … on the stage were exhibited pantomimes of various kinds, mummers, snapdragon, a pantomime, including harlequin and columbine and the clown in proper dresses.

The scene would have continued an hour or so longer had not the mob outside, consisting of all the cotton devils in the neighbourhood, begun to be clamorous to get in and wanting ale. The Man (her husband, Lord Stanley) – who only wants some evil inspiration to aim at popularity with a mob, and acquire it – talked to them and received three cheers; and when the monster was once patted down and quieted, he thought the best way out was to divert its

## Whose Side?

Unlike some dedicated guides, I rarely led visitor parties, but in 1994 I took a tour group of Belgian sales staff to see the Waterloo Barn, explaining that its name commemorated the great victory over Napoleon in 1815. It suddenly occurred to me (to my horror) that I had briefly forgotten which side Belgium had been on in the battle. I admitted my failure; and waited nervously.

'Yours, of course!' They roared with laughter.

attention, and shut up the barn. He told them first, in a very big tone, that he would have order; and having made them stare, he said two barrels of ale were sent to the bonfire for their own property, and that he was sure that Cheshire men could conduct themselves with propriety and no tumult … you may guess at the number assembled when I tell you 320 gallons of ale were drunk, and no drunkenness ensued; at least, not that extreme of intoxication which causes men to be swine.

## Loggia

The quiet rose garden at the top of the Water Garden is overlooked by the First World War tea pavilion or loggia, formerly an open heated summerhouse in the form of an Italian colonnade until it was converted into a visitor's dining room. To its right, the Doric arch with triangular pediment was probably how visitors entered the Garden. The pillars of both have been replaced. The original colonnade of the loggia was made with oak from the burnt-out stables of the old Alderley Hall. A tradition of rubbing the sandstone pillars of the Doric arch for luck was not lucky for them, as it led to them being worn to hourglass shape!

Loggia and Lord Stanley's bay tree

Ice House with snowdrops
PHOTO ASTRAZENECA

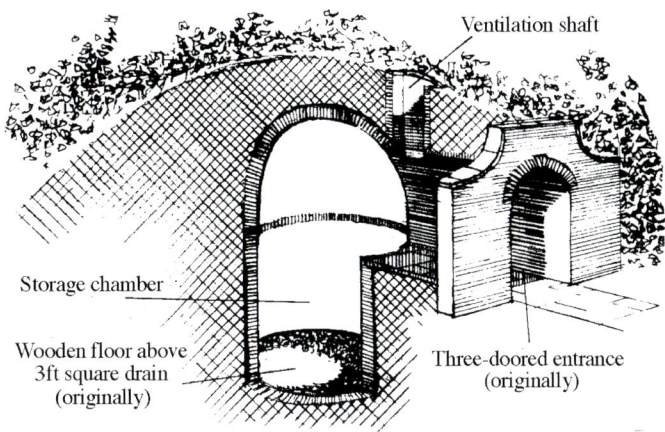

Ventilation shaft

Storage chamber

Wooden floor above
3ft square drain
(originally)

Three-doored entrance
(originally)

## Ice House

The interesting and superbly engineered Ice House was built in the early nineteenth century to store ice for the Stanleys' domestic use. It was carefully restored by staff from Mereside, working in their spare time in 1973, after one of them, Richard Ernill, rediscovered it from an old map and located it buried in the undergrowth.

A long, barrel-vaulted passage leads to an inner chamber with a remarkable domed roof thirteen feet in diameter. The ice, collected locally by sawing lake ice into blocks, was stored with salt and straw, and taken to the house as necessary for use in many household purposes including cooling wine and preserving fish, meat and game. It was probably stored on a wooden floor covering a thick sand draining layer. Water drained through a metre-square hole into a sand drain leading to a now drained pond. The corridor leading to the chamber originally had three doors, either airlocks or to control ventilation. There is evidence of a line of hooks along the corridor which may have been used for hanging game. Fresh vegetables may also have been stored here.

Ice houses were said to have been introduced to Britain in the time of Charles II around 1660. This example is mentioned several times in the Alderley Park estate farm accounts compiled during the nineteenth century, and in other records. Around the 1850s, skating was popular in winter on Radnor Mere. However, in November 1880 some of the estate's

The Stanleys' Ice House
COURTESY MSP (IMAGE ASTRAZENECA)

Ice House cutaway
COPYRIGHT ASTRAZENECA

Ice House Wood snowdrops
PHOTO GEORGE B. HILL

## And in the Ice House!

A fence now protects visitors from any danger at the pit edge. This was put in place after one ICI visitor walked in carrying a pint of beer – and fell into the pit! He was trapped for some time before being rescued, unhurt but soaked in beer!

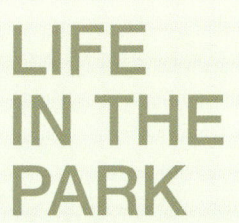

LIFE
IN THE
PARK

labourers were employed in 'protecting the Mere from skaters' as well as in 'filling the ice house'. With care the accumulated ice could last several months or perhaps even a whole year, due to the insulation provided by the building's shady site, thick walls and a deep earth cover with trees such as yews planted over it.

## Church Lodge

All visitors arriving at the north (Mereside) entrance of Alderley Park pass Church Lodge. Built in 1817, it is an unusual, irregular Gothic-style building consisting of one, two and three-storey blocks. In the estate's heyday it was home to the gamekeeper.

Interestingly, most of the real windows in it (some are fakes) are turned towards the estate; it is said Lord Stanley did not like his tenants overlooking the countryside, as they should concentrate on his estate.

The curious weathervane on the Lodge was crafted in the 1950s in the ICI site workshops. The vane is said to represent a human louse and reflects the high level of interest in pediculocides (anti-lice treatments) at that time.

Bollington Lodge, the former Matthews buildings, and Eagle Lodge are also from the same period.

# Old maps of the Park

### Sir John's map of 1798

Several old Ordnance Survey maps exist showing the Park, one of 1872 being the most detailed. But the main large working map of the Stanleys' whole Alderley estate was one of 1787 drawn by William Crossley. A description and details of it appear in *The Story of Alderley*, including the roads around the Park before and after the Stanleys re-routed them.

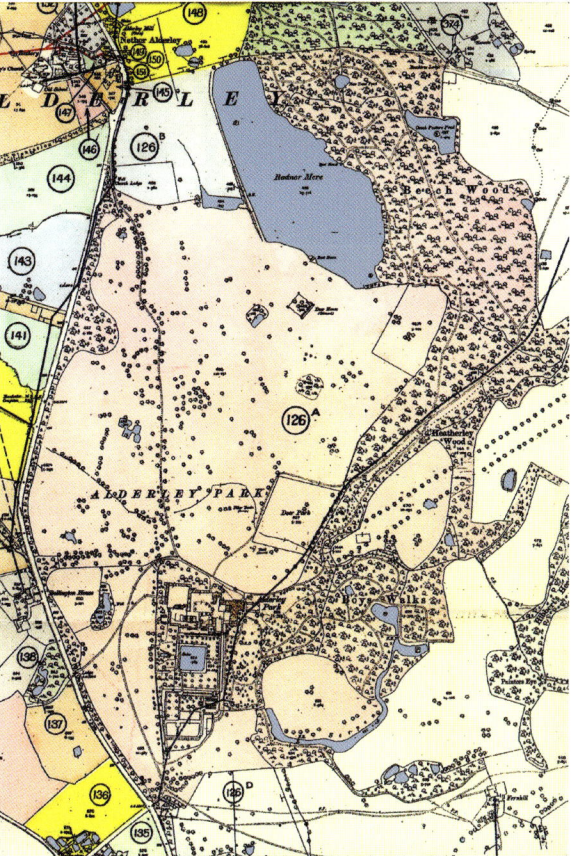

The first baron's
1798 estate map
CHESHIRE ARCHIVES
AND LOCAL STUDIES Ref
DSA 241/1A

The 1938 auction
map
COURTESY MSP (IMAGE
ASTRAZENECA)

This clearly hung in the estate office until at least 1850 and was the source for the more portable copies of parts of the map in the various estate books.

The finest of these is the hand-drawn map of Alderley Park in the Estate Book of 1798–1800. I was told of this by former Alderley Park Site Manager John Robinson, who came across the original while researching the history of the Radnor Mere dam at Chester. When I went to see it for myself, I discovered to my delight (only having seen a black-and-white photocopy) that it was a beautiful, slightly larger than A4, watercolour. It was drawn by Sir John Stanley (later the first baron) who was an accomplished draughtsman (not by the sixth baronet as has been suggested previously).

The map also shows the earlier (smaller) Radnor Mere that existed at that date. It also shows the main carriage drive to the north as being the principal highway; the A34 was then a wandering track.

## The 1938 estate auction map

Another classic map of Alderley Park is the 1938 auction map. All the copies of this I had seen were black and white. So I was again pleased when Audrey Cooper kindly showed me the official copy in the company archives – which was also attractively coloured.

The temptation to set the two maps in direct comparison was irresistible! When the 'Purple Route' Alderley Park Nature Trail guide was printed, the 1798 and 1938 maps were presented side by side, as a remarkable duo, and appear again here. The 1798 map shows the smaller Radnor Mere that was not enlarged until 1826.

Gateway of Alderley Old Hall.

Alderley Old Hall.

# Nether Alderley: Old Hall and Church

Outside the modern Park stand other notable buildings that were once part of the Stanley's core property, including the Old Hall, the Church and Nether Alderley Mill. Only short notes on the first two appear here as all three are described fully in *The Story of Alderley*, which includes Mike Redfern's definitive account of the Mill. (In 2003, when the Mill's water wheel required major restoration, AstraZeneca made a donation towards the cost.)

## The Old Hall

The present Old Hall stands in a hidden position behind and to the left of Nether Alderley Mill; its rear is visible from the end of the Radnor Mere dam. Encircled by a channel of the Moat (the millpool), it is a fine substantial Jacobean-style brick house.

This is the only remaining fragment of the imposing Alderley Old Hall and is a Grade II* listed building. Alterations were made in 1912 by Edmund Warre and the building remains occupied.

## Nether Alderley Church

The parish church of St Mary, Alderley, was originally dedicated to St Lawrence and remained so for 200 years; the reason for the change is unknown. The approach to the church is singularly beautiful, along a lane flanked with cottages beneath great beech trees. The church buildings form a splendid little group, with the seventeenth-century schoolhouse on the left leading the eye to the church.

The oldest parts of the church itself are the fourteenth-century nave and porch; the squat tower is sixteenth century. All are built from sandstone quarried in the parish. The new chancel, finished in 1856, was paid for by Lord Stanley. The church organ was presented by Lady Stanley in 1875.

The first rector, Robert Byran, died in 1328; he and several of his successors were presented to the living by members of the Arderne family, mentioned earlier, who were Lords of the Manor before the Stanleys.

The first two Lord Stanleys and their wives are buried, as described earlier, in an underground crypt; while the impressive Stanley Mausoleum, constructed by the fourth baron at the bottom of the churchyard, holds his and his wife's ashes.

Among many interesting features in the church are nine shields of arms representing the Stanleys and other local families. Over the south aisle, reached by steps from outside the church, is the beautiful seventeenth-century Stanley Pew, with its Jacobean panelling. On the south side of the sanctuary is a reclining figure of the first Lord Stanley of Alderley, while the second Lord Stanley lies on the north side. They and all their families would have often heard the bells of 1787 ringing in the ancient tower, one of which bears the daunting inscription:

I to the Church the living do call,
And to the grave do summon all.

# Wartime activities and noise!

Alderley Park's life changed considerably during both World Wars.

The use of the Tenants' Hall and the former Stanley Arms as hospitals has already been mentioned. During both Wars some of the Park was ploughed up for corn, being restored again to grass afterwards. After the Great War, estate manager Lewis Williams expressed relief that wartime farming was nearing an end: 'It will ease things considerably when all the land is 'grass side up' again.'

## Cadets and Scouts

During World War II, classified information was held in the Park. Part of it was used for battle exercises, including thunder flashes and blank ammunition. At one exercise, former ICI engineer, Harry Ainsworth, then Commanding Officer of Army Cadet Units at the Park, 'heard the whistle of live bullets above our heads'. The Cadets who had fired them claimed they were shooting at woodpigeons, but the officers, who were not popular, were understandably suspicious!

Scout camps were set up by Dr John Wilkinson in 1919. He organised the first camp only weeks after receiving his Wood Badge from Baden Powell, founder of the Scout movement. He was one of 19 Scouts who received personal instruction from 'BP' and were then asked to set up training centres around the country.

## LIFE IN THE PARK

### Evacuees

Wartime included some lively events, as the following exchange from the Oral Archive of the Alderley Edge Landscape Project reveals:

[Doug Poyner talks to John Ecclestone]

'We had evacuees like, at the school. We had some lads from Wallasey, young lads from Wallasey, when I was at school. And we went up there fishing on this mill pool, like. These lads … I couldn't swim a stroke, because we never got a chance to learn to swim, like. The war was on. They never took you from school.

'They spots these boats on the island. So these lads, they stripped off starkers, swam across, fetched one of these, rowed one of these boats back. And they carried it across the fields to Radnor Mere, which was further over. And they were all in the dining room watching all this going on! And he comes running out and shouting. Well, the lads are running like mad with the boat on their heads with nothing on! Nothing on! They was proper daredevils like, you know … It really got bombed at Wallasey like, the dock area you know. Very tough, very tough characters they was. You only fought one of them lads once. They were good scrappers.'

*(Reproduced with kind permission of the Manchester Museum, Manchester University/National Trust)*

The wartime Park
from south
ANDREW BROOME,
GEORGE B. HILL.

A programme of courses for 1922 describes the camp as: 'situated on the edge of the great beech wood, close to the Radnor Mere, in the park.' Camps were held in Alderley Park until 1950. 'We used to have a marvellous time,' Dr Wilkinson recalled of the earlier camps. 'Lord Stanley gave us full access to the park. It was magnificent. In those days it had red squirrels and a lot of other wildlife.'

## Wartime farming

The Alderley Park farmland was successfully farmed by Walter Curtis from 1940–1952, as a tenant of Hambling, Crundall & Co. His daughter, Mary Watts of Durham, recalled

Alderley House site
during WW2
ANDREW BROOME,
GEORGE B. HILL

the family living in the old farmhouse next to the Tenants' Hall. The tenanted estate farm
at that time:

> … was a mixed arable and dairy farm with a herd of pedigree Ayrshires, including some prize-
> winners, and the barns, shippons, hay lofts and stables were well used. The high standard is
> indicated by the fact that every year a party of student nurses was brought by a tutor from
> Manchester Royal Infirmary to observe the milking and dairy procedures, and were then given
> tea and biscuits by my mother in the farmhouse.

**Plane landed!**

**Mary Watts recalled a surprising event:**

'On one occasion a US light aircraft landed in the deerkeep field, so that the pilot could "take a look at the castle"!' (He meant, the Deer House).

## War in the Park

Mary also remembered that the war had only a limited effect on the Park:

> The Tenants' Hall was occupied by the American Red Cross to store equipment and my father was issued with a tin hat to guard it! I don't think the contents were ever moved until the war was over, but we did have both Italian and German prisoners of war working on the farm at different times and one of the Germans made wooden toys for us children.

## Family fun

She also recalled that:

> The park, beechwood and mere made the farm a paradise for children growing up and I have very fond memories of a den in the wood, a raft on the mere, the pear tree on the wall outside my bedroom window and games of rounders, French cricket and hide-and-seek on the green by the dovecote. The granary [more recently the Stanley Arms] was occupied by several families in succession at this period, and there were children in the two farm cottages to the right of the second archway on the approach to the farmhouse, so we three children had companions, increased at times by evacuees.

She also recalled that the Pony Club at that time had a track with jumps in Beech Wood.

> Mary's family left Alderley Park in 1952, and Mary records that:

> It was a bitter blow when we knew that the land was to be sold to ICI but my father was able to buy land in the Cotswolds, so the story had a happy ending.

After purchase by ICI in 1950 the parkland was divided into fields; these have now been merged together again, so that the parkland appears very much now as it would have done 150 years ago.

Pigs were also bred during wartime. The ClubAZ Bridge Section possesses a trophy engraved as the 'Pig Club' cup, said to have been bought with some of the funds raised.

# PART II
## Natural Alderley Park

'There is nothing in which the birds differ more from man than the way in which they can build and yet leave a landscape as it was before.'
*Robert Lynd*

# 5 Nature recording in the park

## Discovering the living Park

It was September 1977, three weeks before I started work as a young synthetic chemist at Mereside, when my first interest in the Park's natural history began. I was birdwatching at Rostherne Mere National Nature Reserve when I bumped into John Dawson, a volunteer bird warden there.

John happened to be the coordinator of regular duck counts across all Cheshire's lakes. It occurred to me he might know about Radnor Mere's birds.

'John? I'm just starting work at Alderley Park. Are there any teal on Radnor Mere?'

He fixed me with a sudden, avid stare.

'I don't know, and I've always wanted to find out. Will you count them for me?'

I started employment. Near the first week's end, I walked at lunchtime to the lake edge. It was *covered* with birds. (October can be very busy on the lake). There began this book …

## Wildlife recording in Alderley Park

Alderley Park is a wonderful place for wildlife. Over 125 acres (50 hectares) of it form two county Local Wildlife Sites and many surveys have been done. Some have continued for many years and may focus on a single species.

### WeBS

The longest-running survey is the British Trust for Ornithology's **We**tland **B**ird **S**urvey, which monitors waterfowl populations across Europe. Formal monthly WeBS 'duck counts' on Radnor Mere were begun by me in 1978 and still continue; the current counter is Paul Flackett.

### Tunnicliffe bird hide

A bird hide dedicated to Charles Tunnicliffe, dedicated 'In memory of the renowned Cheshire ornithologist and artist who watched and sketched these waters in the 1940s' was opened at the south end of Radnor Mere in 1997. It faces northerly straight down the Mere, with good lighting which allows accurate counts of species that are difficult to count, like ducks that frequently dive.

### Heronry counts

The most spectacular wildlife sight in the Park is Radnor Mere's heronry on the lake's one island. It is perhaps the most easily watched (and easily disturbed) one in Cheshire. A count of occupied nests is done every year.

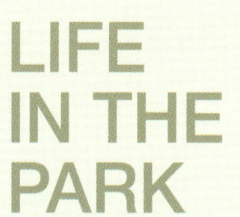

**RADNOR MERE AND THE ISLAND**

The density of the Grey Heron nests means that a three-dimensional map of the colony has to be created each year in order to give an accurate figure. The colony began around 1975, though odd birds had nested in the nineteenth century despite persecution.

The birds repair the same old nests very early in the year, with experienced pairs laying eggs sometimes by mid-January. Early sitting herons sit low and can be amazingly

### Sinking heron!

I was puzzled one day to see what appeared to be a long-necked duck swimming below the island. My confusion was shared by a nearby nesting coot, which leapt from its nest in alarm, displayed at the intruder, then jumped on its back and attempted to sink it! It was in fact a young heron, which protested loudly, before escaping onto the island.

**LIFE IN THE PARK**

inconspicuous. The colony becomes dramatically more relaxed once the bill-clacking sound of youngsters in the nest becomes audible, usually in early March.

Herons rarely fish on the Mere but the colony reached a peak of 53 nests on the island in 2009 before hard winters and tree loss reduced numbers to the present figure of around 30 pairs. One single large alder tree on the island has held an astounding 16 nests. Thankfully, a plan to build houses on the car park near the heronry was abandoned in June 2015.

## Rookery surveys

Many **rooks** nest in the woods, next to the Alderley House site, near the estate farm and along both sides of the A34. Counts have been made annually since 1978. The nest structures are counted each April. The total for the Park (and its close environs) exceeded 200 nests in the early 1980s, but averages half that number now.

## House martin nest surveys

Since Mereside was built, **house martins** have favoured its buildings. Interestingly, there are no old records of house martins at Alderley Park: Edward Stanley made no mention of this species in the early 1800s.

Nest numbers at Mereside peaked at nearly 60 nests in the early 1980s. At that time there were also up to 35 nests around the lower courtyard buildings, though increasing disturbance means there are now few there. Sadly, numbers have also declined to around 10–15 nests at Mereside, mostly on Block 24, though this population appears healthy.

## Censusing small birds

A Common Birds Census in the 1970s and 1980s studied the birds of Beech Wood. One discovery was of the striking density of nesting **wrens** in the wood,

The heronry in 2009, at its peak of 53 nests
PHOTO GEORGE B. HILL

Heron nest
PHOTO GORDON YATES
BY KIND PERMISSION

An Alderley Park heron
PHOTO PAUL FLACKETT

Rook
PHOTO GEORGE B. HILL

which plummeted after a hard winter. A map of the population shows a drop from 27 to 11 in the two example years shown.

In later years, the Census was replaced by more general Breeding Bird Surveys across the estate, done by Hugh Pulsford. These continue to show year on year that Alderley Park is a rich habitat for songbirds! The table shows the results for some species for 2012.

| Species | Nesting Territories in the Park |
|---|---|
| Blackbird | 27 |
| Blackcap | 23 |
| Bullfinch | 7 |
| Chaffinch | 22 |
| Chiffchaff | 18 |
| Coal Tit | 22 |
| Dunnock | 24 |
| Garden warbler | 2 |
| Goldcrest | 8 |
| Goldfinch | 8 |
| Great spotted woodpecker | 6 |
| Greenfinch | 14 |
| Grey wagtail | 2 |
| Mistle thrush | 6 |
| Pied wagtail | 5 |
| Robin | 22 |
| Song thrush | 22 |
| Treecreeper | 10 |
| Willow warbler | 4 |
| Wren | 24 |

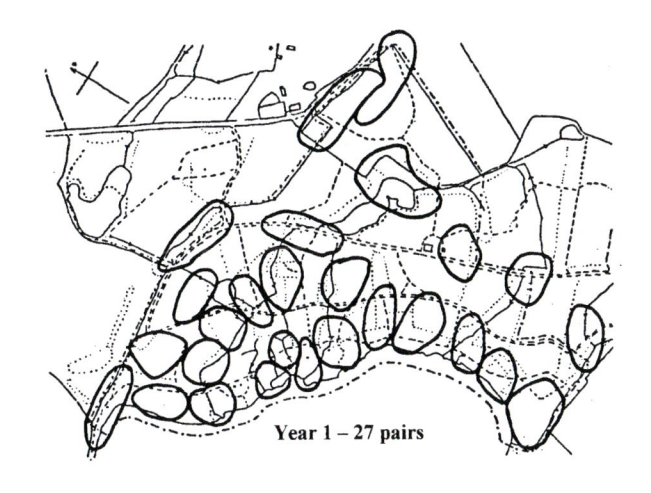

Year 1 – 27 pairs

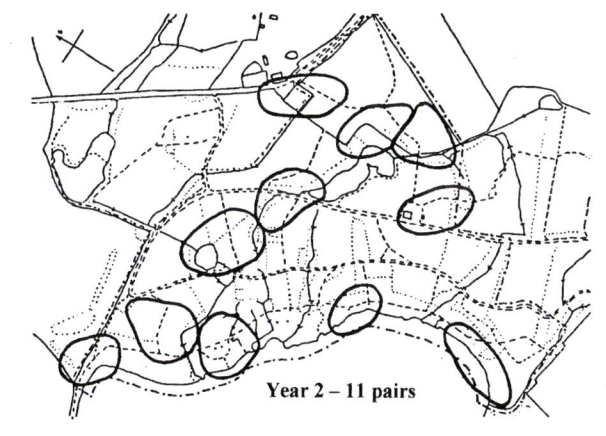

Year 2 – 11 pairs

Wren population crash in Beech Wood after a hard winter
GEORGE B. HILL

## Nestbox surveys

Nest boxes (with 32 mm diameter holes) have been erected around Alderley Park in large numbers since 1989, in a project also run by Hugh. About 140 squirrel-proof 'woodcrete' boxes are currently available and are visited at least twice during the season

The most frequent user is **blue tit**, which has occupied up to 64 boxes, followed closely by **great tit**. A few pairs of **nuthatch** nest and **coal tits** and **long-tailed tits** occupy a box or two in some years. The latter build a superb ball-like structure of moss and cobwebs, sometimes with many feathers. In four years, **pied flycatchers** have occupied boxes, the last time in 2007.

Larger boxes are for the local **little owls**, **stock doves** or **jackdaws**, though only the latter regularly use them. In 2007, six boxes suitable for **barn owls** were added. Six

Nestbox no. 52 in Boys Walks
PHOTO GEORGE B. HILL

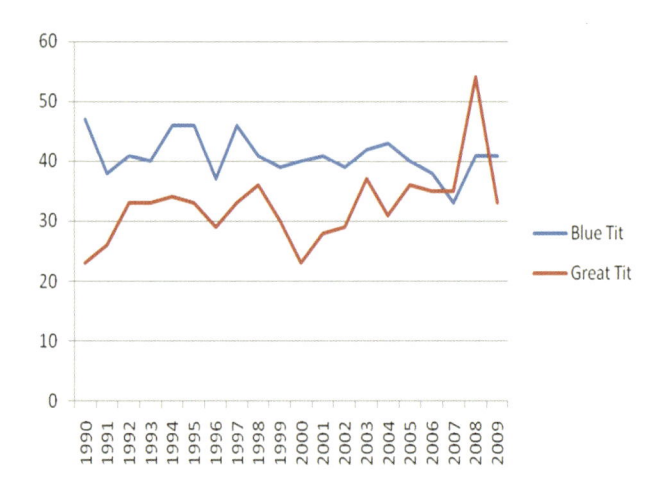

boxes near water are for **mandarin ducks.** A pair by Radnor Mere were the first to use a nestbox in NW England; however, squirrels may attack the newly hatched young.

## Bird ringing in the Park

In the nest-box programme, pulli (unfledged youngsters) of some species are ringed each year. At Alderley Park's 'Farm Open Weekends' in recent years, a more concentrated bird ringing effort by the farm has proved very popular with many families, who invariably form a fascinated audience. **Brambling** (a species rarely seen by most people) and **great spotted woodpecker** have generally been the stars of the show. All ringing is done by licensed members of the South Manchester Ringing Group.

## The annual Alderley Park and Radnor Mere Wildlife Report, 1978–2011

Many other sorts of wildlife have been recorded in the Park. After one or two informal reports were produced by naturalists working on the site in the mid-1970s, records of all this wildlife began to be collected each year. An annual in-house 'Bird' Report began in 1978. Later it became the 'Wildlife' report of the Natural History Section of ClubAZ. It included a full set of checklists in its last paper version in 2011, since when a simple online list has been kept.

## History reports

From 1989, many historical snippets and articles also began to be included each year in an Appendix to the Annual Report. Publication of these constantly led on to further information that might have been forgotten or overlooked.

Much of this book might otherwise not have been written; and the reader is therefore invited to enjoy the results of this process of accretion!

# Biodiversity in Alderley Park

Biodiversity loss around the world is alarming.

In 2006, an AstraZeneca project was launched to create a Biodiversity Action Plan (BAP) for Alderley Park, assisted by the Cheshire Wildlife Trust (CWT). This was part of a strategy aimed at conserving, and if possible increasing, local biodiversity on and around the company's properties worldwide.

Alderley Park was an ideal place to begin a feasibility study. The estate was already managed sympathetically using traditional methods involving, for example, livestock grazing rather than resorting to grass cutting or the use of herbicides. The fields that had remained 'unimproved' (over several years) already held four times as many grass species as those still receiving light fertilizer. Moreover, lots of historical data about Alderley Park already existed.

## FacilityCenter™

The project was managed for AstraZeneca plc internationally by Ross Brown at the AstraZeneca Brixham site.

An exciting new way of conducting such a study turned out to be the use of 'FacilityCenter™', a database/GIS tool already in use within AstraZeneca's Facilities Management group. This was designed to drill down to identify 'People, Assets, Rooms and Buildings'. It was easily modified to monitor 'Species, Habitats, Land Types, and Green

Biodiversity areas map
COPYRIGHT ASTRAZENECA

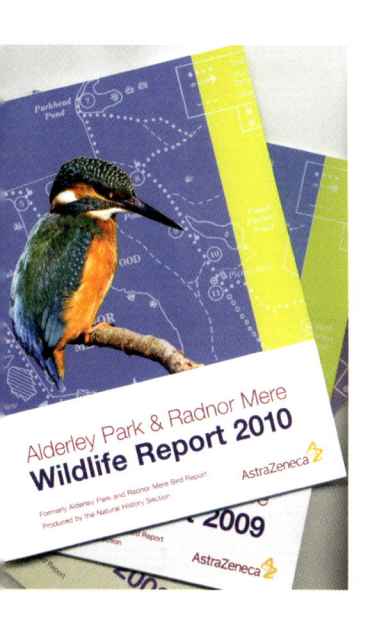

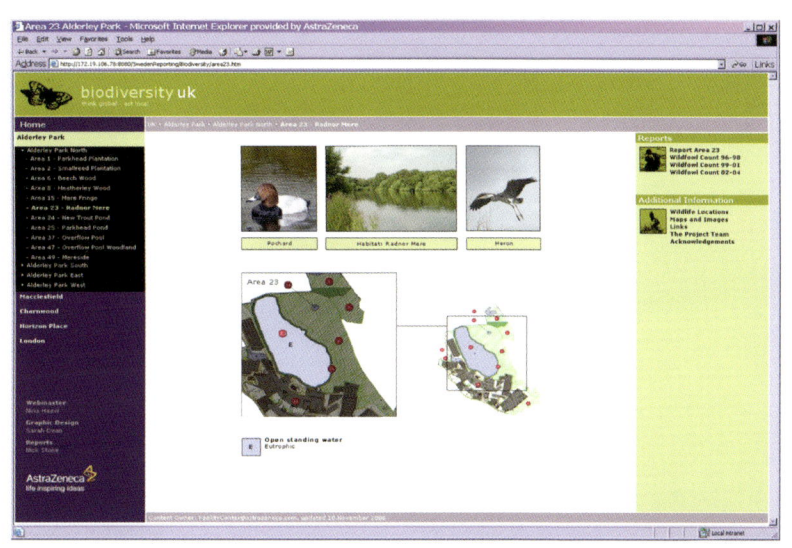

Annual Wildlife Reports

Biodiversity recording software
COPYRIGHT ASTRAZENECA

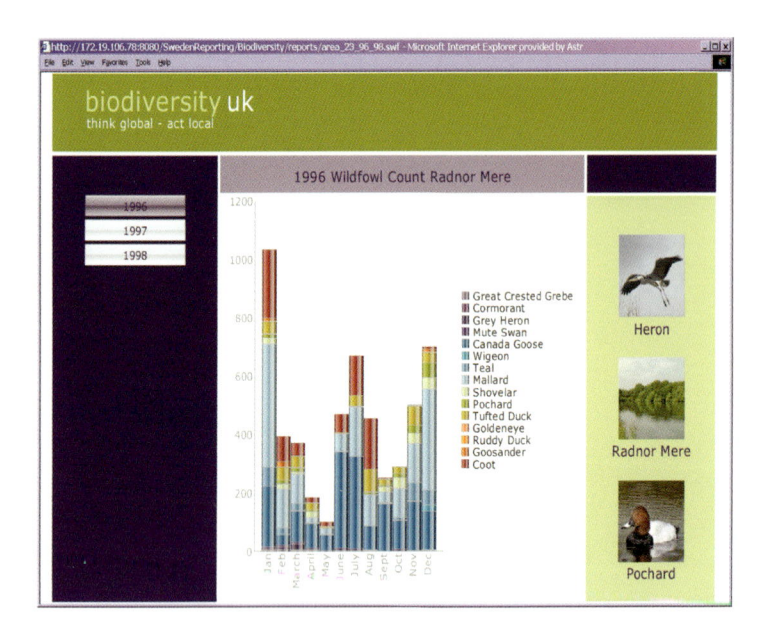

Space'. This unconventional adaptation allowed varied and impressive ways of manipulating data.

In fact, this methodology for collection, presentation and display of data was totally without precedent; and it was commented that AstraZeneca's work had put them light years ahead of the vast majority of other companies in the North West. The new AstraZeneca Biodiversity Database was launched on the company intranet in October 2007.

## Surveying

In 2006, the Park was divided into 50 discrete areas. Armed with numbered site maps and portable waterproof tablets, volunteers in May and June recorded many species found in each area. This meant that the prototype version of this novel use of 'FacilityCenter' had now been validated. It was a great success.

## The AZ Global Biodiversity Scheme

Sadly, extension of the project overseas was limited for obvious reasons, though not before projects had run at Dunkerque in France, at Gärtuna in Sweden, at Canóvanas and Carolina in Puerto Rico, at Charnwood in Leicestershire, and at Avishkar and Yelahanka in Bangalore, India.

These surveys revealed a wide variety of habitats and species, as many AZ sites lay within mega-biodiversity areas. These included rare species of insects, birds and mammals, some of whose populations have declined dramatically in recent decades, such as in Europe, the **red-backed shrike** (*Lanius collurio*) and on the Indian sub-continent, **slender loris** (*loris tardigradus*).

## Awards and accreditation

Alderley Park has won awards as an industrial site for wildlife. In 2012, the Alderley Park application was completed for official biodiversity accreditation from the National Wildlife Trusts. Alderley Park has been an example both locally and globally in this. We hope that the lesson will endure and that the global spark ignited here may burn on both here and elsewhere.

## External liaisons

AstraZeneca became accepted as a sponsor by several outside nature bodies. The Cheshire Badger Group used the company logo on their publications, while both the Mammal Society and the Cheshire Wildlife Trust used Mereview Restaurant for major conferences, with speakers including TV naturalist David Bellamy.

The Cheshire Wildlife Trust, while pressing the company hard to create more 'green' habitats, nevertheless commended the ClubAZ Natural History Section's survey data on Alderley Park's wildlife site as an example for other county landowners to follow.

## Manchester Science Park (MSP)

Further environmental surveys commissioned by new owners MSP in 2014 are building on the data store, as its owners continue to protect the Park's wildlife.

# 6 Species of the park

## Birds of Alderley Park

Alderley Park, then, is a rather good place to watch birds!

The first written bird 'record' from the Park was composed by the first Lord Stanley in 1791. Since then, many ornithologists – including a few famous ones – have observed the birds of Alderley Park.

A total of 171 species has been recorded in the Park since 1791, 148 of which were genuine 'wild' birds. This is a selection of interesting ones.

### Waterfowl of Radnor Mere

Radnor Mere at most seasons has only a few waterfowl. Big flocks very infrequently appear.

The **mute swan** nested regularly, probably for centuries until the population collapsed in 1978, due to lead poisoning from the weights used in angling. Numbers soon recovered but nesting on the Mere is irregular now because the reed-beds provide insufficient nest material.

Flights of true wild geese pass (very high) overhead only rarely in winter. But a flock of feral **Canada geese** is long established; the field below Radnor Mere which later became the Radnor Car Park was known in 1948 by artist Charles Tunnicliffe as the Goose Field:

> I hoped to approach the Mere unseen today, but as I walked up the pasture two watchful Canadas, which had been grazing on the bank, lifted their heads and gave tongue, telling all and sundry of my purpose.

Canada geese family at Radnor Mere by Charles Tunnicliffe

FROM 'MERESIDE CHRONICLE', 1948. COPYRIGHT TUNNICLIFFE FAMILY

**CANADAS WITH WELL FLEDGED FAMILY**

### Fun with geese!

The Canada geese have long been a source of humour, and there are many records of geese winning battles against people. More than one jogger has leapt into a bramble bush to avoid a chasing goose. One site security officer who went to investigate a nest was seen running away, at full speed, with an enraged goose flying behind his shoulder blades!

On another occasion, an extremely bolshie bird took off and flew right at my head. I ducked, and the bird landed behind me; and for a couple of seconds neither of us – as we stood back to back – knew where the other was. Then we both turned round. The gander's reaction, when he found me standing between him and his sitting mate was, to put it mildly, not warm.

## LIFE IN THE PARK

Canada geese breed (or try to) all along the open shores of the Mere and a few ponds. They may even nest in the balconies of Mereview Restaurant, from which the goslings on hatching must leap to escape! Later, large numbers of Canadas gather for the annual moult in summer by the Mere, after which the birds depart for local stubble fields, returning later in the winter especially when flocks are disturbed from other local waters.

A pair of feral **barnacle geese** that raised one gosling by Radnor Mere in 1978 was the first for Cheshire. Various escaped waterfowl turn up on the Mere, often causing puzzled or excited comment from non-birdwatchers.

A showy **mandarin** that appeared in 1981 was the first of a veritable army, for this small feral duck species is now one of the Park's commonest. They are little noticed because they shyly hide in overhanging bushes and trees, but can build up into big flocks in winters, and as mentioned they now breed.

**Teal** were first recorded on the Mere in 1791. They were once common in winter but

Wigeon

A female goldeneye
PHOTOs GEORGE B. HILL

| Waterfowl on Radnor Mere since 1791 (* = rarely seen) | | | | | |
|---|---|---|---|---|---|
| Species | Number of records if rare | Largest visiting flocks on record | Species | Number of records if rare | Largest visiting flocks on record |
| Barnacle goose* | | | Moorhen | | |
| Bewick's swan* | 2 | | Mute swan | | 21 in 2002 |
| Black-necked grebe* | 1 | | Pintail* | | |
| Canada goose | | a staggering 525 in 2011 | Pochard | | 139 in 2008 |
| | | | Ruddy duck* | | 47 in 1982 |
| Common scoter* | 1 | | Scaup* | 1 | |
| Coot | | 532 in 1977 | Shelduck* | | |
| Cormorant | | | Shoveler | | 97 in 1979 |
| Gadwall* | | | Slavonian grebe* | 1 | |
| Goosander | | 36 in 2005 | Smew* | | |
| Great-crested grebe | | | Teal | | 313 in 1979 |
| Greylag goose | | 132 in 2008 | Tufted duck | | 180 in 1997 |
| Little grebe* | | | White-fronted goose* | | |
| Mallard | | 648 in 1982 (probably including released birds) | Whooper swan* | 4 | |
| | | | Water rail* | | |
| | | | Wigeon | | 80 in 1923 |
| Mandarin duck | | 154 in 2012 (a record for NW England) | | | |

are now usually few. **Mallard** are the most numerous duck. They may nest in weird places – once on a roundabout, which caused chaos when the ducklings hatched! Another pair nested on the roof of the Sir James Black Conference Centre, off which its ducklings had to jump. A few **wigeon** and **shoveler** winter on the Mere but large flocks may appear on rare occasions. At all seasons **tufted duck** are present and occasionally produce scurrying little black families on the Mere, but the similar **goldeneye** is a scarce winter visitor.

Far rarer still was a **velvet scoter** sitting on the ice on the frozen Mere in 1979, only the third inland record for Cheshire. Visiting ornithologists have studied the Mere for many decades: a **great northern diver** was on the Mere in 1901, for example.

Some birds are obvious. A regular flock of large and very visible **goosander** is a striking sight on the Mere in winter. In summer**, great-crested grebes** nest with varying success and are a popular and very visible resident. **Mandarin duck** are now a feature and a flock of 154 in 2012 was a record for NW England.

**Coot** are always present and occasionally form huge flocks which eat all the Mere's aquatic growth (followed by very low numbers the following year). Two or three **water rails** visit the Mere's reed-beds in winter, but are almost never seen, although their strange squeals may puzzle anglers or walkers!

## Wading birds, gulls, terns, and other visitors

The **heronry** was described earlier; **grey herons** visit at all seasons. But perhaps the most celebrated recent birds of Alderley Park were relatives, in the form of two **bitterns** that

## Wild Claims

One hazard of being known as a birdwatcher is being approached with outrageous claims. On one occasion I was asked about 'the flock of razorbills on the Mere' – they were actually tufted ducks! (Razorbills are seabirds). But I was even more suspicious about a peacock reported 'standing outside the door of the Stanley Arms at closing time', yet that one (escaped from a nearby property) was true!

# LIFE IN THE PARK

occupied reed-beds beside Radnor Mere during the winter of 1997. They caused great interest when it was realised they could be watched from the newly opened Tunnicliffe bird hide, which became standing room only. Later, I was able to watch one while eating rhubarb and crumble in Mereview Restaurant!

In 1964 an **avocet** – a rare bird nationally at the time – was spotted by site manager Noel Cusa. **Oystercatchers** occasionally overfly the site and **common sandpipers** visit the Mere on passage. But there is little habitat on the estate to suit other small waders.

The most mysterious waders of Alderley Park are its **woodcock**. A few of these still appear in quiet areas of Beech Wood each winter, in its young plantations and most impenetrable parts, although very few walkers see that beautiful chestnut form leap up and jink away through the trees. They probably breed. Once they were common, as described by Rev Edward Stanley in 1835:

Heron in flight over the island
PHOTO PAUL FLACKETT

> It was a favourite amusement, in former days, to catch Woodcocks, by dozens, of a night, in places where now not a dozen could be taken in a whole season. Large openings were left, or rather made, in woods which, at night, were filled up with side-spreading nets, fastened by pulleys to tall branches. A man stood concealed on one side with a rope running through the pulleys, who, the instant he felt a cock touch the net, let it go and the net falling over the bird, secured the prize. In the fine old beech wood, which we have more than once referred to, numbers were formerly taken in a wide space, still known by the name of the Woodcock-glade, where many a Winter's night might now be spent unprofitably and possibly without meeting a single bird.

In about 1880 a 'pale dove-coloured' woodcock was shot in Alderley Park and preserved by Lord Stanley.

**Black-headed gulls** visit the Mere regularly but **terns** are very rare, probably because the Mere does not lie in a valley or natural flyway. Two remarkable marine visitors to the Mere were a **Leach's petrel** in 2006, and a juvenile **gannet** that dived spectacularly in the Mere in 1980.

Woodcock
PHOTO GORDON YATES
BY KIND PERMISSION

Red kite
IMAGE BY
PETER SMITH,
PETER.NATURE@
TALKTALK.NET

## Birds of prey and larger ground birds of the Park

**Ospreys** are rare passing visitors over Radnor Mere, with five known records, though one flew past a window as I was chatting to colleagues in a chemistry lab! **Common buzzard** now lives up to its name, and nests and flies over the Park at all seasons. But records of eagles are treated with great suspicion, especially when perched on stone pillars! A **hobby** now occasionally hunts the house martins spectacularly over the Mereside buildings.

The most famous bird observation from the Park (and one of the most historic in Cheshire) appears in a letter by the first Lord Stanley in 1791 to his fiancée, mentioning the **kites** of Beech Wood. The subsequent near disappearance of kites from British skies was little noticed at the time, which makes our record interesting. His record almost certainly refers to **red kites**, although the Continental **black kite** may also have occurred in England then. (A red kite flew over the Park in 2004, the first for over two centuries.)

On the other side of this Mere the eye rests on a thick venerable wood of beech trees above a hundred and forty years old, planted by one of our great grandfathers on his marriage. There are no trees so large in the county – that is, in the beech – for the oaks, alas! are gone. The finest gloom is caused by the blended branches of the wood, and the silence that reigns there is only broken by the shrieks of the large kites, which constantly build their nests in the neighbourhood, and the calls of the teal and the wild ducks to each other …

A strange record was of a confused **corncrake**, calling noisily in 1990 in the vegetation

| Birds of prey and larger ground birds of the Park | | (* = rarely seen) | |
|---|---|---|---|
| Species | Number of records if rare | Species | Number of records if rare |
| Common buzzard | | Merlin* | 2 |
| Corncrake | (2) | Osprey* | 5 |
| Grey partridge* | 1 | Peregrine* | |
| Hen harrier* | 2 | Pheasant | |
| Hobby* | | Red-legged partridge | |
| Kestrel | | Sparrowhawk | |
| Little egret* | 3 | | |

below Mereside Library. This was remarkable, as the species had not been recorded here since 1883 when it nested in the Park's fields.

## Parkland and larger woodland birds

**Woodpigeons** are abundant and flocks of many hundreds may develop. But **cuckoos** are now hardly ever heard. At one time, an escaped **sulfur-crested cockatoo** resided in the woods for a few years. It was occasionally mistaken for a **barn owl**, a species formerly rare but now a feature of the site roosting in nest-boxes on the farmland. Along the Old Drive, **little owls** still nest but are rarely seen.

**Swifts** fly over the Park from May until August; the old courtyard buildings now include 'Swift Bricks' as artificial nest holes. Attempts to get **sand martins** to nest in a specially built wall by the Mere have failed as the species declines nationally. **Kingfishers** are generally only glimpsed, but have nested and in one year laid three clutches. **Great spotted woodpeckers**, first mentioned nesting in 1902, are common in the woods and at feeding stations. The **house martin** and **rook** colonies were mentioned earlier; in winter flocks of 500 **rooks** and 1500 **jackdaws** are a familiar sight to Park staff in woods near the old buildings at dusk.

**Ravens** are rare visitors but in the early 1800s Rev Edward Stanley recorded that ravens nested annually at the top of a ninety foot beech tree in Beech Wood, known 'from time immemorial' as the Raven Tree:

> At one extremity of this wood, a noisy troop of Jackdaws have long been accustomed to rear their progeny unmolested, provided they venture not too near the sacred tree of the Ravens – in which case, one or other of the old birds dashes upon the intruder, and the wood is in an uproar, till the incautious bird is driven off. Few have dared to scale the height of this famous tree; but the names of one or two individuals are on record, who have accomplished the perilous undertaking, and carried off the contents of the nest.

Raven
PHOTO GEORGE B. HILL

## Small songbirds of the Park

**Wood warblers** summered in Alderley Park in the 1920s but only once in recent decades, while **willow warbler**, common in the 1970s, became scarce but has now recovered a little. **Garden warblers** are rare nesters but **blackcaps** sing throughout the woods in summer; rarely one may turn up in winter. The song of **nuthatches** is a familiar sound.

**Blackbirds** are ubiquitous and more birds appear from Scandinavia in winter but are far outnumbered by wintering **redwing**, which have numbered over a thousand. **Spotted**

| Parkland and larger woodland birds (* = rarely seen) | | | |
|---|---|---|---|
| Species | Number of records if rare | Species | Number of records if rare |
| Barn swallow | | Little owl* | |
| Carrion crow | | Magpie | |
| Collared dove | | Raven* | |
| Feral pigeon | | Rook | |
| Great spotted woodpecker | | Sand martin | |
| Green woodpecker* | | Stock dove | |
| House martin | | Swift | |
| Jay | | Tawny owl | |
| Lesser spotted woodpecker* | gone | Turtle dove* | 3 |

**flycatchers** were once not uncommon but now rarely nest in the creepers around the former *Stanley Arms* or along the Water Garden wall. **Pied flycatcher** have nested four times in the woods; but a migrating **nightingale** in Lord Stanley's bay tree in the Water Garden in 2002 was a big surprise!

Migrating **black redstarts** have been spotted around the Mereside buildings three times in spring. **Wheatear** had not been seen in the Park since 1923, until three appeared right outside the bird hide in 2010.

The Park's **grey wagtails** are an attractive and obvious feature as they feed and breed around the open shores of the Mere and Water Garden and on nearby buildings. **Pied wagtails** are always present around the buildings and roosts of up to 100 have been noted in winter. **Goldfinches** are often to be seen and charms of six or ten may be seen along the Mere dam, feeding on alder tree seeds. In some winters flocks of **siskins** do the same. The rare **hawfinch** is known to have been present in the 1920s; more recently they were glimpsed only twice in the 1990s.

Some small birds have disappeared. Both **willow tit** and **marsh tit** have vanished from the Park in recent decades, and **tree pipits** no longer nest in the woods. **Lesser whitethroat**

Young nuthatch
PHOTO GEORGE B. HILL

Hen blackbird carrying peacock butterfly caterpillars
PHOTO GEORGE B. HILL

was last recorded nesting in 1945. **Reed warblers** have not visited the Mere since the 1970s.

The great **starling** roosts of yesteryear are also only memories. In 1835, Reverend Edward Stanley wrote the following:

> (The mere's) western margin is bounded by an artificial dam, which as the water is upon a much higher level, commands an extensive view over a flat, rich country, the horizon terminated by the faint outline of the first range of Welsh mountains. This dam, on the finer evenings of November, was once the favourite resort of many persons, who found an additional attraction in watching the gradual assemblage of the Starlings. About an hour before sunset, little flocks, by twenties or fifties, kept gradually dropping in, their numbers increasing as daylight waned, till one vast flight was formed, amounting to thousands, and at times we might almost say to millions.

But he later notes that:

> It is now a rare thing to see a passing flock of even fifty, where in years gone by, they mustered in myriads. Their favourite dormitory of reeds, indeed, has dwindled gradually away, since the dam was raised [the new Radnor dam was built in 1826] and the depth of water increased, which may partly account for the diminution; but still reeds are left in sufficient abundance for the accommodation often times the number that are ever assembled ...

Pied flycatcher – a rare nester
PHOTO PAUL FLACKETT

### Small songbirds of the Park  (* = rarely seen)

**Species**    **Number of records if rare**

| Species | | Species | | Species | |
|---|---|---|---|---|---|
| Blackbird | | Grey wagtail | | Skylark* | |
| Blackcap | | Hawfinch* | 3 | Spotted flycatcher* | |
| Black redstart* | 3 | House sparrow^ | | Song thrush | |
| Blue tit | | Lesser redpoll | | Starling | |
| Brambling | | Lesser whitethroat* | gone | Treecreeper | |
| Bullfinch | | Linnet* | | Tree sparrow * | |
| Chaffinch | | Long-tailed tit | | Waxwing* | 2 |
| Chiffchaff | | Marsh tit* | gone | Wheatear* | 1 |
| Coal tit | | Meadow pipit* | | White wagtail* | |
| Common redstart* | 3 | Mistle thrush | | Willow tit* | gone |
| Common whitethroat* | | Nightingale* | 1 | Wood warbler* | 2 |
| Crossbill* | 2 | Nuthatch | | Wren | |
| Dipper* | 1 | Pied Flycatcher* | | Yellowhammer* | |
| Dunnock | | Pied wagtail | | Yellow wagtail* | |
| Fieldfare | | Reed bunting | | | |
| Garden warbler | | Reed warbler* | gone | | |
| Goldcrest | | Redwing | | | |
| Goldfinch | | Robin | | | |
| Great tit | | Sedge warbler* | | | |
| Greenfinch | | Siskin | | | |

## Escapes

Many escaped bird species have been seen in the Park. One was an escaped eagle owl that terrorised the rook colony during 1999 – and which was reported by one startled member of the night staff to be three feet tall!

*Escaped bird species in the Park since 1900*

| | |
|---|---|
| Bar-headed goose | Muscovy duck |
| Black swan | Red-crested pochard |
| Budgerigar | Ruddy shelduck |
| Cape teal | Sacred ibis |
| Common peafowl | Snow goose |
| Egyptian goose | Sulfur-crested cockatoo |
| Lady Amherst's pheasant | White stork |
| Lesser white-fronted goose | |

# Mammals of the Park

Excluding bats, a total of 23 species of mammal could be claimed as seen around Alderley Park (though this includes an escaped **brown bear** in the sixteenth century!). Also, no **red squirrels** have been reported in the Park since Dr John Wilkinson saw lots in Beech Wood in the 1920s, when he recorded that there were few greys present.

Mammal prints
GEORGE B. HILL

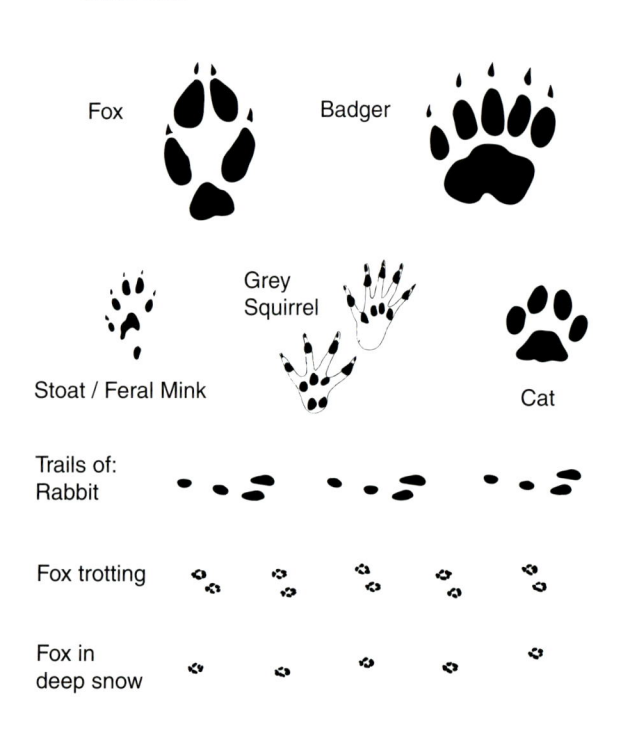

Fox

Badger

Grey
Squirrel

Stoat / Feral Mink

Cat

Trails of:
Rabbit

Fox trotting

Fox in
deep snow

## Mammals

**Hedgehogs** were formerly scarce and confined to one small area due to predation, but are now more common. Small mammal surveys of **voles**, **mice** and **shrew**s were done by Manchester University in 1996 and 2004, when five species were recorded.

As night staff know well, huge numbers of **rabbits** emerge after dark. Myxomatosis is rare and the population appears healthy. An odd black one is seen, but more common are the well-known 'golden' rabbits that started to occur around 2005 in the colonies around the Mulberry's area.

**Hares** are not now seen, although in the 1970s I used to find them sheltering in the plantations near Hocker Lane. On one occasion, a tiny leveret was found in the field below

Mereside Library and was rescued when it became clear it was abandoned – though it was thought at the time to be a rabbit kitten.

In 1850 Maria Josepha, wife of the first Lord Stanley, commented that in spring 'the hares seem *very* mad.'

**Grey squirrels** are widespread, with up to 9 counted in the Arboretum at times. The first grey squirrels in Britain were released into the wild at Henbury Park, less than two miles away, in 1876. Rodents are scarce around the buildings, discouraged by the feral **cat** population which has survived for many years, with the help of sympathetic humans.

**Badgers** are long-standing residents. At the top of Beech Wood, a hole through the base of the nineteenth-century drystone wall may well have been left for them when it was built. They follow their own trails and cross the estate and local roads without paying much attention to us, sadly sometimes to their peril.

**Stoats** are rare, although one once met me half way across the footbridge over the Serpentine in the 1980s, to the astonishment of both of us! **Weasel** sightings are nearly as rare, although in 2000 a still live and very sharp-toothed young one, caught by a cat near Bollington Lodge, was proof they breed on site. **Polecats** appeared in Alderley Park in 1998 as the species recolonised Cheshire, but are rarely seen, though in 2002 I saw a young one carrying a dead baby rabbit – on the same footbridge as above.

**Mink**, rarely seen, are the enemy of the anglers and take valuable fish; on one occasion, a mink robbed eggs from a moorhen's nest in full view of a whole office block full of distressed secretaries at Mereside.

The **red fox** is a dominant species on the estate, to judge from the tracks that may be found in snow in winter along all the tracks and paths of the estate, even after snow has lain for only a day or two. In winter they may appear on the ice of Radnor Mere. In 2005, a family bred in an earth in a sandy rabbit warren in the field in front of Alderley House. The vixen and three cubs took little notice of nearby sheep (foxes will not attack a healthy lamb), although estate manager Andrew Broome watched his lambs trying to play with the cubs.

**Fallow deer** (though not strictly free to roam) were kept on the parkland until the 1920s, where they may have been preserved since the 1400s.

In 2007, there was a dramatic sighting of a large, full-antlered **red deer** stag, when Andrew Broome was startled as it leapt a hedge from Hocker Lane onto the estate fields, ran across one of his fields, and departed in the direction of the Birtles Estate!

Young wild rabbit
PHOTO GEORGE B. HILL

Grey squirrel
PHOTO GEORGE B. HILL

In 1990 there were reports of a strange dark animal 'as big as a large dog'; and a cat-type print 2.5 inches across was found. However, no further evidence was found and the mystery of what it could have been has never been solved.

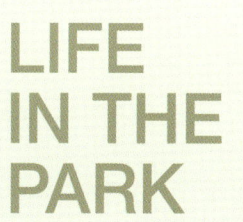
**LIFE IN THE PARK**

# LIFE IN THE PARK

One site manager once commented wearily to me that, despite the immense legal burden of keeping a huge industrial site on the right side of the law in everything from pollution to traffic regulations, the nearest he had ever come to finding himself in court was over a handful of newts …

## Non-mammals

A very small colony of **great-crested newts** was discovered in the Coach Pasture Pond in the early 1980s. When essential work was required, the colony was officially surveyed, but only the same two or three newts were found each night. In fact, the pond, surrounded by woodland, is not very suitable and they also occur in more suitable waters including the Larchwood Field Pond.

Snakes are probably no longer present, but in 1965 it was recorded that **grass snakes** up to a yard or more in length were frequent in long grass areas in the Park. They fed on frogs, toads and newts in Radnor Mere. In drier areas, poisonous **adders** also occurred at that time and 'have been observed in the grass bordering the laboratories'. Both species at that time occurred 'quite abundantly in the grounds'. The last record of an adder came from near Parkhead Pond in the 1970s. In the 1920s, **slow-worms** were recorded in the Park by Dr John Wilkinson.

*Animals observed in the Park since 1900* (* = rarely seen)

| Species | Dates | Species | Dates |
|---|---|---|---|
| Badger | | Red deer* | 2007 |
| Bank or field vole | | Red fox | |
| Brown rat | | Red squirrel | 1920s |
| Cat (feral) | | Roe deer* | 2006 |
| Common shrew | | Short-tailed vole | |
| Fallow deer | 1920s herd | Stoat* | |
| Grey squirrel | | Water shrew* | |
| Hare* | 1980s | Weasel* | |
| Hedgehog | | Wood mouse | |
| House mouse | | Adder* | 1960s & 70s |
| Mink* | | Frog | |
| Mole | | Grass snake* | 1960s |
| Muntjac* | 2014 | Great-crested newt* | |
| Pigmy shrew* | | Slow worm* | 1920s |
| Polecat* | | Smooth newt | |
| Rabbit | | Toad | |

# Bats of the Park

Alderley Park is an important location for bats, with all seven of Cheshire's commoner species having been recorded around the estate, and one rare one. Ultrasonic 'Anabat' bat detectors are used to identify the different species by their call. Around fifty bat boxes have also been erected across the Park.

An exciting recent discovery in the Park was a rare **Brandt's bat**, confirmed by DNA analysis of bat droppings in 2012. A species that can be found far more easily is **Daubenton's bat**, also known as the 'water bat'. On mild evenings up to ten may be seen flying close to the surface of Radnor Mere, where licensed bat workers can follow them cautiously with a torch beam – one powerful enough to see them without disturbing them. The photos were taken during bat surveys in the Park.

The **whiskered bat** occasionally turns up in strange places. The first one was found in 1994 hanging up (after feeding) in the cycle shed at Alderley House.

The loft of the old coach houses in the lower courtyard was known for many years to hold bats. In 1993 these were identified as the uncommon **Natterer's bat.** This was at the time the only known roost of this species in the county. After the buildings were turned into offices it was no longer possible to enter the loft, but special bat bricks were installed to allow the bats access as before, and it is thought likely they are still there.

High above Radnor Mere, or over the old courtyards and Water Garden, are the places to look at dusk for the Park's biggest bat, the **noctule**. (Occasionally in spring these fly above the Mere at the same time as 'roding' woodcock doing their dusk territorial flights above Beech Wood). Up to 5 have been counted. In 2008, a large noctule got into the Alderley House building and startled a security man who found it filling more than a little of the corridor as it flew straight at him!

At the old Alderley Park Farm (near Parklands) there formerly existed an intermittent colony of the most familiar bats to most people, 'pipistrelles'. In the 1990s over a hundred gathered in late summer, in a nursery roost to which the females came to give birth before moving away in autumn. Other similar roosts undoubtedly exist around the site. Actually there are three species of **pipistrelle** and these call at different frequencies. Two occur here and the **soprano bat** is pictured here. A third, rare species not yet found here is Nathusius' pipistrelle, which may well be here as it occurs at Tatton Park and Marbury Park.

Bat Boxes near Parkhead Pond

Daubenton's Bat
PHOTO STEVE PARKER

**LIFE IN THE PARK**

Natterer's bat
PHOTO STEVE PARKER

Noctule
PHOTO STEVE PARKER

Soprano bat
PHOTO STEVE PARKER

Brown long-eared
bat
PHOTO STEVE PARKER

ALL THE BATS WERE
PHOTOGRAPHED IN
ALDERLEY PARK

**Bats of Alderley Park** (* = rarely seen)

| | |
|---|---|
| Brandt's bat* | Natterer's bat |
| Brown long-eared bat* | Noctule |
| Common or Alto bat (pipistrelle) | Soprano bat (pipistrelle) |
| Daubenton's bat | Whiskered bat |

Detective work by the Cheshire Bat Group in 1994 found the Park's first **brown long-eared bat** in the same roost as the Natterer's, and this species is also thought to be still present, though scarce. The surveys done so far have proved Alderley Park to be the best bat site in the area – perhaps because of its many water bodies – although surprising there are no significant bat records from the mines on Alderley Edge, for example.

# Butterflies and other insects

The butterflies and other insects of the Alderley Edge district are described in *The Story of Alderley*. The Park does not have much rich butterfly habitat, although 22 species have been recorded. But there are gems to be found if you know where.

One is the tiny **small skipper**, which first arrived in 1992, while the **large skipper** has been around for slightly longer. A more obvious and lovely sight are the **orange tips** which are common around May. In 1999 a colony of the uncommon **purple hairstreak** appeared in the top of an oak tree at the edge of Boys Walks but none have been since.

The most resilient species is the **red admiral**, which has even been found in a Mereside office in mid-winter. The **comma** first appeared in 1982 and is now frequent.

A **dark green fritillary** that appeared in the very hot summer of 1990 was a most unexpected visitor, and it could have been a vagrant from the nearest colonies in the Derbyshire dales.

Brown butterflies can be numerous. The **speckled wood** arrived in 1986 and is now found on most woodland rides. The **gatekeeper** has become widespread in the Park since 1991 but is exceeded by its larger brother the **meadow brown**, which can number hundreds in hay meadows in midsummer.

Small skipper butterfly
PHOTO GEORGE B. HILL

A male orange tip
PHOTO GEORGE B. HILL

Small tortoiseshell
PHOTO GEORGE B. HILL

Comma
PHOTO GEORGE B. HILL

Speckled wood
PHOTO GEORGE B. HILL

## Other Insects

Bumblebees are common around the Water Garden and in the Park's many garden areas. All seven of the common British species are present, of which the most striking is the **large red-tail**. Eight or nine species of ladybirds have been recorded, while striking large beetles that have been seen on occasion include the **devil's coach-horse** and in water the **great diving beetle**. The **alder fly** is a common sight around the edges of the Mere in summer and **mayflies** are occasionally seen.

# Dragonflies of the Park

Alderley Park's pools are rich in damselflies and dragonflies. The former can be abundant on Radnor Mere but the best hunting grounds for both are the Larchwood Field Pond, Coach Pasture Pond and Darter Pond, plus the water lily pads on the Water Garden.

## Damselflies

One beautiful resident is the **azure damselfly**, which can be very numerous in early summer, after which the **common blue damselfly** replaces it.

Azure damselflies mating
PHOTO PAUL FLACKETT

Azure damselfly
PHOTO PAUL FLACKETT

## Rare focus!

The two images of flying dragonflies in this book were taken in Alderley Park, by keen naturalist Paul Flackett … eventually! It took him five hours at the Larchwood Field Pond trying to photograph them – during which his camera autofocus only 'locked on' twice. But his emperor (here) and southern hawker images made it all worthwhile.

The uncommon **red-eyed damselfly** was first found in the Park in 1992, but in most years large numbers now appear on the lily pads at the Coach Pasture Pond and the Water Garden. It is surprisingly easy to spot despite the fact that it prefers lily pads out in the pond, as it almost always rests on the outermost pads of each water lily bed. The **emerald damselfly** is seen rarely at Darter and Larchwood Field Ponds.

### Dragonflies

The big dragonfly most walkers notice is the **southern hawker**, which may be found patrolling over water or along many woodland trails. This is an insect with an insatiable curiosity and it may well come close to look at you.

The magnificent **emperor dragonfly** moved into Cheshire in the 1990s and was first found in the Park in 1998; two or three of the estate's small waters usually have patrolling males in midsummer.

The **four-spotted chaser** may buzz madly over the water, often at Darter Pond. The pond itself was named after the Park's sole record of a **black darter**, which appeared in 1997 just as the refurbishment of the pond had finished.

In late summer in Alderley Park, the little orange **common darter** is abundant in places, constantly flying up from the feet of walkers passing along sunny paths near water, or sitting on handrails, benches and other surfaces. They may be approached quite closely and provide a wonderful illustration of how dragonflies can constantly surprise us with their life, colour and vivacity.

Darter Pond

Emperor dragonfly female laying eggs watched by a common spotted damselfly
PHOTO PAUL FLACKETT

Emperor dragonfly in flight in Alderley Park
PHOTO PAUL FLACKETT

| | |
|---|---|
| **Azure damselfly** | **Broad-bodied chaser*** |
| **Blue-tailed damselfly** | **Brown hawker** |
| **Common blue damselfly** | **Common darter** |
| **Emerald damselfly*** | **Common hawker*** |
| **Large red damselfly** | **Emperor** [1998] |
| **Red-eyed damselfly** [1992] | **Four-spotted chaser** |
| **Banded demoiselle*** | **Migrant hawker*** [2009] |
| **Black darter*** (one record only, in 1997) | **Ruddy darter*** |
| **Black-tailed skimmer*** (two records only) | **Southern hawker** |

# Plant records and orchids in the Park

A huge number of flowering plants and trees have grown in the Park. The estate checklist includes 392 wild flower or tree species and at least 360 introduced ones.

## Orchids

Four species of wild orchids occur. At least one, **broad helleborine**, is anciently native, growing mostly near the Mere, even in lawns and garden beds. Its spikes may reach 18 inches tall, though its flowers are not showy.

The **common spotted orchid** was once sporadic on the site but became an attractive feature after a wild flower slope was approved by site manager Chris Jump in the centre of the Park in 1988. Numbers climbed rapidly and the Slope held over 200 flowering spikes by July 2013. The **southern marsh orchid** appears sporadically on tracks in Boys Walks.

Perhaps the most exciting find was of a **bee orchid** in 2006, in grassland between Mereside and Parklands. It vanished the following year but thereafter a colony with very fluctuating numbers appeared, reaching a peak of 48 of the diminutive but stunning flower spikes in 2012.

Broad helleborine
orchids near Radnor
Mere

Common spotted
orchid

The native bluebell
PHOTO GEORGE B. HILL

Red campion
PHOTO GEORGE B. HILL

## Native species

The woods are not rich in unusual flowers, but hold beautiful extensive areas of **native bluebells**. They also hold significant colonies, below the Serpentine and in Boys Walks and Heatherley Wood, of **wild daffodils** – a species further planted during the Stanley era but already native to the Alderley Edge area. These are replaced later in the summer by swathes of **red campion**. The little white **wood sorrel** is another abundant feature, while blue **bugle** is less common but **snowdrops** carpet the woodland floor near the farm, and in the Ice House Wood, where 5,000 may bloom.

Some species occur irregularly. Both the pretty **slender St John's wort** and **imperforate St John's wort** have appeared more than once in Beech Wood then vanished again.

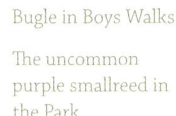

An estate speciality is the scarce grass **purple smallreed**, which also vanished for years when the plantations grew too shady, but reappeared when they were felled, in a hollow midway between Coach Pasture and Parkhead ponds. The uncommon little fern **black spleenwort** clings to cracks high in one wall of the old walled garden behind Mulberry's.

Along the streams, **opposite-leaved golden saxifrage** grows in places. **Common whitlow grass** grows on top of an old wall near Eagle Lodge, as it has probably done for very many years. The rare grass **orange foxtail** has been found on exposed mud by one of the parkland ponds in extremely dry summers.

## Strange 'weeds'

A busy place like Alderley Park also interests botanists for its alien arrivals, although non-botanists may at this point prefer to say weeds! One never knows what might turn up next. Strange visitors have included the first Cheshire record of **bristly hawk's-beard**, plus **Deptford pink**, **field pennycress**, the curiously named **weasel's snout** and **flixweed**, the last two being the first Cheshire sightings for 115 and 65 years. A delightful arrival was **chicory**, that ephemeral and mysterious traveller, which appeared near Radnor Mere. **Slender trefoil** was a Cheshire rarity when it was found being trodden underfoot in the old cricket pitch turf in 1968; it still thrives there and beside the Churchill Tree.

## Fungi

Alderley Park has been an excellent place for fungi due to its limited access. Studies by the North West Fungus Group and others have identified

Fly agaric in the park
PHOTO ANDREW
BROOME

240 different species in the Park. The ones most noticed by passersby are the **shaggy ink-cap** or **lawyer's wig**, a common species alongside the site roads, and in three or four places the white spotted red **fly agaric**, which is very photogenic – and has led to one or two trespassing photographers getting into trouble!

In 1951, when Alderley Park was first bought by ICI, penicillin chemist John Burns searched the Park for interesting fungi for anti-biotic research.

## Fish in the Park

Most of the larger waters hold good fish stocks.

Some, such as the Serpentine, Coach Pasture Pond and Parkhead Pond are fished regularly; and other species have been added to supplement the natural population. The New Trout Pond, described earlier, was briefly used for fly-fishing for **trout** during ICI's time. The Water Garden and the Mere overflow pool hold mainly ornamental species. At one time the Water Garden held two huge **golden orfe** of which at least one would probably have broken the British record if angling had been allowed in the pool.

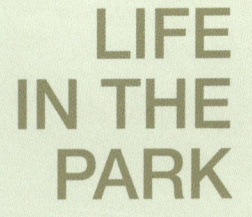

VIEW THROUGH THE DERELICT BOATHOUSE

APRIL 27TH. RADNOR MERE

I HOPED to approach the mere unseen to-day, but as I walked up the pasture two watchful Canadas, which had been grazing on the bank, lifted their heads and gave tongue, telling all and sundry of my presence. Canadas are almost as bad as Redshanks for sounding the alarm.

## Fish in the Mere

Radnor Mere is only lightly fished and has a more natural species distribution. The lack of much stocking and then only of large fish has allowed these to grow to remarkable size at times.

**Pike** have always been present and some have grown very big, the largest on record reaching 29lb. Very old **bream** (to 13lb), **tench** (to 9lb) and **roach** (to 3lb) have also been present in the past, though the bream and big pike have probably all gone. In some years, predation by cormorants massively reduces stocks, mainly of the roach, of which a thriving population estimated at 20,000 was largely wiped out. **Carp** were introduced in 1990 after half a dozen were rescued from Capesthorne Lake when it was polluted.

Radnor Mere in 1948, View through the derelict boathouse by Charles Tunnicliffe
FROM 'MERESIDE CHRONICLE', 1948.
COPYRIGHT TUNNICLIFFE FAMILY

One of the ancient tench
PHOTO PAUL WOODCOCK

| *Fish species found in at least one water in the Park* | |
|---|---|
| Barbel | Pike |
| Bream | Roach |
| Carp | Rudd |
| Chub | Silver Ide |
| Crucian carp | Tench |
| Ghost carp | **Three-spined Stickleback** (in the streams) |
| Golden Orfe | **Rainbow trout** (probably none left) |
| Goldfish | **Wels catfish** |
| Koi carp | |

# PART III
## Green Alderley Park

'The green earth, say you? That is a mighty matter of legend,
though you tread it under the light of day!'
Aragorn to Eomer in *Lord of the Rings*
by J. R. R. Tolkien

# 7 Farm and estate

## Historical farming and use of the land

Detailed records of the early estate and its farm were kept in the estate books of the first Lord Stanley.

### Field changes

The Stanleys made big changes in the early nineteenth century. By 1842, many field boundaries had changed. Small fields were combined, though one of the former field names was usually retained in each case.

Oats or barley was grown on some smaller fields and much hay was harvested. In early 1799, four haystacks held a total of 46 tons of hay. Also in 1799, Sir John Stanley's list of livestock included 15 cattle, 9 horses, 78 sheep and 11 pigs. At that time the open parkland was confined to the east side of the long drive, where the uneven nature of the ground meant that much of it was left unfenced and occupied by cows, horses, sheep and pigs. The west side was all enclosed fields, but by 1899 this too was park.

Lower Courtyard
stables c.1925
COURTESY MSP (IMAGE
ASTRAZENECA)

### Brickmaking

The 1798 map shows some Brick Kiln Pits near the centre of the Park, which used locally sourced clay and red sand. Around 1800, brickmaking was a major enterprise and records show that large numbers were used or sold locally. Around 2,000 bricks a week could be produced in kilns heated with locally harvested timber and gorse; the fierce heat produced by the aromatic leaves of the latter was ideal for firing the brick kilns. Records show that estate workers spent many days harvesting gorse from Alderley Edge. Not surprisingly, gorse is now very rare here!

The brick kilns were large, open structures filled from the top, using the gorse to flash fire them to a high temperature. A longer burning fuel, such as oak logs, was then placed on top of the bricks, followed by more layers of bricks and fuel. After firing, the kilns would have stood for weeks before they could be opened and the bricks collected. By 1823 production had mostly ceased and sales of sawn timber from the estate sawmill became the largest source of income; the brick kiln itself was moved out of the Park to Little Clays field across the (old) A34 before 1882.

## Woodland: mature and managed

After the best timber was felled in two World Wars, only a few good trees remained, mostly along the shore of the Mere. An ICI description of the estate was blunt:

> All useful timber had been felled and rhododendrons had taken over, so that there was an impenetrable mass 10–20 feet high with what could be described as non-commercial beech trees sticking through it in places.

Sixty years of work followed. In addition to huge plantings, some **rhododendron** has been removed to allow native plants to regenerate.

### Mixed woodland

Away from the buildings, over fifty tree species and some ten shrubs now occur in Alderley Park. The former include thirteen locally native broadleaved trees, twice as many introduced ones and seventeen conifers.

Beech woodland in spring by the Serpentine

### Changing of the guard

The story of the Park's **beeches** was told earlier. A few fine beeches still survive all around the Park, for instance standing amid wild daffodils below the Serpentine. But now they are being supplanted by thousands of youngsters which fill the woods with fresh green foliage in spring.

Wild daffodils in the Park

Young Beech foliage

# LIFE IN THE PARK

## Going lime

The historic avenue is still lined with *white* and *red horse-chestnuts*, one or two with arms like octopi. Sadly, both are suffering from bleeding canker disease. So the avenue will slowly change from a Chestnut Avenue to a Lime Avenue as *common limes* replace them.

## Hardwood species

Numerically, other species now dominate beside the Mere, particularly fine **oaks**, along with the 'forester's weed', **sycamore**. Oaks occur here and throughout the estate as mature trees – including the well-known one growing from an old water trough in the upper court-yard. A fine **wych elm** stands near the Ice House.

A large **walnut** and other specimen trees are features of the parkland. Fine mature **sweet chestnuts** mostly grow in the Arboretum and by the Serpentine. Only two or three mature **hornbeam**, a species once common in the Park, remain. **Black Italian poplar** stands have been planted in the damper woods.

## Damp areas and smaller trees

Mature trees line the Mere's shore and other pools. These are **crack willow** and **alder**, which form an important habitat and in places create a small marshy carr, favoured by

*Oak from the water trough*

*Damp silver birch woods*

*Southern Hawker dragonfly*
PHOTO PAUL FLACKETT

nesting waterfowl. In other damp areas, **silver** and other **birch** is plentiful. Many **rowans** have been planted and some **holly** but **hazel** is self-sown.

## Tulip trees

A well-known feature that once stood beside the old mansion was a great **tulip tree** over 150 years old, at the time one of the largest in the north of England. After it fell in 1975 it was replaced by another, planted below the lower courtyard buildings, by Percy Hammett, head forester at Alderley Park, on his retirement in 1982. This flowered well for the first time in 2002 and is now a fine tree.

## Anniversary

Another commemorative planting was the Alderley Park 50th Anniversary copse, planted in 2007, comprising fifty saplings of seven indigenous species on the grassed area opposite Bollington Lodge.

## Mature conifers and woodland restoration

Percy's tulip tree

Conifers that walkers encounter commonly along the woodland tracks include **Japanese larch; Scots pine**; **Norway** and **Sitka spruce**; **Corsican pine** and the elegant **noble** and **grand firs**.

The Park's woods are generally classed as ancient replanted (i.e., not primeval) woodland. The last old planting was in 1908, but younger trees dominate. In 1950, new owners ICI decided the woodlands should be rehabilitated. Undergrowth was cleared and residual trees were clear-felled before replanting. This was for environmental and privacy reasons as well as for financial return. A resident forester, G. A. Atkinson, was appointed in 1958 to

Beechwood regenerating; interplanted larches were removed earlier

Track with mixed trees

manage the forest operations. Most of the woodland was to be commercial plantation but with an unusually diverse tree selection.

It was thought that the soil had been left too poor for pure deciduous woodland, so from 1953 quick-growing conifers plus some deciduous trees were put in, the intention being eventually to fell the conifers to leave hardwood forest.

Hardwoods also lined the woodland tracks, forming useful habitat corridors along which deciduous-loving bird species could travel to new areas.

In a few areas **poplar** was planted, to be felled in thirty years when about sixty feet high, for the matchboard, chipboard and turnery trades. Occasionally **Norway spruce** was also grown and sold to employees as Christmas trees.

The plantations formed a patchwork so that there would always be a variety of tree ages across the wood. By 1976 at least 150,000 trees had been planted, including 6,000 young beeches planted to restore the ancient beechwood, as ICI to its great credit looked ahead fifty years; and there were 149 acres of woodland in the Park, 40% of its total area. Over a quarter of a million trees were eventually introduced – though voles and other small mammals reduced the population somewhat.

## Not a motorway

One planting was unnecessary. In 1969/70, a screen of poplars close to the Park's western boundary was planted to line the southern approach to the proposed 8.5 mile Wilmslow spine road. This motorway, due in 1974–6, was never built and the new bypass is a much reduced version, though with the same new roundabout. The southern dual carriageway approach road to it had been expected to slice away a four metre strip down the Park's western boundary and consume Bollington Lodge. This would have necessitated a new access road across the Park from the roundabout, and would have left some Park trees in the central reservation.

## Recent times

More recently, a new era of 'green' management has made the woods an amenity area, with timber sales merely financing woodland improvements, and also extensions: a new 11-acre woodland area was created on newly purchased land next to Heatherley Wood and became known as the New Heatherley.

Major thinning exercises have been carried out. Trees with unsuitable growth habit are removed, along with some of the commercially valuable timber; about 1000 tons of timber was recently taken. Then the clear-fell areas are 'mulched', grinding and smashing up the remaining stumps and brash ready for replanting in early spring. Forestry Commission grants, annual woodland maintenance payments, and the sale of extracted timber have made the recent programme so far completely self-financing. And the trees grow for free!

# Matthews and flowers

The late Fred Matthews set up his first nursery in Cheadle Hulme in the mid-1920s. During the Second World War he moved to the south end of Alderley Park, where in the early days he ran a market garden within the walled garden. Fred was intimately involved with the early development and restoration of the Park.

Planting for new woods

## Water Garden restoration

He was the man behind the restoration of the historic Water Garden laid out by the first Lord Stanley, which had badly deteriorated with brambles to the water's edge and self-seeded oaks growing actually out of the encircling path by the Pool. Work began in 1962 and continued until 1977; in the early days, orange trees were planted around the pool, later replaced with Spanish laurels, now themselves replaced with conifers. An eagle statue was placed in the Garden to commemorate Fred's work. The self-seeded oaks were allowed to remain as features around the Pool.

Few of the original plants remained, but some venerable treasures were preserved including Lord Stanley's bay tree by the loggia and Lady Margaret's avenue of blue-green Lawson cypress leading down from the decorative Adam gates. Striking new ones included the *Gunnera* and the New Zealand Flax.

## Matthews Garden Centre

In 1964, the southernmost section of the Park, leased by Fred Matthews, became Matthews Garden Centre. The Centre's staff helped maintain the gardens in all of Alderley Park for many years. After Fred's death in 1977 the business in the Park was taken over by his son-in-law Ian Urquhart and kept in the family. It thrived and was popular with many Alderley Park staff shopping in their lunch breaks.

### Water Garden guide

In 2007, Ian and others assisted me to prepare the guide to the Water Garden's plants, described earlier. Although this is now out of date, it painted an amazing picture of what Fred and his family had by then achieved.

### Relocation

In 2005, facing the expiry of their lease, Matthews Garden Centre applied for planning permission to relocate across the A34 road from their original site at the south end of the Park. They were granted special permission, to expire in April 2010. In the meantime, however, Matthews in 2006 started relocation to Somerfood Booths where they no longer operated a garden centre business, and so never moved their site at Alderley.

When they applied for an extension, the authorities noted that Matthews now had an option to sell the relevant land, and had been trading from a different site. The council considered that Matthews could potentially sell their permission to the Tesco's subsidiary Dobbies, who hoped to open a major garden centre at the end of the new Alderley Edge bypass. They refused.

In 2008 the lease expired and Matthews was very sadly closed, marking the end of a colourful era. Part of the old nursery became the new cricket pitch.

## Estate farms old and new: sheep!

After ICI's arrival in 1950, an experimental farm was created on the site of what is now Parklands, but became obsolete after animal health research ended in the Park.

Agreement was given to relocate the farm and construction was completed in autumn 2001. The farm had been little visited by staff but on its relocation it was decided to open up the farmland area to staff to enjoy the parkland.

### New farm

The new farm buildings stand south of Church Lodge, next to the woodland that screens the site from the A34. It has proved an excellent location. Buildings will be buildings, and cannot be made invisible, so much thought was given to the new design, which was deliberately put on an accessible site, away from most of the 'developed' site (and completely

Sheep and Parklands
PHOTO ASTRAZENECA

Sheep building at the new Alderley Park farm

hidden when the trees are in leaf), but next to the central parkland/farm area. And no trees were harmed in the making of the new farm!

Materials were important. Red brick, with the distinctive blue diamond pattern of the Waterloo Barn, was used to face the office and workshop/storage building, the one most visible from the site road. The main sheep building was clad in dark brown timber boarding and the dark blue roof blends in well with the skyline and backdrop of mature trees. And rooks, resident red-legged partridges and a tawny owl are all to hand.

### Sheep

The main purpose of the sheep flock is to trim the grass. The farm is self-supporting in that lamb sales cover the cost of production. During the foot and mouth epidemic of 2001 when the sheep were absent, mechanical mowing of the meadows cost a surprisingly large sum.

Previously to that, as many as 550 breeding Welsh half-bred ewes with around one pedigree Texel ram per forty ewes had made the farm reputedly the biggest sheep farm in Cheshire. After 2001 it was reduced to around 350, to allow less intensive grazing with less fertilizer use and animal waste, fostering the development of a more natural flower-rich grassland.

The ewes are given good grazing in later summer prior to introduction of the rams in early October for about 6 weeks. Ewes then stay in the fields until the New Year, when they are housed at the farm. They are pregnancy scanned in early January then penned according to how

Today's Alderley Park farmers at work – Andrew Broome, Graham Evans and Richard Hughes
PHOTO ANDREW BROOME

many lambs each ewe is carrying, since twin or triplet bearing ewes have much higher nutritional needs. After lambing ewes are penned with their young for at least 24 hours, then put in larger pens for a few days before being taken out to the fields. Weaning is at about 3 months and shearing is in early summer.

### British white cattle

A very popular addition to the site's livestock was a small group of British white cattle.

These are descendants of the ancient indigenous 'wild white cattle' of Britain; there were herds at nearby estates and parks as long ago as the sixteenth century (perhaps even at Alderley Park?). With their black eyelashes, doe-eyes and pure white coats they have become a great favourite (especially the young calves) with all passers-by. Although they would live happily outdoors all year they are brought into the barn in winter to protect the wet fields, and in readiness for calving. The calves are born in in winter or spring and stay in the barn until the weather improves. The calves are weaned the following winter, when bull calves are sold and heifers retained for breeding at around 2 years of age.

The first four of the Alderley Park herd bore the names Alderley Firefrost, Moorside Scottish Mist, Moorside Fleur-de-lys and Moorside First Lady; although the last two were usually known just as 'Flower' and 'Lady'!

Redwing
PHOTO GEORGE B. HILL

## Making the Park greener

Access to the parkland has been made much easier in recent decades. The many dividing fences that for years had crisscrossed the Park fields were removed to give a truer parkland feel. (In 1882, the unfenced parkland had extended from Bradford Lane all the way to the Birtles estate). Later, footpaths were created and new feature trees planted. Persuading the staff to live more green lives also became a popular theme!

### Green transport

Staff members have always been encouraged to use green transport, and the 'Drivers for Change' project of 1997 was an early initiative by the Green Commuting team. A shuttle

**LIFE IN THE PARK**

**Green noise!**

Some managers were known for their notorious lack of interest in green ideas. But on one occasion, it was different. In a hot meeting room with tightly closed windows, the subject of wild flower meadows came up, at the bottom of a long agenda.

One very senior manager suddenly snapped:

'I think letting the flowers grow is a great idea,' he announced, to general astonishment. 'If it stops the infernal racket of those lawnmowers outside all my meeting room windows!'

Mike Baker at the wheel of the site bus
*PARK LIFE* June 2003

### Green enthusiasm

Some people tried hard to catch the site bus. Mike Baker, the driver, recorded that on one occasion a gentleman (suited and briefcased) made a dash across soft ground but sank to his knees, lost his shoes, fell face down and slithered some yards – before recovering his shoes and composure, to loud applause from the hysterical passengers!

# LIFE IN THE PARK

bus ran for some time between Mereside and Alderley House, to encourage people who frequently worked at both places not to run their cars back and forth. Other buses operated as inter-site shuttles, and subsidised bus services to the Park commenced around the same time. For many years, there were debates about whether some financial incentive could be applied to reward staff who chose not to arrive by car in the Park, but sadly none was ever offered. Car sharing had been popular in the early days, but declined as staff found their working days becoming much more unpredictable.

Cycling has always been encouraged. On one occasion, the company's top management argued against the cost of a new bike shed that would take a few dozen bikes. But when they were told that the cost of it was the same as providing just two car park places, they changed their mind immediately! I cycled throughout my career – although one bicycle snapped in two after hitting six years' worth of road humps.

Parkland and paths

# People in the park

## Alderley Park Farm open days

The annual spring open weekend at the Alderley Park Farm is popular with those employed on the estate and their families.

The first was held in 2003, 18 months after the opening of the new farm at its current location. It started merely as an invitation to staff and their families to visit the farm at lambing time to view the newborn lambs. Around 400 visitors were anticipated, so everyone was pleasantly surprised (and slightly overwhelmed) when around 1,500 turned up. As they say, the rest is history. Annual attendance at the weekends can rise to some 3,000 visitors over the weekend – especially if the sun is shining!

### Date prediction

The date of the open weekend is set in stone when the rams are put out with the ewes in autumn – the gestation period of sheep is 145 days, so farm staff know almost exactly to the day when lambing will start – and once they start there is no stopping them, as around 80% of the 300 sheep lamb within the first 'breeding cycle' of 16/17 days. The open weekend is timed for the end of this cycle – when the main rush of lambs is over but whilst there are still some ewes left to lamb. Calving is similarly planned. Until the day, the farm looks like a very different place – it is, after all, a working farm, and by necessity all the preparation, organising, sweeping up and polishing happens just a few hours before the gates open.

### Activities

Outside groups arrive to give fascinating demonstrations of live bats and bird ringing. A telescope is manned at Radnor Mere for bird watching. Various farm animals, from shire horses to goats and rare breed cattle, visit for the weekend. Tractor and trailer rides operate;

Lambing shed
PHOTO ASTRAZENECA

British White cattle

although the first time this was tried the trailer, fully loaded with people, sank deep into the wet ground on the first run, so everyone had to be unloaded, and the trailer pulled out with the help of another tractor. The visitors were not at all perturbed at all by this; but the tractor remained on hard ground thereafter!

Many visitors can pet the newly born lambs, and hopefully even see lambs being born. Long may it continue.

# The Park's History and Nature Trails

The history and wildlife of the site have always fascinated people. Various guides have been popular.

### Estate History guide

Perhaps the most beautiful of all was the *Alderley Park Estate History* booklet, published as a compilation from the series of estate history articles published in *Park Life* between May 2001 and June 2004, written by Paul Woodcock.

### First two booklets

The first two History and Nature Trails were laid out in 1988–1990 with a combined length of around 3 miles. The texts for the 'Red' and 'Blue' Trails were written by me with John Rayner's help following a suggestion by Keith Faulkner and keen support by site manager Chris Jump. David Richards led creation of the route and Heather Green laid out the booklets. Paths were chosen to show off the wildlife without disturbing it.

### Then there were five

They were so popular that in 2003 site manager Will Spinks agreed I should rewrite the two into five, now with colour illustrations, helped by Helen Hayes. The project included the creation, overseen by Andrew Broome and his team, of major new paths.

This led to the *Blue* (Water Garden and Serpentine), *Purple* (Parkland and Old Buildings) and *Gold* (Ice House and Trees Walk) guides (using the new AstraZeneca company colours), which covered circuits around the southern part of the Park. These were followed two years later by the *Red* (Radnor Mere and Beech Wood) and *Green* (Parkland and Woodland) routes around the northern Park.

The new guides were provided free to staff although a donation to the company charity of the year was invited, which raised a substantial sum.

Natural images of the Park included 'before-and-after' pictures of the Larchwood Field Pond during its refurbishment, when it was transformed from a choked, water lily-smothered soup to a rich mixed environment. Other nature pictures taken for the guides appear in this book.

Estate History booklet

Original two guides

N

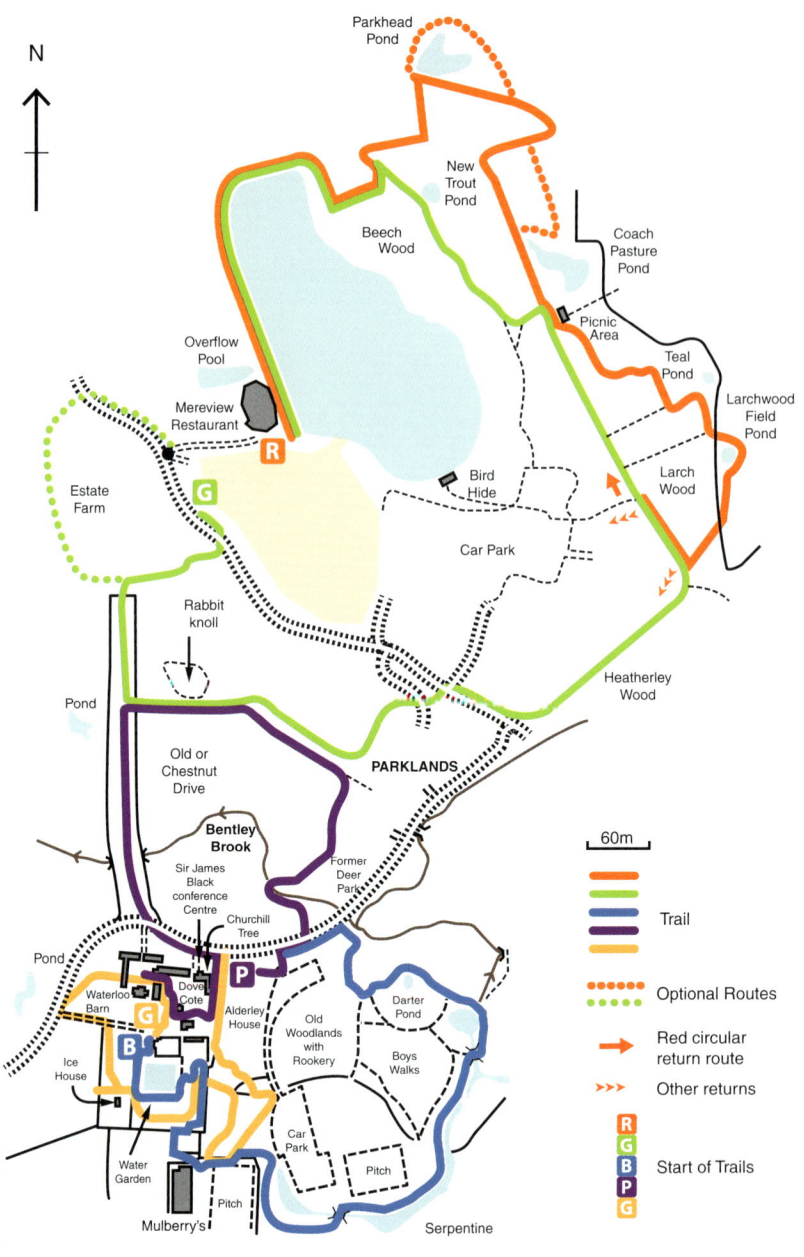

Parkhead
Pond

New
Trout
Pond

Coach
Pasture
Pond

Beech
Wood

Picnic
Area

Teal
Pond

Overflow
Pool

Larchwood
Field
Pond

Mereview
Restaurant

R

G

Bird
Hide

Larch
Wood

Estate
Farm

Car Park

Rabbit
knoll

Pond

Heatherley
Wood

Old or
Chestnut
Drive

PARKLANDS

Bentley
Brook

Sir James
Black
conference
Centre

Former
Deer
Park

Churchill
Tree

P

Pond

Waterloo
Barn

G

Dove
Cote

Alderley
House

Darter
Pond

B

Ice
House

Old
Woodlands
with Rookery

Boys
Walks

Water
Garden

Car
Park

Pitch

Pitch

Mulberry's

Serpentine

60m

Trail

Optional Routes

Red circular
return route

Other returns

R
G
B
P
G

Start of Trails

Five AstraZeneca guides

History and Nature Trail marker

Water Garden and Serpentine

# Icicals to ClubAZ

## Icicals to ClubAZ

The ICI Pharmaceuticals social club at Alderley Park and Macclesfield was founded in 1949 as 'Icicals'. It was a core of site life for four decades.

In 1994, following the demerger, Icicals manager Linda Mainwaring announced the club would be renamed Zeneca Leisure. After the formation of AstraZeneca it morphed again into the huge operation of ClubAZ, which currently has 8,400 Cheshire members enjoying a whirl of activities revolving around www.clubazonline.com, its packed website. In 2010, ClubAZ was incorporated as a separate company, ClubAZ Ltd. It recently opened its membership to local people.

The early Club began with what facilities it could find; the Tenants' Hall was used for badminton and the estate ponds for fishing, for example.

## Sections

The Club has always operated numerous Sections. All were allowed a budget for operating costs and equipment, normally provided they raised a similar amount themselves. Much sports equipment and fine events and facilities – even a bird hide – were thus provided for staff.

Sections involved in less active sport or other activities included the highly organised Angling Section (the largest), together with and sections for hikers, ramblers and walkers, archers, and everything from wine to bridge, singing and drama to films, and a staff Christian Fellowship. All sorts of groups got together. The activities of the Natural History Section have been described earlier; it is now defunct along with others due to staff reductions, but some of its wildlife monitoring work in the Park continues.

## Events, activities, charity and benefits

Many events are part of the social life of the Park. An elegant (though not inexpensive) summer ball was formerly a feature, while a ball at Christmas has more recently been one of the Club's flagship events. Many charity events are supported. All are well reported and many staff appeared, sooner or later, in the report on some function or other, in the social club's own publication (Zeneca Leisure News and ClubAZ Magazine in recent times) and in company staff publications.

A wealth of activities continues throughout the year, from bike maintenance to white-water rafting. The discounts offered on production of a ClubAZ card save money at a wide range of local businesses including restaurants, shops, professional services and child care, whilst also offering great savings at theme parks and tourist attractions across the UK. Theme park offers and similar attractions are popular, as are vouchers for childcare.

## Too high a bid

Larger projects were considered. But the club committee was startled when enthusiasm got a little out of hand and one group started a Flying Section. They put in a hopeful budget request to buy a plane … It was too high!

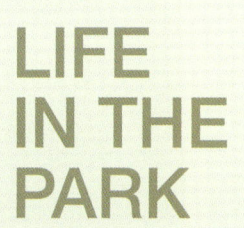

**LIFE
IN THE
PARK**

The ClubAZ lottery also remains popular, with the monthly draw giving away prizes from £100 to £15,000. Over the years it has delighted thousands of lucky members with wins small and large, including cars and holidays.

The Club also operates a shop at Mereside (and one formerly in the upper courtyard) to sell its gifts. In the past a garden shop also sold ICI products from Agricultural Division. A different sort of discount is provided by the sale of bedding plants each May, which pleases almost everyone, apart from the poor ClubAZ staff member left at the end of collection day, desperately trying to contact the owner of the one last forgotten tray of lobelias that has been left behind!

Holiday and summer Kids Camps became a well-established feature, popular with parents of children from ages 4 to 14. In recent times activities have involved everything from sports, dance and drama to learning circus skills.

## Stanley Arms

The older history of the building which became the Stanley Arms was told earlier. This welcoming facility was opened in 1964 and for thousands of staff provided many a lunchtime pint as well as hosting barbecues on summer Fridays.

Everyone was sad to see the Arms close (at the request of AZ) in 2007 due to changing attitudes towards serving alcohol in a working environment. Members of the Club staff are still frequently asked when they will re-open 'the Stanley'.

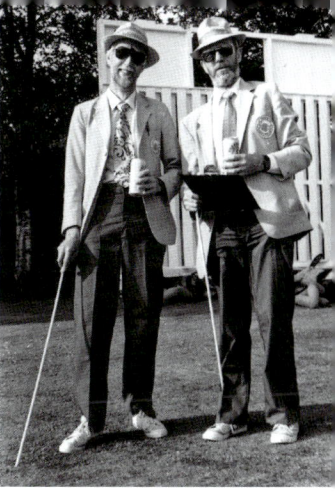

## Annual Fancy Dress Rounders and It's a Knockout

A key event on the staff calendar was always the Annual Rounders Competition. This was a fiercely contested (by some) team fancy dress event, which was one of the most colourful aspects of life in the Park.

The first match had been played in 1958 by two teams on the grass in front of the Stanley Arms, after being launched by a group who transferred to the Park from The Frythe in Hertfordshire. In 1978, it became the Fancy Dress Rounders. By the 1980s, as many as 600 people were taking part, in 64 teams, in an event that was a highlight of the year for many staff.

On Rounders day, all sorts of strange figures could be seen wandering around the laboratories at Mereside, as staff created remarkable costumes from whatever they could find. Old redundant laboratory stores could be turned into the most amazing outfits if required, impressing everyone – except, on at least one occasion, some bewildered Far Eastern business visitors who happened by. A large sum was raised for charity; many comic photographs were taken, to be printed in the next edition of *Scan*; order was supposedly kept by umpires with very dark glasses and white sticks, or (Stu Mills) with a toy parrot on his hat; and a good time was had by all. The Competition ceased for a few years but has now returned each May.

Another popular event is the 'It's a Knockout' in July, a time of year when all sorts of wet things can be done without getting too cold! Jokers can be played!

## Long Service

Long service with the company was always valued and recognised. The award dinners to celebrate 20, 30, 35 and 40 years were lavish events at which old colleagues were able to reminisce about the good old days, and to do so more freely as the wine flowed.

Since 2000, ClubAZ has organised the long service dinners, which are still as prestigious as they always were and greatly enjoyed by the recipients and their partners.

The dinners were also an interesting opportunity to hear senior managers speaking a little less guardedly than usual. They appreciated the opportunity, knowing that this audience had little time for corporate cant and much time for genuine ability (and the ability to distinguish it).

The respect shown for long service did not, however, mean that fools were suffered gladly. At my 20-year award dinner, I arrived wearing a tartan tie. It was a tartan that my wife, Scottish born, was entitled to wear. Unfortunately, however, I was not.

I had just drunk a glass of champagne on an empty stomach and was consequently feeling a little dizzy, when I was spotted by no less a figure than the most famous Scot at Mereside, Sir Tom McKillop. Having once been my section manager, Sir Tom knew that his subordinate's veins held not a molecule of Scottish haemoglobin. He approached rapidly, and I was seized by my tie.

I managed, before my chemistry career could cruelly be cut off, to gasp an apology. It was accepted; but it was only then that I was informed grimly that the insult was a double one. Unknown either to me or my beloved, the tartan in question was not only one linked to *her* family line. It was also, by grave mischance, that ancient pattern treasured since time immemorial by the McKillop clan themselves.

*John Patterson (left) presents the author with a 30-year long service award in 2007*

## Amazen

Not part of the Social Club but often holding events at Mulberry's is Amazen (from Alderley/Macc/Zeneca). Previously called the ICI Pharms and then Zeneca Pensioners' Association, first founded in 1977, this is the pensioners' association for Alderley Park and Macclesfield. It organises all sorts of walks, outings and meals as well as coffee mornings, and keeps in touch with retired staff all over the country.

## LIFE IN THE PARK

### What a load?

Various long service award gifts could be chosen. But one not on the usual list was chosen by retiree Anthony Borrow in 1981, when he requested and received five lorry loads of animal manure from the Alderley Park farm – and had it delivered to his house!

One debate was what to call the new sports centre. Because of my work on the estate's Nature Trail guides, I was asked to suggest the names of wild flowers that grew nearby. I suggested Wild Daffodil or Bluebell Centre, both of which went down like lead balloons.

But the new building overlooked the spot in the centre of the (old) cricket pitch where in 1611 the mulberry trees sent to Alderley Park by King James the First were planted. I pointed this out (perhaps others did also), the committee added the ''s', and Mulberry's became the official name.

## Mulberry's and Sport

The bulk of the Social Club's activities always revolved around sport. A sports centre was initially created next to the old walled gardens, which were soon converted into a slightly small cricket pitch. With its lovely old brick walls adorned with fans of ivy the latter was popular, except when an odd ball shot over the building into the car park!

In due course the old sports centre became dated and the new Mulberry's centre that replaced it was opened in 1996. As the ClubAZ sports and fitness centre in the Park, it holds a large sports hall as well as state-of-the-art gym facilities that are in constant use; in 2015 there were 2,400 gym members. The future may bring changes (current plans mean Mulberry's would move half a mile), but a decline in popularity is not likely to be one of them.

On its opening, two small drooping mulberry trees were supplied by Ian Urquhart of Matthews and planted at the front of the building by Barrie Thorpe and Linda Mainwaring, with the help of 'Warrior' from the popular 90s TV show *Gladiators*. They were appreciated by all – except by those wearing white trainers, who discovered the wonderful colours that ripe mulberries can stain footwear when trodden underfoot.

Mulberry's gym
PHOTO ASTRAZENECA

Mulberry's sports hall
PHOTO ASTRAZENECA

'Warrior' helps Barrie
Thorpe and Linda
Mainwaring to plant
a mulberry
PHOTO ASTRAZENECA

Mulberry's tennis
courts
PHOTO ASTRAZENECA

Mulberry's goalposts
PHOTO ASTRAZENECA

Cricket scoreboard
PHOTO ASTRAZENECA

## Sports and leagues

The central facility is the sports hall itself, a focus for fitness and for Sections for all sorts of competitive sports. Among the biggest has always been the Badminton Section. Several other racket sports are strongly represented, not least by the Tennis Section on the outdoor courts within the nineteenth-century walled garden. Squash is also popular. A vast number of staff over the years became involved in a wide network of leagues.

The football pitch in front of the sports centre is also popular – with rabbits, too! A second football pitch was created above the Serpentine, though a football too far there could become a wading job. The pitches are also used for activities from frisbee to archery; the latter Section also created a woodland course. (The pitches may move soon.)

The Cricket Section renamed itself as the Alderley Park Cricket Club in 1999, to avoid future name changes. The cricket pitch within the walled garden opened in 1967, and was replaced by the new pitch for the 2009 season.

Some sports Sections use even more of the Park, particularly the Running Section which explores its many paths. An entertaining history of the Section was written by Brian Morris.

The stories, achievements, successes and disappointments of the other sportsmen and women of the Park are far too many to describe here; their fascinating stories must be told elsewhere.

But the fitness they all gained was one of the best legacies of the Park for more than a few.

## Annual Site Race

The Site Race began in 1974 and takes place over a 2·3 mile course each winter. The overall record stands at just over 11 minutes, though there are different categories.

The thundering mass of runners passes the windows of Mereview Restaurant, so a popular alternative to running is watching it from a table! But the winners are always assured of a great reception, and the last home gets applauded, too.

# Estate in the local community

Alderley Park and Nether Alderley have always been very aware of each other and their common history. A village history produced by the local Womens' Institute around 1951 includes a pointed comment:

> A large section of Alderley Park Estate has recently been purchased by Imperial Chemical Industries for their Pharmaceutical Branch, on the understanding that they keep the rural amenities as far as possible.

An ex-ICI employee who was a Nether Alderley Parish Council member when the news first became known, remembered the concern:

> In the great days of the Stanleys, many people were employed on the estate. When it was first sold – to a property developer – all these people were sacked, and, with the local farmers unable to take them on, they had to leave the area. And so we welcomed the prosperity that was likely to result from the coming of ICI. On the other hand, we wanted to keep Nether Alderley as it was – at least as far as possible. Our greatest worry was that ICI would want to build hundreds of houses for their employees – altering the character of the place completely. The Company met us, and gave assurances that this was not their intention.

## Sound relations and noise levels

Feelings have not always, of course, been warm with all. An analysis by the local traffic authorities found that the stream of vehicles leaving the site when AstraZeneca went home was so large that it distorted all the traffic patterns in the district, with the result that traffic lights were installed at Church Lodge in 1991.

Closer to home, noise from the research buildings (which have huge fans) also concerned residents; and many trees were planted over the years to alleviate it. Frank Booth, former engineering maintenance supervisor, recorded how one team spent all night shutting down items of plant one by one to locate the source of the noise. But eventually they found that:

> … overwhelmingly the loudest noise recorded during the night had been caused by the rustling of the beech trees under which they stood. 'We could always put hair nets on the trees I suppose,' grunted a tired technician, rubbing a hand over the newly-formed stubble on his chin.

### Community in the estate

Only on rare occasions did outsiders find their way into the Park. Now and then, trespassers or simply lost walkers aroused attention. More amusingly, someone would think that the Park was a public space and turn up at an entrance wanting to pay for admission!

A different sort of intrusion into the Park's space happened every year, to the pleasure of most staff. This was the overflight of planes for the Woodford Air Show, which often flew low over Mereside as they arrived in the days before the show. The Battle of Britain Flight, led by its Lancaster, was a regular. But I was taken aback one summer by a huge Vulcan bomber thundering over me; and Concorde also startled staff by flying low over Mereside on occasion.

## Image, Communications and charity

The global issue of public image, for ICI and its successors over the years, is not a major Alderley Park story, as the company head office was never in the Park. Nevertheless, two public relations issues have always been a matter for what latterly became a distinct 'PR and Comms' function.

### Every enquiry!

One of these is the duty to respond to any and every enquiry from the public that might conceivably be justified. And the public can ask almost anything!

Many odd queries could be cited; but perhaps the oddest I heard of was when one enquirer asked if he could photograph the Deer House – on top of which Mereside had been built in 1957! The answer was no.

### Support, please

The other very public-facing task is to work out how far a business can possibly go to meet the requests and pleas that flood in – 'You are a multinational company. We are a good cause. You have lots and lots of money. Could we have some of it, please?'

Immense numbers of charitable requests have always arrived each year, many of them clearly very deserving. Moreover, many came from, or via, members of staff who were deeply interested – and who knew who to speak to, as well. An employee Matched Funding scheme begun in 1999 was well supported as a way of persuading the company to support 'my' charity. And endless Gift Aid donations were made by staff themselves.

**LIFE IN THE PARK**

With my background of studying the Park, I was often asked strange questions myself. But I was unimpressed when one past Alderley Park site manager, who shall remain nameless, commented cheerfully of me that: 'If I have no idea what else I can do with an external query, I send it to George!'

The company's main support area is medical charities. In recent times, local causes with medical links became favoured, especially if staff were involved. This became more focused in 2003 when a formal Adopt-a-Charity scheme was set up, to point staff to charities the company would particularly help them to support if they wished.

The wider pressure became so great that the company also decided to focus its direct support on two chosen charities each year. Staff were invited to nominate and vote on what they should be, with the first being the East Cheshire Hospice and Macmillan Cancer Care in 2003. In 2009 AstraZeneca in the North West alone donated £67,000.

## Big gifts

Big single gifts have also been frequent. Notable donations included £50,000 to build the East Cheshire Hospice in 1985, with the same amount again to support the first five years of operation; and £40,000 to the local scanner appeal in the 1990s. In 2000, 100 staff volunteers worked together to upgrade facilities at Park Lane Special School in Macclesfield. Yet these are just the top headlines of a news story that is sixty years long.

## World gifts

Probably few people realise how many tragic situations around the globe have been eased over the last couple of generations by consignments of company medicines, particularly antibiotics and treatments for developing world diseases, following every major earthquake or similar disaster. Very many tons of free aid have travelled from the company's Macclesfield works or other warehouses; £25,000 of antiseptics to UNICEF for the children of war-torn Rwanda in 1994, £170,000 worth of antiseptics and anaesthetics to Bosnia, and £180,000 worth of medical supplies to Romanian orphanages were just three of the larger gifts.

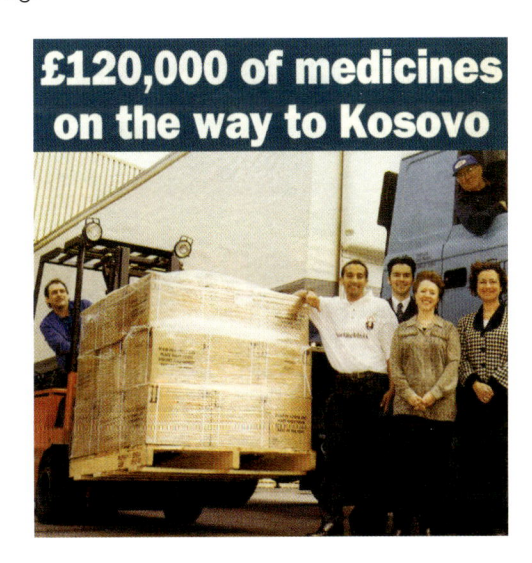

Donation to Kosovo
*SCAN*, 30 April 1999

In more recent years, the company's product range has been less suitable for emergency situations, so after the Asian tsunami in 2004 well over £300,000 in cash was donated through the Red Cross. The Haiti earthquake and breast cancer screening in Cambodia are other recently supported causes.

This side of the work of a multinational company is perhaps more understated than any other. Yet even the fifth (at its peak) largest pharmaceutical company in the world can do only so much. Until it does more.

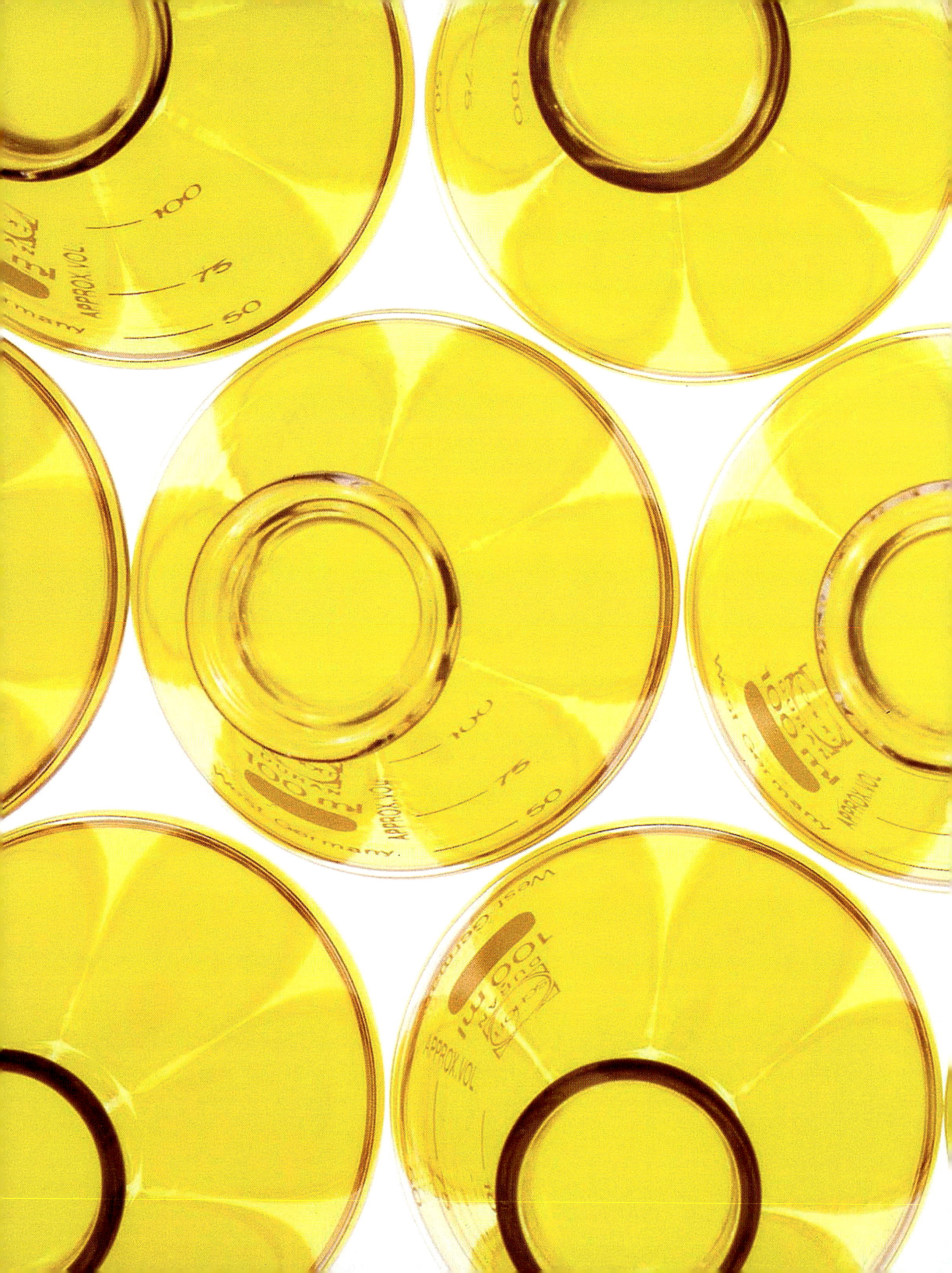

# PART IV
## Global Alderley Park

'If we knew what we were doing, it would not be called research, would it?'
*(reputedly) Albert Einstein*

# 9 Company beginnings

## 'Pharms' – the early history

The scientific history that led to Alderley Park is fascinating.

The branch of Imperial Chemical Industries (ICI) that became Pharmaceuticals Division opened in Alderley Park only in 1957, but its origins are worth tracing.

### Blackley and dyes

In the eighteenth century, dyes were the business. The desirable hue was a fiery dye known as Turkey Red. An emigrant French industrialist, Angel Delaunay, first made this in England. He set up in north Manchester a dyeworks that much later became ICI's Blackley site (pronounced 'Blakeley').

In 1856, the young William Perkin discovered his famous aniline dye, **mauveine ('Perkin's Mauve')**. Perkin and Son was the direct ancestor of ICI Dyestuffs Division and the synthetic drugs that are the mainstay of modern medicine stem straight from Perkin's research. The original Blackley site later grew into an international business under a German emigrant, Ivan Levenstein.

Delaunay's 1830 works
ASTRAZENECA ARCHIVES

Blackley, 1900s
ASTRAZENECA ARCHIVES

Ivan Levenstein
ASTRAZENECA ARCHIVES

# LIFE IN THE PARK

By 1918, the 90-acre site (producing 169 different products) was manufacturing rudimentary pharmaceuticals, mostly made from coal tar, developed after researchers found that certain industrial dyes could affect bacteria. The yellow dye **acriflavine** had been used from 1916 for disinfecting war wounds. It and another antiseptic, **proflavine**, were both later made as medical agents by ICI. The Levensteins also produced **procaine**, a local anaesthetic discovered in 1905.

## Dyes, drugs and ICI

All this attracted attention and, in 1917, the *Manchester Guardian* reported that 'dyes and drugs must be thought of together. Whatever serves the modern dyemaker directly serves national health.'

In 1926 the composite company Imperial Chemical Industries was formed from Nobel Explosives, British Dyestuffs Corporation, Brunner Mond and United Alkali, with a total of 33,000 employees. With government encouragement, the aim was to counter the might of Germany's IG Farben, and DuPont in the US, with a British company in the front rank of the world chemical industry. The ICI 'roundel' was an adaptation of the Nobel company logo. The new ICI Dyestuffs Division became central to its business and several of its dyes were supplied to existing pharmaceutical firms.

In the German laboratories of Bayer, bacteriologist Gerhard Domagk discovered the antibacterial action of a red azo dye, **'Prontosil'**, which the company had patented in 1932. His landmark publication of that news in 1935 led to it becoming the first commercially available antibiotic, a discovery for which he received the 1939 Nobel Prize (and with which he also saved his daughter's arm from amputation). Franklin Roosevelt's son was another patient.

It was soon realised that 'Prontosil' changed to a more active chemical once in the human body. Ernest Fourneau at the Pasteur Institute in Paris determined that this was **sulfanilamide**, which was not covered by industrial patents, unlike 'Prontosil' itself. This was the spark that alerted ICI – despite the fact that sulfanilamide itself was relatively toxic.

ICI's first board meeting in 1927
ASTRAZENECA ARCHIVES

## Cecil's prodding

Interest was awakened by Bayer's discovery. Cecil T. J. Cronshaw, at that time ICI Research Director, did some 'prodding'. So did the company's academic advisers, who included chemist Robert (later Sir Robert) Robinson of Oxford University and bacteriologist Carl Browning of Glasgow who had worked with the famous Paul Ehrlich in Germany. A decision was made in 1936 to set up, at the research labs at Blackley, a Medicinal Chemicals Section within Dyestuffs Division.

## New Section

The new Pharmaceutical Section began under Section Leader Mr Sam Ellingworth, the only scientist with previous medicinal experience. (In 1935 ICI's medicinal work had been merged with that on pest control.)

Sam Ellingworth had already, in 1934, made a sample of the first compound to be marketed by ICI as a medical product of its own. This was cyclopropane, a highly flammable as well as volatile product, at that time a newly introduced anaesthetic. Ellingworth personally carried to London, for clinical trials, a steel cylinder holding over a pound of cyclopropane under his train seat on the 'Comet' – no doubt contravening a swathe of L.M.S. railway regulations! Later, it went into production at Blackley then at Grangemouth, Scotland.

ICI was a chemical group rather than a biological one like Glaxo, founded the previous year, or the established Burroughs Wellcome group, and its organic chemistry based business dictated its early growth. It entered the pharmaceutical world relatively late and speculatively; the budget given to the new Section on 11 June 1936 was £15,000 p.a. for the first five years – a small sum, even then. If nothing was discovered, the Section would be disbanded.

At Blackley, Ellingworth took charge on 1 October of a team including six with PhDs: I. E. Balaban, W. R. Boon, D. G. Branscombe, H. C. Carrington, F. H. S. Curd and F. L. Rose, (all but the last two were new recruits) plus Mr H. C. Lumsden. W. A. Sexton headed the pest control section. The brief was not only to manufacture known drugs for other companies to market (as Dyestuffs had already done) but also to use a 'research approach' to develop new medicines. Ellingworth wrote that:

> Our aim and hope is to contribute to the progress of chemotherapeutic development, where nothing approaching finality has been reached.

The new research block at Blackley was shown to the Press and public, although there was little yet to display. Various exhibits had been acquired for the opening. The centrepiece image in most newspapers was of chemist Frank Curd, staring at a huge model of a tsetse fly, borrowed from the London School of Hygiene and Tropical Medicine.

## First steps

A panel of biologists and chemists initially sent novel compounds for testing by researchers in the labs of various university medical schools. From 1937, ICI started to recruit its own PhD biologists to work in a multidisciplinary team, and built labs for them in 1937–8. Led by Charles M. Scott, these were A.R. Martin, A.L. Walpole, W. Lees and J. Raventos. Raventos (in anaesthetics) and Walpole (in cancer therapy), were to become famous, as described later.

Much of the early research focused on sulfonamide drugs. An important lead was May & Baker '693', or **sulfapyridine**, which famously saved Winston Churchill from pneumonia during the war. 'Frank' (Francis L.) Rose was an azo dye chemist who had joined ICI in 1932 and transferred to the new Section in 1936. Rose and his colleagues created a route in 1940 to an analogue (a chemical relative) of sulfapyridine – one with a pyrimidine ring in its structure instead of a pyridine. The new drug, **'Sulfamezathine'** (**sulfadimidine**), was an anti-infective and weak anti-malaria agent free from sulfapyridine's nausea and kidney problems. It was rapidly adopted by British doctors. It is commonly known now as **sulfamethazine**; its inventors applied for this more appropriate name but had to use the other on first launch.

## Wartime

The start of war had a profound effect on ICI's research programme, when it moved immediately into production of drugs formerly imported from Germany. The British government arranged with several key companies that it would pay for the cost of plant devoted to war production. Through this, ICI played an important role in wartime research, becoming the government's largest industrial agent. The pharmaceutical programmes to produce synthetic anti-malarials and penicillin were prioritised.

### Schoolboy chemistry

Rose was a scientific prodigy. As a choirboy, he was supplied with chemicals by a friendly churchwarden who was a pharmacist. (His stock of white phosphorus nearly burnt the house down when its water containment evaporated while he was on holiday.) 'How I escaped serious accidents I don't know, but even at 8 to 11 I must have been a fairly careful laboratory worker. By the time I was 11 … I knew what I wanted to be.'

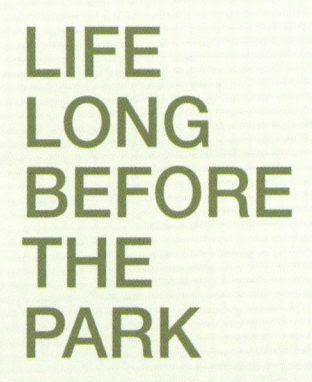

# LIFE LONG BEFORE THE PARK

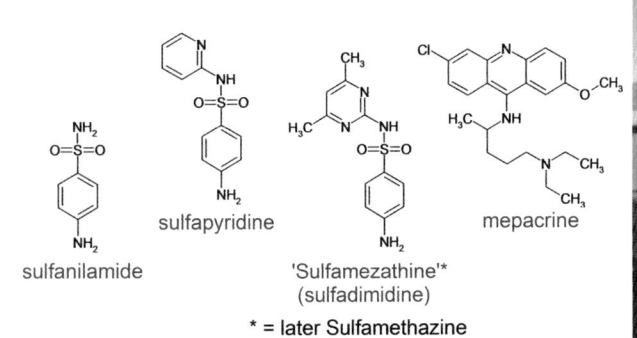

sulfanilamide

sulfapyridine

'Sulfamezathine'*
(sulfadimidine)

mepacrine

\* = later Sulfamethazine

Early sulfa and anti-malaria drugs
GEORGE B. HILL

Mepacrine production at Billingham, c.1941
ASTRAZENECA ARCHIVES

Curd and Rose in the lab
HILARY FORD

ICI and other British firms (Glaxo, BDH, Boots and May & Baker) were asked to 'crack the German patents' and create manufacturing processes for German drugs. The immediate targets were the anti-malarials **pamaquine** and **mepacrine** (**quinacrine** in the US), for which the Allied military medical authorities asked ICI to find a replacement. Pamaquine, whose structure was revealed by Bayer in 1928, had some value. But mepacrine could sometimes suppress the malaria parasite wholly, achieving a complete cure. It was later conjectured that without mepacrine to combat malaria, or penicillin to counter infectious disease, the casualty figures from disease would have exceeded battlefield ones in WWII.

The chemical structure of mepacrine was probably (contrary to some accounts) already known in the UK but Bayer's process was secret. In 1939, a team led by Frank Curd found in six months a new manufacturing route and it was already in production by the time war started. The work was done in great secrecy, under a committee chaired by Harry Hepworth, later Pharmaceuticals Division managing director, and was not even recorded at first in ICI's main board minutes.

Mepacrine was far from ideal; it did not prevent infection or relapses, allowed patients to become anaemic and was hazardous to make. It also had another problem – as Rose remembered well:

> It was quite a common sight in port towns like Liverpool to see people walking about with their skins completely yellow.

Yet it proved invaluable when, in 1941, access to the East Indian plantations which were the source of quinine (from cinchona bark) was cut off by the Japanese invasion. This was reversal of history; in the previous war it was Germany whose quinine supplies had failed, a problem that had been a driver for research there. (German attempts to reproduce the anti-malarial properties of quinine in a synthetic drug dated back to the work of Ehrlich in

# LIFE BEFORE THE PARK

the 1890s.) By war's end ICI was making at Grangemouth two billion tablets a year, for all the armed forces in the Far East. The drug also became important in Europe, where malaria was still endemic throughout the Mediterranean – one reason why so few people then took summer holidays there.

## IC (Pharmaceuticals) and Pharmaceuticals Division

ICI was asked to work on some twenty essential drugs. Because of this and the rising profits from the sale of anti-malarials from 1942, ICI decided to establish (mainly for tax reasons) a new selling company, Imperial Chemicals (Pharmaceuticals) Ltd, with 'an intention to transform it, at some indefinite future date, into a manufacturing company with its own plant'. Its first Chairman and Managing Director were A. J. Quig and T. M. Willcox; but P. A. Smith held both roles by the end of 1947. Global turnover in 1942 was £247,000. (Practically all wartime sales were made to government departments.) In that year the company moved into veterinary medicine, acquiring Stamford Lodge near Wilmslow in 1943 along with its 41 acre estate, to assist that work.

Subsequently on 1 January 1944 a full Pharmaceuticals Division was set up, marking the beginning of an era that lasted for fifty years, although it dealt initially with only commercial and service activities. Dyestuffs were strongly represented on the 'Pharms' board at first and the fledgling Division did not break away fully until 1957 when Alderley Park was opened. The first headquarters of the Division, with its commercial department and first Biological Laboratory, was at Fulshaw Hall, Wilmslow, which during the war had been used for training Special Operations Executive agents. Product distribution was originally based at Earlsfield, London but centred on Grangemouth (on the Forth) from 1951, with a new plant there in 1953.

The new pharmaceutical world created problems for non-scientists on the staff. One typed report described the remarkable disease 'psyphillis'. Also, one Medicinal Products Panel meeting became celebrated for a certain mispronunciation of 'psoriasis', which perhaps should not be explained further …

IC (Pharmaceuticals) created
ASTRAZENECA ARCHIVES

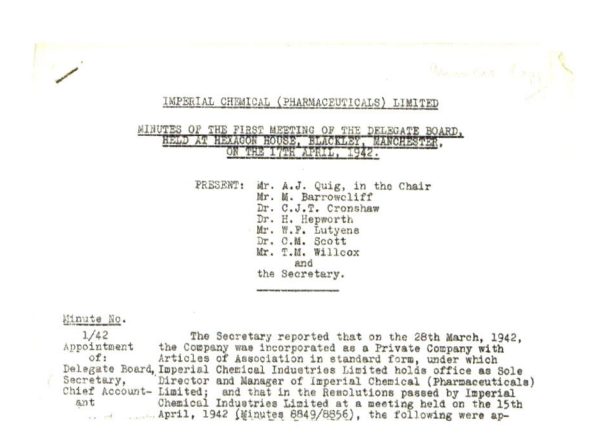

## Alone

The company had originally intended to take over an existing business. But various approaches made during 1939–42 all fell through. A previous agreement with BDH Ltd to market medicinal products made by Dyestuffs Division prior to 1936 was ended in 1940. Eventually, Cecil Cronshaw became determined tho company should depend on its own resources:

> We're not going to become another company selling Glauber salts – we're going to live and profit by our own discoveries.

So the Division deliberately avoided licensing products or buying information; and its early growth was organic. As W. J. Reader, the official ICI company historian, asked in 1975: 'This was magnificent, but was it wise? It ran clean contrary to accepted practice in the pharmaceutical industry …' ICI had been designed in 1926 for the heavy chemical business. This bias, and accidents of war, meant that even when the importance of fine chemicals was grasped, the company structure was never, before Alderley Park days, properly adapted to this new world's needs.

## 'Paludrine'

Following the Japanese invasion, a vast US-UK anti-malarial research programme began. ICI was a major player to develop a new anti-malarial drug that would be easy to make. The drug also needed, like mepacrine, to act as a prophylactic (stopping the early stages of the protozoal infection before the malaria parasite invades the bloodstream).

Frank Rose replaced Ellingworth as Medicinal Chemicals section leader in 1942 and led the effort in a programme drawn up by Frank Curd earlier that year. The parasitologist D. Garnet Davey was recruited in 1942 to devise a biological screen; despite his reluctance to work in industry he agreed to join the company in order to support the war effort (and was not the only useful top academic gain for ICI from the war).

The team also included another 1942 arrival, Alfred Spinks, who had been recruited to work at Imperial College for ICI on £350 per annum two years earlier, and now moved to Blackley. Spinks was given a tour of the company and worked briefly in patents, whose head described him caustically as 'not only ignorant but insolent', a story Spinks later enjoyed telling. He started work on pyridylpyridines but was soon asked to tackle a new topic.

Frank Rose had been intrigued and puzzled by differences between the antibacterial activity of their compounds *in vitro* (literally 'in glass', in the test-tube or other vessel) and *in vivo* ('in life', in living animals or tissue). With biologist A. R. Martin, he developed methods for measuring the concentration of experimental drugs in blood, now a standard procedure but then a novelty. At first, ICI's management objected to the huge amount of time spent on Rose and Martin's studies. But Rose's promotion allowed him to ask Spinks to continue; and Spinks's personality and skill then ensured that the study of the pharmacodynamics of sulfonamide and anti-malarial compounds continued. (In simple terms, pharmacodynamics is about what a drug does to the body – as distinct from pharmacokinetics, which is about what the body does to the drug.) A Royal Society biographical memoir of Alf Spinks records that:

> Almost from that point onwards, consideration of chemical structure-drug absorption relationships became an important concept in new drug design within the Section, and this at a time when there existed almost no formal pharmacological academic teaching on this aspect.

## On the bus

The science theory was accompanied by practical ingenuity, as Alf Spinks overcame wartime restrictions to accumulate his equipment. The first use of radioisotope-labelled compounds (radioactive ones) was a particular challenge:

> Because of the all-prevalent danger of air raids, not to mention the hazards from the 'flying bedstead' ironmongery sent up from the local Home Guard anti-aircraft unit (of which Spinks was a member), the management at Blackley wisely refused to house the radiochemical unit in the new (medicinal) laboratories that had just been erected on the site.
>
> Entirely of his own volition, Spinks overcame this setback by negotiating the purchase of an obsolete double-decker bus he had seen advertised for sale by nearby Leigh Corporation. This was installed in a far corner of the car park and equipped as a mobile laboratory, with chemistry downstairs and biology up aloft. In the event of disaster, it was hopefully intended to tow the bus away. It served, however, to get work started, and profit resulting from later disposal of the massive lead shielding with which it had been fitted was reputed to have more than paid for the entire venture!

Amongst other things Spinks was able to suggest the best size and frequency of dosage to malaria patients – at a time when one of his colleagues had just stopped grinding test drugs up with biscuit powder! Spinks later continued his pioneering work in pharmacodynamics at Mereside along with another colleague of Rose's, South African Jeff Thorp. Spinks was made an FRS in 1977.

The combined efforts of the team led, after preparation of 1,400 new chemical compounds, to compound ICI 4,888, named **'Paludrine'** (**proguanil**). This was discovered when chemists making sulfamezathine-like compounds decided to swap the pyrimidine ring these contained for an open-chain 'biguanide'.

Very opportunely, it proved possible to prove 'Paludrine', first made in November 1944, on a large scale in army personnel in Australia by the following April. It was used extensively in the Malayan campaign, was made by the ton for the Pacific campaign (which never happened because of the atom bomb), and became one of the most widely used anti-malarials

The Medicinal Chemicals Section in 1943 with Spinks in uniform
HILARY FORD

proguanil oxidizes to cycloguanil in the body

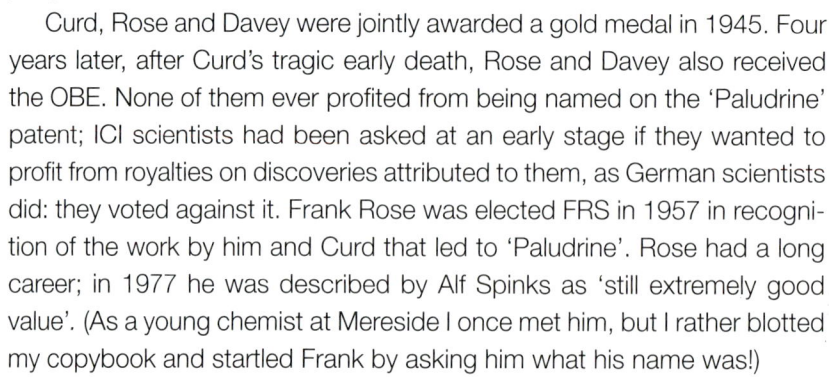

in the world along with the Bayer drug **chloroquine** – now sold by AstraZeneca as **'Avlochlor'** – with which it is still taken in combination. Unlike mepacrine it also prevented infection, and controlled fever in all of the several forms of malaria. It is very safe, as an overdose makes you sick – Rose believed no-one had ever died from using it. Ironically, later work showed that the biguanide did not make the drug active! Paludrine's biguanide is converted back into another sort of ring, a cycloguanil, which gives the drug most of its anti-malarial property.

Curd, Rose and Davey were jointly awarded a gold medal in 1945. Four years later, after Curd's tragic early death, Rose and Davey also received the OBE. None of them ever profited from being named on the 'Paludrine' patent; ICI scientists had been asked at an early stage if they wanted to profit from royalties on discoveries attributed to them, as German scientists did: they voted against it. Frank Rose was elected FRS in 1957 in recognition of the work by him and Curd that led to 'Paludrine'. Rose had a long career; in 1977 he was described by Alf Spinks as 'still extremely good value'. (As a young chemist at Mereside I once met him, but I rather blotted my copybook and startled Frank by asking him what his name was!)

The experience of making, testing and selling anti-malarials had a considerable impact on the postwar development of 'Pharms'.

## Penicillin

ICI became the first industrial firm in Britain to mass-produce **penicillin**, after its 1929 discovery by Sir Alexander Fleming had finally borne fruit.

Sulfonamides acted against streptococcal bacteria but penicillin worked against the common staphylococci which caused many infections. In August 1941 the *Lancet* printed what it later described as 'the most important paper it ever published', describing penicillin's development for human use. Anti-malarials were thought more important for the war effort at the time but in 1942 ICI was approached by the paper's author Howard Florey (who became a consultant to the company), then working on the drug at Oxford. It was expected that the structure of penicillin would soon be learnt, allowing its easy manufacture, like the sulfonamides. William Boon, another of the original Blackley team, set out to synthesise the drug. (Making penicillin from scratch never became industrially useful.) But the emphasis within ICI soon switched to fermentation production as the company joined in a UK-US collaborative effort to make penicillin in bulk.

Following a small production unit at Blackley, ICI set up a penicillin plant at Trafford Park, Manchester. Manufacture was in milk bottles, all that was available in wartime! These were plugged with cotton wool and tilted to increase the surface area; later, deep fermentation, a process strongly recommended by Robert Robinson, was used. The plant was visited by Sir Alexander Fleming, and each day's output of 5g was sent for use by the armed forces.

In 1943, a project set out to discover penicillin's structure; ICI supplied the brilliant Dorothy Hodgkin with penicillin's rubidium salt and Charles Bunn's team at ICI's Northwich labs worked using X-ray crystallography with her, to show in 1945 that the drug contained its famous (fragile) 'beta-lactam' ring. In 1944, a famous ICI film included interviews with all of the penicillin inventors Fleming, Florey and Cheyne.

In 1944 the penicillin pioneers were knighted, and supplies of the 'wonder drug' were parachuted into France, only six hours after the first seaborne landings, to assist wounded D-Day soldiers. By 10am on D-Day, the first sample had already been administered to a young soldier. 'In the last war,' declared Fleming, 'there were many soldiers whose wounds remained open for six months or more. Now we hope that wounds will not become infected at all.' It was not publicly mentioned that penicillin also enabled thousands of troops to return quickly to battle after contracting venereal disease.

ICI never did well out of penicillin. Their independent company mindset led them, by the late 1940s, to become 'probably the only manufacturers in the world who were trying to make penicillin from their own knowledge alone.' When other UK companies sought American help and the world price began to drop rapidly, ICI soon lost any useful profit margin.

Antiviral research also began in 1944, but proved difficult.

Penicillin production in wartime milk bottles
BARRY SMITH

penicillin
(R = various)

Penicillin structure
GEORGE B. HILL

## The poetry of science

The startling revolution in pharmaceuticals began to change the modern world. An anonymous versifier in *Hexagon Courier*, ICI's company newspaper, summed it up near war's end:

'Hail chemotherapy – set your mind at ease!
　　Do we approach the conquest of disease?
A group of drugs the fatal reaper foils,
　　Controlling pneumonia, tonsillitis, boils.
Staphylococcal sepsis has no power to shock us
　　A murrain on the wily streptotoccus!
Doctors are overjoyed and won't send any bill in,
　　Thanks to sulphonamides and penicillin.

# LIFE BEFORE THE PARK

## Planning for the future

In 1943, Charles Scott from ICI visited pathology professor James Yule Bogue (who used his middle name) at the Nuffield Institute in Oxford and invited him to join the medicinals group at Blackley, as a physiologist and head of biological sciences. An inducement offered was that Yule Bogue should be involved in its post-war expansion. From 1946 he headed the planning committee for this. His remit included the planning and design of new laboratories. In a seven-month tour of the USA and Canada in 1946–7, he visited 26 companies and 25 academic labs to study American pharmaceutical and chemical research.

In 1947 the 32 research chemists employed were split into three sections under Curd, Rose and Boon. In the same year, a Chemotherapeutic Research Committee was set up under Scott with Sam Ellingworth as secretary, to advise the company on research matters. In 1948, this committee laid out the research programme for that year, which included new projects into amoebiasis, helminthology, bacterial and virus work, tissue growth, anticonvulsants and many others.

ICI's pharmaceutical business grew in several ways: manufacturing competitor's products; utilising chemistry the company was already doing; and discovering and testing new products. In time the third became the dominant approach, and Alfred Spinks played a key part in this.

Alf Spinks was posted off to Oxford University in 1950 to do a course in physiology. Hypertension was identified as a specific target in 1951. A Pharmacological Section was set up at Blackley in August 1952. In March 1953 after his return, Spinks put forward a new list of fields for research, this time based on quantitative data rather than just a literature search, and predicted that the market for drugs to combat hypertension could soon reach £100,000 per annum.

Scott and Bogue accepted the proposals drawn up by Spinks, whose influence grew in a career that eventually took him to the ICI Main Board as director responsible for 'Pharms'. They agreed there should be a push toward diseases of organic dysfunction, in particular of the heart and central nervous system. (An early result of the latter was the company's first CNS product, **'Mysoline'**, for the control of epilepsy.) In 1955 a new unit was set up under Jeff Thorp, to study atherosclerosis, with Spinks doing much of the work.

It was also clear the new research push needed to be unfettered. Archie (W. A.) Sexton, lecturing as the Division's research director in 1954, pointed out that 'pursuit of <u>unexpected</u> observations can lead to results of considerable moment.' This was a prescient comment on what was to lie ahead. His further advice has by no means always been followed by his successors:

> The important thing is that the general organization of scientific work must be so conceived that an <u>opportunist</u> policy can be pursued when necessary.

## Out of Blackley – a new site

A move from Blackley became inevitable, driven partly by the unpleasant physical atmosphere there, which Frank Rose described as:

> 'Absolutely filthy – all the smoke from Manchester, all the coal, the perpetual smell of naphthylamine. You couldn't open a window for black smuts. It was just like a Lowry picture, but it was convenient for the textile industry, which is why it was there.' The atmosphere was polluted with zinc and arsenic; and frequent viral infections made it difficult to keep laboratory animals alive.

Yule Bogue had gathered much data from his tour. He began designing biological laboratories, experimental animal facilities and support services, drawing up plans with a much increased focus on biochemistry and histopathology. H. C. Carrington was responsible for design of the chemical labs, modelled closely on those at Blackley.

Bogue and Carrington began work … and the end result was a plan for a new site for the new research direction. This site would become Alderley Park.

# ICI's choice; building begins 1955–64

## The search for a new site

J. Yule Bogue, later Managing Director and Deputy Chairman of the Division, wanted the new laboratories to lie within the heaviest concentration of medical research and practice, namely the Oxford–London–Cambridge triangle. (Recent events have shed an interesting light on this view, with AstraZeneca now moving its R&D in 2016 to Cambridge for exactly this reason.) A possible site was located there.

## Cheshire

But the ICI Dyestuffs board decreed that the site should be near the heart of the chemical industry, i.e. not far from Blackley. Cecil Cronshaw ruled it should be in Cheshire, if possible.

The aim was a site of not less than 250 acres with no public rights of way across the land. Yule Bogue and Tony Cartmel from the Dyestuff's secretary's department drove around Cheshire looking for sites. There were several possibilities, but all were traversed by public footpaths. Finally, Cartmel made a visit to Macclesfield Borough Council and came up with the suggestion of Alderley Park.

They considered it an ideal site and it was probably Harry Hepworth, the new Division's managing director, who signed the purchase agreement on ICI's behalf.

## Application and purchase

On 25 October 1949, ICI applied to build research laboratories in Alderley Park. An outline request was filed for a Biological Research Station in Alderley Park, with an administrative office block envisaged later. The initial planning application map (which I have studied) shows a comb-like building design quite different from the one eventually built.

The purchase of the historic estate itself, from property developers Hambling, Crundall & Co., was announced in national newspapers on 22 June 1950. The price was £55,000 for 350 acres – including the copper mining rights. The London estate broker A. E. Becheley Crundall had paid £40,000 for the estate when it failed to sell at auction in 1938. Outline planning permission was officially granted on 12 October 1950.

In 1949, the Division also established a very different lab

In the *Manchester Guardian*, 22 June 1950
*PARK LIFE*, March 2002

Mereside as first planned and as in 2005
COPYRIGHT ASTRAZENECA

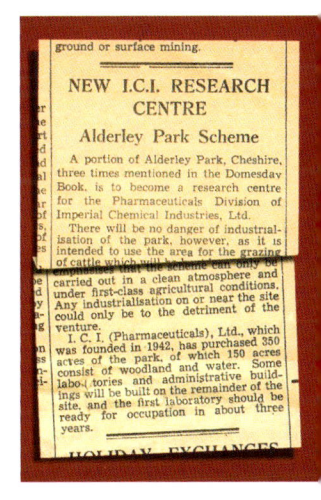

ground or surface mining.

**NEW I.C.I. RESEARCH CENTRE**

**Alderley Park Scheme**

A portion of Alderley Park, Cheshire, three times mentioned in the Domesday Book, is to become a research centre for the Pharmaceuticals Division of Imperial Chemical Industries, Ltd.

There will be no danger of industrialisation of the park, however, as it is intended to use the area for the grazing of cattle which the scheme can only carried out in a clean atmosphere and under first-class agricultural conditions. Any industrialisation on or near the site could only be to the detriment of the venture.

I. C. I. (Pharmaceuticals), Ltd., which was founded in 1942, has purchased 350 acres of the park, of which 150 acres consist of woodland and water. Some laboratories and administrative buildings will be built on the remainder of the site, and the first laboratory should be ready for occupation in about three years.

RADNOR MERE

Former Deer House

52
26
50
13
8
7
11
3
19 Pond
35
33
23
24

■ Eighteenth-century Deer House (now gone)
■ Pond (now gone)
■ 1949 planned laboratory buildings
⋯ Existing Mereside buildings 2005

## Touring Britain

**In 1952, the pharmaceuticals business launched its two new promotional vehicles. Staff were amused to learn that these would display and proclaim ICI's products at country fairs and shows the length of the country, 'from Wick to Tavistock'!**

– in the grounds of the main prison in Nairobi, Kenya! A willing supply of imprisoned volunteers and patients were tested with various anti-infective products. The lab closed in 1955.

## Changes and design

Little seems to have happened rapidly in the Park, although the Division's business expanded vigorously through its other sites. The farm passed into ICI's hands, and woodland restoration also began, with the first new conifers planted in autumn 1950.

Final plans for a 'centre for speculative chemotherapeutic research', at an estimated cost of £1,250,000, began to be made by March 1953.

In the same year, the Chemotherapeutic Research Committee, despite the fact that minor improvements to existing products was known to be a profitable approach, stated that a flow of new products was required for the company to prosper. To that point, research expenditure on pharmaceuticals had represented roughly 10% of the company's turnover. It was expected that by 1957–60, with a likely turnover then of £5 million, about £500,000 per year would be spent on research.

## Construction

Construction of Mereside was begun in February 1955 by John Laing & Co. The final cost proved to be £800,000. The architects were Harry S. Fairhurst & Sons. An ICI history written around 1972 describes the challenge:

> The problems were how to fit modern laboratory buildings into this very rural site without ruining it further, because underneath the neglect the parkland was landscaped in the best English tradition. It was probably the work of Repton and so was potentially very fine. Then, having done that, how to get the site back to something of its former magnificence without bankrupting the Company and how to maintain it at a minimum expense, for we are a commercial organization with responsibility to our shareholders.

The dilapidated eighteenth-century Deer House on the site was first demolished. The first completed building was the (old) Mereside Restaurant, with its 'crush hall' with aquarium and fountain, shop, and barber's shop, a main hall (in part of which waitress service became available for an extra shilling or so), and a directors' dining room (with bar), to which directors drove each day from Alderley House, in a Ford Escort.

The research complex comprised chemical and biological labs, an animal breeding unit and animal houses, and a wash-up unit and store.

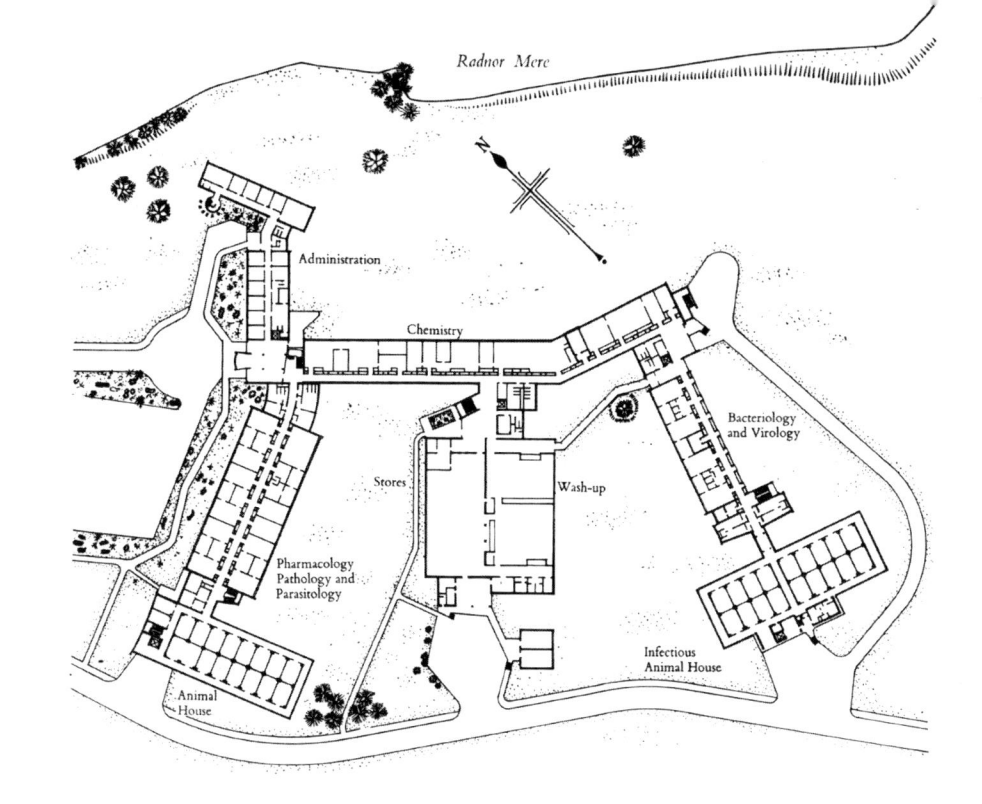

Mereside as first built
*PHARMACEUTICAL RESEARCH IN ICI 1936–1957 (1957)*

One of the first chemistry labs
*PHARMACEUTICAL RESEARCH IN ICI 1936–1957 (1957)*

An early pharmacology lab
*PHARMACEUTICAL RESEARCH IN ICI 1936–1957 (1957)*

A histology lab, 1957
*PHARMACEUTICAL RESEARCH IN ICI 1936–1957 (1957)*

Animal room in 1957
*PHARMACEUTICAL RESEARCH IN ICI 1936–1957 (1957)*

Layout of an early biological suite
*PHARMACEUTICAL RESEARCH IN ICI 1936–1957 (1957)*

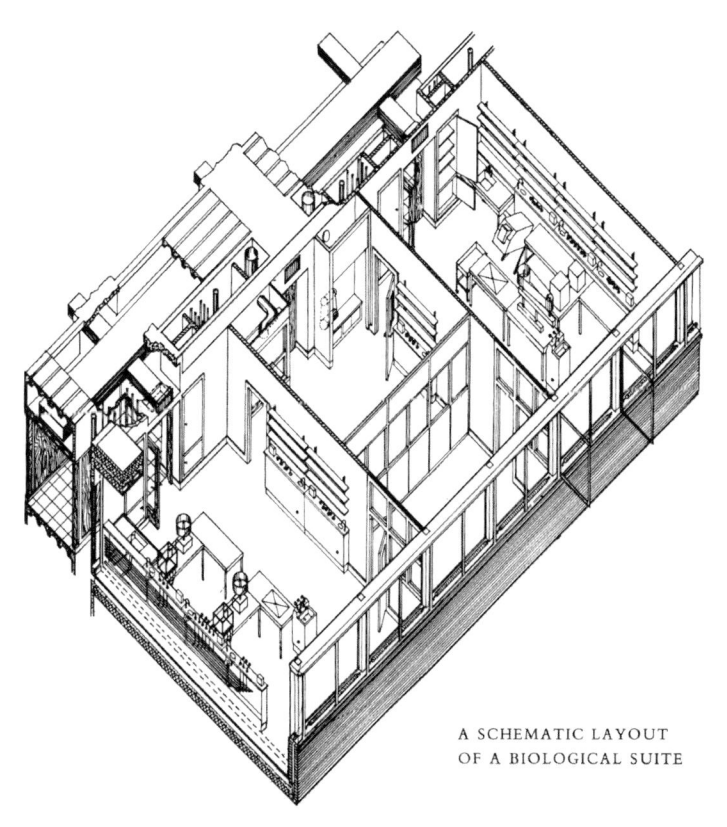

A SCHEMATIC LAYOUT OF A BIOLOGICAL SUITE

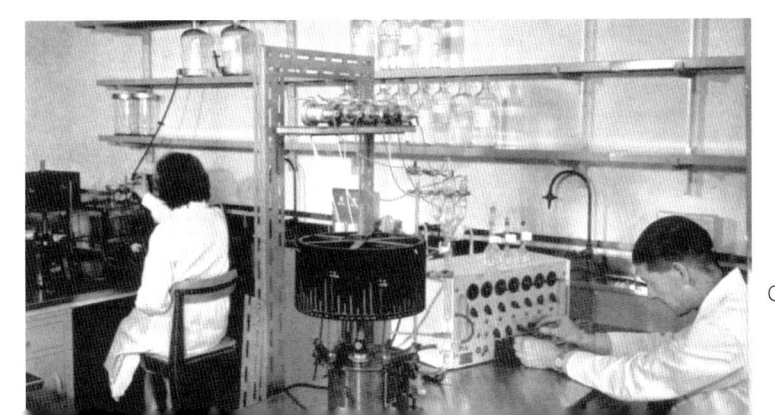

The biology labs required two distinct types of facility. The infectious diseases wing (dealing with viral, bacterial and fungal diseases) had special safety precautions for the *in vitro* work with live infections carried out there. The non-infectious wing dealt mainly with physiology, pharmacology and pathology, plus protozoa and fermentation processes.

The new labs held a staff of 450, 100 of them graduates of whom more than half were chemists. The latter synthesised some 2,500 new compounds a year. Many new staff began arriving, including later research director Brian Newbould. The labs were occupied some time before the formal opening, under research director Archie Sexton, with Bogue and Rose as biology and chemistry managers.

## Mereside opened

Mereside was officially opened by Lord Waverley, a director of ICI and former Chancellor of the Exchequer, on Tuesday 1 October 1957, an event of sufficient local importance as to fill four pages in the *Manchester Guardian* of that date. He was welcomed by the Division Chairman, Philip (P. A.) Smith, on his arrival with Chairman of ICI, Sir Alexander Fleck, FRS.

Some 250 distinguished guests, including representatives from foreign embassies as well as government and industry, watched Viscount Waverley unlocking the main doors. The visitors were deeply impressed by the quality of the facilities. Afterwards, the 'enormous gold' ceremonial key and banner were presented to Yule Bogue by Harry Fairhurst. (In 1987 Bogue offered to return the key to Mereside from his home in California, but its whereabouts is now unknown.) The staff settled happily into their new home. Pharmaceutical research at Blackley ended that year.

The reason for the choice of the name 'Mereside' is unknown. It was probably not suggested by the name of Charles Tunnicliffe's then recently published local book *Mereside Chronicle*.

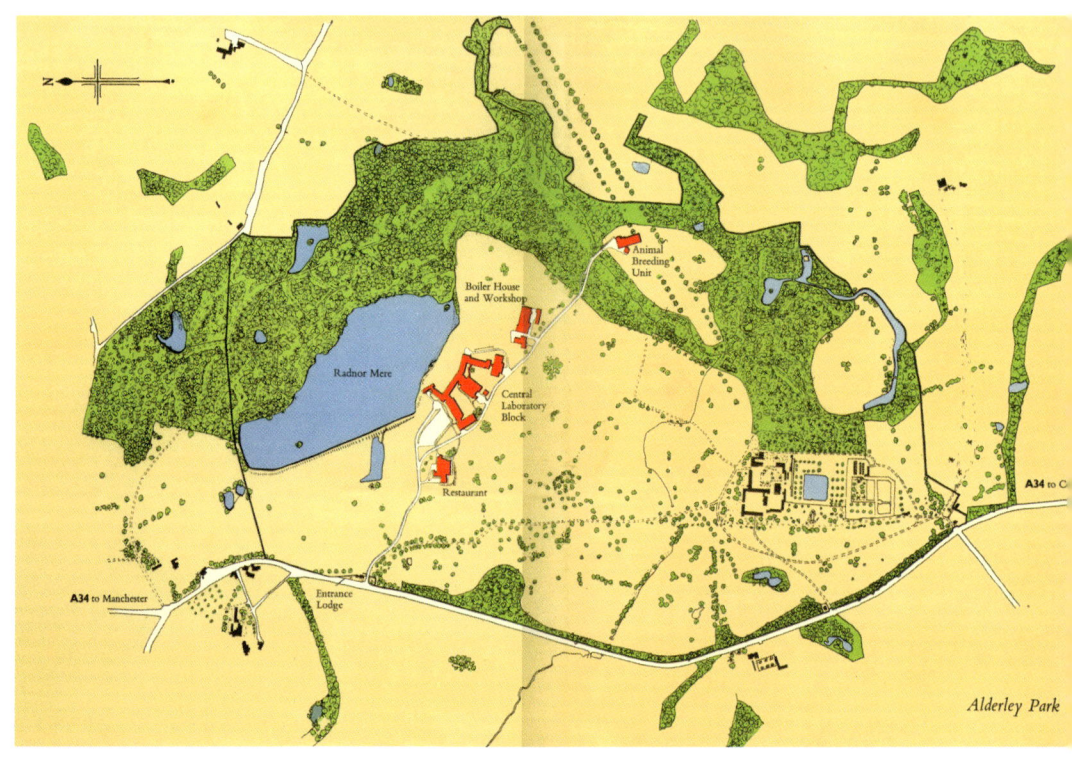

Alderley Park

Mereside original
car park
*PHARMACEUTICAL
RESEARCH IN ICI
1936–1957, 1957*

Mereside in 1961 –
one car park
ASTRAZENECA ARCHIVES

Brass ornament from
light standards

## The Park

The rest of the Park was at first little disturbed. Radnor Mere became a nature reserve, with no fishing, boating or bathing. In winter there were 'often 2 or 3000 duck of all sorts on the water.' The best chestnuts and oaks were repaired by tree surgery; and the parkland was restored to support large numbers of sheep and cattle.

A Site Road, lined with elegant lamp standards, later created to link Church and Bollington Lodges via the site buildings.

## Butlins!

Audrey Thompson, who co-ordinated international clinical trials of 'Casodex' in early prostate cancer before retiring in 2003, was the last working member of staff who remembered Mereside's opening. She recalled a very different company. One feature of those early days was the overcrowding of the bike sheds! A bus service entered the site at Church Lodge and left it at Bollington Lodge, causing confusion amongst the non-ICI passengers, especially when the bus conductor took to shouting 'Butlin's Holiday Camp'! Hardly anyone had a car at that time; when Audrey cycled to work, as ladies weren't allowed to wear trousers, she had to change for work behind a wall before cycling up the drive!

# LIFE IN THE PARK

Alderley House under construction
ASTRAZENECA ARCHIVES

Alderley House in recent times
PHOTO ASTRAZENECA

## Construction of Alderley House

Soon offices were needed. In December 1961, outline permission was granted for ones beside Radnor Mere, where Mereview Restaurant now stands. A new 'Mere Canteen' was to be included overlooking the Mere, at the south end of the current Radnor Car Park.

In 1962, the planning authorities changed their mind and suggested the site of the demolished former mansion at the south end of the Park. The company agreed, and in 1963 construction of the new Alderley House, to a little-changed design, began. It was completed in 1964 at a final cost of £2.5 million, and moved into by sales and publicity staff from Harefield Hall among others.

Alderley Houses in 1930 and 2013
COPYRIGHT ASTRAZENECA

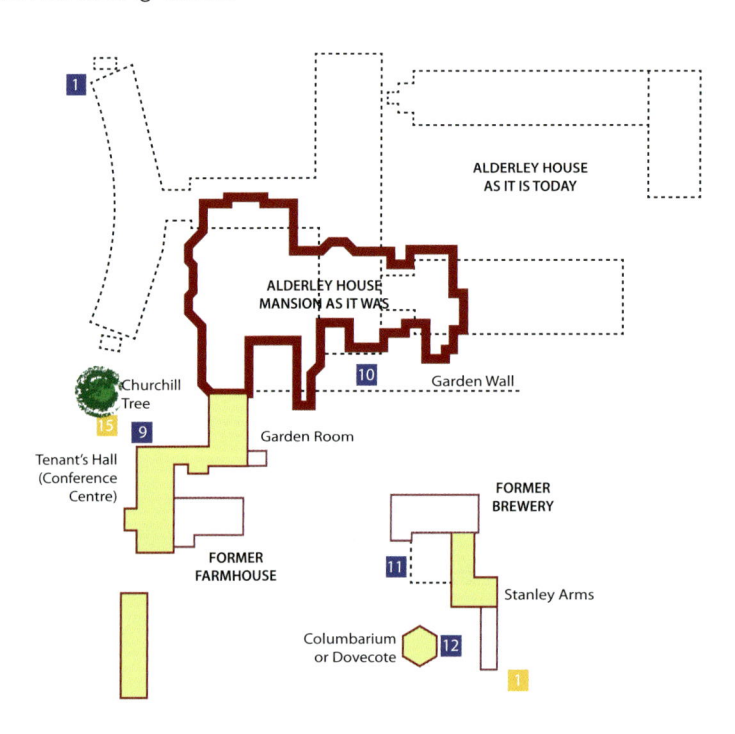

## Offensive?

After Alderley House was completed, estate manager Brian Owen chose one day to muck-spread the nearby farm fields. But the wind changed direction. Shortly, Brian had a phone call from the chairman himself. 'Have I done something to offend you?'

# LIFE IN THE PARK

Restoration of the historic old buildings (some into offices), and the building of the Water Garden Restaurant (opened in July 1964), also began. Fred Matthews in 1962 started the renewal of the Water Garden mentioned earlier; and the last of King James I's mulberry trees was felled!

# IHRL and CTL

In 1948, ICI founded an Industrial Hygiene Research Laboratory at The Frythe, Welwyn, in Hertfordshire. The 1930s and 1940s had seen a growing understanding of the potential dangers of chemicals, particularly to those manufacturing them. The IHRL was created to help protect the physical well-being of the company's workforce.

Water Garden Restaurant in winter

A move was mooted from the old wartime huts of The Frythe from 1958; and in 1961 the ICI board approved £436,000 for new buildings at Alderley Park and construction began of an IHR laboratory there, which was not a part of Pharmaceuticals Division and reported direct to ICI's Head Office. The publication of Rachel Carson's book *Silent Spring* had brought public concern about pesticides to a level where major investment in their safety was widely approved. In September 1963 thirty staff from Welwyn transferred to the new G block facility at the Park.

## Later history of the Central Toxicology Laboratory

The name was changed ten years later to the Central Toxicology Laboratory or CTL, a title that better reflected its main role of testing potential products for toxic side-effects.

The vision for CTL was largely inspired by its director Andrew Swan in the 1960s and 70s. It is difficult to overestimate the contribution CTL has made to the practice of toxicology throughout the world. CTL tested for safety immense numbers of chemicals, including ones that now meet us in every sphere of our lives. It assessed the effects of ICI pesticides and agrochemicals, and other products, both on humans and on the environment. It became recognised internationally as a major centre of excellence, as ICI's key provider of toxicology advice and testing to ICI businesses worldwide. Legislation in the US from 1972 to regulate pesticides increased its significance. By 1975 the overall cost of CTL had risen to £1 million.

## New smoking material

In the same year, CTL hit the headlines due to a project operated jointly with Imperial Tobacco, with the aim of creating a safer smoking material. The experiment was encouraged by two government departments (which, it should be noted, produced the list of tests that had to be done). But the work became highly controversial when an unauthorised photograph taken by an undercover reporter appeared in *The People* newspaper. ICI dropped the research soon afterward.

Protests in later years reached a height in April 1984 and as a result two people were imprisoned. The protests shocked staff (who merely stood by and watched; thankfully, no-one on either side was ever hurt). There was one moment of black farce. Reputedly, one ICI engineer, isolated deep in the parkland, was sitting peacefully in vegetation beside his equipment, measuring building sound levels, when he heard voices. Rising from his sunny post, he saw a group striding past him wearing balaclava helmets. This was before the days of mobile phones; after less than a millisecond's reflection, he dropped like a rock back into his hidden seat, and stayed there.

## Alternatives to animal testing

The CTL labs became a focus for animal rights demonstrations until CTL closed in 2007. Meanwhile, demonstrations against the medical research of Pharmaceuticals Division and its successors were rare. Everyone knew that, in the mind of the public, cigarettes and industrial chemicals (despite the safety requirements of governments around the world) were one thing; heart disease and cancer were very much another.

But CTL itself devoted considerable resources over many years to finding alternatives to animal testing. Along with Pharmaceuticals, they supported extensive work through the Humane Research Trust and other bodies. In 1992, CTL's work on alternative ways of assessing acute toxicity was recognized by the OECD, the international Paris-based economic and trade body. And it is striking to note that in 2000, Europe's premier laboratory animal welfare prize was awarded to a CTL scientist.

## Major achievements

By the mid-1970s, the CTL chemical compound toxicology reference index was considered one of the best in Europe, and resourced the entire ICI group. The computer system that later handled the growing database, ARTEMIS, was launched in 1979.

Specialist sections dealt with toxicity issues for chemical compounds and products, particularly acute toxicity, reproductive mutagenicity and inhalation toxicity. Analytical biochemistry was a major area including blood and tissue studies. A particular success of the 1970s and 1980s was in understanding the problem of **paraquat** poisoning.

Fermentation research at CTL led to the protein food product **'Pruteen'**, work which later helped in testing leading to the launch of the important meat substitute **'Quorn'** in 1984. 'Quorn' was produced by Marlow Foods, a joint venture between ICI and RHM, later sold by AstraZeneca for £72million in 2003 and now worth over £500million.

In the 1990s, a spin-off of work at CTL resulted in a treatment in children for the rare but fatal inherited condition of Tyrosinemia Type 1.

CTL's work was vital to enabling the launch of many Agrochemicals products. It was particularly commended by ICI chairman Sir Denis Henderson after its work leading to the herbicide **tralkoxydim**, launched in 1998.

## Staff and community

CTL staff played a keen part in the wider life of the Park, including social events such as the Annual Rounders. CTL itself held Open and Family Days. A strong community ethic also meant that external good causes benefited from donations of many tens of thousands of pounds.

## Changes

The £8.5 million biotechnology/research Block I opened in 1988. By 1993 staff numbers had risen to 375 and at the demerger in that year CTL was providing a toxicology service for the new Zeneca, although Zeneca Pharmaceuticals did its own toxicity testing and Zeneca Agrochemicals was its major customer for at least 50% of its work, along with some work from Zeneca Specialties and also some external customers. In the late 1990s new investment created high-tech bioanalytical labs. In 1999 control of CTL passed to AstraZeneca.

On 13 November 2000, AstraZeneca merged its agribusiness (still known as Zeneca Agrochemicals) with that of Novartis to spin off the new company of Syngenta, the first global group focusing exclusively on agribusiness. CTL became part of Syngenta's Health Assessment Group. The largest international research centre of Syngenta is the former ICI Agrochemicals Division research facility at Jealott's Hill, Berkshire, founded in 1927.

## Closure

In September 2006 CTL was earmarked for closure. Some 4.5 linear miles of records were archived in the Deepstore salt mine facility (itself once owned by ICI). In 2007 the last Syngenta staff left Alderley Park and the labs were mothballed. A highly illustrated history of the laboratory and its staff was published on its closure, under the title *60 Years of Scientific Excellence: Memories of the Industrial Hygiene Research Laboratory (1948–2008).*

In July 2015, work began to refurbish the old CTL buildings, as part of the 'science for life' park.

# Foundational products

The company discovered or was influential in a range of treatments for humans (and in some cases animals) in the era leading up to Mereside's creation, in a range of therapeutic areas. Some became famous, despite ICI's late start and small sales team compared with its competitors:

| Brand name | Generic name/identity | Number | Purpose | Discovered (launched) |
|---|---|---|---|---|
| 'Kemithal' | (thialbarbitone) | | intravenous anaesthetic | 1939 (1946) |
| 'Gynosone' | triphenylchloroethylene | | synthetic oestrogen | (c.1939)* |
| 'Sulfamezathine' | (sulfadimidine)† | ICI 1,988 | sulfonamide antibacterial | 1941 (1942) |
| 'Paludrine' | (proguanil) | ICI 4,888 | anti-malarial | 1944 (1946) |
| 'Lapudrine' | (chlorproguanil) | ICI 5,943 | anti-malarial | 1946 (1958) |
| 'Mysoline' | (primidone) | ICI 8,827 | anti-epileptic | 1949 (1952) |
| 'Hibitane' | (chorhexidine) | ICI 10,040 | general purpose antiseptic | 1950 (1954) |
| 'Fluothane' | (halothane) | | inhalation anaesthetic | 1952 (1957) |
| 'Fulcin' | (griseofulvin) | | oral anti-fungal (composition) | (1956) |
| 'Etisul' | (ditophal) | ICI 15,688 | leprosy treatment | 1957 (1958) |
| 'Tenormal' | (pempidine) | | anti-hypertensive | (1958)* |
| 'Atromid-S' | (clofibrate) | ICI 28,257 | lipid lowering | (1958)* |
| 'Epodyl' | (ethoglucid) | ICI 32,865 | cytotoxic agent in bladder cancer | (1962)* |
| 'Peptavlon' | (pentagastrin) | ICI 50,123 | gastric monitoring agent | 1964 |

† = now better known as **sulfamethazine**    * = first chemical sample not made by ICI

ICI also encouraged development of **dapsone** for leprosy, marketing it as **'Avlosulfon'** from 1947.

It was not uncommon for products to graduate through different discoveries and creatures; not listed above is **'Lorexane'** (**lindane**), which was first made (as a mixture) by Faraday in 1825. In 1942, ICI discovered its insecticidal properties against turnip beetles. It became used widely in agriculture and then against scabies in sheep by the 1950s. By 1965 it was a standard treatment for scabies and head lice in humans. It is also an example of a treatment now mostly superseded and disfavoured, due to side-effects once thought far less important than a troublesome medical problem but now all too well realised.

Of products not already described, a few stand out as world-class medicines used for many years.

## 'Hibitane'

After the discovery of 'Paludrine', related compounds were found to be powerful bactericides. The first sample of **'Hibitane'**, a topical antiseptic used in disinfectants, cosmetics and pharmaceuticals, was made in 1950 by Frank Rose, the full patent application was filed in 1952 and it was launched in 1954. It long remained one of ICI's top-selling products.

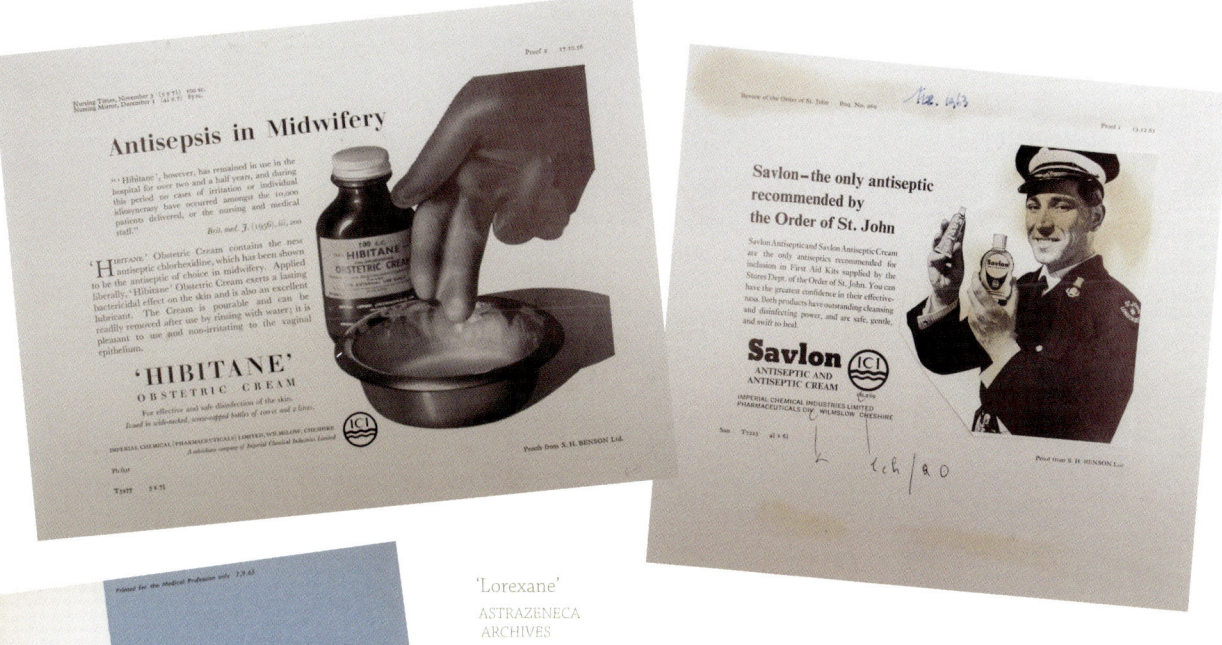

'Lorexane'
ASTRAZENECA ARCHIVES

'Hibitane' poster
ASTRAZENECA ARCHIVES

'Savlon' poster
ASTRAZENECA ARCHIVES

'Hibitane'
(chlorhexidine)

'Atromid-S'
(clofibrate)

'Hibitane' and
'Atromid-S'
structures
GEORGE B. HILL

In the 1960s and early 1970s formulation work was done leading to **'Corsodyl'** mouthwash from 1976.

## 'Cetavlon' (cetrimide) and 'Savlon'

Another ICI antiseptic discovery, **'Cetavlon'** (**cetrimide**), is not one compound but a mixture of organic ammonium salts effective against both bacteria and fungi, launched in 1942. In 1955, ICI launched a mixed antiseptic solution of hibitane and cetrimide as **'Savlon'**, which later as a well-known ointment became one of ICI's most familiar products.

## The griseofulvin ('Fulcin') controversy

One early product led to a sharp debate with another company, Glaxo. They had done most of the work to develop the natural antibiotic **griseofulvin**. But ICI held some patent rights through work by A. R. Martin; and when his work seemed overlooked he angrily insinuated misconduct by Glaxo. But an enquiry cleared them; and friendly joint marketing of the product was restored (by ICI in 1959), the sales of which were enormous, totalling $6 million by 1966 in the US alone.

Test tubes
COPYRIGHT
ASTRAZENECA

Flasks
COPYRIGHT
ASTRAZENECA

### 'Atromid-S' (clofibrate)

An insecticide developed by ICI in 1954, phenylethylacetic acid, caused illness in French farm workers. Their blood cholesterol levels were found to be extraordinarily low. Jeff Thorp at Mereside realised the potential of this sad event and screened some similar compounds that had been made as plant hormone mimics, then went on to synthesise an analogue, CPIB, a compound first made in Italy in 1947. It was given the name **clofibrate** and patented with Glynne (W. G. M.) Jones, Thorp and W. S. 'Sam' Waring as named inventors. After a paper was published on it by Thorp and Waring in 1962, a flurry of research by other companies eventually led to safer analogues. In 1963 ICI began selling it as **'Atromid-S'**.

It was not known at first how clofibrate worked, though it was known to reduce some lipoprotein levels in patients. Worryingly, it was toxic in mice. Only after 30 years of intensive research did Isseman and Green, two ICI researchers at CTL, identify a new sort of steroid hormone receptor, known as PPAR. This revealed the secret of how 'Atromid-S' worked – and as expected it acts differently in different species. Fibrates became the main sort of drugs used for lowering cholesterols until statins were developed. Clofibrate itself was discontinued in 2002.

## Animal medicine

Animal health research was going on at the Park as soon as it opened, into anthelmintics and treatments for coccidiosis and liver fluke.

Compounds first made elsewhere such as **'Promintic'** (**methyridine**), proven and developed at ICI by Arthur Broome and others, were launched. **'Zanil'** (**oxyclozanide)** for liver fluke was the only new animal medicine discovered in the Park's earliest years.

### With a bang

Jeff Thorp later caused a different stir, after his death in 1992, when he had his ashes blasted spaceward in 28 rockets from Kerridge Hill. Planes were diverted away from the colourful display.

## LIFE IN THE PARK

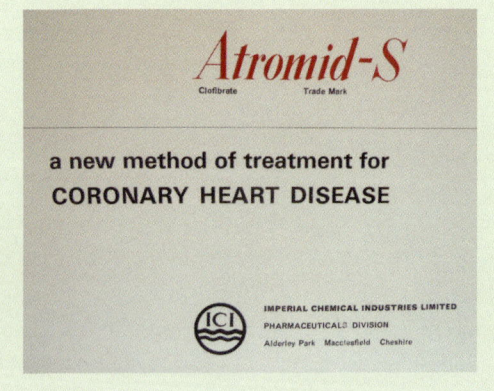

a new method of treatment for
**CORONARY HEART DISEASE**

Jeff Thorp
ASTRAZENECA
ARCHIVES

'Atromid-S' cover
ASTRAZENECA
ARCHIVES

Alderley Park's farm became part of 'Pharms' Division in 1959; and the open farmland on the Alderley Park estate was then used for nearly 25 years for ICI's thriving animal health business. Discoveries included:

| Brand name | Generic name/identity | Number | Purpose | Discovered (launched) |
|---|---|---|---|---|
| 'Antrycide' | (quinapyramine) | ICI 7,555 | veterinary trypanocide | 1947 (1949) |
| 'Helmox' | (dictycide) | | anti-lungworm | 1957 |
| 'Aimax' | (methallibure) | ICI 33,828 | gonadotrophin inhibitor | 1959 |
| 'Promintic' | (methyridine) | | anthelmintic | (1961) |
| 'Zanil' | (oxyclozanide) | | anthelmintic | 1963 |
| 'Estrumate' | (cloprostenol) | ICI 80,996 | luteolytic agent | 1970 (1975) |
| 'Equimate' | (fluprostenol) | ICI 81,008 | luteolytic agent | 1970 |

Arthur Broome
PHOTO ANDREW BROOME

## 'Antrycide'

A successful series of products prior to Alderley Park days had culminated in **'Antrycide'** (**quinapyramine**), a cattle treatment for types of trypanosomiasis (African 'sleeping sickness' carried by the tsetse-fly). In a joint project between ICI and the British Government, this had been the brainchild of ICI's Frank Curd and Garnet Davey.

An insectarium at Mereside, 1957
*PHARMACEUTICAL RESEARCH IN ICI 1936–1957 (1957)*

Trials of 'Antrycide' in Africa apparently mostly went well. But one was interrupted, when Garnet Davey had to climb a tree, to escape a 'high velocity rhinoceros'!

# LIFE IN (OR RATHER, FAR OUTSIDE) THE PARK

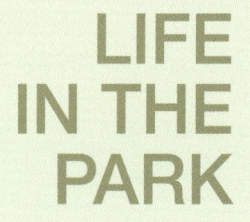

# LIFE IN THE PARK

## Dropping off

One early research field was anaesthetics for horses. These were tested on a placid old horse called Charlie. He did not seem to object to these repeated treatments. In fact, the suspicion arose that as soon as Charlie saw the biologist approaching with the usual equipment he began to prepare for his regular sleep – and would probably have 'dropped off', even without treatment!

## Frank Curd dies

Curd, who had first worked on possible trypanocides for ICI in 1934–6, and was later one of the co-discoverers of 'Paludrine', was a brilliant chemist who managed to identify 'Antrycide' as the active impurity in another compound. Walter Hepworth then worked out its structure.

Frank Curd would undoubtedly have been a key player in the development of Alderley Park's research had it not been for a tragic train crash on the Stockport railway viaduct in intense fog in November 1948. At the time, the discovery of 'Antrycide' was still on the secret list (it was announced by the Colonial Office the following year). Curd was carrying home papers from a meeting detailing recent promising results.

His injuries appeared slight but in fact proved fatal; before he lost consciousness, he asked for the papers to be located in the wreckage. The death of Curd at 39, a devoted family man and keen churchgoer, left his wife with three young children.

His colleague Frank Rose was due to catch the same train but was diverted on meeting his sister. Later, Rose 'was always keen to mention his especial debt to Curd, his close friend and scientific partner'. Alf Spinks could have offered the men a lift but to his great distress never thought to, probably due to the fog, said to be the worst in memory.

The Park's story might have been very different had events taken another course and another brilliant mind been available to direct its research.

Frank Curd
COURTESY HILARY FORD

## 'Aimax' (methallibure)

'Aimax', sold in the 1960s to pig farmers, was an influential discovery. J.K. Walley, a colleague of brilliant biologist Arthur Walpole's, was testing prospective treatments for a chicken disease (coccidiosis) when he found one compound caused their wattles and combs to shrink. These are organs with sexual significance: could this be a hormonal effect? Walpole's team quickly showed the drug was inhibiting the work of the pituitary gland in the brain (it was acting as a 'gonadotrophin inhibitor'). This affected the reproductive organs which needed gonadotrophin to produce sex hormones. Yet it was *not* acting as a hormone itself. This unprecedented find meant such a compound had potential uses in both sexes. Chemist Dora Richardson modified it to give the more potent **methallibure**, which was tested in human contraception and prostate cancer. Human use proved impossible and it was later marketed as 'Aimax'. But its striking properties made Walpole interested in gonadotrophin inhibitors … but that is a later story.

## Prostaglandins

Prostaglandins are naturally occurring compounds with a range of biological effects. In the late sixties, a speculative research programme was started in Alderley Park with the aim of discovering analogues of these compounds with potentially useful properties. Of particular interest were analogues of **prostaglandin F2α** (**dinoprost**) which is produced by the uterus when it is stimulated by the hormone oxytocin and can be used to induce labour. It is also a luteolytic agent involved in the regulation of the oestrus cycle and it was this effect which was one of the research targets.

Very potent analogues of prostaglandin F2a were quickly discovered using laboratory tests set up by biologist Mike Dukes. Testing in large animals was done by Mike Cooper of Animal Health and Husbandry Department who realised the commercial potential of synchronising oestrus in farm animals and was responsible for much of the successful development work. Two compounds reached the market place. ICI 81,008 became **'Equimate'** (**fluprostenol**), with Jonathon Hutton as named inventor; and ICI 80,996 was **'Estrumate'** (**cloprostenol**) with Jean Bowler, who made the first sample, and Neville Crossley, named as inventors. 'Equimate' was used in mares but 'Estrumate', launched in 1975 and now sold by Schering-Plough, is used more widely in horses, cows and (especially in Eastern Europe, where it became very popular) in pigs and sheep.

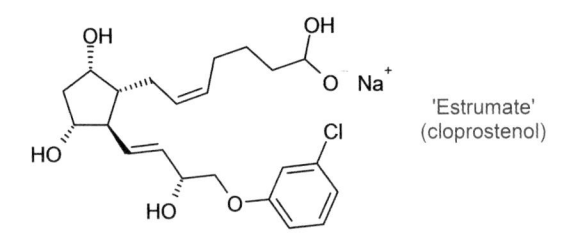

'Estrumate'
GEORGE B. HILL

'Estrumate'
(cloprostenol)

Even further east, one ICI biologist was sent to arrange an adventurous clinical trial – to test 'Estrumate' on giant pandas in China! It is also used on elephants.

Their discovery represented a significant scientific success for which ICI received in 1979 (for cloprostenol) the Queen's Award for Technological Achievement. They are remarkable compounds not least because of their potency. For 'Estrumate', a single dose of half a milligram is sufficient to bring about luteolysis in a cow – an amazingly small amount considering the size of the animal.

Mike Dukes, Jean Bowler and Fred Brown are congratulated by Brian Newbould on news of the Queen's Award for 'Estrumate'
SCAN

They were the first wholly synthetic prostaglandins to be marketed. Although they are expensive, because of their potency it has only ever been necessary to make kilogram quantities. Nevertheless, the manufacture of these complex molecules using a 27-stage process also represented a significant achievement by chemists in Process Development Department.

## Back to farming

In 1982, experimental research with farm animals ceased in the Park, and the Alderley Park farm reverted to its original purpose as a livestock farm. In 1984, ICI's animal health business was merged with that of Burroughs Wellcome to become Coopers Animal Health, a joint venture then sold to an American company in the late 1980s.

# Historic milestones

## Heart Drugs 1: 'Inderal' (propranolol)

In 2012, 31% of deaths worldwide resulted from heart disease. Even when not terminal, it can make life profoundly hard. Its obvious form, heart pain, was first described in a Sanskrit text in the sixth century BC. More specifically, angina was defined by William Heberden in 1768: 'They who are afflicted with it are seized while they are walking (more especially if it be uphill and soon after eating)...'

Similarly, hypertension (high blood pressure), which was known as 'hard pulse disease', was known from ancient times and was treated by bloodletting or the use of leeches.

### Drugs

The first proper medical treatment was the herbal drug **digitalis** (derived from foxgloves), discovered by William Withering in 1775, which can control heart rate.

Little more progress was made in the eighteenth and nineteenth centuries. In the mid-twentieth century, diseases of organic dysfunction began to be recognised as a serious and worthwhile target for research. These included hypertension; the latter became a specific ICI target in 1951, though a broad programme set up in 1954 had to compete with resources allotted to protozoal and bacterial diseases for several more years. But in 1958 it was realised that ICI's sales of anti-malarials, on which (particularly 'Paludrine') it had long relied, were falling; and other markets needed to be developed. And at Alderley Park, a new leader for the project was about to be recruited: a Scot, James (later Sir James) Black.

### Adrenaline

The discovery of the neurotransmitter **adrenaline** (epinephrine) was reported by William Bates in 1886. It was first isolated and purified by Napoleon Cybuski in 1895 and was patented as a mixture in 1901 by Jokichi Takamine, who made a fortune from it under the name '**Adrenalin**'. At times of stress, adrenaline surges into the bloodstream, causing the heart to beat more strongly and faster and increasing blood pressure. In a diseased heart, that can bring on angina or even a heart attack.

Its activity was investigated by two well-known scientists, H.H. Dale in the UK and Walter Cannon in the US. Dale speculated that something in the cells of certain organs had an affinity for amines like adrenaline. But he concluded that this theory of 'receptors' – areas on cells that specific chemical molecules attached to – struggled to fit with the wide range of chemical structures he had investigated. For example, early anti-adrenaline drugs like **phenoxybenzamine** were found to reverse rises in blood pressure without affecting heart rate. Dale's subsequent silence on the topic and his great prestige meant others also ignored the possibility.

In 1939 Cannon came up with a different and rather fanciful theory for the action of such compounds, the 'sympathin hypothesis', which did not involve receptors, despite the fact that the theory of receptors actually dated back to the work of Paul Ehrlich and of the British physiologist John Langley around 1900.

## Ahlquist's theory and James Black

In the 1940s, the little-known American pharmacist Raymond P. Ahlquist proposed two different receptors to explain the various effects of adrenaline. He is today hailed as a visionary scientist. He gave them the labels of alpha (α) and beta (β).

His theory received little attention, perhaps because he unfashionably used mathematical modelling to explain medicine. His first attempt to publish it in a scientific journal was dismissed because of what Sir James Black later called the 'long shadows' of the work of Dale and Cannon (although Black also acknowledged a crucial observation by Dale on reversing the pressure-inducing effects of adrenaline). But in 1948 Ahlquist's paper was accepted by a different journal, the *American Journal of Physiology*; and it revealed what is now seen as his famous distinction between alpha- and beta-adrenoceptors.

'Ahlquist's theory' explained Black later 'was that a cell responded to different substances because of the decoding machinery it had on its receptors'. It was a paper before its time. Remarkably, Black himself only came across Ahlquist's theory in a 1954 edition of a textbook, *Drill's Pharmacology*, which he was using to prepare lectures for students. But he later admitted his work would not have begun but for Ahlquist's idea.

## Dichloroisoprenaline, beta-blockers and angina

Ahlquist's theory received a boost with the discovery in 1958 by Powell and Slater, scientists at Lilly, of **dichloroisoprenaline (DCI)**, a chlorinated relation of adrenaline that blocked some of its effects and was the first clinically usable 'beta-blocker'. Moran and Perkins in Atlanta showed that it also blocked cardiac response and linked this with Ahlquist's theory. But neither they nor Lilly recognised the potential of the beta-blocking effect itself, even though Moran claimed that the compound belonged to Ahlquist's 'beta-adrenergic' type.

Black, researching in physiology at Glasgow University, thought they were wrong to dismiss it. He later suggested they were influenced by a theory of Cannon that the central nervous system was also involved; blocking this would be a bad idea.

Black had been studying angina. In 1928, Chester S. Keefer and William H. Resnik had shown that the cause of angina was oxygen starvation of the heart. It was commonly treated by increasing oxygen flow to the heart using nitroglycerin (a nitric oxide source) to dilate the blood vessels.

James Black doubted drugs could significantly dilate stiff, diseased coronary arteries. He raised a question: if a small amount of oxygen works, would a small decrease in the *need* of the heart for oxygen do the same thing? The death of James's father from a heart attack following a car crash made him focus on the role of stress in producing adrenaline and precipitating a heart attack. Black believed that oxygen consumption in the heart is determined both

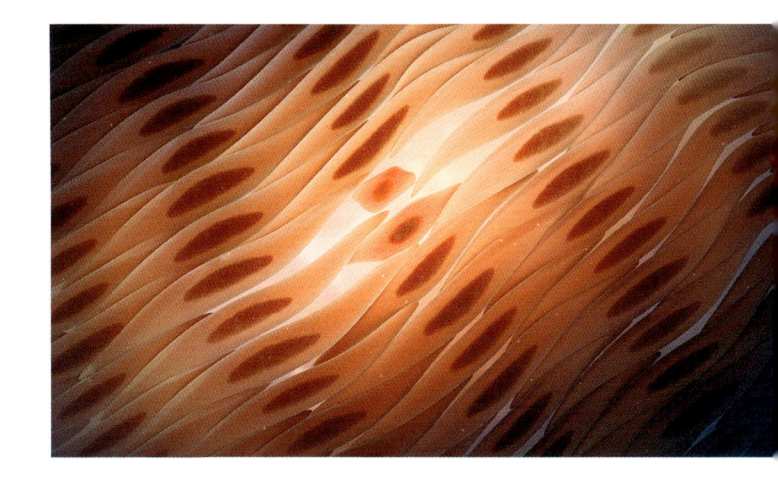

Cardiac regenerating muscle cells
COPYRIGHT ASTRAZENECA

J. Yule Bogue

*SCAN*, January 1966 & 25
January 1991

by arterial blood pressure and heart rate. It would be dangerous to disrupt the former; but reducing heart rate, which is controlled by the nervous system, should have the required effect.

If the action of adrenaline on the heart was thus blocked, might that prevent angina?

## Black comes to the Park

Yule Bogue had researched since the late 1920s into cardiac physiology and believed it a promising field. In 1958 he was appointed managing director and chose Garnet Davey as the new head of biology at Mereside on the strength both of his management skills and of his intimate understanding of the key relationship between chemistry and biology. Meanwhile, Black approached ICI with a request for funding. They were interested in his work in Glasgow and Davey went to meet him. They startled Black when Bogue (through Davey) offered him a job, promising that Black would not be a mere tester of speculativo chemicals.

James Whyte Black (despite his remarkable name!) joined ICI in July 1958. He already had a clear goal: to find a β-receptor blocker to treat angina. In early 1959, he came upon the discovery of DCI and then Moran's report on it. Black seized on the report, interested in the effect of catecholamines like adrenaline on the heart. Might blocking their action prevent potentially damaging increases in the rate and strength of its beat? In 1959 he proposed a project (though it is not clear whether work started just before or after Moran's report became known at ICI) and DCI was made and tested but was found to stimulate the heart, which was undesirable. The search started for a beta-blocker without this effect.

The initial chemistry research at Mereside was headed by Bert (A. F.) Crowther. A new animal test showed that DCI blocked the effect of adrenaline – but stimulated only *some* heart tissues. In his Nobel lecture, Black recorded: 'We were astonished … nothing [had] alerted us to expect that the agonist activity of these compounds could be so tissue dependent.'

## Chemistry

It was chemist John Stephenson (who Black considered brilliant) who in January 1960 had the idea of replacing the two chlorine atoms in DCI with another phenyl ring, to make a naphthalene – the naphthyl analogue of isoprenaline. In only 30 days, lab assistant Les (L. H.) Smith had made ICI 38,174. This new compound caused no stimulation in any heart tissue examined and Black knew instinctively that it was what they were looking for. By June he and his colleagues were sure. It was named **'Alderlin'** (**pronethalol** or **nethalide**) after Alderley Park and perhaps Nether Alderley.

## 'Inderal'

'Alderlin' was effective against angina and was later given a restricted licence. But things had already moved on. Testing had shown it produced tumours of the thymus gland in mice, probably because of conversion in the body to a carcinogenic aziridine. After the thalidomide disaster in 1962, the general public had become highly concerned by drug side-effects; 'Alderlin' was never used clinically.

Garnet Davey never put pressure on Black to market 'Alderlin', as they realised it was merely a stepping-stone. To everyone's surprise, Stephenson left; but in 1962, after two years making and testing hundreds of compounds, ICI 45,520, soon to be named **propranolol** or **'Inderal'** (not quite an anagram of 'Alderlin') was synthesised in Lab 8F3 by Smith working for Crowther. The latter recorded that David Dunlop, who had just tested the compound, bounded upstairs to the chemistry lab and crashed into him: 'We've done it – it is twenty times as active as pronethanol!'

The key structural modification, which was carried through to essentially all subsequent beta-blockers, was the insertion of an oxymethylene group into the structure of 'Alderlin', which gave the potency boost. This also eliminated the carcinogenicity of 'Alderlin'.

Like 'Alderlin', 'Inderal' is chemically a mixture of two enantiomers (mirror image structures). Other workers led by chemists Ralph Howe and pharmacologist Robin Shanks, formerly Ahlquist's research assistant, later discovered that only one enantiomer of the drug was a beta-blocker (the one related to natural catecholamines). Bill (W. A. M.) Duncan developed a novel method for analyzing blood levels. An army of others then set famously to work to make the new drug a world success.

'Inderal' caused a reduction in angina symptoms corresponding to the dose, and after 3 years patients treated with beta-blockers had four times fewer heart attacks than others.

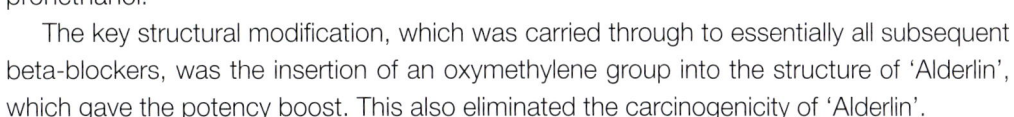

'Alderlin'
(pronethalol)

'Inderal'
(propranolol)

Structures of 'Alderlin' and 'Inderal'
GEORGE B. HILL

Sir James Black
KENNEDY, ICI, *THE COMPANY THAT CHANGED OUR LIVES*

'Inderal' team members (in 1989): Bernard Langley; Bert Crowther; biochemist Peter Bond; chemists Bernard McLoughlin, John Stephenson and Ralph Howe
PHOTO BILL ROUSE

Les Smith
*SCAN*, 15 January 1982

## Out of stock

The key 1-naphthyl group in the structure of 'Inderal' was a serendipitous find. Les Smith actually went to the Mereside stores to obtain a bottle of the 2-naphthol; but this was out of stock, so he used 1-naphthol instead – and made history. The 2-naphthol compound would have been no more active than 'Alderlin'...

LIFE IN THE PARK

Many ICI drugs were injected into Alderley Park staff, as some of the first volunteer humans treated. A rather gory poem from the 'Inderal' days was composed to recall the suffering involved! Provided by Bernard Lancaster, the last verse reads:

'Just one pious hope, and we hope it like hell,
('Inderal', 'Alderlin', Alderley Park)
That after this trouble, the damn stuff will sell,
(With [various] and old Wam Duncan and all …)'

## Novelty?

Later it was realised that 'Inderal' had already been made by another company, Boehringer, shortly before ICI made 'Alderlin' – but they had not recognised its potential and had not patented it!

ICI's first sample of 'Inderal' was registered on 12 October 1962 and the patent application was filed in November 1963 with Crowther and Smith as inventors. Broad patent filing consolidated ICI's place in beta-blockers and restricted competitors.

## Changing the face of medicine

Clinical trials in humans began in summer 1964, only 32 months after the drug was first given to an animal; 'Inderal' was launched in July 1965 as the first beta-blocker to reach the market. It became a hugely successful product for ICI, improving the quality of life for countless thousands of patients suffering from angina and becoming the world's bestselling prescription drug of the time. As a recent writer put it: 'Since then, propranolol has changed the face of cardiac medicine.' (Modern surgical techniques are now able to produce similar life-prolonging effects, but at a different sort of cost.)

It was not approved in the USA until 1973, much later than other countries. In the 1970s, the conservatism of the US Food and Drug Administration delayed approval of many drugs developed outside the US, particularly cardiovascular ones – an indication of problems to come for the industry.

'Inderal' has one odd side-effect. Because it suppresses the effect of adrenaline in times of trauma, it dilutes memories, removing the strongest emotions. So it also lowers anxiety levels. Snooker players, violinists – and even research scientists nervous about giving talks on it! – have found such treatment useful.

In 1991, a novel use for 'Inderal' was reported in *National Geographic* magazine: the Saudi Arabian wildlife commission had used it to sedate seabirds requiring clean-up after the Gulf oil disaster.

## Hypertension

Remarkably, the bulk of its sales proved to be for a different medical problem, hypertension (high blood pressure), which ICI had not expected; and in which it was shown to be effective by Professor Brian Prichard in a London hospital in 1964. By 1970 this use was well established; Garnet Davey, who became research director at the Park in 1969, called the anti-hypertensive activity of 'Inderal' 'absolute luck'.

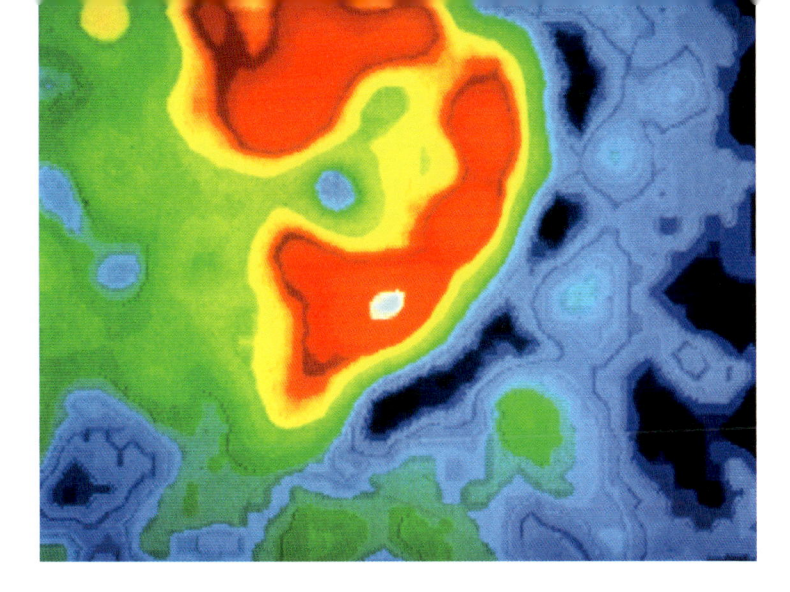

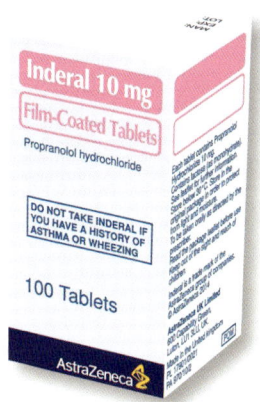

Heart cells
COPYRIGHT
ASTRAZENECA

'Inderal'
COPYRIGHT
ASTRAZENECA

It remains an outstanding example of a brilliant theory carried through to brilliant effect, as well as an example of Archie Sexton's determination as Research Director to exploit serendipity. In 1969, the Division was presented with the Queen's Award for Technological Achievement for its development of drugs for ischaemic heart disease by Lord Lieutenant Viscount Leverhulme at the Water Garden restaurant.

# Heart Drugs 2: 'Tenormin' (atenolol)

### Jimmy Black leaves

Black had already left ICI, in mid-1964:

> When a company has made the discovery of a new compound, what they want the scientists to do is to help them promote the drug. So [ICI] wanted me to go to meetings and give papers … and they didn't want me to stop working on beta-blockers. I wasn't interested in marketing. And I got interested in [histamine receptor antagonist], so I said I wanted to do this and they didn't want me to do. So I said, 'Alright, I'll go.'

According to Yule Bogue:

> After Crowther's synthesis of propranolol, [Black] became unhappy with later developments. He came into my office in Fulshaw Hall in 1964 to tell me he was resigning. This was a great disappointment to me as his approach to solving scientific problems was both original and stimulating. I had visions of him as our future research director.

Black was now fascinated by the prospect of blocking the histamine receptors in the stomach which produce gastric juices. Famously, (or infamously, for those who believe ICI should have kept him) Jimmy Black joined Smith, Kline & French where he developed his second major drug, **cimetidine**, an H2-receptor antagonist: a drug acting in a related manner to propranolol but to treat stomach ulcers. It was launched as **'Tagamet'** in 1975 and soon outsold propranolol to become the world's largest-selling prescription drug. Later it was followed by Glaxo's even bigger **'Zantac' (ranitidine)**, which had the same mode of action. If only …

Sir James Black – Nobel Prize
*SCAN*, 21 October 1988

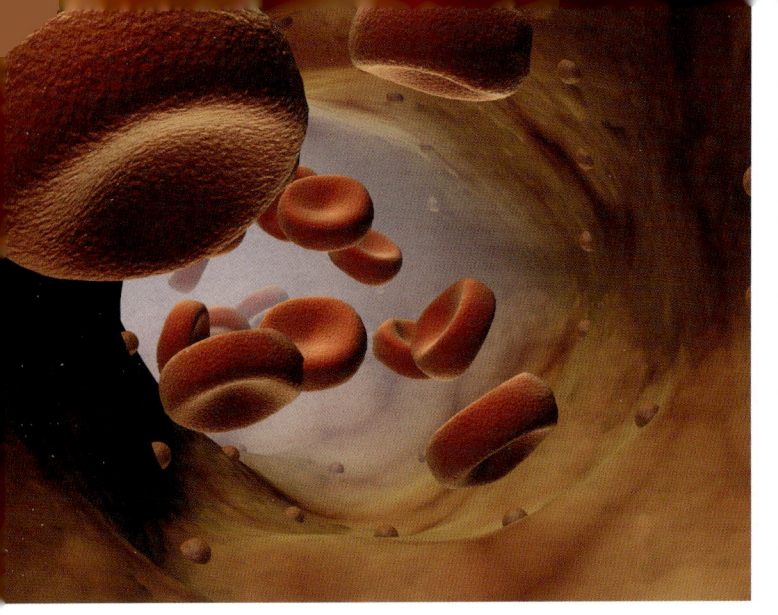

Blood vessel
COPYRIGHT
ASTRAZENECA

Replica of Sir Jim's
Nobel medal at
Mereside
PHOTO GEORGE B. HILL

Black was made FRS in 1976, was knighted in 1981 and was awarded the Nobel Prize in 1988 for work leading to the development of both propranolol (commemorating the team effort during 1958–1964 at Mereside) and cimetidine.

## A bad setback

'Inderal' was still being marketed successfully but had one problem: it affected not only receptors in the heart but also bronchial ones, and so caused problems for asthma sufferers. A better drug would avoid this.

Two other effects needed avoiding. A new drug should not be a 'partial agonist' (i.e. a drug that worked but also slightly blocked its own effect). Also undesirable was 'membrane stabilizing activity' or MSA, which makes a drug act as a mild local anaesthetic. Inderal was not a partial agonist but did have MSA.

A different compound from the research programme was found, unexpectedly, to be selective for receptors in the heart alone, unlike 'Inderal', and had no MSA, although it was still a partial agonist. The new compound, inspired by the structure of **sotalol**, a Mead Johnson compound, was given the name **'Eraldin' (practolol)**. It was launched in 1970, but later was immediately restricted to hospital-only use and then withdrawn in 1975, when serious and wholly unexpected side-effects showed up in patients after long-term oral use, with eye damage that was sometimes irreversible and tragically even led to blindness in some cases.

There were also potentially lethal abdominal problems. These were discovered only after many thousands had used the drug; and they could not be reproduced in the laboratory. This caused deep distress to the patients affected and was a bitter and disastrous lesson for ICI (who gained credit for setting up their own compensation scheme) but was a milepost in understanding the need to monitor all 'Adverse Drug Reactions', which undoubtedly helped to avoid other disasters across the industry in later years. John Patterson considers that without the wisdom of medical director Colin Downie, and of senior medical adviser John Waycott who handled the whole issue, it might have destroyed the company. It was a moment when ICI needed to (and did) react rapidly and so show its highest quality.

Structures from
'Inderal' to
'Tenormin'
GEORGE B. HILL

'Inderal'
(propranolol)

'Eraldin'
(practolol)

*(cardioselective, toxic)*

'Tenormin'
(atenolol)

*(cardioselective)*

## Research to 'Tenormin'

Most of the research team was still working on the 'Eraldin' project (before its problems were known) when chemists Roy Hull and David LeCount started to discuss the nitrogen atom attached to the 'Eraldin' ring. An improvement on both 'Inderal' and 'Eraldin' was desirable. The new drug should not be a partial agonist like 'Eraldin', and must be selective with no MSA, unlike 'Inderal'. Another big improvement would be to make the new drug much more water-soluble than 'Inderal'.

Hull suggested an alternative structure to LeCount, but that proved hard to make. LeCount decided to try a phenylmalonate derivative, a structure with a carbon in place of the nitrogen and with a second acid group. The easiest way to make this would be to have the carbon there first. The second acid group was to be added later – although this turned out to be unnecessary.

The obvious starting material was 4-hydroxyphenylacetic acid. (This later became a serious issue, as no manufacturer could be found to supply it when the new drug became successful, so ICI had to make it themselves.) This was converted to its ethyl ester. Synthetic chemist Chris Squire was asked to attach to the molecule the chemical chain that should make it a beta-blocker. But he obtained a bis-amide, as the ethyl ester had also reacted with the amine. But this compound, when isolated, proved to be active! Nonetheless, it was not likely to be water-soluble, so LeCount asked him to make the corresponding compound with only a primary amide group; and ICI 66,082, **'Tenormin'** (**atenolol**), was born, with its first registration on 18 December 1968. Remarkably, it was the only compound in its chemical series that was active as a drug. Pharmacologists A. Mike Barrett and J. 'Des' Fitzgerald (later research director) led the testing; and the patent application was filed in January 1970 with Barrett, John Carter, Hull and LeCount as named inventors.

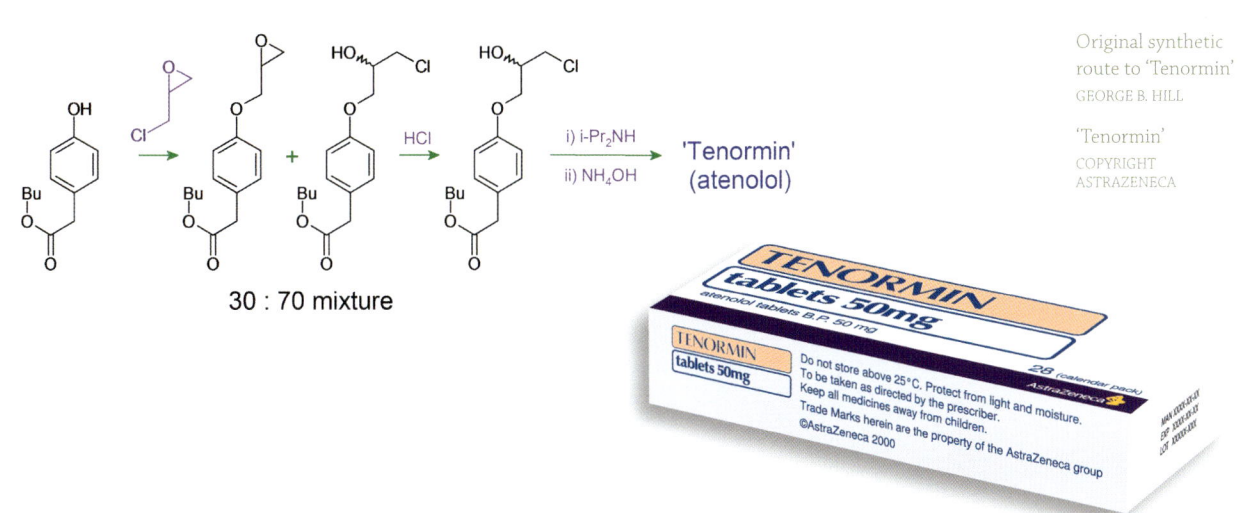

30 : 70 mixture

## Just what the doctor ordered

Within two months of its synthesis, 'Tenormin' was found to be a cardioselective beta-blocker, as hoped. Surprisingly, it was also found to have neither partial agonist nor MSA activity. So the compound unexpectedly achieved all the required targets at once.

Bert Crowther had in 1966 become the first manager of the newly-formed Process Development Department, based in laboratories at the new £7 million production site in

Macclesfield (opened in that April by Lord Florey). Crowther led the very difficult challenge there of scaling up the laboratory synthesis of 'Tenormin' to manufacturing scale.

Although no more potent than 'Inderal', 'Tenormin' was longer acting, required dosing only once a day, was not metabolised by the liver, and (similarly) had no long-term side-effects. Since it did not bind strongly to proteins in blood like 'Inderal', it gave much more stable blood levels. The market for beta-blockers was considered to be saturated, so 'Tenormin' was marketed primarily for hypertension.

It went into development in 1972, was launched in 1976 and became a huge seller, at its peak becoming the best-selling beta-blocker in almost every market and the third bestselling drug in the world. In 1987, 'Tenormin' and its related products had sales of £500 million, half the world market. In 1991, three years after the patent expired, it was still in the top five with sales of more than a billion dollars annually. At the demerger of Zeneca in 1993 it still accounted for 25% of the new company's sales. The company's position has remained secure ever since (with its more recent heart drugs having been invented outside Alderley Park); in 2013, AstraZeneca was number one in the world for cardiovascular drug sales.

By 1989, it was estimated that 100,000 potential beta-blockers had been synthesised by chemists around the world, and 5,000 papers and 2,000 patents had been produced on them. But the potential impact of such a wealth of research on patients around the world is beyond estimation.

## 'Nolvadex' (tamoxifen) – discovery

**'Nolvadex'** (**tamoxifen**) is one of the most famous modern medicines.

Its generic name is so well-known that, as with the beta-blockers, I have included 'tamoxifen' in the title of this chapter. Both names refer to one of two 'isomers' of the compound. ('Nolvadex' is the citrate salt of that isomer).

It was estimated in 2003 that more than 400,000 women were alive and millions had lived longer and better because of it. In 2004, tamoxifen was still the world's largest selling hormonal drug for the treatment of breast cancer, a disease first recorded in Egypt in 3000 BC. In 2008, it was described by Harvard cancer expert Harold Burstein as 'the most important drug in the history of medical oncology'. Treatments using or inspired by it are partly why almost two thirds of women diagnosed with breast cancer in this decade are predicted to survive their disease for 20 years or more and 80% for ten years, compared with only half that when 'Nolvadex' was first launched in 1973.

This is the early inside story of **'Nolvadex'** (**tamoxifen**). It originates from two different areas of research.

## Hormones and oestradiol

The first area is the question of why breast cancers differ.

Pioneering cancer surgeon Sir George Beatson noted that Scots farmers could extend lactation in their cows by castration (i.e. ovary removal). After animal experiments of his own, in 1896 he surgically removed the ovaries of three women with breast cancer, one of whom went into complete remission (he also applied thyroid therapy). This suggested that the ovary was a source of some sort of 'chemical messenger', a subject already of growing interest to scientists. In 1905, physiology professor Ernest Starling coined the word 'hormone' (from the Greek 'to arouse or excite') for such messengers which pass between organs of the body.

Beatson's operation became common for breast cancer patients until the 1960s; and it gave researchers the first clue that female hormones, known as 'oestrogens', of which the ovary is a source, were involved in the growth and development of breast cancer. But the drastic treatment only worked in 37% of patients – and there was no way to tell which they would be.

## Hormone therapy

Interest in hormone-related medicines rapidly developed. By 1935, scientists in Germany and America ensured that **oestradiol**, the main female hormone, had been isolated and identified. The 'first orally effective oestrogen', a soluble ester of oestradiol called **'Emmenin'**, had already been introduced in 1930 by Collip and Ayerst Laboratories. But it was obtained from the late-pregnancy urine of Canadian women – not a useful industrial source! In 1936 Albert Doisy isolated 25 milligrams of pure oestradiol – but he did it from four *tons* of pig ovaries. And making a complex steroid-like oestradiol from simple chemicals would not be scientifically possible for another forty years.

So scientists searched for new sources of oestrogen to experiment with. In the 1930s, scientists identified and doctors tested a variety of simple synthetic chemicals that might mimic the effects of oestrogen while being easier than steroids to manufacture, i.e. they were 'non-steroidal' oestrogens. A key one was **'Stilboestrol'** (**diethylstilboestrol**), developed by Cancer Research UK scientist Professor Sir Charles Dodds in 1937 – after being discovered by pure chance. Synthetic as well as natural oestrogens were shown by radio-labelling to influence the reproductive organs. But most such compounds had side-effects too severe for widespread use in medicine.

## Opposite effects

In any case, treating a breast cancer patient with oestradiol appeared (in principle) to be the wrong thing to do. If the cancer was one that responded to oestrogen, surely it would flare up? In women a related drug, but one which had the *opposite* effect, was what was needed.

This was quite a different area of research. And although scientists were working on it, it was not with cancer in mind. This remained the case for a remarkably long time. In 2000, Professor V. Craig Jordan, who later helped to make tamoxifen world-famous, said:

> Tamoxifen was born into a world of indifference in the 60s, when the focus of the research was on contraception. It grew up in the 70s, in a world where chemotherapy was king and hormonal therapies were perceived as non-starters in the quest to cure cancer.

So how was it discovered at all?

## 'GYNOSONE' A NEW SYNTHETIC OESTROGEN

'Gynosone' brand triphenylchloroethylene may be employed in all the conditions which normally respond to the administration of natural or synthetic oestrogenic hormones. It is particularly useful in the symptomatic treatment of the menopausal syndrome, in atrophic conditions of the vagina and for inhibiting lactation. Given by the mouth, the duration of action of 'Gynosone' is similar to that of stilboestrol, but its administration is characterised by the almost complete absence of unpleasant side-effects such as nausea and vomiting. 'Gynosone' for oral administration is issued in tablets of 0.5 gramme in packings of 10, 100 and 500.

*Further information will be supplied on request.*

### IMPERIAL CHEMICAL [PHARMACEUTICALS] LIMITED
THE RIDGE, BEECHFIELD ROAD, ALDERLEY EDGE, MANCHESTER

Ph.151*l*

## ICI's cancer research

ICI's Arthur Walpole, based in the company's Biological Laboratory at Wilmslow, had collaborated around 1940 (with work in ICI's own labs from 1943) with cancer research pioneers Edith Paterson and (the later Sir) Alexander Haddow. They used two ICI synthetic (nonsteroidal) oestrogens supplied by Walpole.

In the absence of any better agent, they treated older patients with high doses of oestrogen. Paradoxically, although this appeared to be the wrong thing to do as a hormonal therapy, it worked! In 30% of breast cancer patients there was a response. It was correctly concluded that this was due merely to toxicity. ('Haddow's paradox' is now understood as the toxicity of excessive amounts of oestrogen particularly towards oestrogen-sensitive tumour cells). After the news was published by Haddow and his team in 1944 the technique became a standard therapy for thirty years until 'Nolvadex' became available, despite its significant side-effects.

Later, in 1949 Walpole tried unsuccessfully to discover why only *some* tumours responded to oestrogen therapy. In the process of his work he became an expert on carcinogens (which were of concern in chemical manufacturing).

He now returned as a reproductive endocrinologist, leading his team under biology head Garnet Davey at Mereside in searching for new contraceptives – but with hormonal cancer treatment at the back of his mind. This was a huge moment, which ensured that Walpole's research throughout his 37-year career would become historic in cancer therapy.

## Alkenes, mirrors and keys in locks

Stilboestrol and the other compounds being investigated are 'alkenes', compounds that contain a double bond between two carbon atoms. More particularly, the double bond they contain is tetra-substituted, with four chemical groups around it, making up a rather flat, rectangular molecule. Some compounds of this sort can have two isomers (mirror images), called the '*trans*' and '*cis*' form (see illustration later)

The molecules of compounds of this sort of alkene can *mimic* natural oestrogens. This means that they may stimulate the body's oestrogen 'receptor' as do normal hormones. A receptor is in one sense like a natural lock, which must be operated to make important things happen. To use a non-scientific analogy, synthetic oestrogens made by the chemist are the equivalent of substitute keys, ones that will enter and turn the same 'lock' as the natural hormones.

But there was a way in which such a molecule might also do the opposite. Perhaps it could act as the 'wrong key' and *jam* the lock! In that case, it might act (as an anti-oestrogen) to stop normal female biology – by blocking all the oestrogen receptors in a woman's cells. It would do this because although it could 'fit' the lock, it could not *turn* it: it could not reshape the protein into the 'active' form of it which the body would use.

## Jensen's evidence

It could, of course, only block the receptors *if they existed*. In 1962, Professor Elwood Jensen solved Beatson's puzzle. With Herb Jacobsen, he synthesised radiolabelled oestradiol and showed that the reproductive organs accumulated it. Jensen proposed this was due to the existence of the oestrogen receptor (or estrogen receptor, ERα) in these. Jensen and Jack Gorski in Illinois in the 1960s obtained samples of the receptor (although its structure was not proven until 1997).

## Anti-oestrogen uses

Anti-oestrogens might do several profound things. Work done in rats in Israel had shown that a high oestrogen level was necessary for foetuses to become implanted and grow – so the opposite effect might give a (later controversial) 'morning-after Pill'. On the other hand, an anti-oestrogen might persuade the body that it is *short* of oestrogen. The body perceives an unusually low level of oestrogen, and makes more. And this can help women who struggle with ovulation (egg production). So, widely different markets could open up. By the late 1950s, pharmaceutical companies were actively hunting for anti-oestrogens.

## Opposite effects

In 1958 Leonard Lerner, a scientist at Richardson-Merrell, described **MER-25**, a compound which had surprised him. Merrell had been making potential heart drugs with alkene type structures when Lerner, realising they resembled synthetic oestrogens, tested one as such. (Drug companies always put any chemicals they make into many different tests.)

But MER-25 did nearly the opposite of what he expected. It was much more an 'antagonist' than an 'agonist' in rats: it was mostly an *anti*-oestrogen (and an inhibitor of implantation).

In humans also, it appeared to have an anti-fertility effect. It also seemed to inhibit breast cancer in one of three patients treated, though it was too toxic to use. (Drug toxicity was at that time assumed to disfavour use of such compounds in most cancer). Lerner's paper aroused great interest.

Further research led to the introduction by Merrell in 1961 of **'Clomid' (clomiphene)**. Contraception was a major business target and the new drug was aimed at that. But clomiphene failed in its planned use as a contraceptive in humans, because anti-oestrogens of its type turned out to stimulate ovulation in women, and are still used for that purpose. It assisted what it was supposed to prevent!

## Clomiphene to 'Nolvadex'

A publication on clomiphene interested Arthur Walpole. Remarkably, this was not for the drug's anti-oestrogen effect but because it was described as a gonadotrophin inhibitor, acting on the brain's pituitary like ICI's **methallibure**, as described earlier. The sex hormones oestradiol, progesterone and testosterone can also work like this; and research elsewhere by Pincus and Djerassi was just then turning derivatives of these steroids into 'The Pill' for women.

Walpole noticed that clomiphene's molecule had a basic ether chain (an oxygen-linked chain containing a nitrogen atom) attached to it. Then he realised the corresponding compound without the chain was one of those he had worked on some twenty years previously, known as **'Gynosone'** and briefly marketed by ICI.

It struck Walpole that clomiphene appeared to have a similar effect to methallibure, less oestrogen effect than the 'Gynosone' series, *and* some anti-oestrogen effect (in fact the first effect turned out to be minor). 'He quickly asked [chemist] Dora Richardson to put the basic side-chain onto the other ICI oestrogens he had worked with earlier.'

Clomiphene contained a chlorine atom. It was already known within ICI from the previous chemistry that if this were replaced with a methyl group it gave a more potent drug. Dora made such a compound, ICI 42,043. It too was an anti-oestrogen in rats; so, aiming at the same contraceptive target as Merrell, a project was started. Walpole's colleague Mike (M. J. K.) Harper successfully devised a test to find compounds that hindered implantation.

But in its turn ICI would learn, too, that its plan for a money-making contraceptive was doomed!

In August 1963, ICI took the precaution of filing a broad patent application, with inventors Walpole, Richardson and Harper, claiming a range of such compounds as potentially useful for control of 'endocrine status' in patients. But the application also claimed – in what now looks like the understatement of the century – that they 'may' be useful against hormone-dependent tumours. For it was in that year that Dora Richardson, in Lab 8S14 at the Mereside chemistry labs in Alderley Park, had herself

Mike Harper in the 1960s
V. CRAIG JORDAN

Structure of clomiphene; 'Nolvadex' (trans-tamoxifen); cis-tamoxifen (ICI 47,699)
GEORGE B. HILL

clomiphene    'Nolvadex'* (tamoxifen)    *cis*-tamoxifen, ICI 47,699

\* 'Nolvadex' is the citrate salt

## LIFE IN THE PARK

### Conviction

Dora Richardson had decided on her life's work when seeing people working in hospital laboratories, while visiting her dying grandmother in London's Cancer Hospital.

Dora Richardson in 1979
V. CRAIG JORDAN

made the first, historic sample of **'Nolvadex'** (**tamoxifen**). It was given the catchy label of M46,474 (the M was a hangover label from Medicinals Group at Blackley) or, for those outside the company, ICI 46,474.

## Isomers!

The new drug was, like clomiphene, a mixture containing some *cis* along with its *trans* isomer: the active medicine was contaminated with a little of its 'minor' isomer. With difficulty, ICI chemists managed to separate the minor isomer from the main one by fractional crystallization. The minor one was given the name ICI 47,699, while the major isomer, ICI's 46,474, was now applied to pure **'Nolvadex'**. This led to a surprise: when tested in rats, the minor isomer on its own proved to have an *opposite* (though weaker) effect – it was an oestrogen, not mainly an anti-oestrogen.

It was clear that the major isomer was the one ICI wanted; the other was a nuisance! Actually, the major isomer was not a *pure* anti-oestrogen but a 'partial agonist' – a drug that acts strongly in one direction and weakly in the other at the same time. But its main effect was powerful.

Yet despite now having the separate isomers, ICI still could not identify the *cis* and *trans* – they did not know which was which! It was only in 1966 that a paper was published by Geoff (G. R.) Bedford and Richardson (after a high-tech study of tamoxifen's dipole moment and NMR spectrum) proving that 'Nolvadex' was the *trans* isomer. Thus the unwanted ICI 47,699 is often called *cis*-tamoxifen, although 'Nolvadex' itself is generally called tamoxifen, not *trans*-tamoxifen. Incidentally, this is presumably why some sources state, misleadingly, that tamoxifen was first synthesised in 1966.

Geoff Bedford
PHOTO ASTRAZENECA

## Patent differences

But ICI's original patent application had a potential 'patentability issue' in that its coverage overlapped with earlier Merrell ones. Nevertheless, when published in 1965 as GB Patent Application 1013907, the separation of tamoxifen's isomers that had been achieved at Mereside enabled the patent to add the innovative – and crucial – observation (described

### Trick question

As a young man in 1967, US tamoxifen expert Craig Jordan worked as a summer student at Alderley Park – having gained an instant interview by imaginatively phoning up from the phonebox outside the Park! There, he met Arthur Walpole, a meeting neither guessed was momentous.

Unusually, in 1972 Walpole was invited by Jordan's university to be the examiner of Jordan's PhD. (The university could find no-one else and grumbled that Arthur was only 'from industry'!)

Walpole asked Jordan *why* the isomers of clomiphene and tamoxifen had different effects. Jordan wryly recalls he answered that it was obviously due to the chlorine atom in clomiphene – 'never considering that clomiphene's makers had mixed up identification of the isomers. The literature was wrong; ICI chemists had correctly identified the isomers in-house – and Walpole knew that!' Jordan still got his PhD!

# LIFE
# IN THE
# PARK

by Walpole and Harper in *Nature* the following year) that the *cis* or *trans* forms of such compounds did indeed have opposite oestrogenic or anti-oestrogenic properties. This later became extremely significant in the quest to get the patent granted (and also ensured that a pure *trans* product could be marketed).

But confusion was still not over. Scientists at Merrell had worked out the biological activity of the isomers of clomiphene for themselves – and in 1967 they inadvertently named them the wrong way round! But X-ray crystallography confirmed ICI's result.

## Into man (or rather, woman)

When Dora's compound was tested by Walpole it was clear that 'Nolvadex' was the best find yet. But what would it do in *humans*?

Arthur Walpole knew very well that since breast cancer was stimulated by the body's own oestrogen, circulating in the bloodstream, an anti-oestrogen should be beneficial. Was this one ICI could use?

The first problem was that cancer treatments were not a corporate priority; ICI was not a 'player' in the cancer therapy community at the time (except for Walpole's earlier **'Epodyl'**, a bladder cancer drug) and did not expect much from the area. In 1957, the company itself had stated that 'our work on the study of cancer … has never, frankly, been expected to yield quick monetary rewards … the cancer research … is likely to reveal knowledge which may be valuable in other directions.' Walpole's main job was in reproductive work and he had first to focus on merely keeping support for studying the compound alive in the face of both company disinterest and the lack of clear patent protection.

A different problem was that the new molecule's effect varied greatly. It was an antagonist (an anti-oestrogen) in rats and monkeys, but an agonist in mice and dogs. Could it save human lives? Its effect varied even between organs – it had the opposite effect in many tissues from the one ICI had hoped for; so by some it was considered a useless drug.

**LIFE IN THE PARK**

### Wallop!

Walpole himself was one of Mereside's great characters, known by everyone as 'Wallop' even to his face. Sandy Todd remembers him as:
'… almost a caricature of an absent-minded professor. He had a habit of standing in the middle of the corridor in 8-block at Mereside, lost in thought, gazing at the ceiling and oblivious of everyone squeezing past him on either side. Bernard Langley [the chemistry manager], well-known as a wag, described him as 'the living embodiment of the Heisenberg uncertainty principle'! He was the most charming man and a brilliant scientist to boot.'

Arthur Walpole
V. CRAIG JORDAN

## A potent agent

It also needed treating with care. 'Nolvadex' is so powerful that male scientists handling it in fine powder form must wear dust masks or they may start (due to its slight agonism) to show female characteristics. In the body it is partly changed into 4-hydroxy analogues which are 30 to 100 times more potent still.

Walpole believed it would be a useful antagonist in humans and was determined to prove it. He became convinced that it *could* be proved in oestrogen-sensitive breast cancer.

Walpole never conducted any basic or clinical studies himself on how 'Nolvadex' might work against cancer. But Garnet Davey and he arranged that it be tried out. In the summer of 1969, Walpole set up a project to test 'Nolvadex' with the oncologist Mary Cole as a partner in the Christie Hospital in Manchester, UK, where 46 women with metastatic breast cancer were enrolled. The results were amazing with, in 10 patients (22%), a tangible reduction in tumor volume and lung metastases, along with a decrease in bone pain and lymph node swelling. Importantly, the researchers noted that the 'The particular advantage of this drug is the low incidence of troublesome side-effects'.

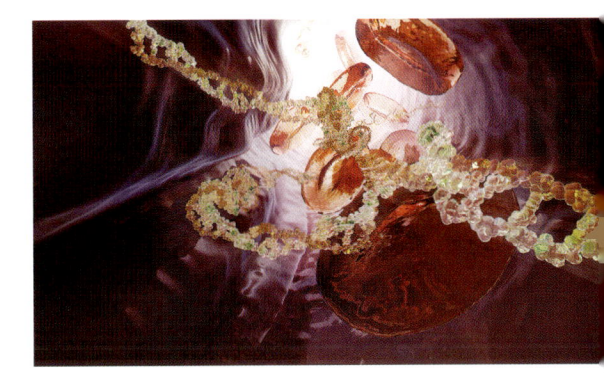

Two less potent competitor drugs fell by the wayside due to side-effects on long-term dosing. (In the aftermath of the **thalidomide** disaster, 'Nolvadex' itself was one of the first compounds checked by ICI for foetal problems in rats; but the only effects noted turned out – after some groundbreaking toxicology – to be due to its own anti-oestrogenic properties.)

## Decision!

Nevertheless, at least two high level meetings, under deputy chairman Eric Hoggarth, a blunt Geordie, then proposed the project be terminated. At the first, Davey and Walpole defended it and obtained a stay of execution pending further results.

Then in May 1972 in a further meeting (without Walpole), this time focusing only on clinical and marketing issues, Hoggarth declared he did not see how ICI would ever 'make an honest copper' out of 'Nolvadex' and again proposed termination. Sales forecasts were very low; it looked like it would no more than repay even the cost of its development, assuming use of it for one year in one third of the few thousand new breast cancer patients in the UK annually. 'The commercial folk at the time did not believe that mere reduction in side-effects would be sufficient to promote the compound at a premium to the existing treatment, diethylstilboestrol, which was cheap, and were not in favour of marketing it.'

It was a critical moment; and the drug faced other dark barriers. There was no granted patent protection in the USA – the US having consistently refused to do this. The company had no facilities to run a cancer programme. If the drug failed, there was no proven 'back-up' compound to it. Moreover, the whole research area was in shadow. Craig Jordan later recorded that 'In 1972 … no-one was recommending anti-oestrogen research as a sound career choice; it was perceived as a dead end.' Several accounts confirm that tamoxifen was close to oblivion.

The financial argument was central. At the meeting Hoggarth metaphorically raised his axe (someone is said to have muttered that he must have had a conversation with the Almighty!); and the fate of 'Nolvader' seemed already decided …

## Walpole's threat

But the voice of Arthur Walpole, *in absentia*, still echoed. He had foreseen the axe's descent and had already made his views forcefully known. He had appealed personally to Division chairman Reg Hoare, and had even threatened to resign his job if 'Nolvadex' was not marketed.

## The ethical view

Then, as Sandy Todd personally recalls, medical director Colin Downie crucially intervened in the meeting, to say that he did not think it would be good for the Division's image if we did not make available a product which was effective in at least a proportion of cancer patients. Todd (who perhaps had an authoritative voice as a descendant of two science Nobel prize-winners!) backed him.

Eventually, Hoggarth dubiously put away his axe. The meeting agreed to market the drug merely as a service to medical science, on a (hopefully) financially 'break even' basis; and a company decision was taken for genuinely ethical reasons to commit resources and apply to market the product against cancer as a public kindness – with few expectations for this failed contraceptive. ('Nolvadex' was also marketed in the 1970s outside cancer, to assist ovulation.)

It is worth noting that if the present UK NICE assessment regime had existed back then, such a product might well not have been recommended for NHS use; and ICI probably would have abandoned further trials.

## The tide turns

But the treatment of cancer was becoming a hot topic, including in the US, where in 1971 President Nixon had declared a 'war on cancer'.

In January 1973, a paper (little noticed at ICI at the time) was published by Harold Ward's team at the Queen Elizabeth Hospital in Birmingham. This confirmed the Christie results but placed even greater emphasis on the low level of side-effects – an excellent result despite the fact that most side-effects were then thought insignificant compared with severe chemotherapy.

## Craig Jordan's proof

But could scientists prove how 'Nolvadex' *worked?*

Elwood Jensen had gone on to reason that since oestrogen worked naturally via the oestrogen receptor, oestrogen levels might also regulate breast cancer. He showed that *some* tumours contained the receptor, and devised a test to tell which these were (in patients this can be determined by a biopsy). He showed in 1971 that patients with such cancers *should respond to a chemical* 'endocrine agent' *of the right sort*.

But was *tamoxifen* blocking the binding of oestrogen to the receptor? Was it really those tumours containing 'positive' oestrogen receptors that were responding to this new drug?

Critically, Craig Jordan, now researching in Massachusetts, showed that it *was*, publishing his results in 1975. (He had been given funds by Lois Trench of ICI Americas to do systematic animal studies, which continued when he returned to Leeds.) With his work the last and most convincing jigsaw piece was fitted.

For the first time in the history of cancer, a drug was connected with its target – the malignant cell – by an excellent biomolecular logic. It was time for it to show its worth.

**Inequality!**

As a historical note, it is worth pointing out that when Dora Richardson made one of the world's most famous drugs for women, she was only on 80% of the salary of her male colleagues …

# 'Nolvadex' (tamoxifen) – launch

## Launch and Ward's paper

**'Nolvadex'** (**tamoxifen**) was approved for the treatment of postmenopausal women with metastatic breast cancer in 1973 in the UK, and launched in September. Chairman Reg Hoare described it in *Scan* on launch as purely an altruistic gesture!

The fine results in Harold Ward's scientific paper were eagerly seized on by oncologists at a number of UK hospitals. The result was that a supposedly 10-year supply of the drug, already made by ICI, was exhausted in 6 months. Company managers suddenly sat up, acutely embarrassed, and took notice. In Parliament in London, the Minister of Health was asked why this important new cancer drug was unavailable, and the Ministry hurriedly asked the Division what it would do about producing more.

## Isomer separation

'Nolvadex' is not easy to make. The early manufacturing process used the method developed in the research labs, which takes six fractional recrystallizations to get the pure drug free from minor isomer. This is a laborious process that I later carried out myself with 4-hydroxytamoxifen (and remember all too well). This slow procedure, and an equally primitive tableting machine, were both early problems.

*Dora at the Queen's Award dinner with Viscount Leverhulme and Peter Cunliffe*
*SCAN*

mixture of *cis*- and (*trans*)-tamoxifen

A different process had just been discovered for making the pure *trans* isomer; and a new tablet formulation was ready. The 'Man from the Ministry' invited an urgent application to save blushes all round, and both were approved in record time – around 2 weeks! (The precise process eventually used to make 'Nolvadex' is, of course, a commercial secret, as it is for all marketed drugs.)

## Success and Arthur Walpole

So the drug that was never expected to make a profit achieved global sales by 2001 of over a billion dollars per annum! Alex Pleuvry oversaw the early marketing of 'Nolvadex'. In 1978 the Division received the Queen's Award for technological achievement because 'Nolvadex' was the first anti-oestrogen for the treatment of advanced breast cancer to reach the market. Lord Leverhulme, Lord Lieutenant of Cheshire, presented it at a celebratory dinner at Mereside. Later, Dora Richardson and Division Chairman Peter Cunliffe from Alderley Park were invited to meet the Queen at Buckingham Palace.

Walpole on stage before the war
PHOTO JACK BAKER, V. CRAIG JORDAN

Tragically, Arthur Walpole, who died suddenly in July 1977, only six months after his retirement, never lived to attend the dinner or see the real success of 'Nolvadex', or of **'Zoladex'** (**goserelin**) in which he was also instrumental. He was both brilliant and highly focused: 'I've always had a commitment to drug-hunting – research for its own sake has no appeal to me.' In 1964, under heavy stress, he had a heart attack, and was told to cut back – so he gave up acting! (He had performed in and directed *Othello*). In 1966, he took only a week off after severe concussion and lacerations in a car crash. He was also exacting towards his staff, among whom Barbara Valcaccia and Rosemary Chester were key members contributing to many drugs.

It seems clear that he might have gained, or shared in, Alderley Park's second Nobel Prize, but for his early death.

## Slow conviction

Throughout the 1970s there still remained a bias against hormone drugs like tamoxifen and a conviction that cancer should be treated with toxic chemotherapy. A major review on anti-oestrogens in 1975 devoted only 14 lines out of fifteen pages to their use against cancer. And 'Nolvadex' still seemed an orphan drug with minimal economic value.

In the US, the rights to 'Nolvadex' were initially held by Ayerst, who sold ICI's drugs in America. They decided not to develop it, whereupon Todd and medic Roy Cotton recommended it to ICI's new American subsidiary Stuart. Ward's paper convinced them and development work began.

The determination of a few key scientists had kept 'Nolvadex' alive. Others now added their enthusiasm, one of them being John Patterson:

I was lucky. I walked into this company in 1975 and they said 'we've got this molecule called ICI 46,474' … which I was asked to look after. I never intended to stay more than a couple of years, and imagined I'd go back to hospital medicine. But I stayed and it turned out to be the five most exciting years of my life.

## US patent problems

The drug was approved in the US in 1977 for use in advanced breast cancer in postmenopausal women. But a huge issue for the company was the patent application covering 'Nolvadex' in the US. The US patent examiner in 1964 had raised various objections. This set in train 20 years of back and forth arguments with the US Patent and Trademark Office (USPTO) to persuade them to grant the patent. ICI Pharmaceuticals Division was repeatedly denied patent protection until the mid-1980s because of the perceived 'primacy' of the earlier Merrell patents and because no 'advance' (in the form of a safer, more specific compound) was acknowledged by the Office. In fact Walpole had earlier shown 'Nolvadex' to be far safer than one of Merrell's compounds which caused cataracts, a result that was eventually the key factor in gaining the patent. But remarkably, 'Nolvadex' progressed for over ten years in the USA without any guarantee of exclusive sales for the company!

Alex Pleuvry, John Patterson and Craig Jordan in 1975
V. CRAIG JORDAN

Failure to obtain a US patent was yet further proof of how unlikely to be useful medical advisers to the industry generally thought the drug could be; but the patent department kept going.

In 1984, ICI sued the US Patent Office's Commissioner for wrongful refusal to grant the patent, and presented all its evidence. The Office's solicitor was ineffectual and called no witnesses; but the judge, who admitted no more than school-level knowledge of chemistry and biology, nevertheless upheld the Patent Office's rejection.

But in the same year 'Nolvadex' was hailed as the treatment of choice for breast cancer by the National Cancer Institute. Then in 1985, when the patent case went to the Federal Appeals Court, the judge's flawed reasoning was exposed and the USPTO was ordered to grant the patent (which it did that August).

This had huge consequence, because at the time (and until 1995), US patents were granted with a patent term starting from the *grant* date (rather than from the filing date, as in the rest of the world). So although the original filing date was 1965, the newly granted US patent did not expire until 2002!

John Patterson records that this was almost 'solely due to the efforts of one man, [ICI patent attorney] Peter (R.P.) Slatcher, who made it his life's work to obtain that patent when all others including his management had given up.' So, at a time when patent protection for 'Nolvadex' in the rest of the world was expiring, the US patent started a 17-year run.

Peter Slatcher
PHOTO ASTRAZENECA

This remarkable situation did not go unchallenged by generic manufacturers. Mike Dukes studied all the written records in great detail to support the patent and legal positions ahead of trials. One of the witnesses called was Justus Landquist, Dora's section manager, who at 76 was, in Dukes' words, 'bloody magnificent'. In 1990 the company was forced to sue Barr Laboratories over 'Nolvadex' (that judge also started the case by admitting he had given up chemistry at 15 and should be assumed to know nothing – and he was right, too!). Barr gained some compensation, and further challenges in the US were made in Baltimore in 1995 and Boston in 1999, but the patent remained the company's property until it finally expired in 2002, as ICI rightly retained legal protection for their pioneering product.

## Palliative to adjuvant

The drug's launch in the early 1970s was as a mere palliative, to alleviate symptoms of late terminal breast cancer. But some more enlightened surgeons realised that more and more drastic mastectomy was not saving more patients and that many would return later when their cancer recurred.

It was guessed that indetectable small islands of cancer cells were surviving surgery, and these were growing later into new tumours. Toxic chemotherapy was applied against these; but ICI then realised that 'Nolvadex' might achieve the same result. The company sponsored studies using 'Nolvadex' in what is called 'adjuvant therapy' (treatment additional to the main one). Craig Jordan at Leeds in 1977 defined the scientific rationale for such long-term adjuvant treatment and in 1983 Mike Baum and others for the first time published proof that this gave real improved disease-free survival, even where patients were not specially selected. But US researchers remained sceptical until an overview of clinical trial data in 1985 proved tamoxifen to be as good as toxic chemotherapy.

## Prophylactic – cancer prevention?

Strikingly, the normal one-year course of treatment with the drug was shown to be ineffective; at least 5 years was optimal. Then in the 1980s, attempts began to prove 'Nolvadex' in disease *prevention* (prophylaxis). In 1973 it had been shown that 'Nolvadex' could actually prevent mammary cancer arising in rats. Trevor Powles at the Royal Marsden Hospital in London started tests in patients with a high risk of breast cancer. Bernard Fisher in the US started a major trial. But giving a drug with slight side-effects to healthy women caused such a furore that he and Patterson ended up having to testify to a US Congressional committee.

## Craig Jordan and SERMs

In the 1980s in Wisconsin, Craig Jordan was studying the effect of tamoxifen as oestrogens in some tissues and anti-oestrogen in others, when he made another major discovery. 'Nolvadex' in humans is a '*selective* receptor modulator oestrogen' (a SERM) – it acts differently in different tissues.

This gave a serious clinical advantage. It meant that the drug (unexpectedly) acts to protect bone, rather than cause osteoporosis, as was expected; it was a leading reason why 'Nolvadex' became a world-beater, and why SERMs soon became the required standard in the field, with five now in global use.

Jordan himself later became the leading international advocate for tamoxifen, encouraging its passage into clinical practice and increasing awareness through papers, books and at conferences; at one, in Cambridge in 1999, tamoxifen was recognised with the prestigious German MMW award of 'medicine of the year'.

For his tamoxifen work Jordan himself was, among other honours including one proposed by Sir James Black, elected to the National Academy of Sciences and in 2002 was awarded an OBE, partly on AstraZeneca's nomination. (Also among his diverse distinctions was, unusually for an American cancer researcher but in fine Pharms tradition, a top-secret security clearance for MI5!)

Yet he never forgot Arthur Walpole, who had arranged for Roy Cotton to send consignments of rats, chauffeured from the Park to Leeds University, for Jordan's experiments for three years; nor Alderley Park, where Craig had first been offered a technician's job at the remarkable age of thirteen (before his mother found out!).

Craig Jordan (left) receives a map of Cheshire from Barry Furr to celebrate his OBE.
V. CRAIG JORDAN

## Illegal manufacture

Popularity can bring problems. The drug must be protected from sunlight, as this could cause a little of the drug to change to the other isomer. Because of this, there was great concern when illegally made tamoxifen from Czechoslovakia was discovered being sold in plastic bags … exposed to light. It was a classic illustration of the hazards of counterfeit drugs.

## Might never have been

After 20 years of use, new experiments showed the drug produced liver cancer in rats. Had this been known earlier, it might never have progressed – but it was also then found that the toxicity does not appear in human patients.

## The gold standard

In 1998, an initial adjuvant therapy trial showed that the incidence of invasive breast cancer in patients at high risk decreased by 49%. So the trial was stopped by the ethics committee, after they noticed the indisputable advantage of the patients taking the drug over the placebo group (who thereafter received the drug too). Other studies were not so clear but the results were enough to make tamoxifen the 'gold standard' drug for prevention of oestrogen-receptor-positive breast cancer.

The rest is history. Tamoxifen, the failed contraceptive, gained the distinction of being the most prescribed hormonal treatment for cancer in the world, having started from its small beginnings in Alderley Park, no less.

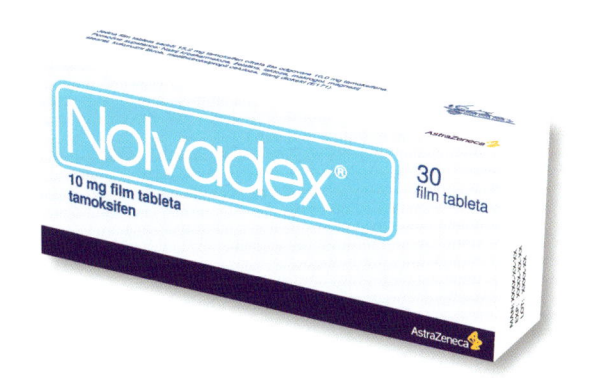

'Nolvadex' AstraZeneca version
COPYRIGHT ASTRAZENECA

'Nolvadex' logo, 2015
COPYRIGHT ASTRAZENECA

It also put hormone cancer therapy on the map with a vengeance. It has been superseded in the US but is still widely used elsewhere in the world. (AstraZeneca itself ceased selling 'Nolvadex' in the US in 2006.)

# The antibacterial search

The Park's research into combating infection continued for many years, until the antibacterial department at Alderley Park was closed and the work moved to Boston in the 1990s.

### Penicillins

In 1958, scientists at Beecham discovered that the core molecule of penicillin was **6-aminopenicillanic acid** (**6-APA**). When it was realised that new penicillins could be created easily from this by varying the culture conditions or by synthesis, it unleashed in the early 1960s a big international hunt for new synthetic antibiotics.

### Crystals

Alderley Park saw its share of the hunt. But one great challenge with these fragile compounds was turning them into stable solids. Only if they were crystalline would they be likely to be stable and pure enough to have an acceptable shelf life. On one occasion, I worked next to an entire team that spent six months trying to crystallize one potent but rather unstable antibiotic. Eventually the team sadly failed and the compound – which was probably simply too unstable for use – was abandoned.

## LIFE IN THE PARK

### Secrets of crystallization

Crystallization was a challenging area from the start; and it also involved some great characters. I once had the privilege, as a young chemist, of meeting the redoubtable 'Tommy' Leigh. I was told then (I know no more) that Tommy was the first person ever to have crystallized a penicillin.

But rumour had it that the secret of his skill lay in the curved pipe and 'Edgeworth' tobacco he smoked (at that time smoking was allowed in chemistry labs!). Difficult crystallizations sometimes need a helpful 'seed' particle around which the first crystal can start to form. And Tommy's pipe was said to be an excellent source of such particles!

Another source of suitable particles is said to be an old chemist's beard …

Tommy Leigh, with pipe, in the old Mereside Library
*PHARMACEUTICAL RESEARCH IN ICI 1936–1957 (1957)*

## The need

Drugs for infectious diseases will probably always be needed. But who pays for them? Unless a disease is widespread, then patients will only take an antibiotic for a few days.

So antibacterials and antivirals are not very profitable therapeutic areas. This adds difficulty to the predicted 'antibiotic resistance time bomb' problem that society now faces; although the principal cause of this is the misuse of antibiotics already discovered.

## Carbapenems

Carbapenems are a class of highly effective but unstable beta-lactam antibiotics. They are derived from **thienamycin**, a product of soil bacteria discovered by Kahan in 1976 and first made chemically by Merck. Thienamycin is structurally different from previous antibiotics of the penicillin or cephem sort and was the first of a new group that was given the name of 'carbapenems'. It attracted global interest but proved unstable and toxic. But it was a crucial pointer. In a number of pharmaceutical companies 'lead optimization' research programmes were rapidly started to find a related but useful antibiotic. An important lead was **imipenem,** invented by Merck in 1980.

Carbapenems sold by ICI were both invented and developed elsewhere, except for two.

## 'Meronem' (meropenem)

One injectable antibiotic, with a structure inspired by imipenem, was not discovered at the Park but in Japan at the relatively small pharmaceutical firm of Sumitomo. **'Meronem'** (**meropenem**) – 'Merrem' in the US – was an ultra-broad-spectrum drug able to treat a wide variety of infections. But some of the rights to it were acquired ('licensed in') by ICI, as the Sumitomo management had decided their scientists needed help to develop it to a marketable drug. As a molecule it can have many isomers (64 in total) of which only one was wanted.

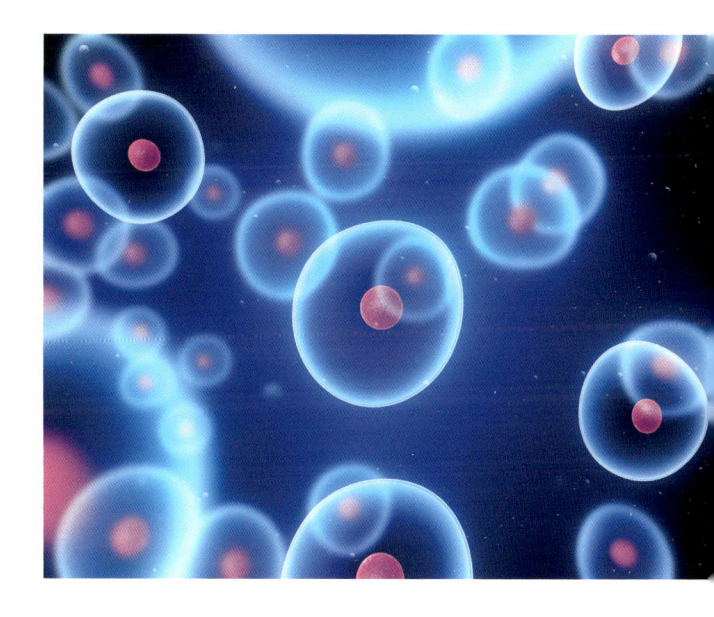

The difficulty of making it on a large scale was solved by research both at Sumitomo and Mereside, where several teams were involved. The aim was a useful manufacturing process with a 25-fold reduction in the anticipated bulk drug cost; the falling costs of key intermediates helped.

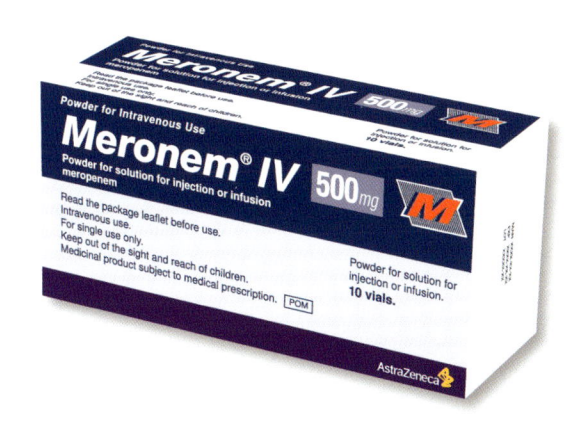

Both the chemical synthesis and the study of its antibiotic spectrum (the range of bacteria it could act against) were tackled in the Park in a project led initially by Terry Hennessey with the teams of Mike Betts on chemistry and Jeff Edwards on microbiology. Afterwards, more work was done on the problem of drug resistance to 'Meronem'.

It was first approved in Italy in 1994 and in the US in 1996. It penetrates well into many tissues and body fluids.

## Invanz™ (ertapenem)

The most useful result for ICI of the work on 'Meronem' was to get the same Mereside scientists started on work leading to a drug of their own making.

The short half-life of carbapenems was an issue. Betts suggested attaching a glycine unit to a carbapenem. This gave an unstable red gum but synthetic chemist Mike Swain proposed reversing the structure and turning the glycine into a proline. This gave stable solids with good activity and half-life. One was selected as a development candidate but in the meantime Gareth Davies recommended making a different isomer. Cathy Darbyshire, a newly employed young chemist at Mereside in September 1991, was asked to make this compound as her very first piece of chemistry on arriving in the company. The result was ICI 251,595, later ZD4433, which was given the generic name **ertapenem** and has a similar core chemical structure to meropenem, with which it shares a key 1-β-methyl group. But medicinally it differs from that and the others in its class. It is slightly poorer in having a somewhat less broad spectrum of activity against bacteria but, on the other hand, has a longer lifetime in the body. This means that it needs administering (by infusion) only every 24 hours instead of eight.

**'Meronem'**
**(meropenem)**

**Invanz™**
**(ertapenem)**

One entertaining note was the subsequent discussion about Cathy's future. One unkind voice was heard to suggest that she should now be sacked, as it would be a double statistical wonder if she ever made another new drug! However, kinder logic very rightly prevailed and she went on to a successful career (albeit in clinical quality assurance).

ICI in turn chose not to develop the drug as its own. So the drug was out-licensed for around £45 million to Merck, from whose thienamycin its precursors had originally sprung. The patent application was filed with Davies, Swain and Betts as inventors in 1993; and the latter, with Ron Galt, led an outstanding campaign (requiring a remarkable 600 chromatography fractions) to make a major batch of this very delicate compound. Swain was sent to the US to give his direct Merck counterparts advice – although he was bemused to find that all the chemists he met there had impressive job titles; not being an assistant deputy vice-president for anything in England, he felt rather left out!

Merck marketed ertapenem as **Invanz™**; and used the slogan 'The Power of One', on the grounds that the dose is one gram, once per day.

Déjà vu

**Ertapenem was clearly a fine drug. Some years later, another ICI department (which shall remain nameless) identified it as an excellent candidate for the company to buy the rights to *from* Merck …**

# LIFE IN THE PARK

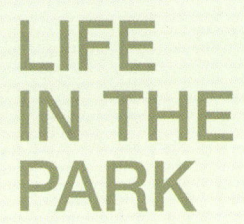

# Anaesthetics – 'Fluothane' to 'Diprivan'

ICI had begun working on anaesthetics early in its history.

Its first new chemical product was the intravenous anaesthetic **'Kemithal'**, made in 1939 and proven to work in 1946. Much more was to follow.

## Old anaesthetics

Anaesthesia was a field in which there had been little progress for nearly a century. Barbiturates (injected intravenously) could easily 'knock patients out'; but they could also endanger them. Inhalants such as **chloroform**, **ether** and **nitrous oxide** were products of the nineteenth century still in use in the mid-twentieth. All had major disadvantages: chloroform can lead to fatal liver damage; nitrous oxide ('laughing gas') was very short-acting and poor except for quick dental procedures (and could be misused); and highly flammable ether had caused many explosions in operating theatres. **Cyclopropane**, a later introduction around 1928, also had the latter problem. **Trichloroethylene** was introduced by ICI in 1941 as **'Trilene'** after a London chemist in an aeroplane factory in 1940 noticed youths intoxicating themselves over vats of it; it was thought safer than chloroform, but obviously had other problems.

## 'Fluothane' (halothane)

In 1930, Thomas Midgley at General Motors was looking for non-toxic refrigerants. He wondered about carbon compounds containing non-metallic elements, and showed that fluorine was the element whose compounds were both volatile and stable (and thus safe).

James Raventos
ASTRAZENECA ARCHIVES

Later, during the Second World War, Charles Suckling, a young ICI chemist at Widnes, began making 'highly fluorinated compounds – I didn't know why but it turned out they were intended for gases to assist in the separation of uranium isotopes.' Only later did he learn he had been doing research towards the atom bomb. (In fact ICI had wanted to build Britain's first A-bomb.)

Similar materials made in America were found to have anaesthetic properties. But in 1950, ICI general chemicals research director James Ferguson was unaware of this when he reviewed the company's range of fluorocarbons. Post-war, ICI was also developing them as refrigerants and in pest control, but their total lack of chemical reactivity convinced him they might work as anaesthetics. The job of investigation was given to Suckling and, in 1951, he and pharmacologist James Raventos from the Pharmaceutical Section of Dyestuffs at Blackley chose strict criteria that led them, with only the sixth compound Suckling made, in January 1953, to a very promising candidate.

# LIFE IN THE PARK

## Knocked down

The first anaesthetic tests were done on grain weevils; but better results were obtained with house flies. As Suckling succinctly put it: 'They either flew or they didn't.'

A Goldman 'Fluothane' vaporiser from 1963
ASTRAZENECA ARCHIVES

'Fluothane' for animals
ASTRAZENECA ARCHIVES

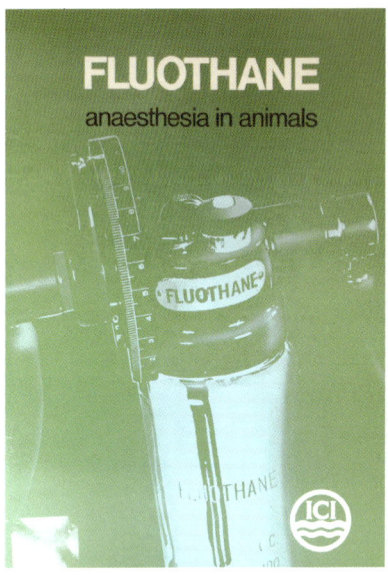

The relatively new technique of gas chromatography was used to test and later purify the new agent, **'Fluothane' (halothane)**. After three years of exhaustive testing it was put into clinical trials by Michael Johnstone at Crumpsall Hospital in Manchester, and was launched in 1957.

It became one of the world's most widely used inhalational anaesthetics. Its many advantages ensured its success even though it cost eighty times as much as chloroform. It was pre-eminent until the 1990s when mild concerns about liver toxicity led to its replacement in most places. Suckling and Raventos received the international highly respected Scott Prize in 1973 for 'Fluothane', and Raventos was also feted in the world of anaesthesiology for changing 'the course of anaesthetic practice through his brilliant work in ICI's laboratories'. Suckling became an FRS in 1978.

Raventos and Suckling receive the Scott Prize
PHOTO ASTRAZENECA (BARRY SMITH)

### 'Diprivan' (propofol)

But Alderley Park would soon have something better.

In the 1970s concerns about hazards, pollution and liver damage among theatre staff triggered a programme to find an intravenous, controllable 'maintenance' anaesthetic.

Intravenous anaesthetics are injected, traditionally to induce (start) anaesthesia – after which gaseous mixtures of inhalation anaesthetics may be needed to *maintain* (control) the patient's level of consciousness and perception of pain. But an intravenous agent would be far more useful if it could also maintain the patient's state by itself.

### An interesting lead

As part of the search for new intravenous agents, samples from the ICI compound collection were routinely tested. In January 1973, a remarkably simple compound, 2,6-diethylphenol, was discovered to have anaesthetic properties in mice. The compound's full 'profile' of properties was not particularly attractive. But medicinal chemist Roger James decided to treat it as a potential 'lead' and arranged to test similar compounds. His chemistry team also started to make further, more complicated, phenols, as well as derivatives with greater water solubility, designed to release the active molecule in the body (as 'prodrugs', which the body itself changes into the 'real' drug.)

Among numerous promising compounds the best, a close relative of the original lead, was 2,6-diisopropylphenol. This was already sold and known as an antioxidant additive for foods and rubber, and the sample of it in the company collection dated from around 1960. The unique profile of it was recognised very early in the campaign (in May 1973) by a bioscience team which included Jimmy Dee, whose years of experience in testing volatile anaesthetics proved invaluable. The team's leader J. B. 'Iain' Glen developed new ways to compare different anaesthetics and his expertise was also vital during the development phase and beyond. Many others contributed to the very complex work.

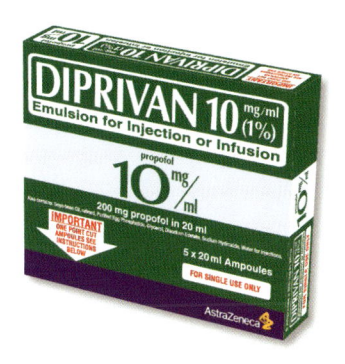

'Diprivan' carton
COPYRIGHT
ASTRAZENECA

'Diprivan' syringe carton
COPYRIGHT
ASTRAZENECA

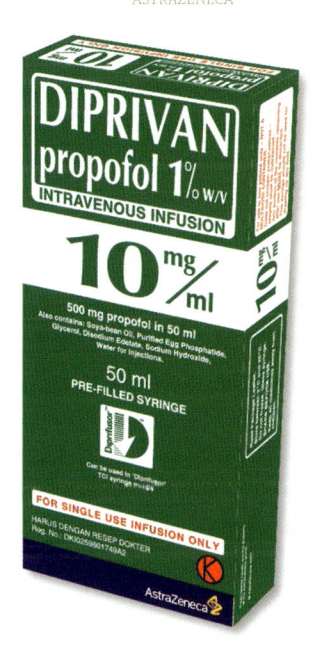

## Expectation

**After his success, Suckling was somewhat cynically amused to be asked by a visiting deputy company chairman, 'Right, what are you going to invent next?'**

## LIFE IN THE PARK

'Fluothane' (halothane)

'Diprivan' (propofol)

fospropofol, a propofol "prodrug"

Structures of 'Fluothane', 'Diprivan' and fospropofol
GEORGE B. HILL

## Huge potential

The new drug's properties were exceptional, producing smooth, rapid induction of anaesthesia, a short period of good muscle relaxation, no cumulative effects on repeated dosing, and then easy recovery. Research later showed that it had a complex mode of action, working in seven different ways at once. Its rapid onset and clear-headed, nausea-free recovery – within minutes even after extended use – made its huge potential clear.

ICI filed patent applications on its use in anaesthesia in 1975, with James and Glen as inventors. **'Diprivan'**, the commercial name, derived from 'Diisopropyl intravenous anaesthetic'. Its generic name was **propofol**. The first clinical trial, reported by Brian Kay working with Professor Rolly of Belgium in 1977, confirmed its excellent properties.

## Difficulties

But the new anaesthetic had a problem. For a drug, it is a relatively lipophilic (fat-like) compound. Human blood is water-based, but 'Diprivan' is poorly soluble in water. So presenting it in a form compatible with blood – that is, one the body will accept – was a major problem. How do we get it where the patient needs it?

A special formulation (mixture for treating the patient with) was needed. One containing Cremophor (a castor oil derivative used previously for getting other drugs into the human body) was tried; but there was concern that Cremophor might produce allergic reactions in patients. A project begun in 1975 to find a better water-miscible formulation failed.

So 'Diprivan' was tested in Cremophor, as what scientists call a 'proof of concept', while another solution was sought. In 1977 the mixture went into trials and proved so suitable that it was tested in one thousand patients; but it was abandoned when the Cremophor content caused anaphylaxis (a form of shock) in a small number of these early clinical trial patients.

Meanwhile, competitor products had hit even greater problems and had been abandoned. But despite much work, another 'Diprivan' formulation using an ICI 'Pluronic' (a soluble polymer) also failed.

At one stage, the commercial potential was also questioned and it was recommended that development be stopped, as the marketing group could not foresee any use beyond mere induction of anaesthesia. But happily, research director Brian Newbould pushed for a re-think. Eventually, a new and effective emulsion form was developed after massive technical challenges were overcome; and 'Diprivan' was formally launched in 1986, for clinical use in adults for short-term maintenance (as well as induction) of anaesthesia.

## A global success

The new anaesthetic became hugely popular – so much so that it even gained a nickname: the opaque, white emulsion was given the apt description of 'Milk of Anaesthesia'. Both patients and doctors were amazed that a patient could come round, look the surgeon in the eye and say 'thank you', on the spot. This meant that it was also particularly useful for day-care surgery, allowing procedures previously impossible with outpatients before this class of anaesthetic was developed. John Patterson thinks that 'Diprivan' was the 'single most changing' of all the Park's discoveries for medicine as a whole. In 1992 the Division was given the Queen's Award for Technological Achievement for it.

'Diprivan' came largely to replace **sodium thiopental** (**'Sodium Pentothal'**) for induction of anesthesia. It also replaced much of the use of gaseous agents for maintenance, because recovery from it is more rapid and 'clear' when compared with halothane. Also, it is not an analgesic, so opioid painkillers such as **fentanyl** can be combined with 'Diprivan' to alleviate pain.

A water-soluble prodrug form of 'Diprivan' was later introduced as **fospropofol**, (**Lusedra**™, initially trialled by MGI Pharma but then sold to Eisai) although this takes effect more slowly and so has been less used.

A growing problem is the supply of the drug. After ICI's patent expired, one major manufacturer was the generics company Teva. But in 2010, Teva ceased production of propofol, citing both process difficulties and several pending lawsuits. This aggravated an existing world shortage of this very popular agent.

## Misuse

The misuse of propofol has led to controversy. Bad practice by anaesthetists in the USA and other countries led to bacterial contamination of the 'Diprivan', which caused problems in some patients in the early 1990s. Christopher Jones and John Platt were inventors on a Zeneca patent application in 1998 for an edetate-containing formulation of 'Diprivan', designed to ameliorate this problem.

In a very different and much darker problem (because propofol when overdosed depresses breathing) the Missouri Supreme Court decided to allow the use of propofol to execute condemned prisoners. It backed down only when Britain banned propofol exports to the US two months later in 2012; and when its own state governor ordered a halt after the European Union also threatened a ban. (Propofol is used some 50 million times annually in the US for medical purposes, with sales there of over half a billion dollars in 2010. It is probably still the most extensively used maintenance anaesthetic in clinical practice.)

More infamously still, due to the 'off-label' misuse of propofol by his personal physician, it was the drug that may have killed Michael Jackson. In August 2009 the Los Angeles County coroner concluded that a mixture of propofol (mixed with lidocaine), plus lorazepam and diazepam, caused the famous singer's death. Clearly this is both a warning against the misuse of such drugs; and a testimony to the research effort and attention to detail that led to 'Diprivan' becoming the world-class anaesthetic that it did.

# 11

# Supporting the work

## Engineering and labs

What have engineers and technicians done in the Park over the years?

### Buildings and global engineering

Research starts with buildings; and in the Park these started with ICI Engineering (in Alderley Park as Engineering Services Group), later AZ Engineering. AZ's construction or refurbishment of new buildings is organised globally. AZ Global Engineering currently handles projects greater than $10million in value, working closely with project management to meet the needs of the business. At first, it just looks like a hole in the ground. But then girders, walls, windows and whole offices and labs appear. The engineers go home; and the staff take over …

### Heat and air

Of course, everyone needs to keep warm. In the early days, heating was by heavy fuel oil that had to be heated before it would even burn. Later gas was used. Heat was transferred by high pressure hot water piped to the Park's buildings and later steam. The earlier Energy Centre beside Block 10 was replaced by 2009 with a new £7.3 million one near CTL.

Oil-fired boilers in 1957
*PHARMACEUTICAL RESEARCH IN ICI 1936–1957 (1957)*

Energy & utilities in 2014
PHOTO ASTRAZENECA

Ventilation of laboratory fumehoods was an even bigger challenge. To keep lab staff safe and ensure vapours were safely dispersed, powerful 'draughts' were needed, changing all the air in a lab in minutes. New fans required on Block 14 were so heavy that the building's foundations had to be strengthened to support them!

Many scientists require supplies of gas cylinders, ice, dry ice and more. Piped supplies of steam and gas, the mainstay of older chemistry labs, are now history; but compressed air, vacuum and nitrogen hopefully never run out. In recent times the latter came from a liquid tank holding the equivalent of a third of a cubic mile of gas.

## Day to day services

Important to all scientists at Moreside and to all Park staff is engineering support on a day-to-day operational basis. In a large laboratory complex, electricians, plumbers, mechanical engineers, joiners and the like are the people who actually keep things going (as they well know). Repair of everything from a plug to a major piece of equipment suddenly becomes critical to someone.

## Electronic damage – expensive

Maintaining thousands of instruments around the site is another huge cost. It is frighteningly easy to damage equipment that cost a six figure sum. Even simple laboratory balances require regular servicing; and as technology grows, it seems that nothing is simple for the instruments and maintenance functions at Mereside.

## Animal services

Medical research must put humans first. But after that few things require more care than the breeding of animals for use in medical research. In this demanding area, all staff were from the beginning profoundly aware both of their responsibilities and the many stern internal and external inspections they would constantly face. The cost was also huge; on purely financial grounds alone, poor treatment of any animal would be grounds for instant dismissal. But without such work, safe medicines would hardly exist today.

### Glass and blowing

Working with metal or wood was one thing; but one of the most evocative places at Mereside until the work was contracted out was – at least for chemists – the glassblowing laboratory.

The glassblowers of what was originally Research Services Group (recently Messrs. Hems and Keenan, under Mereside's resident Dutchman, Wim Van de Bospoort, universally known since he began in 1964 as 'Van de Blowpipe'), knew they were a breed apart. They flamed and melted and blew at 900°C in ways that youngsters thought were incomprehensible, and which awed older chemists who had tried it themselves. In their spare time, to hone their skills, the glassblowers would occasionally make stunning pieces of craft that would have sold for large sums in a tourist shop. But the urgent needs of grovelling suppliants with broken equipment were always treated with kindly condescension.

## LIFE IN THE PARK

### Fatal flaw!

The merest flaw in the glass itself was not tolerated. The slightest chance of a dangerous chemical being spilt due to a 'star-cracked' flask was enough to doom many expensive (and favourite) pieces of laboratory ware.

On one occasion, I took down to John and Larry in the glassblowing lab a five litre flange-neck Pyrex glass flask (the size of a football). Replacing it would cost dear; and it had one tiny star-crack in it. But it was worn. The glassblower looked at it, while holding it over an empty steel waste barrel; then he opened his fingers and let it drop. The sound and the horror together were enough to lift me several inches off the ground. 'No. Too scratched. Order a new one.' Sympathy – none. We soon learnt to take no chances.

## Facilities and services

Who provided all those other facilities for us?

Many more services on the Alderley Park (and Macclesfield) sites were (and still are) provided by departments whose names over the years have included Research Administration, Site Services, Facilities Management and Business Services.

These allowed staff to do their jobs by providing a quite enormous range of support. By 2000, more than 400 staff occupied these roles alone in the Park.

A description of all their activities would probably take up the rest of this book. But they include everything from supply of water and energy, food and printing, through to caring for the environment and, of course, cleaning, of both offices and science areas. Video conferencing is there as well. And staff needed finding, for everything from secretarial services to drivers.

### 'Media'

Right from the opening of Mereside, a particular challenge was always meeting the hunger of the scientific labs for clean washed glassware and biological 'media'. Even by 1963, 1,650,000 items of glassware were being washed each year – and the washing up just

Media compounding
and dispensing table

grew from there. In fact, every medicine that came out of Alderley Park started with you, folks! Thanks a million.

## Food

The histories of the restaurant buildings are described elsewhere. The scale of provision by their friendly staff was enormous, as everyone on the site well knew. In 1996 an innovation was the cashless Klix Key system. In later times, numerous other eating places were provided, at Parklands and elsewhere, as 'Grab'n'Go' facilities and Costa™ outlets, perhaps the most elegant being the Cedar Café below the great tree at Alderley House. The year's biggest event was always the sold-out Christmas Lunch.

Parklands Café
PHOTO ASTRAZENECA

The Restaurant and many other services were later contracted out and operated successfully by Sodexo Holdings Ltd, to which many staff transferred, as AstraZeneca found it necessary to focus on its core business.

## Phones

Unimaginable numbers of conversations have helped run the business throughout its history. At one stage, the company even had its own private phone cable across the Atlantic – until a ship dragged the cable away in mid-conversation. In 1985, Alderley Park moved to a computer controlled phone system. The previous rotary dial phones were replaced with push button handsets. Later, direct dialing arrived. Telephones proliferated to become eventually more numerous than staff (though fax and teleprinter, once essential, became words hardly remembered). But humans still answer, even now, for those who need them!

## Office printing

Photocopiers eventually became on hand for all staff, and were networked to personal computers, so that one could send one's document straight to a nearby machine (logging on at the machine with one's cardkey) and print it directly – if there was paper, ink and no jam! The huge budget for paper rapidly declined in any case, as electronic storage became standard.

But in earlier days, the Office Printing Section was a monster operation, producing everything from publicity materials and internal reports to posters for exhibitions. It was based at Alderley House and later at the farm, but by 2000 it had found a home in the former Division Garage in the lower courtyard, where company vehicles had formerly been kept ready for staff.

## The E of SHE: energy conservation

The Safety and Health parts of the familiar acronym SHE are mentioned elsewhere. But a key focus in recent times has been the environment, which in business focuses on energy conservation, now a massive concern in Alderley Park both for financial and global reasons. The Alderley Park Energy Committee was formed in 1987; all significant decisions taken in the Park probably were shaped by its advice.

## Deliveries – on the trolley

One key role that affected all buildings in Alderley Park was the well-oiled electric trolley delivery system operated by logistics staff, who collected 'requisitions' from hundreds of delivery points four times a day, and delivered anything and everything as well.

The trolley men (and ladies) were always popular with their customers, as mistaken orders and confused deliveries could be sorted out rapidly if one was on good terms with the trolley operator. The Stores themselves are mentioned later.

Mereside entrance
in 1957
*PHARMACEUTICAL
RESEARCH IN ICI
1936–1957 (1957)*

Radnor Reception
in 2014
PHOTO ASTRAZENECA

## Visitors

In addition to meeting the needs of staff, the handling of visitors to a multinational company is an industry in itself. Immense numbers of people have visited Alderley Park over the years, perhaps from most countries on Earth. Welcoming the welcome, directing them to a plethora of destinations, and looking after them, their vehicles and their safety, is no small task. An online visitor management system now watches over them all.

## Guided tours

This was another huge industry, run by regular staff tour guides. Everyone loved to see more of Alderley Park, from journalists to scientific and sales visitors – occasionally with multiple translators, too!

Once, I was asked to show the Water Gardens to a party of top Far Eastern hospital presidents. I was told they spoke little English; and an interpreter with them had never visited the Park before. When I asked for backup, I was given a mobile phone (the first I had ever held) and shown two buttons on it.

'Press that one in an emergency,' I was told. 'Not the one that says 'Mum' by it.'

I started by explaining that photography was not allowed, then watched helplessly as they all produced cameras to record the magnificent Cedar. Hurriedly, I led them away from the buildings down to the foot of the cypress avenue by the pool. Facing the pool, I started talking, but was largely ignored. An odd noise puzzled me; when I turned round, I saw a large hydrangea bush. It was shuddering violently, and pointing out of the top of it was a very small pair of shoes.

A tiny lady had climbed up on the wall for a photograph and had fallen headlong into the bush. My first reaction was to wince, as my entire professional career flashed before me. Then I thought of the mobile phone; and wondered to call for urgent assistance … or whether 'Mum' might be of more comfort.

By now the group had rescued their casualty and had closed ranks firmly. It was made *very* plain (to my great relief) that any publicity would represent an unthinkable loss of face. Despite my attempts to wipe her down, the little lady shot off like a rabbit into the next part of the tour, still smeared with lichen and slime. And I declined to act as a guide thereafter.

## Office supply

Office equipment could mean almost anything, from pencils to photocopiers, from chairs to carpets. Furniture became more and more health oriented as ergonomic factors became a key focus of Occupational Health – watch that posture, and don't watch that screen for too long!

## Meeting rooms

Booking of meeting rooms was always an issue in the Park, but became more and more difficult as the population increased, despite the fact that the buildings were also spreading.

Management of meeting rooms was also a challenge, one that became much harder as technology invaded them. Policy dictated that every meeting room should be set up with state-of-the-art projection and other facilities. But each time a new meeting room was opened or refurbished, the art had changed! So the technology differed even from room to room.

## LIFE IN THE PARK

### No more!

**But there had been bigger changes in the past. In 1962, a shocking edict was issued by the Alderley Park Site Manager. Cigarettes were no longer to be stocked in meeting rooms – but for economy, not for health reasons! All smoking on site was banned in 1995.**

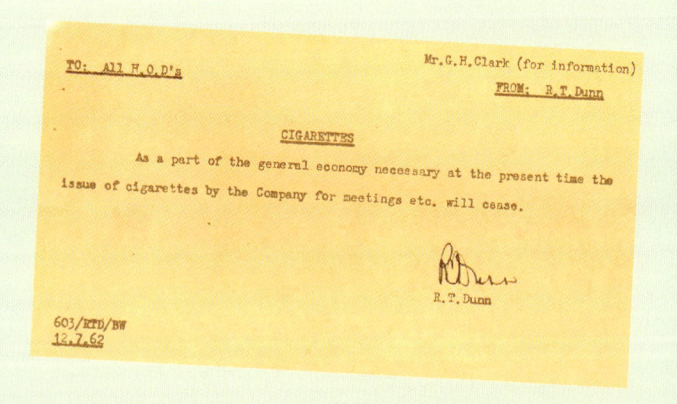

No cigarettes!
ASTRAZENECA ARCHIVES

## Security

Visitors as well as staff and property need to be kept secure. And security is a never-ending job. When everyone else goes home, or switches off, quiet eyes are watching and observing.

Sometimes the observations are not quite what one expects – like the badger that used to roll past Church Lodge gatehouse each evening, on its late patrol, to the amusement of the officer on duty, or the odd weasel or polecat or owl passing by. Rarely, challenges were more serious; but in an uncertain world, there were always those ready to meet them.

## Fire

The site firemen were old hands, experienced and passing on their experience to generations of staff, particularly ones like chemists and electricians whose work could hold unpleasant surprises. Regular training sessions – involving real fire – were run, where extinguishing with water, powder, $CO_2$ and foam was undertaken by those willingly stepping forward (and sometimes by those not!). Fire was always a huge concern in Alderley Park (and in the woods in summer, too). The chemistry labs were, as is common modern practice, always on the top floor – so everyone beneath could escape easily. They were full of fire extinguishers, but the message was always, laboratories are replaceable; people are not.

## Travel

In a multinational company, people travel incessantly. The foreign travel budget of a global business, with endless international and academic contacts to maintain, can be the biggest single item in some departments' *entire* budgets. Someone must manage it.

# Office life

Who has worked in offices in the Park? An immense number were occupied.

## Changing culture

Office life changed considerably over the years. In earlier days, offices were largely spaces private to managers.

A Park-wide change came from the 'open door' policy that became standard. An even bigger and (at the time) surprising change in office culture came with the building of Block 11 at Mereside, whose walls were moveable. Offices could be reshaped and resized as the business required.

They also became transparent! Glass-walled offices, even for senior managers, were a startling change. Until staff became used to working in goldfish territory, they had to take care in respect of things written on whiteboards or posted on walls visible from outside. Offices for scientists were alongside their labs.

Offices and chemistry labs, Block 33
PHOTO ASTRAZENECA

In recent times, electronic working and 'hot desking' have reduced the sense of privacy even further, with many staff having no desk storage space of their own at all.

## The tea-ladies

One much-lamented loss were the tea-ladies who used to appear outside all the offices and labs before self-service machines replaced them.

In the early days, heads of departments had offices with solid wooden doors – which were a message to the tea-lady that tea should be served to *them* in cups! Everyone else gathered in the corridor to be served in mugs – and to aim for the nice, moist piece of toast at the bottom of the pile.

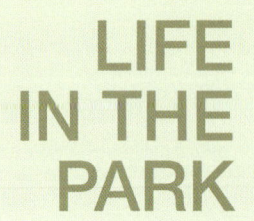

**LIFE IN THE PARK**

### Strange brew

The tea was usually excellent. But on one occasion it tasted a little off. However, no-one knew why; until the tea-lady was emptying out her urn at the end of the day and found in it the dishcloth she had mislaid that morning.

# Computers – starting from one

The first computers at the Park were loaded with manually generated punched tape. In 1970, Pharmaceutical Division possessed just one mainframe computer (an ICL1900 series) in the new Computer Centre (Block 18) at Mereside; and was just putting in a second one, a Burroughs. Burroughs mainframes became the mainstay for a quarter of a century. They became bigger and bigger; in later times the Computer Centre had to have a separate fan unit purely to keep it cool! They were used for all Division activities, including accounts, commercial matters and production, as well as R&D.

A few other smaller computers around the Park then existed; but what were originally described as 'minicomputers' were on the way. PCs bred like rabbits, and networked like spiders. By 2010, the company could call (for big problems) upon whole 'server farms' with hundreds of computers in each; while AstraZeneca itself had seventy thousand plus PCs globally.

## Searching for a compound

How can a computer store the 3D structure of a complex chemical compound? How do you talk to a computer in 3D? With difficulty!

A Company Compound Centre was set up (originally by ICI Head Office, using an IBM machine at Piccadilly Plaza, Manchester) to store the identities of all compounds made by 'Pharms' (and also Agrochemical and Organics Divisions). This was done using the rather clumsy but efficient Wiswesser Line Notation (WLN), which was a way, invented in 1949, of turning chemical structures into strings of text symbols. ICI's Canadian subsidiary invented a new software system, called CROSSBOW, which allowed this database to be searched and the results printed in 2D. Both the Company Compound Centre and CROSSBOW moved to Mereside in the early 1970s and the system continued into

**Keyboard gender**
It was a shock when the first bulky desk computers had keyboards that were used by men! Until then, female secretaries had been the only typers.

the 1980s and beyond. A different and far faster chemical naming and search system, SMILES, then became standard in what became known as 'computational chemistry', but was only slowly adopted elsewhere.

Individual computers later brought a step change. This was more than just a technological shift; ICI's scientific computing had begun as an adaptation of corporate accounting systems which were not easily tailored to the needs of research information systems with their different ways of operating. Introduction of more suitable technology was, in the earlier period, slowed by this.

## Chemical drawing packages

Early systems were not suitable for use by non-specialists. The epochal shift for most staff began when the MACCS™ molecular access system of Molecular Design Limited was introduced to scientists at Mereside. When fully implemented, this allowed storage of chemical structures in a user-friendly database (IsisBase™) which was accessed using MDL's fairly simple 'chemical drawing package' of IsisDraw™.

It was a revelation as well as a revolution when ordinary lab staff found they could draw their own chemical structures and run their own searches, and (with a licence) that they could run as many as they needed! Everyone was happy – except new recruits and students who had already learnt ChemDraw™, a different system used by universities. Oh, and IsisDraw™ has now been replaced by Symyx Draw™.

Although apparently simple, chemical drawing packages automatically decide how the structures must be drawn to obey the rules of chemistry, which are occasionally very far from simple. They became immensely important when all standard chemical records became electronic; the modern chemist uses them constantly.

## Notebooks

For many years, scientists at Mereside recorded all their findings in elegant maroon or other coloured hardback notebooks that were preserved as the ultimate original store of company discoveries.

In later times, these had to be signed as each page was completed. This was a major legal necessity; on one occasion, two other companies almost simultaneously discovered a valuable antibiotic. The original notebooks had to be consulted, and it was found that one set of researchers had discovered it only a week before the other! Months of patent litigation and tens of millions of pounds hung on that proof.

For a long time, the stupendous implications of a mistake meant that the paper notebooks continued in use, as a last throwback to a more elegant age. But in the new millennium the computer age finally caught up, as the Electronic Lab Notebook (ELN) became the permanent (and digitally signable) store of all work in AstraZeneca's labs.

The first ELN used in AstraZeneca was by Intellichem, operated in PR&D at Macclesfield, and at Södertälje, Charnwood, Avlon and Mölndal. (Intellichem was later bought out by Symyx.)

ELN came to Mereside in 2008, when a different ELN system was chosen. A big investment by the chemistry and bioscience functions was made in the Cambridgesoft system, which became the standard tool in research labs. This was the final link in the chain, as all information across R&D became available to any staff in the world – if they wanted it.

## Registration, ISAC and HiTS

All new compounds always had to be 'registered' into the company's collection of compounds. Details were originally stored on file cards; I was once fascinated to be shown the original file card for 'Nolvadex', number 46,474.

All data on these was later computerised. Then, on the formation of AstraZeneca, databases from both Astra and Zeneca were harmonised in the International Samples And Compounds database, ISAC, which held chemical and biological data on all compounds. This allowed scientists to discuss and physically exchange samples with colleagues worldwide.

Registration of new compounds into the database had now, of course, to be done by rules agreed globally. Exact (and complex) rules for naming and identifying a chemical compound had to be followed by everyone.

This huge information store made possible the large-scale introduction of High Throughput Screening (HTS, with its Mereside database known as HiTS), in which ten thousand compounds might be tested for activity against a biological target (such as an enzyme) in a single day. The sheer volume of data that can be generated by this (for hundreds of enzymes and hundreds of thousands of compounds) is almost inconceivable, particularly to older biologists who would have been pleased to test thirty compounds in a day. Biology manager Mick Turnbull helped drive its development.

## Test data

Data had previously been the province of the Data Services Section, responsible for both data acquisition and retrieval.

But biological test data was now available to all authorised staff, as the Excel-based systems that underpinned it became standard when personal computers were rolled out across the Park to most staff. Comparing data is the heart of research; and research now had the tools for it.

## C-Lab and E-Lab

The study of compound properties was similarly revolutionised. C-Lab provided every imaginable tool for comparing the 3D shape and 'measured properties' of all AstraZeneca's chemical compounds, and matching them to the enzymes they were intended to interact with. Key parameters could now be checked in moments.

The new world of gene and protein studies was a particular growth area. By 2000, 'bioinformatics' had become the name of the game, as recording, studying and analyzing the data became an industry. The AstraZeneca portal was given the name E-Lab; by 2001 it had become one of the most heavily used intranet sites in the company, with some two million hits a year.

## Global computing and GEL

Alderley Park's computers stretched their tentacles around the world. A particularly large project was GEL, the Global Electronic Library. This was implemented in the early 2000s, to author and produce drug submissions to health authorities around the world. It became used in over 100 countries and had some 10,000 users, connected through 200 back-end servers via around 10,000 personal computers.

## Hardware and the cloud

From the early 2000s, the dominant database for just about everything became Oracle. Whether it was documents or data elements, Oracle could handle it.

The older generation of hardware similarly underwent a sea change. ICL, Burroughs, VAX and PDP were now fading into memory. Sun servers and Blades operating UNIX were the new guys on the block. The chemistry function moved into high-end processing using Linux and Linux clusters. And then came the Cloud …

The Cloud?
PHOTO GEORGE B. HILL

## Captain Kirk?

Imposing computer systems on scientists was quite a challenge in the early days. One unexpected problem arose during computerisation of the main chemical labstock at Mereside, which comprised of many thousands of bottles in dozens of different laboratories. The names of all the chemicals in each lab's file index had to be entered in the new database; but it was found that not all of them even existed!

Over forty years at Mereside, unknown chemists in moments of boredom had tested the system by inventing several outrageous chemicals. When the computer team tried to put them into the system, helium chloride was immediately picked up as suspicious (the element helium forms no chemical compounds). Pentafluoromethane failed when someone tried to draw its (impossible) structure: a carbon can have four but not five fluorines attached to it! Sodium dichloride and others were less obvious but similarly non-existent. One or two others had probably better not even be named here. Yet they all had file cards in different laboratories suggesting a real bottle was shelved there.

Even more curious were genuinely organic-sounding fakes like 'eneaminamine' and, obviously from an old time music fan, 'cumene-2-D-gardene-maud'! After the Apollo missions there were also several bottles of moon dust!

A file card in one lab claimed a bottle of 'Dilithium Crystals'. But anyone searching for the bottle found that it had been signed out on the file card by a certain J.T. Kirk (with a meaningless star-date). Presumably Captain Kirk had borrowed it in order enterprisingly to boldly go on powering the most infamous split infinitive in the galaxy!

All of the 'ringers', including the starship fuel, were eventually removed by non-chemistry staff, a fact that has been claimed as evidence that chemists are the only scientists with a real sense of humour …

**LIFE IN THE PARK**

## Online security

But who can access the data? A big company's data is immensely valuable. An in-house computer network, the 'intranet', was created which could communicate with the outside world but was protected by a series of powerful firewalls. It is hard to see how anyone who penetrated it could possibly know what data to steal; but they could do enormous damage. The decline of AstraZeneca's R&D at Mereside and the arrival of many smaller companies in the BioHub has pushed computing at Mereside into a complicated new era. But it must remain safe. We trust it will.

## Office and presentations

Microsoft 'Office' changed, of course, the world as we know it. The plainest example was in presentations (made by all staff to their peers). These varied from simple talks to overhead projector slides and whiteboards, and finally to smart boards and the inevitable 'Powerpoint'. But was the work *done* any good?

## Training

And, of course, everyone needed *training*. Courses for everything, which is yet another story. Just make sure that you can remember how to do it when the course is over …

## Challenged staff

Quite a few company staff members have faced serious physical challenges at work over the years; and the company has always demonstrated not only a very supportive and inclusive employment policy but also a willingness to help in practical ways.

A familiar figure at Mereside, Ricci Downard, brought a Braille peripheral with him when he arrived from Runcorn as a blind computer programmer in 1992. This gathered dust and so was often hard to read. So when Windows arrived, Zeneca paid for an access suite called WindowEyes, a laptop and a Braille printer – although, as Ricci recalls, 'this made a noise like the Kennedy Space Centre at Cape Canaveral'! He required numerous speaking devices, so the company also bought him a mixing desk!

Ricci had never seen Windows (which was not then very reliable) and often struggled when helpdesk staff advised him to 'click on the menu' and so forth. The company paid for the much better JAWS (Job Access With Speech) system; but when Ricci's programming language Focus was decommissioned he had to tackle JAVA, HMTL, Business Objects and Oracle. He became the only blind Oracle programmer in the British Computer Association of the Blind, though submitting queries proved impracticable for him and he has had to do much research of his own to support his job as an IT project manager. He remains very grateful that he has repeatedly been given 'chances to try out expensive solutions without carrying the financial risk myself'. He is, however, less appreciative of some of AstraZeneca's modernist architecture:

> I should mention Simon, a receptionist @ Radnor. When I biffed my head on one of the diagonal steel buttresses on the front of that building, he straightaway had it covered in expanded polystyrene padding. Now it looks a bit scruffy, but it's much much safer as I zing underneath it in the morning!

# Finance, budget, investment

The finance function at Alderley Park was during ICI's time the Accountancy Department. Its activities included budgets and forecasts, a cashier's section, accounts, invoicing and salaries. More recently it became part of the Operations area. (It was from an early date a heavy user of programming and computing, though this later fell within Information Systems). In either guise, it was little seen by most of the staff, until they wanted to receive or spend something.

## Budgets

But the budget was always a matter of public debate. How do we slice the cake? Who gets what? The questions and answers were keenly contested at all times. Why has our consumables budget been cut? But why has our capital and equipment budget got more money in it than we can spend? Can't we switch funds from one to the other? No! But why not? Because, that's why. If we let you, the company would fall down. Perhaps it would.

## Authorisation

Authorisation of expenses was only small financial change but another headache to all staff. In earlier days employees would be refunded in cash amounts, large or small. (One parasite expert who went to a worm conference in Egypt once made a *very* small claim – 6d!) The tortuous electronic systems which were used in later days were so complex that many staff suspected (perhaps not entirely without reason) that they were actually intended to deter all but essential claims.

## Investment

The huge international finance management behind the company's global operation was certainly beyond the sight and mind of most employees, as it handled the accounts in billions of the company's sales and profits. The City undoubtedly watched it closely; but it was a mystery world, far above most people's knowledge.

### Ping pong

Sometimes the electronic forms that had to be filled in on screen admitted of no logical explanation at all, or could not be filled in by any strictly truthful means whatever. Legend has it that the company's top leaders even fell foul of the latter problem on one occasion. At a prestigious event on another continent, a couple of hundred top business customers were provided with a meal. The company CEO discovered that he had forgotten his company credit card; so another very senior manager had to use his instead.

The latter returned home to Cheshire and was presented on his computer screen with the form to explain his huge bill. Perusing it, he discovered that it could only be completed if he inserted the names of every single guest at the meal! One way was to name them Ping and Pong alternately to the end of the form ....

LIFE
IN THE
PARK

But even from afar it affected Alderley Park immensely. The biggest financial challenge was and is, of course, the cost of achieving every pharmaceutical company's aim (and that of any novel science organisation in Alderley Park) – new discovery. Older statistics suggested that around only one in every 5–10,000 compounds made by chemists might reach the market. These odds represent what is known in the industry as the 'attrition rate'. The figure is now far higher. Perhaps a hundred or two of these will show sufficient promise to need testing further. Five to ten of these may qualify for tests on humans.

But if only one changes thousands of lives …

Once a lead compound has been identified through the process of drug discovery, bringing a new pharmaceutical drug to the market involves whole armies of staff, state-of-the-art facilities – and huge resources. A 2010 study suggested that the cost of marketing a major new drug had by then reached £600 million of upfront costs, plus twice that amount in additional capital costs over time. These sorts of investments are only possible if investors are confident their money will be well handled.

# Procurement and consumables

The procurement department at Alderley Park was throughout its history at the heart of the physical operation. In ICI's time it was known as the Division Purchasing department and was divided into purchasing of chemicals, office supplies, packaging, engineering supplies and others.

It handled everything from the smallest requisition for a pencil or rubber (although preferably not, since handling the requisition itself would cost far more than the goods!) to the largest piece of capital equipment. In principle anything could be ordered from anywhere. I once bought some spatulas that turned out to have come from Liechtenstein. It was a point of honour that *anything* was possible – if there was a business need for it.

## Consumables and the stores

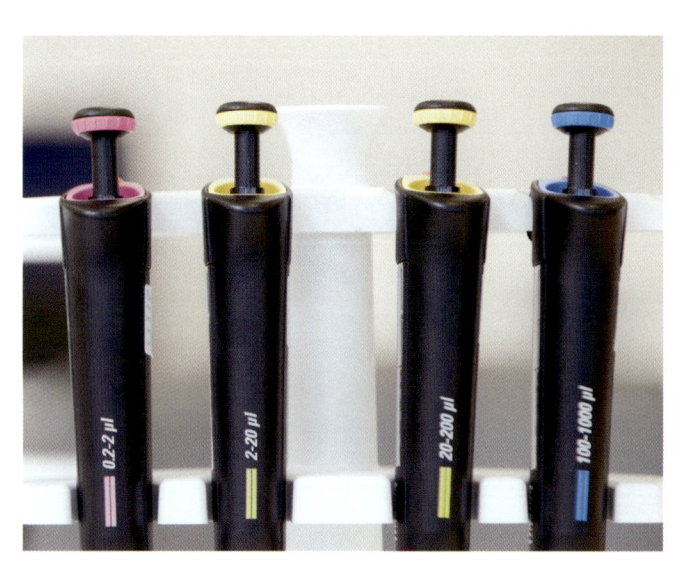

The consumables budget of the Mereside laboratories alone increased to several million pounds over the years. This was a huge sum when everything supplied by the stores had to be ordered on paper requisitions filled out in duplicate.

In more recent times, efficient computer software, sometimes linked with Kardex storage systems that might extend up through four storeys of a building, and later with a purpose-built master store on the site of the old ABU building, allowed stocks of consumables to be kept to a minimum.

In earlier times, an immense stock of consumables was maintained in each set of buildings. At Mereside in particular, the old Research Stores was a cavernous hoard in which it seemed possible to find almost anything useful to man or woman in a practical job.

### From the Post Office!

One sort of consumable that the synthetic chemists needed in large quantities was balloons. They were used in every laboratory for creating atmospheres of hydrogen or inert gas over chemical reactions. Unfortunately, they soon perished in the vapour-laden atmosphere inside the chemists' fumehoods, so stocks were always running out.

Chemical suppliers were a possible source; but sometimes deliveries were delayed. But the Mereside Stores enterprisingly discovered that balloons could be obtained (and more cheaply) from Alderley Edge Post Office, to which one of the stores staff could be sent during the lunch break. Once or twice, the Post Office ran out of standard balloons. The chemists were delighted by the result: soon their fumehoods were decorated with long wiggly-worm-shaped balloons, and ones covered with pictures of birds.

### Just in time

For less commonly used items, the so-called Just-In-Time system became standard across industry; and however much the staff puzzled, over why a van had to arrive just to deliver a trivial item to one of them, the economics did work out better than keeping the huge consumable stocks maintained heretofore. Or so they were confidently told.

### The Stores counter

On occasion, nevertheless, when something was urgently needed, a scientist would walk down themselves to see if they had managed to guess correctly the opening times at the Stores counter. If so, they were always welcomed – except at tea break.

### Young and unwise

The trip to the Stores counter was also used when any rather bumptious newly arrived young member of staff needed taking down a peg or two. An older colleague in the labs would declare that they required a certain piece of support equipment, and would send the green recruit down to the counter with a suitable chit in his or her hand.

'What can I do for you, son?'

'I was told to come down here for a long stand, please.'

'A long stand you shall have, then.' The stores man was in on the joke and would leave the youngster to deduce at their weary leisure why he had walked away from the counter with a smile and not come back. (An order for some 'benzene rings' was another favourite.)

# 12 Route to the patient

## Scientists and sciences

A rainbow of sciences has filled Mereside through its history. Some of the finest minds have eschewed the managerial staircase and have chosen instead to follow the 'science ladder' which led to influential positions in the company's research.

The company's first medicinal scientists would have struggled to understand the later huge science departments as much as a 1930s pilot would have struggled with a modern Airbus. Nor can they be more than sketched here. Nevertheless, the traditional sciences still survive; although chemistry and biology (or rather, now, bioscience) now have many forms.

### Chemistry

Chemistry involves the hunt for 'small molecule' drugs. This in turn involves 'synthetic' chemists who make the drugs; 'medicinal' chemists, who decide what to make and how to find out if it has led to a useful, biologically active product; and chemists concerned with physical chemistry, spectroscopy, purification, 'computational' chemistry and other forms. (At least 700 permanent staff chemists worked at Mereside between 1957 and 2014.)

Chemist at work 1957
*PHARMACEUTICAL RESEARCH IN ICI 1936–1957 (1957)*

Chemist at work, 2006
PHOTO ASTRAZENECA

Practical chemistry begins on research scale. Synthesis of a new compound is followed by its purification, in earlier times by crystallization but now always by Liquid Chromatography. The latter was done manually for many years (or even at one time by clockwork!) until the first electrical LC pump at Mereside was introduced by Martin Joseph in Gareth Davies' team in 1977. LC then became a central high-tech operation in every lab, latterly often being given to specialist groups. The synthesis is then scaled up repeatedly before leaving the Park to go to the Macclesfield Works and its process and manufacturing functions.

Since chemicals are all chemicals, most chemists can move fairly rapidly from one sort of drug or project to another. So, many chemists are easily redeployed (which is good) but can be replaced (which is bad!). Many Alderley Park chemists moved to new disease areas every few years and were encouraged and keen to do so, to broaden their experience.

The job changed massively in more recent times. Medicinal chemists and lead bioscientists were once the practical experts of their teams, leading by example. They are now more the wielders, at extraordinary speed, of complex 3D graphics and charts and 'Excel' spreadsheets of monstrous size. (My supervisor Nicola Heron was the fastest operator I ever witnessed.) The computer gaming world would have a big shock if their talents were diverted there!

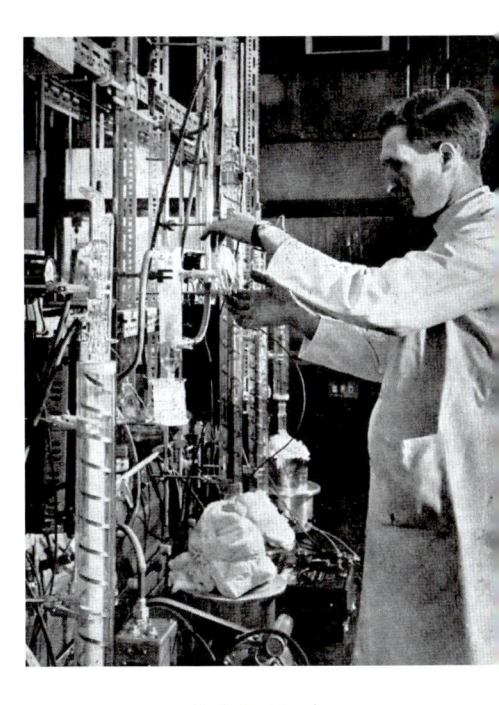

Early fractional distillation
BARRY SMITH

## Bioscience

The Park's bioscientists, who often specialised in an area for life, inhabited a quite different world. On one occasion, Mereside Library put up a display listing the different kinds of

A bioscientist at work
COPYRIGHT ASTRAZENECA

bioscience jobs at Mereside. It totalled fifty-three! This means that bioscientists are less easily moved, and have more irreplaceable but less transferable skills. There are *in vitro* biologists, working, as the Latin phrase suggests, with enzymes and cells literally 'in glass'; and *in vivo* and *ex vivo* ones working with whole organisms or materials from life. The work varies greatly with the disease area; bioscientists working with disease bacteria inhabit a different world from ones working with (very hazardous) human cancer tissue, or on non-life-threatening conditions.

Like chemists with their compound collection, bioscientists also have material archives. At Alderley Park one example of this was the Global Cell Bank, which ensured cell 'lines' used in drug testing and research around the world would be identical and give similar results for different scientists, who received them through the Cell Line Tracker database.

## Interactions

Interactions between the different scientists were always interesting – and sometimes prejudiced! From time to time there were charming skirmishes of disparagement between the bioscience and chemistry departments, each of which considered the other's members as well-meaning but faintly intellectually misguided.

Bioscientists think chemists are mad, because chemists read (or nowadays, download) ancient manuscripts. All bioscientists generally ignore any science publication over five years old. But chemistry does not change through time, so chemists may look up and follow the methods of others who lived decades or even a century before. The oldest scientific paper I ever had to find was dated 1866. (It is interesting to note that the first Lord Stanley was a friend of the brilliant chemist Sir Humphry Davy.)

Chemists can have equal disdain. Once, two unwary junior biologists from the floor below came up to our chemistry lab to ask for advice.

## Ivory towers seen from industry

Pharmaceutical researchers are, of course, united in one thing: their fine dismissal of the naked scorn shown to them by some academics. One of my colleagues, coming to ICI from research at Cambridge University, was mocked as he left. 'You're going to work on FLAT molecules?' But academia, proud of its castellated molecular monsters, loves to demonstrate amazing chemistry using well-behaved compounds mostly of little use as medicines. Chemists at Mereside had to aim for compounds the human body might recognise (particularly ones of the 'heterocyclic' variety), ones that ignored polite rules – and this convinced them that they alone did *real* synthesis for the pharmaceutical world.

'Er – could you help us? We need to find out how to make dilute sodium hydroxide solution.'

We chemists smiled benevolently.

'Oh, that's really easy. You make up some sodium hydroxide solution. And then …' we paused for effect '… you dilute it.'

## Sections

There were many reorganisations over the years.

Both sciences were organised in sections, under a section manager. In 1961 there were just four in chemistry, with Bert Crowther over heart and central nervous system work, plus Roy Hull on speculative chemistry, Justus K. Landquist on antibacterials, and Glynne Jones on animal health. (At university in 1976, I had my first lectures in medicinal chemistry from the latter two, who were far more interesting than most lecturers, if a little prone to anecdotes!)

After some staff were transferred to the Park from The Frythe at Welwyn, a Natural Products section under Brian Turner entered the scene, as did peptides (mentioned later). Natural products chemistry continued for many years, without leading to any major new drug, despite interesting results from extracting all sorts of plant material. (One incidental result was a nice colony of the pretty water plant bogbean, which flowered in Radnor Mere for several years.)

By 1971 there were ten chemistry sections. Biology had six sections encompassing a range of biological areas, under Arthurs Walpole and Broome, Des Fitzgerald, Terry Hennessey, Alan (G. A.) Snow, and John Lowe – the latter covering Biochemistry, which was made a distinct department under Trevor Franklin the following year.

Trevor Franklin, first head of Biochemistry
*SCAN*, 8 September 1972

## Two departments, then four TAs

In 1980, research was split into departments I and II (under Dick Clarkson and Des Fitzgerald) and the number of projects was cut by around a third. After a review from Boston Consultants, there was further improved focus, though more was still needed and many chemists had more than one project they were pursuing.

In 1993 on the formation of Zeneca, the whole Research function was restructured into four 'Therapeutic Areas': Cardiovascular under Mike Turnbull, Cancer under Dave Julian, Vascular Inflammatory & Musculo Skeletal under Barry Furr, and Infection under Graham Boulnois (all under David U'Prichard).

## Three RAs then 'i'

Following the merger with Astra, Barry (B.J.A.) Furr, who had led all Zeneca's late-stage discovery work since 1997, was appointed Chief Scientist. Research was divided into three 'Research Areas' (RAs) which all had global identities. These were Cancer and Infection (CIRA) – of which the Infection part had moved to the new company site at Boston and also at Bangalore in India; Respiratory and Inflammation (RIRA) – of which Respiratory was at Wilmington and at Charnwood in Leicestershire; and Cardiovascular and GastroIntestinal (CVGI) – of which the GI part was in Sweden. (RIRA later closed altogether).

Later yet, another bit of terminology arrived as the fashionable letter 'i' crept into AstraZeneca as everywhere else, and the puzzled staff had to learn what an 'iMed' was … to go with their iPod, iPhone and iPad …

As an indication of the new company's research world, a list of key research techniques was printed in the 1999 AstraZeneca Annual Report:

| Biology | Chemistry | Informatics |
|---|---|---|
| comparative genomics | structural chemistry | bioinformatics |
| pharmacogenomics | compound management | cheminformatics |
| toxicogenomics | compound collection growth | advanced statistical science |
| gene expression tools | targeted chemical libraries | complex systems analysis – modelling |
| human genetics | high throughput screening | intranet delivery of discovery information |
| | | combinatorial chemistry |

## Lab life

Another steady change was to lab life, as kettles and mugs and eventually desks were moved out of labs, which became pure working areas. By around 2000, all lab staff were moving into nice clean offices, with computers (and email!)

This created a clash between the habit of strict cleanliness (which extended later even to 'clean' and dirty' cupboard door handles), and the urgency of the science. Scientists were strictly required to record all data as experiments were done, for both legal and safety reasons; but paper notebooks were still needed for legal reasons. Yet when carried back and forth between lab and office the books were a contamination hazard.

Eventually, when ELN e-notebooks finally arrived, so did 'CrystalView' workstations in labs, which 'mirrored' the office computers of staff, allowing input to switch instantly between in or outside the lab. This made life much easier for all, although it was a huge cultural change. But non-lab staff visiting laboratory areas were less impressed. As IS expert John Jones put it:

Desks were still in the labs in 1986

Research and development
COPYRIGHT ASTRAZENECA

*'Before* [ELNs and 'Crystalview' arrived] *it was very easy to talk to chemists, as they spent 80% of their time in the office. Afterward, they spont 80% of their time* [out of easy reach of visitors] *in the labs. An absolutely astonishing change.'*

It was, of course, nearly a reversal back to the early days at Mereside when many staff inhabited only labs – although the kettles and tea-breaks have not yet returned into them!

## Global science

It was in the science departments, perhaps more than any other, where the global nature of the company was most obvious. We realised this one day when, as an Englishman working for Scot Nicola, I was in the lab alongside Australian Chris and Frenchman Cyril. Our friend Santosh, on secondment from our lab in Bangalore, wandered into our area and started looking in a cupboard for something to borrow.

Australian Chris protested. 'Hey! What are you after, you cowboy?'

I looked up, and spoke without much thought on a point of order. 'He's not a cowboy, he's an Indian.'

No work was done for the next ten minutes, until the paroxysms subsided; and Santosh himself was still chuckling when I met him again a year later.

## LIFE IN THE PARK

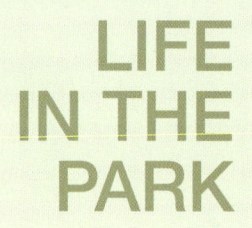

**Braveheart**

Lab chemists co-operated because everyone knew theirs was a hazardous profession – in more ways than one. Two Scots chemists were nervously synthesising a chemical very sensitive to shock. An oblivious English colleague, ambling past them with a heavy lab notebook, chanced to shut it with a loud bang, just behind them.

William Wallace would have been proud of the verbal result.

# Analysis, spectra and drug properties

When a drug is made, it must be analysed and its structure proven! How?

## Physical methods

The Physical Methods function had more than one home. It was originally No. 7 section of the chemistry department, but was felt later not to belong in chemistry and by 1971 was a scientific services group within Research Administration. Used at first almost solely to 'characterize' (identify) chemical products, it later moved back into chemistry when reaction planning and drug design became important in research, as described later. It used several quite different types of technology.

## Microanalysis

Elemental analysis is the traditional way of analyzing a chemical. It was important in Mereside's early years, though much less used now.

A sample was first burnt; and from the gases and ashes its contents were determined, as the percentage of each element. At one time, all further work paused until the analysis could be 'got right'. Only then could the drug be tested. As a young chemist, I used to dread the appearance of an analyst who had walked up two flights of stairs to present me with an analysis that was 'out' – and, adding insult to injury, to demand *why*?

## Infrared

In earlier days at Mereside, infrared spectroscopy (IR) was important. Samples were ground up with 'Nujol' mineral oil and the resultant film placed between transparent 'IR plates' – discs (or cells) made of pure salt – which were transparent to the IR radiation, and long lasting, except when one chemist dropped his in the sink!

## Mass Spectrometry and Chromatography

Also as a young chemist, I was sometimes presented with a miniature roll of wallpaper! This was a 'mass spectrum' of my compound. Not surprisingly, fast forward thirty years and the capability of the modern mass spectrometrist opens up a whole new story.

It was always very useful to combine 'Mass Spec' or MS with a chromatography technique. In the earlier years at Mereside, this was either Thin Layer Chromatography (TLC) or Gas Chromatography (GC). TLC was the mainstay of generations of chemists, but both of these have become far less used since the arrival of a science that has changed the world: MS with Liquid Chromatography. The combination of the two, LC-MS, is now known everywhere because of its ability to detect fantastically small traces of materials in the body fluids of Olympic athletes and many others. All modern chemistry and biochemistry depends on it.

The history of MS itself in ICI began in 1947 when physicist John H. Beynon first arrived at Blackley: 'I … arrived for my first morning, was shown to my desk, and on my desk there was a note that said, 'Remit for Mr. Beynon, build a mass spectrometer', which came as a huge shock to me, because I didn't really know what a mass spectrometer was.' Beynon (later made FRS), though he never worked in the Park, helped put MS on the map in the UK as well as in ICI and at Alderley. Later the technology grew beyond recognition from such small beginnings.

Larger scale chromatography facilities, immensely important in 'scaling up' new product production, are mentioned elsewhere.

## A case of contamination

In chemistry labs, huge precautions are necessary against contamination, because intensely potent agents may be in use. One example was the well-known antiseptic **triclosan**, a frequent object of study at Mereside over the years.

In 1998, Zeneca learnt (before it was published) the 'mechanism' of how triclosan worked, which was by blocking the enzyme EACPR. It proved *astoundingly* potent (for the experts, down to picomolar level; for the non-scientist, think of a teaspoonful in a swimming-pool). The company compound collection was checked for other similar agents, and Neil Hales asked me to examine a sample of one compound that appeared to act like triclosan. But it seemed an unlikely candidate; and we came to suspect the material was *contaminated* with an infinitesimal amount of triclosan itself – perhaps having been made decades before in a lab where triclosan was also being handled. Yet the finest spectroscopy could not find it. Eventually, Neil asked me to run a TLC (a thin layer chromatographic 'plate'). This, as expected, showed no visible trace of triclosan. But ingeniously, Neil then had the plate placed face down on top of a bacterial culture plate – in a brilliant combination of chemistry and biology – with the result that a patch of the bacteria stopped growing – *just* where we suspected there was a trace of triclosan on the plate so small that no chemical analyst could possibly detect it. Our case was proved!

AstraZeneca's NMR lab in Block 33
PHOTO ASTRAZENECA

## NMR

Rapidly increasing in importance throughout Mereside's history was NMR, perhaps the modern chemist's finest tool. It was also at Blackley that ICI placed one of the first three commercial NMR machines in the UK, a Varian 40MHz that arrived there in 1957. By the mid 60s as described earlier the isomers of tamoxifen had been identified by research at Alderley; this used a machine at Macclesfield Works, but later entire NMR laboratories operated at Mereside.

An immensely powerful magnet is cooled (using liquid nitrogen and, later, liquid helium at near Absolute Zero). The atoms of the compound are forced to show whether the magnetic field affects them. (The medical equivalent today is MRI). This became a huge industry, and the magnets in their cold tanks also became huge. They are able to stop heart pacemakers and access to labs is thus restricted.

The spectra they produce are, to the scientist, as informative as a painting to an art critic, or a fingerprint to a forensic expert. Without NMR, much of modern chemistry would not be possible. By 2010, 400MHz machines were standard in chemistry, and 1000MHz in protein NMR. Simple NMR studies the element hydrogen in molecules (medical MRI maps it as water molecules in the body), but bigger machines later covered carbon, fluorine, phosphorus, nitrogen and others.

## 'Physprops'

The physical properties of the drug must be found out, particularly its solubility in water and its crystallinity (using X-ray). A compound that looks promising will be studied intently, particularly to make sure it is stable and will not break down chemically or precipitate physically from a solution.

## LIFE IN THE PARK

### Private lives

The spectroscopy and analysis staff controlled an amazing world. Their abilities seemed uncanny; like Sherlock Holmes, they could build certainty from the tiniest clue.

They were the servants of many, and were always in demand. Often, in business, when someone is doing work for you, it is easy to forget the people. Sometimes, therefore, I asked what people's hobbies were. The spectroscopy staff were particularly interesting; in his spare time one helped to run steam trains: another, a theatre.

But I was taken aback when I tried Kath, the little grey-haired lady who did microanalyses for me. I expected knitting, or cooking as her hobby. But she pulled out a large photograph of a mountain track. She was clearly visible – several feet above it. Wearing a hard smile, she was strapped beside a grim-faced driver in the navigator's seat of a rally car, one with all four wheels high in the air, and flying toward the camera.

Writing out analysis forms was never quite the same again.

## DMPK

A much bigger question is what a drug will do inside the body. This involves two questions: what does the body do to a drug? And what does the drug do in the body?

DMPK – the study of 'Drug Metabolism' and 'PharmacoKinetics' – is at the heart of making a new medicine successful. The body may 'metabolize' the drug – break it down chemically. Or the drug may become useless, if the body holds onto it physically and prevents the 'free' drug from moving around the body.

Determining a medicine's 'duration of action' is perhaps the most important task of all, once it is proved to work. A drug that acts for so short a period that it would have to be dosed far too often is useless; one that stays in the body so long that it could act in other ways and cause harm, is dangerous. But this is very difficult to predict. Animal tests (in rats or mice) are very important and give a good idea of whether a drug might be safe for humans or not. Later, tests in human volunteers must be done to determine the 'duration of action' in man.

Comparing results
with different
potential drugs
COPYRIGHT
ASTRAZENECA

## Comparing result with different potential drugs

Several critical measurements must be made. How fast is the drug absorbed? Some of an organic compound will bind to protein in the bloodstream (and so become medically useless); how much protein binding is there? How fast does the body 'clear' (get rid of) the drug? The latter rate is measured as the drug's 'half-life'. When in the body, how does the drug distribute itself among the body's organs? There is little point treating a patient for a brain condition with a medicine that never reaches the brain.

These are questions affecting both the choice of the right drug to develop and the safety of it as a medicine. Block 24 was designed as a purpose-built Drug Kinetics facility. This is another topic far too deep to describe here, but it became one of Mereside's most amazing abilities. DMPK was always among the most closely watched areas, as the results of its work could weigh the balance down finally for, or against, a possible drug. (In the Park, it was part of the drug kinetics and safety of medicines functions until the merger, but moved more into research in AstraZeneca's time.)

One of the biggest scientific changes in Mereside's long history was the knowledge brought by this area of study, whose current abilities would have seemed positively uncanny to the company's first scientists.

## Radiolabelling

Closely linked to DMPK is the use of radiolabelled drugs (weakly radioactive versions for research). After early efforts at Blackley (working in Alf Spinks's double decker bus 'laboratory'!), John Burns started a radiochemistry lab in 1962 at Mereside, then in 1964 led work to create $^{14}C$ radiolabelled 'Inderal'. $^{3}H$-labelled 'Nolvadex' followed.

Early analysis used TLC. Better labs were developed and hundreds of labelled compounds had been created by 1980.

A new radiosynthesis lab was opened in 1989 and saw labelling of 'Faslodex' with four different elements. As drug metabolism studies became increasingly complicated, labelling became routine during development work, and allowed quantitative measurement of changes at drug receptor sites, amongst other things. In the late 1990s, mass spectroscopy-based methods using LC-MS revolutionised analysis. John Harding published a history of the work in 2007.

John Burns
*SCAN*, December 1981

# Samples, the Collection and Weighing

A sample once made and identified must be 'registered' into the company system, as described earlier, and placed in the Company Compound Collection before it can be studied. This is essential, because otherwise there is no proper way of storing the rapidly growing amount of data associated with each new compound.

All compounds made by the chemists entered a battery of tests done by various biologists in different therapeutic areas. This 'primary screening' usually produced nothing; but just once in a while the scientific prospector would spot a gleam of gold at the bottom of his or her pan … and look for more of interest. Projects costing several millions of pounds a year and potentially producing world class medicines have resulted from a few very interesting 'leads' of this sort.

## The Collection

The actual physical Collection was (and still is, for even if the whole company vanished its collection would still be immensely valuable) the combined store of all of the chemical samples ever made by the company, at Mereside and other sites, that has not been already used up by testing. For example, there are undoubtedly still samples in it that were made during WWII. Today around two million compounds are registered in AstraZeneca's collection, half of them unique to AZ, though physical samples only exist for a fraction. In November 2015, a huge swap of over 200,000 each of these was agreed between AZ and a competitor, Sanofi. Nowadays it is the province of robots, of which more later. But 'twas not always thus.

## Weighing and sample preparation

Samples for the biologists were weighed on very accurate balances in a strictly controlled environment, in a repetitive but very important job. As little as two milligrams could be enough to do a whole series of tests once dissolved in the commonest test solvent, dimethyl sulfoxide (DMSO).

## LIFE IN THE PARK

### Orange boxes!

In the early years at Mereside, sample bottles were stored in much more primitive containers. Perhaps the most famous were the orange crates that were stored in an underground repository near the Mere that was so damp that it frequently flooded; some chemists declined to put their precious compounds in it at all!

Later, the bottle numbers became huge, with hundreds of thousands to keep track of. In about 1993, a bar-coding system was introduced by Judith Provis. She records that she was inspired to the idea one day while shopping in Tesco!

Samples
COPYRIGHT ASTRAZENECA

In recent years, a project was therefore carried out to 'solubilize' a part (not all) of every sample in the Collection. This involved dissolving small amounts in DMSO (of all that were soluble) and freezing the result. Samples could then be dispensed very rapidly, as required by state-of-the-art testing using High Throughput Screening – already in solution!

# Patents and Intellectual Property

Patents! In the Park, the word is pronounced 'patt-ent', not the English 'pay-tent'. But everyone – in every business in the Park, not just AstraZeneca's – knows how important the word is.

## IP (Intellectual Property)

Global Intellectual Property (until 1988 it was called Industrial Property) and its associated legal function is in AstraZeneca the department that protects the innovation that is AZ's lifeblood. The patent department has been essential since the first days of Alderley Park, in protecting *innovation* deriving from the discovery and development of the company's medicines. (When the Division was first founded in 1936, things were a little different. Only process patents were then allowed in the UK, not ones for compounds). Its main purpose is to protect AstraZeneca's 'IP' so that the business can secure an appropriate return on its investment. This allows it, critically, to be able to invest in the next generation of innovative drugs.

The length of duration of a patent is critical. The 'innovators' of a new drug, by global agreement, are granted a monopoly of a certain number of years to prevent others from exploiting their invention, allowing them to recoup (and reinvest) their huge investment in research. Once the monopoly is over the drug becomes 'generic' and can be made by anyone (and sold at a much lower price). The fact that drugs have two names – the company's 'brand' name and the generic name (International Nonpropietory Name, INN), assist with this crossover.

The brand name or 'Trademark' is itself another form of intellectual property which is applied for by the company's Trademark department. This remains the property of the company and may not be used by anyone else without permission. Nonetheless, generic companies are free to use the INN on their packaging – so that prescibers are clear on what they are prescribing!

## Global

Protection must be worldwide, for patents have to be applied for (and hopefully granted) in every country where a drug is to be sold. They must be written with extraordinary care; and one initial patent application needs to deal with the foibles and complexities of the law, potentially in more than 60 countries.

Yet a patent does not give you the right to sell your product! Sometimes a deal needs to be arranged with another company in one or more 'territories' to allow you to exploit your invention.

Patents must also be *understood*. So an army of translators around the world turn AstraZeneca's patents into the local language when needed. Local patent firms advise the company's businesses in every country.

**Oops!**

Translations can go wrong! One common patent expression, 'pharmaceutically acceptable diluent or carrier' was once transmogrified into 'pharmaceutically acceptable dishonest courier'...

Patenting the actual chemical compound or 'drug substance' usually gives the strongest protection; but innovative 'formulations' used to dose it (as with **'Diprivan'**, **'Zoladex'** and **'Faslodex'**) and occasionally the device used to deliver it (such as the 'Zoladex' 'protected needle') may be protected by patents as well. The brand name and any related designs or images are also valuable commodities and must be protected by trademarks (which may need to be different in different countries).

Other key IP tasks include working out if a patent's life can be extended and, if appropriate, applying for these extensions. In the pharmaceutical (and agrochemical) industry, patent term extensions of up to five years, in certain countries, are permitted if the time taken to obtain regulatory approval takes longer than a certain period. This further assists with recouping costs and investing in the next generation if the drug was very difficult to develop and bring to market. The most striking recent example of lengthy development was **'Faslodex'**, discovered in 1983 but was not launched until 2002, just before its 'substance patent' expired. All of these issues require a team of dedicated professionals; many of those within AstraZeneca have worked in the Park where the group will remain for the foreseeable future. As patent law has become more complex over the company's lifetime, the need has become more exacting. It has, as the earlier record of **'Nolvadex'** also showed, always been a crucial – and occasionally a protracted – story.

## Detail and records

Patent scientists have to like detail. In Alderley Park, the early patent scientists were all chemists, as are nearly all the current ones, many of whom originate from AZ's own labs.

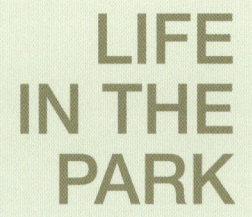

A good brand name must also first be chosen! This can be a very important point. When ICI and Merck were selling the same heart drug, as 'Zestril' and Prinivil™ respectively, Merck admitted that ICI's version may have sold better, because heart patients liked the idea of gaining a little zest!

Keeping of records is an interesting point. This is not indefinitely required by law, except for data that is under 'legal hold' due to litigation (usually in the US) – which may be substantial; the 'Nolvadex' patent cases required photocopies literally by the lorry load. When AstraZeneca's site at Södertälje in Sweden closed (a site half the size of Alderley Park), a total of eighteen linear *kilometres* of paper had to be archived. But this is costly. Nevertheless, patient safety remains an issue in perpetuity: AstraZeneca still gets occasional questions about **'Paludrine'** (invented in 1944).

# Drug Development and Safety

The huge world of the drug development process at Alderley Park is not something this book can do more than point to.

## SOM

The Safety Assessment or Safety Of Medicines department at Mereside, latterly based in Block 23, was from a patient safety aspect perhaps the most important department of all. Its immense programmes, supporting safety studies to determine any possible toxicology problems or side-effects with new drugs, are not even within sight of the scope of a book like this.

The department had originally been part of the Biology department at Mereside until it gained a separate identity. By 1971 it was a major business function, divided into toxicology, pharmacology, toxicological pharmacology (!) and teratology. Needless to say, this was only the start of an expansion that saw it grow out of all recognition.

Safety has a profound cost; and that price is well-known, for the animal testing done in the Park has always been the greatest cost of all. But huge numbers of patients have benefited from it and the work of those committed to doing it. Alderley Park would have produced no medicines to benefit society at all, without them and their work.

## Timeline

The three key phases of drug development are pre-clinical research (involving microorganisms and then animals); clinical trials (on humans); and the obtaining of regulatory approval to market the drug.

## The pre-clinical phase

Alderley Park had the ability to do all of these three, as well as further work after drug launch. But they could only begin once the chemical compound (that would hopefully become the new drug) had been identified. A novel small molecule is what is called a New Chemical Entity (NCE) and is probably a chemical that has never existed in the history of the world before. (A newer term which includes biological drugs is New Medical Entity or NME.)

Drug development timeline

No chemical compound is completely safe. The pre-clinical safety evaluation itself has three elements.

Firstly, work must be done to try and uncover any potential *human safety issues*. These may be identified at any time during the drug's development, and can stop it absolutely when they are.

## Formulation

Secondly, the drug must be made to work to its full potential. This involves *optimising its efficacy* in man by finding for the drug the best salt or best 'formulation' (generally, the best liquid mixture for transferring it into the bloodstream). The formulation department was mostly based at Macclesfield Works but was important in the Park in the early stages of some drug development.

# LIFE IN THE PARK

Not all humans always took the safety issues as seriously as they should. In the very early days, this included one scientist working on artificial sweeteners who used to taste all the compounds he made. Unfortunately, he did not try the right one! Aspartame, 'discovered' by Searle in 1965, had previously been made at Mereside, but no-one there thought to taste it; and it made billions – but not for us!

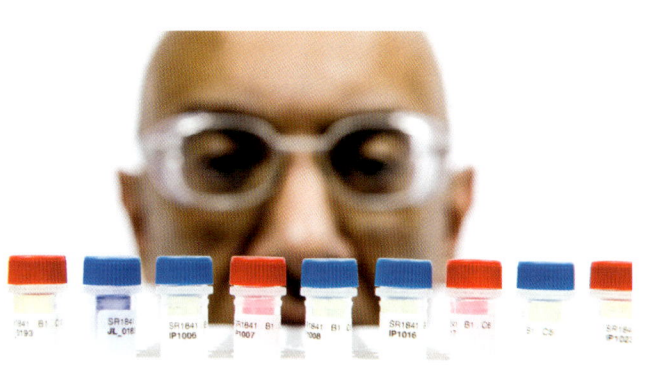

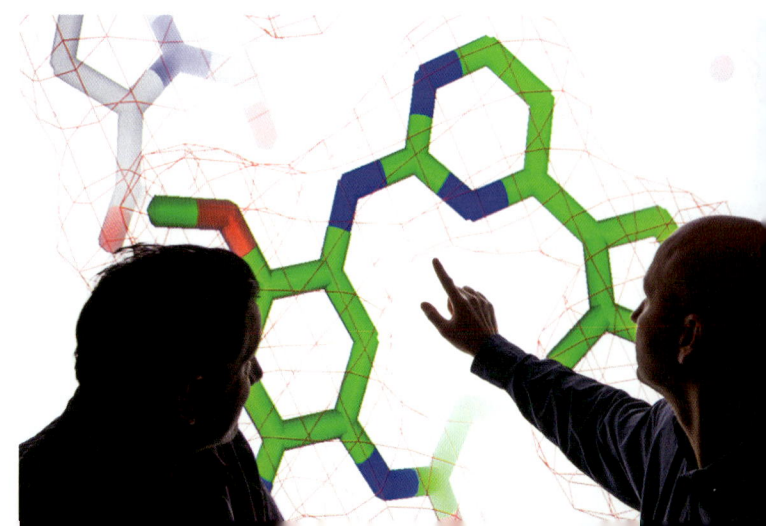

Pharmacy is the study of how to present active principles to patients. Formulation, with some Alderley Park drugs, particularly **'Diprivan'**, **'Zoladex'** and **'Faslodex'**, has been the hugely expensive factor that has turned a difficult chemical into a historic medicine. A dedicated formulation department is at the heart of every pharmaceutical company.

## Process development and packaging

Thirdly, work is essential to ensure the *quality* of the medicine. This will include purification, scale-up of its synthesis, process development, and stability studies – the last being critical for antibiotics and similar fragile molecules.

Packaging is also critical both to a drug's safe storage and its appeal. This was never an Alderley Park function but began as a distinct operation around 1950 in a lab at Harefield Hall, Wilmslow, before moving to Macclesfield. This book does not describe Macclesfield's work.

## Patient safety

Throughout, the drug's safety is paramount. To ensure this, it must be absolutely understood. How will it behave in the body? *'If we put this drug into a human being, how do we dose it, how much of it do we dose, and how long will it work for?'* Could it be dosed orally as tablets, or by injection, or by infusion?

And what will the *patients* think of that?

# Clinical Trials and Regulatory Affairs

Clinical trials are the immense test track where the rubber meets the road!

The Medical function of the business began with the appointment of a single physician, R. P. Liston, in October 1942. The first offices were in Oxford Street, Manchester, and then at The Ridge, in Alderley Edge. When Dr Liston was asked to help organise clinical trials, it rapidly became evident that many other aspects of medical services would have to be taken up, including maintaining close contact and goodwill with bodies such as the Medical Research Council and the Department of Health. In 1947, a consultant anaesthetist, L. B. Wevill, took charge. At least one tropical disease expert was also then always included in the department. New offices at Harefield Hall, occupied in April 1949, were still the headquarters of Medical Department when Mereside was completed. By 1971 there were medical products planning and medical services departments as well as ones devoted to different parts of the global market, as the business developed to cover all needs.

## Clinical Trials

The medical arm of the business is called in as soon as the first clinical trial on a new drug is contemplated.

Clinical trials include tests done on both healthy and sick volunteers, managed by the Clinical Research, later Clinical Pharmacology, function. Under the latter's manager John Nicholls, the Clinical Pharmacology Unit (CPU) was built for that purpose in 1977, as virtually the only in-house pharmacology unit in the country, as 'the philosophy of ICI was that volunteer studies should be done in-house if at all possible'. Healthy staff members were

recruited as volunteers to undergo early trials. These were designed to determine the dosage levels at which bigger clinical programmes could be done.

A growing issue in the pharmaceutical industry is transparency of information gained through clinical trials, to benefit, among others, both patients and science researchers. One fact revealed by AstraZeneca's commitment to openness was that by 31 December 2013, the company had carried out 2,241 registered clinical trials, and had posted on public websites reports or summaries of more than half of them.

The company has always had a reputation for strict ethical control of its trials. This had many aspects. A patient must have the trial process explained to them in their own language. And great care was taken to make sure that patients fulfilled the criteria for the trial. Thus, doctors' notes were carefully examined by clinical trials scientists, to make sure that the patient was, for example, suffering from mild to moderate (rather than severe) heart failure – if that was the trial's remit. No pressure to take an unsuitable patient (and thus benefit his or her sponsor) was countenanced by the company, both for ethical reasons and to avoid confounding the results and hindering the drug reaching the market.

The clinical trial programme was in recent times handled by Global Clinical Development department, based at Parklands. This was responsible for planning clinical trials, for running them in co-operation with hospitals and doctors in many countries, and for statistical analysis and publishing of results. The presentation of these was critical, for the regulatory authorities in every potential customer country would take a close interest both in how the trial was carried out and what it had revealed.

Trials usually comprise three or four steps:

### Phase 1 trials

These, undertaken in volunteers or patients, start with very small single doses in individuals. The first will be very few and in closely controlled conditions. Phase 1 trials are designed mainly to see how the body tolerates the drug. They have the added ability to give early evidence of therapeutic activity in some cases.

### Phase 2 trials

Traditionally these are designed to obtain the first evidence of safety and efficacy in patients. Importantly, they aim to find the range of doses that should be tested in the large Phase 3 studies to come: that is, those doses that are effective and well tolerated over a period of some weeks.

### Phase 3 trials

Phase 3 then takes these findings and applies them in controlled studies against the appropriate standard treatment over longer time periods. These can be large, or even gigantic, pivotal tests in very big numbers of patients. This is designed to reveal how the drug will work in general use.

Trials can be done anywhere in the world, though care must be taken to make sure that they are fully and carefully recorded wherever that may be. The choice of where to do them can be complicated. The EU's Clinical Trials Directive of 2001 required approval to be obtained in each individual European state, whereas only one approval is needed to allow trials involving any of the population of the whole of the United States or other big nations.

## International Registration – what the law requires

The above is required to meet the regulatory requirements of drug licensing authorities. These aim to reveal all ways in which a new drug might conceivably cause unexpected problems in a human being, and also to confirm the drug has the desired effect. They do not merely test whether the compound might damage single cells in the body, but also how it might affect major organs, particularly the heart and lungs, brain, kidney, liver, digestive system and skin. This is why many of the earlier (pre-clinical) tests can still only be done by using experimental animals, since only an intact organism can reveal the complex interplay of effects that could hurt a patient if not controlled; and why the later tests can only be done on human beings.

One-page trial proposals by Arthur Walpole in 1944 – a different world!
MIKE DUKES

When all is ready, the Marketing Authorisation Application (in Europe) or the New Drug Application (in the US), is submitted. An application must be made in every country where the drug is to be sold.

One problem in earlier decades was that different countries strictly controlled sales. Initially, ICI did all its trials in the UK. But later in some, such as France and Japan, a new drug could not be sold unless much of the research and clinical trials had been carried out in that country. This greatly increased the global complexity and cost of the trial process.

## Approval

If a compound after anything up to ten years of testing looks good, then it can be submitted for marketing approval. When a drug gained 'approval for launch', it was an occasion in the Park to put the flag up!

## Phase 4 trials and early disease

Many medicines might also work in *early* disease – rather than serious 'advanced' disease – but discovering this can take a long time. And it may need studying in many patients.

This leads to a difficult choice. A major decision for all innovative R&D companies (ones launching brand new compounds) is whether to launch their drug rapidly as treatment only for severe disease – in a relatively few patients in whom the drug's effect will be obvious – or to do long and expensive studies to prove it has wider value. But these will delay the drug's launch; and the clock of the drug's patent life is rapidly ticking! (A patent generally lasts for 20 years from the formal filing date of the patent application – which for substance patents is about a year after the time of the drug's invention – *not* from when it begins to be sold.) If the company gets it wrong, they may spend a fortune on research and then other companies will make the profit from it – ones that did not invest in the research at all.

Doctors can also try drugs for purposes for which they have not been approved – though on their own responsibility.

## Tamoxifen

The classic example of difficult study choices is tamoxifen.

Initially, in the early 1970s, **'Nolvadex'** was developed for treatment of advanced breast cancer in women. A few hundred patients with it, who were expected to live for less than one year, should show an effect within 6 months – and did.

This was promising, so 'Nolvadex' was approved in 1973 in UK, and sales began world-wide. But more studies then began.

Patients likely to live for 5 years were then treated; but how long would their trial need to run for? During such a long duration many coincidental things can happen – anything from heart attacks to being run over by a bus! So many more patients are needed to make the statistics reliable – a few thousand, say.

So with all the complexities, this trial could (and did) take up to ten years. When it proved the point, it greatly increased the number of patients that should be treated (and the sales).

And much bigger trials followed. But who should pay for *these*? By now, the drug is not, or soon will not be, covered by its patent; and when the patent expires and most doctors prescribe cheaper tamoxifen rather than 'Nolvadex', profits plummet. But ICI went on paying for the new trials, the costs of which increased massively.

Could we not dose the less ill patients from the beginning? No, that would be unethical, if we had not properly proven that the drug worked. Could we not refuse to pay for the new trials? No, that would be unethical, too!

Do you see the dilemma?

Very many staff at Alderley Park were needed to support all this work. This was, for over half a century, perhaps the most influential business of the Park and of those working there.

## Process, Manufacture, 'Macc. Works' & Purity

Away from Alderley Park, the new Process Development Laboratories at the Hurdsfield site in Macclesfield began operation in 1966. 45 acres of the 80 acre site were developed at first, with a mere dozen staff arriving ahead of the expected initial complement of 900. (This book does not contain their story.)

Planning had started in 1961 with an estimated cost of £4·5 million. Prior to this, pharmaceutical production had been based at Regent Works, Linlithgow, from 1949, also at Huddersfield, with later a small plant at Blackley.

Later, the £3.5million sterile wing and warehouse completed in 1971. Many further expansions eventually filled the entire site and an adjacent 19-acre Ciba site was purchased in 1997.

Some ingredients used at 'Macc Works' were also made at Severnside, Bristol (now the Avlon site), which was approved in 1957 and closes in 2017.

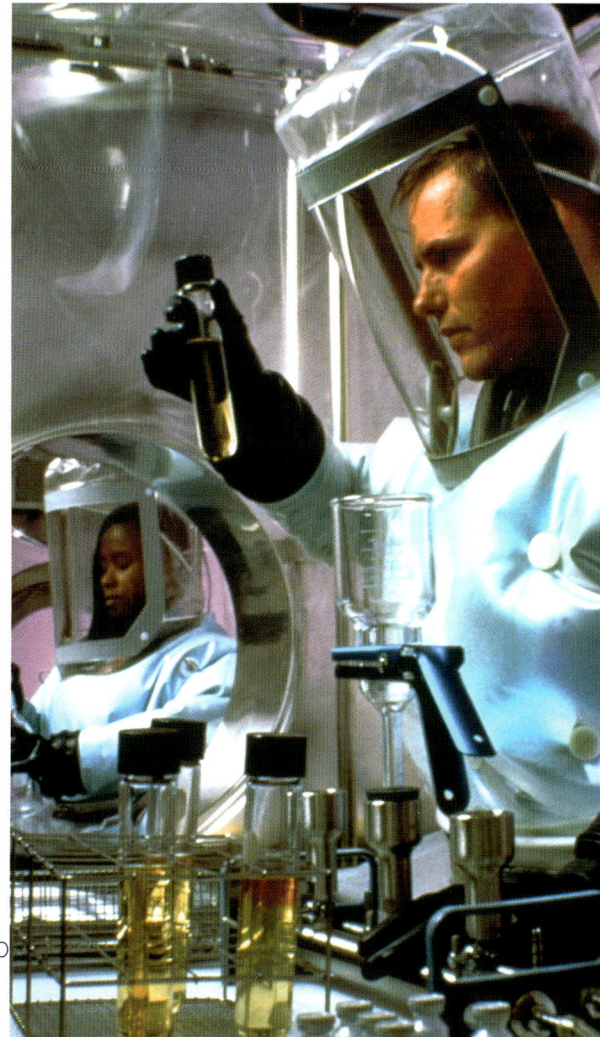

## Another book?

The great stories of drug process development and manufacture by the company outside Alderley Park, at Macclesfield and elsewhere, cannot be further covered in this book. Not least because the record could form a book by itself (though parts would be restricted for legal or business reasons).

Nevertheless, a huge amount of interaction between Alderley Park and 'Macc Works' – and movement of staff – was part of the world of Alderley Park – and few Macclesfield staff would not have been at least a tiny part of the Park's story at some stage.

## Production and quality

But Macclesfield was where the business faced the real world; and applied very different criteria to the work. The original chemical synthesis of every new drug was mainly chosen for speed, and potentially to make multiple similar analogues for testing. Now safety, economy, supply, environmental impact and many other factors must be considered.

Also at the heart of medicine manufacture is one of the most critical functions of all: quality assurance. With a forward-looking agenda, ICI Pharmaceuticals was one of the first companies to introduce during the 1970s a department dedicated to the new philosophy of 'scientific practices'. Alderley Park's work in this, led by Ray Broad, was at the forefront of the drive to ensure that new drugs were safe and met international standards. This became best practice in the industry worldwide; and now the research and safety of new drugs and medical devices is tightly controlled.

# Launching – World Market – Patient Safety

## Early commercial

The company's commercial history began with the creation of IC (Pharmaceuticals) Ltd as an ICI subsidiary in 1942, with H.V. Askham as the first commercial manager and J.A. Cochrane as his sales manager. Technical and sales teams were separate at first but after strong debate about this policy of 'segregation', these were combined.

ICI already had associated companies in several major territories and some of these undertook pharmaceutical sales in India, Africa, Australia and New Zealand. Others in the US and elsewhere chose not to; and Cochrane travelled widely to appoint suitable agents, laying the foundations of the sales forces of later decades. In some territories, bulk drugs were at that time processed on the spot.

## Launch, distribution and marketing

Alderley Park has seen the launches of a gallery of groundbreaking medicines into clinical use in hospitals and by doctors around the globe. Launch is an immense process, often in different countries at different times, as doctors globally have to be given information about the new product's properties and use. It is also strictly controlled by industry and government bodies.

After launch, an international sales team was set up from the Park for each drug, with responsibilities shared among staff members who looked after every part of the world. Its activities took place around three 'Commercial Regions': North America; Europe and

International; and Japan, formerly the company's second largest market until overtaken by China in 2014. Each drug has a set of global product directors, who are responsible for all sorts of interfacing with the outside world, including media as well as customers. Global working, accountability for product strategy and delivery, and contact between the internal and external world, are key. And a precious drug deserves strong promotion.

One big event of the year in product marketing was the glittering presentation of the 'Moscars', the marketing awards for the best teams. Apart from prizewinners, probably not many of the global sales team ever set foot in the Park itself – although good performance was sometimes rewarded with group visits which were especially popular with overseas staff – but the Park always kept a close watch on the pulse of the international business. Many who worked in the Park itself were employed in doing that.

The economic value of the company's pharmaceutical exports has been nationally recognised, with Queen's Awards for Export Achievement in seven different years as ICI and two as Zeneca.

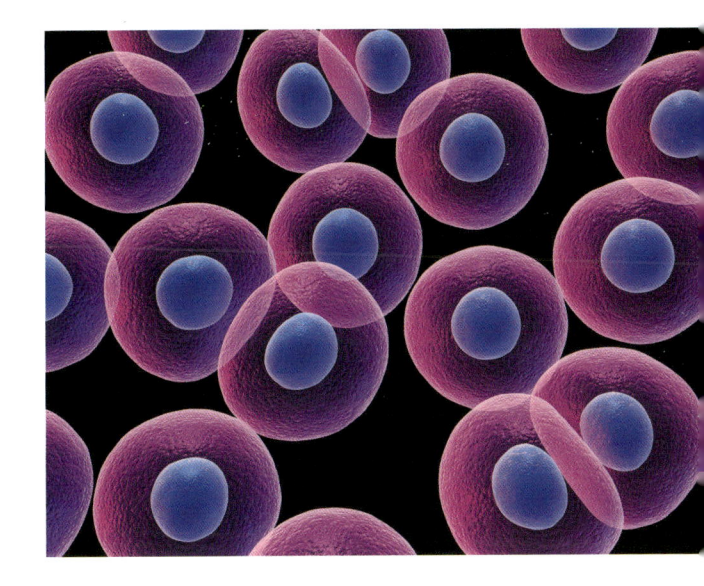

Healthy cells
COPYRIGHT
ASTRAZENECA

## Pharmaceutical industry

The company was from the beginning a major player in the industry. This was recognised at many levels. In the UK, Peter Cunliffe, Till Medinger, John Patterson and Chris Brinsmead were four who were recognised as they each became president of the Association of the British Pharmaceutical Industry.

## Code of Conduct

A particular emphasis became placed on employees' awareness of the company's Code of Conduct, undoubtedly because of the massive commercial cost of a bad mistake as well as

Till Medinger
*SCAN*, May 1993

Chris Brinsmead
PHOTO ASTRAZENECA

for ethical reasons. Many Alderley Park staff, while not leaving the site for business, might meet external parties; so their good instincts were enhanced by a surprisingly hard set of questions rolled out by way of an interactive test which every member of Alderley Park's staff had very strictly to complete every year to ensure their knowledge was current.

This certainly helped the company's reputation. From ICI days the company had been known for doing things properly; although some bench scientists who rarely left the lab were prone to wonder why they needed to know how not to take a bribe in a dodgy restaurant in some distant country!

## Patient safety and website

But patients were always at the heart. Once a new medicine is launched, a critical part of the business, the Patient Safety function, comes into action. It is very important to listen to every report of anything unusual happening to patients. A whole section of the business exists to trawl through medical reports and health records to watch for any trends or problems starting to arise.

From a different angle, patients themselves are encouraged to understand and manage their own conditions. A new innovation was the Patient Health International website, launched in 2004.

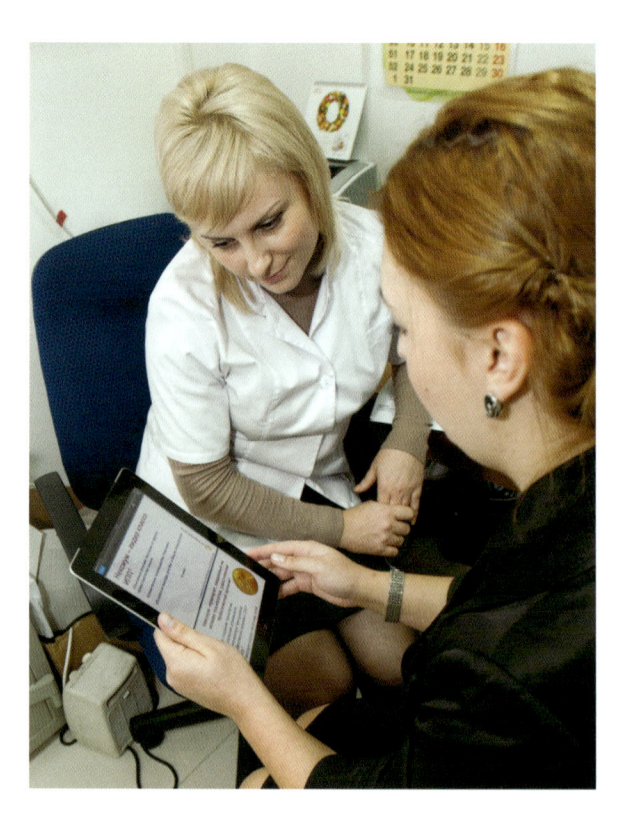

George's wife Christine in 2015, a patient treated successfully with **tamoxifen** (1998–2004), **atenolol** and **lisinopril**.
PHOTO GEORGE B. HILL

Providing information
COPYRIGHT ASTRAZENECA

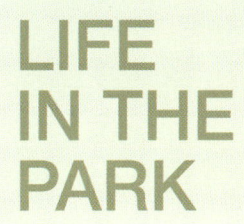

## Caution!

Over-caution was endemic at Mereside itself. On one occasion, I heard an electrician complain that ICI had demanded that so many safety systems be fitted into one electrical circuit system, that there was not enough room for them all in the box! But the answer must be, 'better safe than sorry!'

## ADRs

Any significant problems are flagged up by doctors and health professionals as Adverse Drug Reactions (ADRs). These might include, for example, an allergic reaction to a drug by a particular patient, of which several kinds might occur.

Equally serious are ADRs which appear in a number of patients. Two general types are the 'augmented' response to a drug – when a patient happens to respond to it rather too well, and should be given a lower dose; and the rarer but more dangerous 'bizarre' response, which is something probably inexplicable but which must be reported. Another way of looking at it is to watch for problems that are dose-related, ones that are time-related and ones unique to a patient.

In any case, everyone will want to know about anything that should be watched for. Most of all, the drug's makers.

# 13 Growing and telling

## Building in the Park: 1965–90

### Land purchases

The original area of Alderley Park in 1950 was enlarged on three occasions up to 2014.

In the 1970s, a plantation beside Hocker Lane, later known as Woodcock Wood, was added to the site's area. At the same time, three fields between the top of Beech Wood and Hocker Lane were also bought, although these were rented by the farmer and did not become part of the managed Alderley Park farm until 1997.

In 1997 the field later known as the New Heatherley plantation was also purchased for a reputed £70,000; it was commented that every farmer around the perimeter of Alderley Park had, not surprisingly, become a keen land agent!

Finally, to stop speculative development and infill after the building of the Alderley Edge bypass, one field on the far side of the A34, anciently marked on maps as Little Clays, was bought in 2006. This curious pasture held pits and holes probably from nineteenth-century clay extraction. This brought the site area to some 410 acres (166 hectares). This maximum area was only temporary, as the ten acres of Little Clays was sold again to Cheshire East county council when the Park was bought by MSP in 2014.

### Building – the bigger outline appears

Major extensions at Mereside (Blocks 12, 13 and 14) had successively been added by the mid 1960s; and in 1966, Alderley House Stage II was built. Equally notably, a new

Mereside, mid 1960s
ASTRAZENECA ARCHIVES

bar was added to the Stanley Arms in 1968! Research expenditure reached £3 million in 1969, with 4,500 compounds being made per year. In 1970, pharmaceutical production began in France, the first of numerous overseas advances by the Division. In 1971, US sales passed £20 million for the first time.

After a pause in the Park, proposals were put to Macclesfield Borough Council for a 50% increase in research on the site and planning permission was granted in May 1970 for the huge Block 19 at Mereside, a development that made headlines all over Europe.

The national power crisis of 1972 meant that gas lamps had to be used in the restaurant, and laboratory glassware was hand-washed!

## Mereside Library

Attached to Block 19 was the remarkable octagonal Mereside Library, which at the height of its importance (as a paper library) was said to be the largest private medical scientific library in the United Kingdom. In 1973, library staff travelled 45 miles in total transferring over 40 tons of books and journals with 30,000 reports from the old library in Block 1 to the new one. The Technical Information Unit also moved to the library building.

A £4 million extension to Block 19 was already complete by October 1973.

## Plantime

In the same month, the growth in staff numbers (which reached 2150 in 1975, when there were 1400 car parking spaces) brought rush-hour congestion to a point where flexible working hours became essen-

Building of Mereside Library begins
ASTRAZENECA ARCHIVES

tial. The Plantime system, remembered by an entire generation of those who worked in the Park, was introduced at Mereside.

Staff soon became used to the notched yellow plastic keys that had to be carried at all times. (The habit was so strong that I forgot to hand mine in when the system finally ended, and was still carrying it years later). Staff had to work 150 hours in every four week period and could carry up to 10 hours 'over', but would get a slap on the wrist if they fell more than 5 hours below their quota. The limits of the working day were 7–30am to 6–30pm (enforced strictly in laboratory departments for safety and insurance purposes); and the hours 10 to 12 and 2 till 4 were compulsory. There were always a few people dashing in at one minute to ten, or one minute to two; and at 4pm the stampede of 'early' working staff began, with a further wave of departure at 5pm from staff whose departments either stuck to a 9 to 5 habit, or who had simply never got out of the habit of one.

There was no incentive for white collar staff to work excessive hours, since neither overtime nor converting excess time into useful days off were allowed, although the latter was a perk for which many staff campaigned long in vain. But gridlock on the A34 was prevented.

## Chemists' shoe leather

The most frequent users of the new library were expected to be the chemists. Four of these, led by Tommy Leigh, made an eloquent plea for all the chemistry items to be collected together, rather than on different floors as first planned: 'The mileage covered by the chemist is usually much greater, *ceteris paribus*, than that covered by other users ... during the course of a library search, shoe leather usage would probably be in the same order ... the [positions of the] library contents should be heavily weighted in favour of the chemists!'

LIFE
IN THE
PARK

### Inside the world of the park

By this time, Alderley Park and the Mereside labs in particular had grown into an entire world. Arriving at ICI in 1977, as I did, was like entering a stately home – not in physical form but in manner. The floors were cork, and the benches were teak and (although I never did it) were still polished at Christmas in the time-honoured manner using wax polish and a Bunsen burner. At one time flame polishing had been a ritual every Friday afternoon.

W. G. M. ('Wigwam') Jones was less a section manager to us than a benevolent uncle, who came round each evening to enthuse his team and to admire the crystals – occasionally of valueless drying agent! – that he thought were valuable products in their flasks. The fact that he was a scientist of considerable repute from the early days of the company (having worked on everything from penicillin to the atom bomb) and a former chair of the prestigious Society for Drug Research was all part of the atmosphere.

# LIFE IN THE PARK

### Happy Christmas!

One legend from past days is not safe to repeat in detail. A lab of chemists, one cheery Christmas, thought up a new way to wish their colleagues in the other nearby labs the compliments of the season. Even though the actual perpetrators are no longer around (it was forty years ago), it is still no longer safe to say what they managed inadvertently to do, given an open window, balloons, bin bags, sellotape, an entire roll of aluminium foil and an unlimited supply of helium …

### On stilts!

A remarkable project initiated in in 1977 and built in early 1980 was the building of the £5.3 million Clinical Pharmacology Unit, constructed on concrete stilts over Radnor Mere. A coffer dam was used during construction to allow deep piling.

The clinical trial volunteers that were later ensconced there nicknamed themselves the 'Heron Club' in 1981 and were awarded special ties, as they suffered in peace.

Block 26 and the CPU

**Fishy story**

On one occasion the peace was disturbed. Dave, a keen member of the site angling section, noticed that some large pike had taken to basking below the CPU windows. He acted accordingly. Legend has it that he was just reeling his snapping, slimy catch in through the ward window when the nurse came into the room unexpectedly, and freaked out!

The CPU itself was toured by many visitors when complete, including in 1991 the Deputy Prime Minister of the Soviet Union, Mr Rudnev.

In 1977, Alf Spinks wrote of Alderley Park that 'we continue to hope that each building will be the last on this site.' In that year the former ballroom, the Tenants' Hall, was converted into a Divisional Conference Centre (previously it was used for badminton).

## RSL

In 1987 the £3.7 million Block 25, the Large Scale Laboratory (later the Research Support Laboratory), at the edge of Beech Wood near the Mere, was completed. It replaced a cramped earlier facility in Block 8. It began with 5 scientists and grew to 20. One of the first compounds scaled up there was **'Faslodex'**. By 2007, 25 Campaign 1 batches of potential drugs and 180 scale-ups were being processed annually there.

The RSL handled up to 400 pressure chemistry reactions per year. High pressure chemical reactors at the back of the building had blast doors opening into a safe zone outside the building, so that any explosive accident would discharge the contents of a bursting reactor without harm. Another interesting point was that most people never realised the RSL was connected by an underground service tunnel to the main Mereside buildings.

The RSL also provided a fully equipped home for the large scale chromatography service, which handled up to 600 requests per year. This was an intermediate scale of drug delivery in between the research lab and the process function at Macclesfield Works. Originally focussing on separations on 'normal' silica, this took on increasing work using water-based and more environmentally friendly 'reverse phase' chromatography, producing new medicines to tight deadlines for clinical trials.

## History made

A major change to the original main Mereside frontage of 1957 came when the £5 million Clinical Data Centre, Block 11, was opened in September 1988. This included a new Reception.

In the same month, the then Lord Stanley of Alderley visited Alderley Park for the second time (the first was in 1977), bringing with him a party

The late eighth Lord Stanley of Alderley (right) at the Park
*SCAN*, 9 September 1988

from the Anglesey Antiquarian Society. Although he had never had a personal connection with the Park (the last Lord Stanley to have lived in the Park had been his cousin, the sixth baron), he was as an amateur historian very interested in everything in the modern Park. He was sufficiently impressed to declare during his visit that ICI had 'done a superb job with the estate'.

A much greater piece of history was made soon after, when the Nobel Prize in Physiology or Medicine was awarded to Sir James Black for his ground-breaking discoveries. This included the work which led to both 'Inderal' and 'Tenormin'. All members of the original beta-blocker team were invited to Millbank in London for a celebration. On 9 June 1989, the Divisional Conference Centre (the former Tenants' Hall) was renamed the Sir James Black Conference Centre in honour of his award and research. A replica of his Nobel medal was displayed in the foyer – appropriately for the company created in 1926 partly from Nobel's own.

Sir James Black (in glasses) is congratulated by directors Sir Charles Reece, Brian Newbould and Peter Doyle
PHOTO BILL ROUSE

Sir James opening the Conference Centre
*SCAN*, 30 June 1989 (RON WINGHAM)

Sir James Black Conference Centre
PHOTO ASTRAZENECA

In 1988 the pleasant Fountain Room below the Water Garden Restaurant was also opened. The first site nature trail, mentioned earlier, was opened that year. March 1989 saw the Park's first major building not built of brick, in the form of the Drug Kinetics building Block 24 at Mereside, which cost £11.2 million and had individual fumehoods for each employee (essential for working with human blood samples). In July 1990 there followed the opening of Block 26, Mereside's first five-storey block, designed by AGP.

## Mereview Restaurant

Much more popular with staff was the construction of the award-winning, sensitively designed (by Cruickshank and Seward) and beautifully sited Mereview Restaurant, also opened in July 1990, at a cost of £4.2 million. It replaced the small old 450-seat restaurant with one able to supply food for 1,500 throughout the lunch period, obviating the need for the regular '12 o'clock rush'. But under manager Trevor Stone it still served (until banned for health reasons!) 'his' jam roly-poly pudding.

Initially there was a brasserie with waitress service. Downstairs was the relaxed Styles café bar for younger staff, with a patio beside the overflow pool. A visitors' dining area with 2–4 variable rooms offered traditional silver service.

From beneath its shallowly sloping roof it provided spectacular views across Radnor Mere. Members of staff who had previously hardly noticed the lake started watching great-crested grebes in summer, goosander in winter, and many other birds from the window tables, which were rarely unoccupied at meal time. (I was once eating lunch when a kingfisher landed on a branch outside the window).

The lovely sketches of Radnor Mere drawn by the famous wildlife artist Charles Tunnicliffe were planned to feature in the restaurant. But an application to the Tunnicliffe Trustees for permission to put these on the lunch trays used in the restaurant was turned down. The Trustees obviously felt that their artistic heritage should not be eaten off!

Beside the picture windows, a whiteboard was left for birdwatchers among the staff to record any interesting sightings on the lake. This accumulated some non-avian extras: for example, mandarin ducks (common on the Mere) were recorded – but so was mandarin orange!

Mereview Restaurant

Mereview Restaurant today
PHOTO ASTRAZENECA

Mandarin duck
PHOTO GEORGE B. HILL

# Staff and message – publications

A constellation of in-house publications have covered every aspect of the Park's life and business over the years.

## Regular publications

When IC Pharmaceuticals Ltd arrived at Alderley Park in the mid-50s, ICI already had eight Division newspapers, with a total circulation at home and abroad of 140,000.

The Park's 'pioneers' were served by *Fulshaw Times*, a quarterly magazine launched in 1950 when the embryonic pharmaceuticals business had its commercial headquarters at Fulshaw Park, Wilmslow. The magazine later included numerous articles about life at Alderley Park – and sometimes its wildlife. One article featured the various snakes then found wild in the Park.

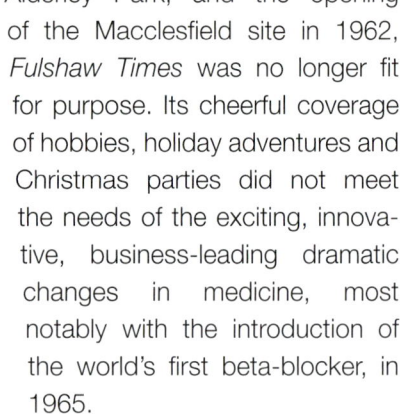

But with the rapid growth of Alderley Park, and the opening of the Macclesfield site in 1962, *Fulshaw Times* was no longer fit for purpose. Its cheerful coverage of hobbies, holiday adventures and Christmas parties did not meet the needs of the exciting, innovative, business-leading dramatic changes in medicine, most notably with the introduction of the world's first beta-blocker, in 1965.

## SCAN

A more professional, business-focused journal was required.

SCAN's
1966 first edition
ASTRAZENECA ARCHIVES

Plans for a monthly newspaper were announced and, with a title suggested by staff, a new eight-page tabloid called *Scan* was launched in January 1966 from a base at Harefield Hall, Wilmslow.

With an initial circulation of 6,000 copies, it aimed to support a well-informed and highly motivated workforce and to comment on every aspect of the business – and its people!

Business and product news appeared next to articles on staff with fascinating hobbies and unusual holidays. Christmas parties and summer sports days were well covered and so too were births, marriages, retirements and deaths.

Bad news was included, such as the demise of the Fermentation plant at Trafford Park, political pressures on the industry, and mishaps involving employees at work or off-site. Sport filled the back page.

But the editorial team always knew that the most avidly-read section was *Scanads*, an entire page of free small ads whose appearance signalled 24 hours of negotiation. On Friday afternoons, when the words 'Scan is out' went round the labs, someone from each office or lab dashed off to collect an armful of copies from the nearest foyer or distribution point, and all work halted for ten minutes. We remember it well!

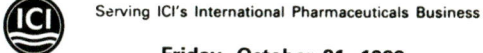

### Editorial freedom

Under the editorship of Roger Langley, who took up the reins in 1971 after 15 years with the provincial and national press, *Scan* became a fortnightly tabloid with a print run of 7,500 copies and a readership of 15,000. Most copies were taken home, if only for the employee and his family to conduct *Scanads* business!

The work of Langley and his team meant that by 1980 it was ICI's only surviving fortnightly house journal. It success was underlined in 1973 when it was named UK House Journal of the Year in its category. The 70s and 80s were the heydays of house journals in Britain; more professional journalists were said to be employed in their production than in all the London-based nationals.

Business needs limited some reports. But *Scan's* editor and team fiercely defended both their mission and their right to freedom in assessing the importance of a story. Roger Langley edited 450 issues of *Scan* in total. Trafford Press printed the journal for more than 20 years.

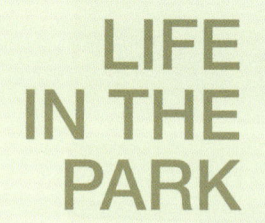

## LIFE IN THE PARK

### No fuss ...

On his retirement in 1996 Roger Langley recalled one notable exception to his editorial freedom. In 1973 when 'Nolvadex' was launched, he was summoned to a meeting with the Division Chairman, who made it very clear that he did not want to see a 'cancer breakthrough' headline across the top of the next front page. He said the business believed that 'Nolvadex' would prove to be a useful addition to the range of drugs available to oncologists treating breast cancer – but no more than that. It would do well to earn £10,000 in its first 12 months on sale.
Of course, he was wrong, and the rest is history!

### Other publications

Internal communications were also served by single-page notices such as the weekly 'Green Sheet' at Mereside.

Until the demerger, the corporate publication was *ICI Magazine*, which began life long before Pharmaceuticals Division appeared on the scene. It was later renamed *Roundel*. CTL had its own business publication, *Focus on Toxicology*, from 1987.

In the late 70s, *Scan* began to publish *Scan International*, for UK worldwide distribution among subsidiary companies and agents in more than 100 countries, including Russia – which received two copies! In 1989, this morphed into *Pharma World*, a full-colour tabloid which subsequently became the corporate magazine of Zeneca Pharmaceuticals. When AstraZeneca was formed, *Pharma World* was succeeded by *Spectrum*, a corporate glossy which gained little attention from staff. A more popular Alderley Park publication was *Park Life*, which was widely read until it ceased production to general sadness in 2004. After that, *Cheshire Life* was a more generic publication with much less site-specific news. It continued until December 2007, after which paper publications became a thing of the past for the Park and company.

### Publicity department

Publications were the internally visible face of the publicity function. Other parts of this, facing externally, included planning and production of publicity, along with photographic and exhibition sections. However, the production of scientific illustrations was a matter for illustration units within the science and other technical departments.

# Publishing the science

The other sort of publication is, of course, the legion of scientific papers that have been published by scientists at ICI, Zeneca and AstraZeneca over nearly 60 years. In 2014 alone, AstraZeneca's medical and scientific staff worldwide (not just in Alderley Park) authored more than 750 publications.

When a scientific paper is widely respected, it is 'cited' in papers by other scientists. It would be quite impossible to list even the highly cited papers from the Park.

### Successful products

The papers behind Alderley Park's key drugs are listed in Appendix 3. But these represent merely a token of the Park's contribution to scientific knowledge, through many hundreds of authors.

### Near misses

Also important were projects that produced no marketable drug but still changed the scientific landscape or blazed new trails. For example, AZD1152 was a potential leukaemia drug that could not be marketed; but the paper on Aurora kinase inhibitors (describing the science leading to it) won the 'most cited' award in that journal for the next several years, perhaps helped by some beautiful graphics.

### Academic milestones – the interferon gene

Some science does not directly lead to company products. The Park's finest example was possibly the paper published in the journal *Nature* in August 1981 on the 'Total synthesis of a human leukocyte interferon gene' by a team led by Mike Edge and Alex Markham. The world of science applauded, as the team showed that a human leukocyte interferon gene of 514 bases could be rapidly synthesised from nucleotides in the laboratory.

At the time this was the largest gene to be synthesised in this way, as interferon was being investigated as an anti-cancer drug. *Nature* commented of the group's achievement that it was 'a little like climbing Everest – known to be easier the second time!'

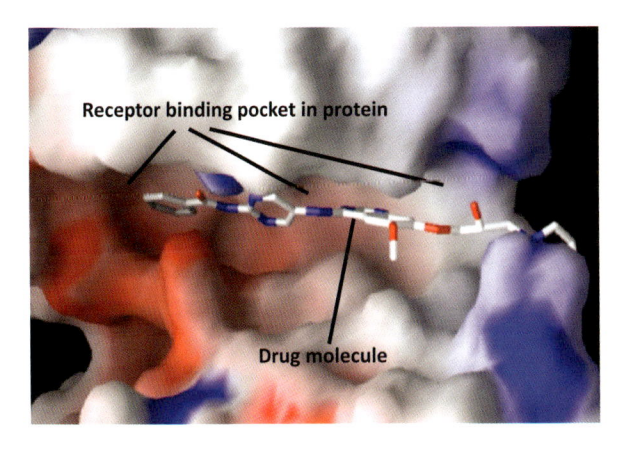

**Receptor binding pocket in protein**

**Drug molecule**

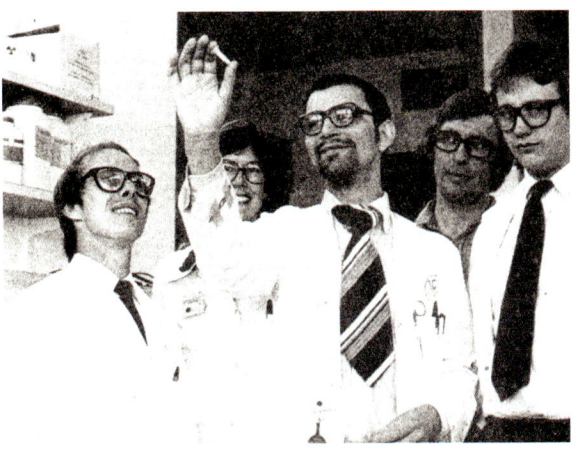

Markham (later Sir Alex) went on to head the ICI Cellmark Diagnostics facility at Gadbrook Park, Northwich (not part of Pharmaceuticals Division), where Sir Alec Jeffreys' 1984 discovery of genetic fingerprinting and profiling was commercialized.

# Dealing with humans …

Occupational Health and Human Resources (or 'HR') are the two departments in Alderley Park mainly responsible for all who work there.

## Body and mind

The Medical Centre at Mereside began its first weeks in 1957 in the barber's shop in Mereside Restaurant – the only place where there was running water! Later it was situated in Block 1, then Block 19 from 1973, then in Block 26, and was a place every employee found their way to, sooner or later. Initially under the care of Mereside's first occupational health nurse Sister Dorothy Bird (who estimated that she had personally known 6,000 people in her 25 years in the post), it was cheerful and straightforward. My own personal medical on first arrival, as a new starter in 1977, ended abruptly when I was asked what age my grandparents had reached. 'The youngest died at seventy-nine.' 'Oh, you can go off and start work, then!' In

later years, it grew into a fully fledged health centre, with a string of visiting health professionals, nearly the equivalent of a miniature hospital.

The focus also changed, as in recent times stress became a growing problem. The phrase 'work/life balance' entered the Park's vocabulary. The CALM programme became the main exponent of this, growing into a company-wide scheme after beginning as an initiative by HR and Occupational Health at Macclesfield Works, inspired by Rev. John Staley, as everyone tried to squeeze more (including lunch) into shorter periods.

## HR

HR's duties include all matters of people protection and the important issue of equality; of finding and employing suitable new people; developing the skills of existing staff, managing the performance system; administering benefits; and complying with employment law. It was responsible for hiring but also, rarely though, for assisting managers who decided that firing was unavoidable for very serious misbehaviour.

Under ICI, it had been the Personnel Department, until it changed its name. The new title of Human Resources was one that became standard across industry – but that moniker never particularly endeared itself to employees! HR's key task was to support the human-power needs of the company.

But that included supporting humans. HR always supported staff and also their families, in all sorts of personal troubles and trials; and much of its invaluable work can, necessarily, never be made public. Its services were deeply appreciated, in times of disturbance or trauma.

It also instituted change, as in the mid 70s when the company abolished separate pay scales under which women got 80% of the male rate! Also, in the 1970s job interviews were relatively informal affairs. A generation later, they had come to include psychometric tests that were intended to tell the management much more about the employee than he or she knew.

## Performance, salary and unions

The hottest personnel issue was, of course, pay. Various performance assessment methods were tried over the years. The Haslam system of the 1970s and 80s was a rigid banded one, in which much of the paperwork fell on supervisors. It was up to ambitious members of staff to 'write up their jobs' with examples of their work, if they wanted to climb higher.

## Negotiation

The actual 'band values' – the salary at each level – were decided by negotiation with the trade unions. This was a necessary procedure, though the company management chafed about it. Many staff were union members and there was a near-formal dance between the company and the unions each year, as offer after offer was voted on, until both felt a point of maximum advantage had been reached. Only on one occasion did the dance break down, leading to Alderley Park's one and only strike in the early 1980s. Support was patchy; and both strikers and union reps were shocked to realise they would lose a day's pay, though ICI after discussion agreed not to deduct it on that occasion. It was an anomaly, as relations with the unions were generally good; but it was an event that startled the company and undoubtedly stimulated its determination subtly to undermine union power thereafter.

This was achieved in any case by the introduction of the 'Hay' system, which abolished fixed salary levels altogether, in favour of 'broad-banding', which made a group-negotiated percentage rise impossible. This was a profound change, for consultation between managers and employees had been a feature of company life since the formation of ICI in 1926. However, joint consultative meetings continued (and became ever more important as AstraZeneca moved into difficult days).

Another revolutionary change was the abolition altogether of required working hours, in favour of an American 'trust time' model. Staff then became assessed on whether they had met their so-called 'SMART' targets (which they had set in agreement with their supervisor). Latterly, another acronym that staff loved to wince at was the famous DPWR, from the Development of People, Work and Reward programme, launched in spring 1989. In contrast, IDPs or Individual Development Plans helped personal long-term planning.

### In passing

At one period, staff members were subjected to a 'second assessment' at the next level up, with their section manager, to see if they had any concerns they had not wished to raise with their direct supervisor. If not, this often became just a business chat, and was not always taken seriously.

But eyebrows were raised when one senior manager 'assessed' his team member on meeting him in the corridor. And Bernard McLoughlin even received one from Bernard Langley while standing in the gents' lavatory!

LIFE
IN THE
PARK

### AZ Advantage

Another big change to terms and conditions came with the introduction of the AZ Advantage scheme, after the merger. This set a value on the benefits and optional perks that employees could select or decline. The main driver for this was so that potential recruits to the company could compare our terms and conditions directly with those of other employees. In fact it also slightly confused most employees as to what their real salary was! But Advantage was attractive in enabling people to pick and choose, and it soon became a part of everyone's thinking.

### Pension

Of more interest to the older members of staff was the pension system. In 1996 the old scheme, despite the ICI Pension Fund having been heavily fenced to deter corporate raiding, was partly closed. In 2000 the break was complete and the old 'defined benefit' sort of pension became a thing of the past for new starters, a trend soon to become more or less universal in the UK. One consequence of this was that employees suddenly became interested in the share prices of other companies besides their own! But they had no idea how soon the pension changes would become much more than of future academic interest for very many of them.

### Management skills

Management training was a world in itself. Every big company tries to spot and nurture its rising stars. A certain managerial toughness is required, a distinctive force of personality …

## LIFE IN THE PARK

### Grooming for management

One rising star I particularly noticed during my career was Andrew. He was clearly marked for higher things. I realised that fact when, while on leave, I sent my supervisor Nicola a pretty holiday postcard. Nicola shared an office with Andrew and when it arrived she held it up in delight for him to see.

But the postcard was from the lovely little island of Lundy; and it bore one of Lundy's exotic postage stamps. And Andrew had a ruthless habit, a true business-style hobby. He was a stamp collector. He reached for the card from Nicola. Taking a firm managerial decision, he gripped it and wrenched off the corner bearing the stamp. Without a word, he handed back the remains.

Nicola was almost tearful, later, as she showed me the maimed missive. We gazed sombrely at the ruined remnant, which should have been brightening the wall behind her desk for many dark winter days to come. The company would be in strong hands in future years; but neither of us would ever put a stamp near them again.

## The Mereside New Year Honours List

The staff were not without their ways of making their feelings known to management. One venerable institution for very many years in the Mereside labs was the mysterious *New Year Honours List*.

Originating somewhere in the chemistry department (its exact source was never admitted), it appeared each Christmas without warning, originally as one typewritten sheet and later several; it only became electronic very late on. It was always topical and some of its content became legendary, such as a response to the introduction of the aforementioned Hay performance assessment system, which appeared as the 'Song of Hayawatha'.

It should be pointed out that I was not involved in it and could not possibly identify those who were. (I was only punished in it once myself, in a mock piece about the new site nature trails I had helped to create). Commonly a copy would simply be found lying on top of a photocopier, and would multiply rapidly, like a virus.

It descended now and then into a scurrilous rag. Nevertheless, it took pains to note whatever events had happened in the company's life during the year, particularly involving the senior management. It could occasionally be generous when credit was felt due to anyone by their peers. Satire, insult and some plaudits were doled out as seemed necessary. Over the years, the *Honours List* became known across the Park, and was known to be read even in the company boardroom.

Some senior members of the company took umbrage, as the caricaturing in the *List* was merciless, especially after the worryingly expensive £15billion purchase of MedImmune, when a reference to buying a turkey for Christmas went close to the limit (and was undoubtedly excessive); but others regarded notoriety endowed by a mention in its pages as a badge of honour, as a blunt judgement on their performance. As the *List* itself commented at the Millennium, 'Some were not thrilled to be mentioned in the Honours List but they endured the experience stoically. Most would rather be spotlighted than ignored … All these people represent the very stuff of life.'

Amongst these was Tom (later Sir Tom) McKillop, when our research head. I well remember him striding into one of his department's labs one Christmas Eve.

'Where's the *New Year Honours List*?'

There was a sudden, uneasy silence.

'Find me a copy. There'll be trouble if I'm not mentioned in it! And mine had better be a GOOD mention!'

Dora Richardson on her retirement

```
"A great step forward," said Tweedledee.
"It will really motivate people -," said Tweedledum.
"To give of their best -," continued Tweedledee.
"Twenty four hours a day," went on Tweedledum.
"Seven days a week," added Tweedledee.
The dormouse opened an eye again, feeling that some explanation was required.
"These two are departmental managers," he said. "Oh really," said Alice as if
she realised that would explain anything.
```

The New Year Honours List reflects on DPWR

## Classic honours

Most of the people given accolades in the *Honours List* over the years could not be named here, not because enough readers would not remember them, but due to fear of litigation. And many prizes only made contemporary sense. Yet it would be a pity not to list a few classic and timeless awards, leaving one to guess who they originally referred to.

Thus, rewards such as the 'Mallory Prize for not quite making it to the top', or the 'Mars Lander Prize for being remote, expensive and you can't get a peep out of him', might have various owners, as might the 'Rentokil™ prize for nitpicking', the 'Golden Thumbscrew Award for torturing visiting speakers', the 'National Trust Award for the preservation of ancient monuments', and the 'Harold Steptoe Award for the most junk submitted to the Compound Collection'.

Projects and functions also came in for stick, with the 'Mary Rose Award for the resurrection of old wrecks', and the 'EuroDisney Award for a Mickey Mouse operation that's never likely to show a profit' being especially cutting. There were campaigns, such as the one to get licensing manager Dudley Earl awarded the OBE (so that he could be described as Dudley Earlobe). Current events featured in articles such as 'The Career Olympics'; and literary treasures included 'Alice in Blunderland'. Quotations were used as elegant put-downs: one I particularly liked was: 'He does not drink at the fount of knowledge, he just gargles'.

More generally, the Montgolfier Brothers Prize was awarded to the Staff Committee for machinery powered solely by hot air; while the England Cricket Cup was awarded to the Senior Executive Team for their triumph of optimism over experience. But the Golden Wonder Award for Crisp Decisions was not presented. Nonetheless, one award was universally approved, to Dora Richardson, on her retirement, of the 'Dulux™ Paint Prize for a brilliant finish.'

# LIFE IN THE PARK

*The Dovecote: no longer for pigeon post?*

# Dealing with e-mail

The revolution came: e-mail arrived …

When I first started work at Mereside as a new graduate in 1977, I often received just one piece of mail (a compound list) each week.

By the time I reached retirement, I had been forced to post above my desk that magnificent epigram of the Duke of Wellington which is the tersest antidote to today's decay of communication that I have found: 'If I attempted to answer the mass of futile correspondence that surrounds me, I should be debarred from all serious business of campaigning.' It seems futile to add anything more.

# Collaborating with academia

Academic collaboration (working with university partners in your industrial research) is hugely important in business.

## Confidentiality

But secrets must be kept! There is no way that business research, jealously guarding news of its discoveries from its competitors, and striving to get legal protection for them, can share all information with university academics whose *raison d'être* is making their research and name famous.

Normally, a few trusted professors are invited to sign confidentiality agreements and are then told any secrets that it seems useful to tell them; while others are only shown generic outlines of work being done. But the interface with academia is the hottest in industrial research.

## Students

Other sorts of university collaborations involved students. Many staff (and figures in academia) had first worked at Mereside as summer students or undergraduates doing a 'sandwich' year in industry, doing company research.

Professor Craig Jordan, well-known for his life-long work on **tamoxifen** (originally ICI 46,474), remembers warmly his first experience as a summer student at Mereside when (later) famous scientists worked and interacted cheek by jowl in Mereside's labs:

> … the telephone call from outside Alderley Park that got me the summer job with Steven Carter. I described how improbable the outcome of that summer in 1967 … Carter (of cytochalasin B fame) was looking at the effects of compounds on mouse cancer cells in culture … Opposite Carter's lab was Walop's [Arthur Walpole's] lab. He had just published a paper on the effects of ICI 46,474 … I went to lunch each day with his technicians and we went to the Rose in Alderley Edge on Fridays. Next to Carter's lab was Mike Barrett's [Barrett led the biology behind 'Tenormin'] … Barrett would be my Professor at Leeds.

Many graduate students came for three months as part of a CASE Studentship (originally called 'Collaborative Awards in Science and Engineering'). These were part of the critical process of introducing the best postgraduate students into industrial labs, where they would use company facilities for three months to carry out their own research, under the supervision both of their academic supervisor and a company one interested in the same work. Usually this was an excellent way of getting young scientists both to energise and learn from older, long experienced staff as they enthusiastically discussed each other's problems. Many settled into great scientific careers – although there were exceptions: one CASE student I looked after abandoned chemistry entirely, to play the bassoon! Was it me?

## Academia

University academics themselves were very keen on links with the Park. Academic speakers frequently visited as lecturers – and enjoyed their well-informed audiences; many world-leading scientists were among them.

Professor Joe Sweeney, a leading academic collaborator and now Head of Chemical

Sciences at the University of Huddersfield, also remembers the rich environment:

One of the distinguishing and admirable features of UK chemistry is (was …?) the very close collaborative interface which exists between the academic and industry chemistry communities.

From the earliest stages of an academic's career, financial, practical and intellectual support was generously provided, especially by Big Pharma; nowhere was this better exemplified than at AP, which was the focal point for many simultaneous PhD collaborations. One would often visit Alderley and bump into more academics than industrialists; and there was a residual permanent air of scientific cut-and-thrust which left a lasting impression on both student and supervisor.

# Education, training and the young!

Industry has profound responsibilities to those not within it.

A huge part of Alderley Park's impact on the local community involved schools. Visits to schools by staff were always popular. Tours of company facilities were even more so.

## Work experience

But actual work experience at the Park was a special privilege enjoyed by few. During ICI's time, an annual Work Experience Conference day tried to give a flavour to a wider group; but there were always very many more young people who wanted a week in a real lab or office than could ever be accommodated. This particularly applied in departments whose work was hazardous, where staff sometimes could not have students shadowing their actual work, and had to set it aside to do safer things.

Another constant challenge was ensuring that students were selected so far as practicable by ability – since most staff had their own children in mind. But it was also fascinating to work with young people who might ask questions from much further outside the box than expected. And if they returned to enthuse their peers about a scientific career, all well and good.

### Safety of students

Safety is always a huge issue in laboratories. But the strongest precautions were naturally taken when looking after work experience students, who could be as young as 14.

When students came, they were first lectured at length on the need for using protective clothing and equipment at all times. Then they were shown the lab. The students were always deeply awed by the array of state-of-the-art technology around them. Generally, they looked up to lab staff with the same awe, and paid great attention to the important things they were told. Then they were usually allowed to do a demonstration experiment, the favourite being to make a sample of the simple drug **paracetamol**. This always impressed – except on one occasion when, after warning three students to take great care with a real drug, I managed clumsily to spill the glistening crystals all over myself!

### Lego™

One of the most popular and colourful of all the events seen in the Park in recent years has been the Lego™ Challenge. Many local schools joined in as AstraZeneca sponsored teams from Cheshire schools and donated the Lego™ kit. The Challenge, a robotic team tournament for children aged 10–16, combined a hands-on, interactive robotics programme with a sports-like atmosphere, designed to trigger youngsters' interest in science and technology. How can I fit these two pieces together?

# Collaborating with colleagues

Working with educated people is always a challenge, for they are always likely to have views as strong as yours – but different. And that applies 'at home' as well! One of the key challenges for management, therefore, was getting research teams to work together.

### Property

The most challenging area in the labs was often the 'ownership' of property. Established scientists tended to accumulate all sorts of favourite bits of equipment and other things over which they believed they had an absolute authority.

If a lab reshuffle meant that they found themselves working next to colleagues with a more communist view of property, everyone knew that sparks would sooner or later fly. Some of the most dramatic moments in lab life could arise as a result, with the thunderstorm that broke over a culprit's head sometimes entering departmental legend.

### Teamwork

The most important aspect of all life in the research labs (and elsewhere) is always teamwork. Staff were constantly reminded of their mutual duty of care.

In some departments, where materials or machines being used are hazardous, this is not merely a matter of morale and galvanising people. You might need to rescue your colleague; or they might save *you*. Teamwork may be a matter of health, or even survival.

### What's Yours is Mine

Some leading chemists kept order and controlled miscreants firmly. One of these was Jean Bowler. Jean had frequent skirmishes over lab behaviour with more casual colleagues.

Jean had several things that she guarded jealously. One was a water spray, acquired from the plumbing department, which she had set up to wash down her chemistry fumehood. One morning, she arrived to find that the head of the spray had mysteriously disappeared. A neighbour in the lab (let us call him) Henry was one of those generous souls who believed that no-one could object to sharing anything if it was in the interests of someone. Jean marched across the lab and accused Henry roundly.

He lived in fear of Jean, but bluffed it out, cheerful as a daisy. 'Oh, I noticed your sprayhead. The one in my bathroom at home is blocked. So I took it home just to check different sizes. I forgot to bring it back, but I'll return it tomorrow.'

Even by Henry's standards this was interesting. The thunderstorm that broke over his head became an instant legend in the department.

Another bane of Jean's was a lively lad called Alan. Jean nearly always won game, set and match against Alan. Then came the day when Alan had the misfortune to spill a chemical solution over his trousers. Following safety procedures correctly, he dashed to the powerful laboratory safety shower and operated it. He had just pulled the contaminated garments to his ankles when Jean walked in. She spun on her heel; defeated, just that once.

One way in which Jean was never defeated though was in her research science. Amongst other advances, she helped lead the follow-up project to 'Nolvadex' which led to 'Faslodex', and also worked briefly on 'Arimidex', thus being connected to three of the most important drugs ever discovered in the fight against breast cancer.

It was, therefore, a poignant as well as tragic moment when at 55 she lost her own personal fight against the same disease. There have been few like her.

MALLARD RISING FROM THE EDGE OF THE WOOD

Mallard rising from the Edge of the Wood by Charles Tunnicliffe
FROM 'MERESIDE CHRONICLE', 1948.
COPYRIGHT TUNNICLIFFE FAMILY

# The big new world

**14**

## Information Systems

As the company grew, the information it stored mushroomed.

Information Systems (IS) is a massive department in the modern company. Its precursor in 1971 was the Management Services Department, which covered hardware and software but had nothing like the all-embracing scope of modern IS. It developed because more and more areas of the business became involved in information handling – in addition to the traditional company activities of managing staff, capital and materials.

The different business areas had various functions but their information needs were not dissimilar, and needed to be linked in various ways. And the new capabilities of modern computing meant that it made more sense to centralise the handling of all that information, rather than each area handling its own.

In Alderley Park this eventually led to an Information Systems (IS) department – not to be confused with Information Technology (IT), or Intellectual Property (IP) departments. It had several parts.

### Science and information

Three of these were largely about the science. The main duty of Discovery IS, perhaps the most critical part of IS in Alderley Park, was to look after data produced by AstraZeneca's so-called 'Crown Jewels' – its chemical compound collection of samples, and everything learnt from it.

An even bigger hat was worn by Development IS, with its huge remit covering clinical trial data both before and after launch. Operations IS was concerned with the physical production and manufacture of drug products.

### More information!

Other areas included Finance IS and Commercial IS, the latter concerned both with sales and planning. Common to all areas were techniques and tools for developing systems, networks and databases. Obviously, IS staff had to keep up to date on computer technology and on business needs, as they bridged between the two.

### Global and division

Perhaps more than any other part of the business, IS has always had a global reach. When ICI and Zeneca split, perhaps

Sample storage
PHOTO ASTRAZENECA

the most difficult and time-consuming job of all (apart from caring for the employees) was dividing the information stores of the companies, deciding who was entitled to what. IS Staff from both companies went through many negotiating sessions as data sometimes decades old was 'claimed'. Then, six years later, came the more friendly but even more complex task of uniting Zeneca's and Astra's computer systems and data. And then there was more and more to store! This is the age when information, like work, seems to mimic Parkinson's law – and fill every space available.

# Information Technology

'Knowledge is power' was Sir Francis Bacon's view in the sixteenth century. In Alderley Park, as throughout industry, it translates also into money. The amounts of information stored by pharmaceutical companies exceed those in most other industries; and who knows which tiny piece of data might be the one that matters? Kings' ransoms have hung on such things. 'Information Technology' is a more modern term from 1958. Now every major firm has its IT department and many people in Alderley Park have worked in it.

## Yours and theirs

In a big company, IT has two parts: your own company's data; and everything else in the world!

These may become nearly the same thing. Chemists working on cancer drugs at Mereside spent many years studying the chemicals called quinazolines. They searched the world for information on them. But eventually, they realised almost all the 'hits' they were getting from their searches were ones that originated from themselves. They already knew almost everything there was to know, because they had written it.

## Origins

The Division's first Technical Information Section was set up 1942, based at Oxford Street in Manchester, to promote interest in the company's products and service them on the technical side. Later, advertising and publicity matters were separated from the compiling of technical information bulletins including literature information and abstracting. By 1956, the latter had itself split into an Information section and an Intelligence one, the latter carrying systematic surveys of scientific, medical and (in earlier times) veterinary information. Searching the company's own data store also became important as it grew.

This distinction continued; in 1971 the Industrial Property department in addition to patents included both a technical information unit and library services, the latter managed by Angela Haygarth-Jackson, who became the first woman president of the Institute of Information Scientists and received an OBE.

But the various distinctions became less important when the electronic age dawned and paper records fell out of use.

## Product enquiries

As early as 1946, the idea of having two bodies of sales representatives, technical and sales, was abandoned. This meant that all sales staff needed to be kept informed of the technical aspects of the company's products. Product reviews thus became an important task for the company's information experts.

Mereside Library
(until recently)

Library interior
PHOTO ASTRAZENECA
(BARRY SMITH)

Angela Haygarth-
Jackson, OBE
*SCAN* September 1972

## Data searching

In later decades, searching out data from the global and company data stores became a Herculean task. The contents of every scientific article published globally are 'abstracted' into huge collections which company scientists (in the early days without, then later with, computers) have to search. The most important summaries for earlier chemists were the colossal hardback collections of *Chemical Abstracts*, listing all the chemistry of the last century and also the German *Beilstein* index. Later, 'Chem Abs' morphed into an electronic equivalent, Scifinder, and Beilstein into Crossfire. Searching by chemical structure was a huge and constantly developing element of all this.

The medical information system MIDAS was introduced in 1972, then replaced by MEDLEY in 1987, when over 70,000 records on company and competitors' products were transferred; 6,500 journals were then being abstracted, with access by telephone to the records which were actually stored in Bern, Switzerland.

Bioscientists originally used *Biological Abstracts* (BIOSIS is the electronic database) and especially *Index Medicus* (MEDLINE, now on the internet as PubMed) and the complementary *Excerpta Medica*.

Data searching has now entered another new world, with application hosting in the 'Cloud' via one's own portable device, for instance in collecting anonymised patient data and biological symptom information.

## In-house databases

All pharmaceutical companies store immense amounts of information. This in theory presents a serious security hazard. What happened if your commercial rival gains access to details of your latest drug?

THE BIG NEW WORLD          **269**

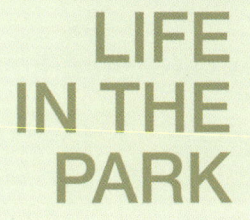

# LIFE IN THE PARK

## Sent back

In fact commercial rivalry is less of a problem than might be thought. Big companies generally do not want hostile neighbours. On one occasion, a discontented scientist at Mereside posted his laboratory notebook, containing details of his work, to another company. He had not expected what happened next: they posted it straight back! And he lost his job instantly.

## Media

And paper has gone! As information grows, the media used to store it changes. Yet the global move from paper to electronic media did not, surprisingly, affect many scientists at Mereside at first. For example, although most scientific journals began to appear electronically, there were two problems: the journal 'back issues' took much longer to become available in electronic form; and – critically – their publishers could not at first guarantee that they would remain so 'in perpetuity'. If a publisher disappeared, how could a crucial scientific article be obtained without extra expense?

## Library's end

By around 2010, both of these issues had become unimportant. That meant that Mereside Library had essentially lost its function in the paper sense, with a role now mainly as an electronic learning centre. With what seemed quite indecent haste, the Library's stores of paper journals and books were rapidly disseminated, as AstraZeneca went entirely electronic.

Apart from the chemists, who decided to keep their own reference collections for a time, all useful textbooks were donated to staff or given to universities; and journals were handed over to an external store or the British Library, from which copies could be obtained electronically as required.

It was the end of a musty but evocative and genteel era.

# Keeping tabs on the competition

Confidentiality is a hot topic when keeping tabs on those doing similar work in other companies. International conferences and the like are outside the scope of this book; but hints and tips brought back by those attending them often had an impact on work done in the Alderley Park labs. Synthetic chemists in particular were sometimes given sketches or images picked up at conferences, and asked to turn them into chemicals that could be tested, to see if 'their' compound was better than 'ours'! (There would be dismay if it was).

Legally, it was quite legitimate to make a sample of another company's favoured compound, as long as the original inventors were told about anything they really ought to know about and may not yet have discovered for themselves (such as hazardous side-effects of a drug they were planning to market). But that could only be done once the sample had been *made*. And doing that depended on finding out what its structure was.

# Sites overseas: links to the globe

The Park history of ICI, Zeneca and AstraZeneca, and the impact on their global company structures are within the story of this book, but the history of the multitudinous colleagues in other countries, and of Alderley Park's employees and the overseas sites where they worked (and also the history of Astra) sadly, is not.

## Many countries

Nevertheless, the key overseas sites themselves that existed at various times should be named. The numbers of research and development staff after the merger at the turn of the millennium are listed below.

| | | | |
|---|---|---|---|
| Alderley Park and Macclesfield | 3300 | Lund, Sweden | 100 |
| (2800 at the Park alone, out of | | Mölndal, Sweden | 1600 |
| a staff total of 4500 there) | | Montreal, Canada | 100 |
| Bangalore, India | 100 | Södertälje, Sweden | 1500 |
| Boston, USA | 250 | Wilmington, USA | 650 |
| Charnwood, Leicestershire | 1200 | | |

The first figure includes the staff at Reims, France.

Production sites are not listed, nor are more recent acquisitions, such as MedImmune's main site at Gaithersburg, Maryland.

## All the same!

On one occasion, I was given a photograph taken at an international conference of an important chemical molecule. It was a photo of a computer image of another company's structure, taken by one of our own people when it flashed up briefly on a lecture screen. I was asked to make a gram or two of that compound.

I magnified the image up until I was sure I had worked out the right chemical structure from it, then invented several steps of chemistry to make the material. When I had finished, the compound was tested. Puzzlingly, it proved to be utterly useless, with no biological activity at all.

A little while later, colleagues were chatting at another event.

'Did you see that structure that [Company X] put up for a few seconds at the last conference? We photographed it on the screen and got one of our chemists to make a sample of it, to see how good it was. But our picture must have been poor, because the compound did nothing at all in our company tests.'

'That's funny, WE did exactly the same.'

'And so did WE …'

'And so did WE …'

Months later, Company X cheerfully announced that they would now show the world the *real* structure of their new potential drug … instead of the image they had shown last time – which had been just an artist's impression!

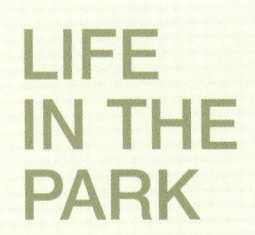

**LIFE IN THE PARK**

# Hanson, demerger & Zeneca

In 1991, ICI was the UK's third largest company. In that year, belatedly, it appointed its first female board member. ICI Pharmaceuticals alone employed more than 12,000 people, a quarter of them in R&D and nearly two thirds outside the UK. During the 1960s, the early policy of relying on the company's own activities had been reversed and since then Pharmaceuticals Division had increasingly established a network of wholly owned subsidiaries and had set up manufacturing plants in various countries. Big acquisitions by the Division had started with Atlas Chemical Industries in 1971 for access to the vast US market, and Stuart in 1972.

The fierce recession of 1980–1981 had then shocked the parent company out of its inertia. Under chairmen Sir Maurice Hodgson (1978–82) and Sir John Harvey-Jones (1982–87), ICI had already contracted painfully, with the loss of some 50,000 jobs globally. ICI had

An original ICI
Savlon tube
ASTRAZENECA ARCHIVES

then pushed strongly into high-value chemicals. By the late 1980s, pharmaceuticals was ICI's most profitable business, and ways of accelerating its growth through acquisitions or alliances were explored, although none came to fruition. By now, the Division had 21 production sites and 150 sales offices around the world, with major research laboratories in the USA and France. Annual pharmaceutical sales passed the £1.8 billion mark and plans were laid for spending more than £3 billion on pharmaceutical R&D over the coming decade.

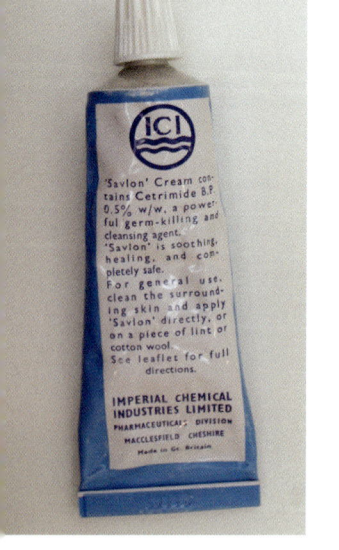

One problem was that ICI's shares had underperformed for much of the 1980s. The older, lower-margin parts of the business kept share prices from reflecting the growing profitability of pharmaceuticals. ICI was a fiercely complex organisation; at one point, it held a startling total of more than 500 legal entities within it!

ICI Pharmaceuticals was already focusing its business. In October 1989 it had announced a move out of selling OTC (over-the-counter) products, including old drugs long out of patent, in order to focus exclusively on ethical (that is, prescription-only) medicines. The OTC business in the US was sold to a joint venture of Merck & Co. and Johnson & Johnson for more than $450 million. Perhaps the most familiar OTC product of all, **'Savlon'**, was transferred in May 1992 to the latter, who relaunched it in the same nostalgic blue tubes, but now minus the ICI roundel.

ICI was not thought a likely target for a corporate predator. But a major improvement in shareholder value was badly needed. At the end of 1990, ICI's chairman Sir Denys Henderson, Harvey-Jones's successor, became convinced that restructuring was critical. During early 1991, the company looked at various ideas: a megamerger; a demerger; or a pharmaceutical alliance.

## Hanson!

Then in May, the landscape changed abruptly as the news broke that ICI had attracted the hostile interest of English industrialist James, better known as Lord, Hanson. Hanson was the British corporate raider with a fearsome reputation for taking over badly managed companies.

His biggest deal had been the 1986 purchase of Imperial Group, a British tobacco conglomerate with a cash-rich pension fund that was his real target. Hanson's team had turned up in Bristol on the morning of that takeover to find the Pension Trustees had closed the fund the evening before, denying him the asset. Lord Hanson had expected to use the IG Pension Fund to subsidise the transaction. In the event, his takeover was funded by selling many of Imperial Group's subsidiaries, leaving him with a profitable business.

In similar vein, he turned his interest toward ICI. In May, the Hanson Trust, which was Margaret Thatcher's favourite conglomerate, announced that it had acquired 2.8% of ICI. The company regarded the move as a precursor to a hostile takeover, and a war of words ensued.

Rumours were wild, and unsettled many staff at Alderley Park and globally. Hanson claimed that ICI would benefit from its management style, but for example there was great concern among the company's scientists when it was said that Lord Hanson would require all company expenditures of over £1,000 to be approved by head office. He clearly had little idea of the massive expenditures involved in pharmaceutical research!

The news caused national shock. Robert Peston, the BBC's business correspondent, described the furore in his blog 16 years later:

> The reaction of the media and fustier elements of the City was one of outrage. It was as though the Tower of London was about to be sacked and the crown jewels sold off to Ratners. ICI was viewed as a national treasure, a nurturer of scientists, a school of management. The British establishment viewed it as little short of scandalous that a bunch of takeover artists at Hanson – whose primary concern was the Hanson share price – should suggest that ICI could put a bit more emphasis on rewarding its shareholders.
>
> In some ways it was the totemic struggle of the age. ICI fought vigorously to undermine Hanson's reputation and credibility. But although it saw off that particular aggressor, it lost the battle – because ICI's executives were forced to recognize, in a way that they never really had before, that their primary responsibility was to the company's owners, the shareholders.

Sir Denys Henderson warned UK Members of Parliament that ICI was not a '*plaything*'. In the end, Lord Hanson's reputation as a ruthless raider worked against him as ICI played on its national Image status as a flagship company and also as it proved as wily as he was.

The ICI Pension Fund was ring-fenced with extra legal protection (in a way that benefited staff for long afterward). Chairman Sir Denys Henderson hired Goldman Sachs to look into Lord Hanson's business dealings. According to *Wikipedia*, they reportedly found that Lord Hanson's business partner was running racehorses at shareholders' expense! Hanson backed down from the takeover attempt. By October, it was clear that Hanson would not launch a bid. But the corporate rain had cleared the air; a dramatic view of the future had now appeared.

## Demerger

Lord Hanson's financial arguments were considered by many to have stirred ICI to what happened next; although it was, in fact, more a case of proving what they had already perceived. The ICI board decided to crystallize the value in the company's fast-growing pharmaceutical arm by spinning it off themselves.

A key figure was John Mayo, a director of ICI merchant banking advisers S. G. Warburg, who had experience in corporate restructuring. An outsider with no emotional ties to ICI, Mayo, who was later appointed finance director of Zeneca, was able to look objectively at the organisation, and became the final catalyst for a bold solution. After long discussion by the directors in early 1992, the final plan was approved and announced in July. An early rights issue was to be launched by the new company, so that most of ICI's cash reserves were expected to remain with the parent one. This would make the latter unattractive for takeover while the new one would become an expensive buy.

The necessary legal and tax changes, and the splitting of information and computer systems, took nearly a year. Some naysayers had doubted that such a feat was possible. But it was achieved.

## Zeneca

On Tuesday 1 June 1993, after the ICI Extraordinary General Meeting the previous Friday, ownership of Alderley Park passed from ICI to the new company Zeneca plc. About 35,500 employees went with the new company globally. It was formed by the demerger of the Pharmaceuticals, Dyestuffs and Specialty Chemicals divisions of ICI. Zeneca Ordinary Shares were admitted to the Official List of the Stock Exchange at 8.30 am – although work to dismantle the world's largest ICI roundel, on the wall of Macclesfield Works, had begun the week before! The old ICI name was retained by the heavy chemicals sectors of the old company, while Zeneca took the fine chemical businesses. Zeneca's new HQ was in London at Stanhope Gate, Mayfair.

David Barnes, who had been a main board director of ICI, became the first chief executive of Zeneca plc under chairman Sir Sidney Lipworth. David Friend, who had been chief executive and chairman of ICI's Pharmaceutical business before the demerger, became chief executive of Zeneca Pharmaceuticals, reporting to David Barnes and the Zeneca board, and also an executive director of Zeneca plc himself. In 1994, David Friend retired and Tom McKillop replaced him in both roles.

In the league of world chemical companies, the effect of the demerger was to move ICI down only from fifth to sixth place. Zeneca entered the top 20 pharmaceutical companies, with its international core at Alderley Park. It was a highly successful result; neither company suffered much loss of status in the eyes of customers or governments. The market value of the combined companies rose from £7.6 billion to £13.2 billion between July 1992 and December 1994; and combined share value rose vigorously.

Sir David Barnes
*SCAN/PHARMA WORLD,*
*4 July 1993*

David Friend

Tom (later Sir Tom)
McKillop
PHOTO ASTRAZENECA

'Zeneca' was an invented name created at the request of Barnes by branding consultancy Interbrand; the first temporary name of the new company had been ICI Bio. Interbrand had been instructed to find a name which began with a letter from either the top or bottom of the alphabet and was phonetically memorable, of no more than three syllables and which did not have a stupid, funny or rude meaning in any language. It is said to have cost £50,000, though no money was spent on advertising as the company merely sent out press releases whenever anything happened, until the name caught on!

The name initially chosen was Zenica, with the letters IC from the old company name. Tragically, this suddenly and by a wild coincidence hit the press as the name of a town in Bosnia near Sarajevo which was besieged and murderously attacked in April 1993 during the Bosnian War. The plans were instantly changed and the new label, though attracting startled comment in places, was accepted immediately nearly everywhere.

Only slight hiccups occurred, as in one territory where a slow uncovering of the name letter by letter at the formal presentation caused brief hysterical laughter, reportedly because in that language 'Zene' was a slang term meaning the equivalent of 'dosh', or 'lolly'!

Zeneca
ASTRAZENECA ARCHIVES

Dismantling the world's biggest roundel at Macclesfield Works on demerger
*SCAN/PHARMA WORLD,* 4 July 1993
PHOTO BARRY SMITH

## Tasty-looking?

The new company of Zeneca plc moved fast, freed from the ponderous constraints of ICI. In 1993 it invested £300 million in research and development, corresponding to 16 per cent of sales. It was recognised from the start that the new company was small and tasty-looking (it was thought to be an obvious target for takeover by a big Japanese pharmaceutical company, for example). In consequence its share price rose rapidly from its starting point of £6. The company immediately moved into corporate acquisition itself, to increase its size. In December 1994 Zeneca agreed the acquisition of half (later all) of Salick Health Care, an operator of cancer care centres in the United States.

## Vulnerable?

**This was a time of re-organisation and also of considerable tension for staff. There was constant discussion about the future. Perhaps the most extreme case came on one occasion when one senior research scientist walked through our laboratory, shouting 'It's Merck! Clear your desks!' Happily, he was wrong. (Roche was another supposed prowler.)**

## LIFE IN THE PARK

In early 1995, the pharma giant Glaxo was looking to swallow up a smaller business. Many rumours suggested this might be Zeneca. The Zeneca executive and board were always conscious that Glaxo might try to do something; and there was plenty of speculation among analysts and the press. But, contrary to some reports, Glaxo never made an approach to Zeneca, who in any event were well prepared to rebuff any overtures. There was never any doubt that the response from Zeneca would be hostile.

But Glaxo turned its attention instead to Burroughs Wellcome. This was 40% owned by the charitable Wellcome Trust, which had a legal duty to maximise its income and thus, as Sir Richard Sykes explained, 'had a weakness that made it an easier prey than others.' When Glaxo made a £9 billion bid for the latter, all the charity's shares had to be voted in favour unless another bidder stepped in. Zeneca itself briefly considered a bid for Wellcome, but the Trust had signed an agreement with Glaxo that made it impossible. Glaxo Wellcome came into existence in April 1995, and Zeneca staff breathed a sigh of relief, for there had been a clear awareness from the start that the similarity of the two companies meant that many jobs would go in any merger.

Zeneca itself was then rumoured to be interested in a bid for its small neighbour Fisons; this never happened but the former Fisons Charnwood site at Loughborough was bought by Astra in 2000, so a marriage thus took place in a different way in later years. Sadly Charnwood was later still a casualty of the same downsizing process that contributed to the eventual earthquake at Alderley Park.

### Narrowing the focus

In May 1996 Zeneca announced the sale of its textile colours business to the German group BASF. In 1997 it acquired a large fungicide business. In 1998, the company announced that it was planning to sell its Zeneca Specialties division, including its biocides, industrial colours, lifescience molecules, performance and intermediate chemicals and resins activities. The sale, for $2 billion to a joint venture by Cinven and Investcorp, was announced in 1999.

# Merger – to AstraZeneca

In May 1998, Zeneca announced that Tom McKillop would replace Sir David Barnes as Chief Executive of Zeneca plc in 1999, with the latter (who was knighted in 1996) becoming Chairman.

Before this could actually take place, the news was announced in December 1998 that Zeneca proposed to merge with Astra AB of Sweden, another pharma giant. Astra had over 23,000 staff, with over 4,000 R&D staff working in respiratory, cardiovascular, gastrointestinal and pain research areas, with four research sites in Sweden, one in the UK at Charnwood, Loughborough, and one being built at Boston, US.

McKillop would become CEO of the £48 billion merger company AstraZeneca. Percy Barnevik, head of a major shareholder in Astra, would be non-executive chairman. David Barnes and Håkan Mogren, the incumbent chief executives of Zeneca and Astra respectively, would be non-executive vice-chairmen.

Despite appearances, Zeneca was not a particularly willing partner in the merger. In retrospect, it appeared to some an obvious move. This was not so at the time. The mid to late 90s were a period of considerable uncertainty for the pharmaceutical industry, particularly following the proposals for major healthcare reforms by the Clinton administration in the US. This had resulted in further consolidation through M&A (merger and acquisition) activity. Zeneca was approached by a number of other pharmaceutical companies who were interested in merging, but after proper consideration these were all rejected; Zeneca preferred to stay independent. It was only after about two years of detailed analysis and highly confidential talks involving very few people, and only after Astra had succeeded terminating an agreement with Merck (one which had given Merck indefinite rights to Astra's products and which the *Daily Mail* said was 'not rated one of the cleverest deals in history'), that a deal was reached for the merger creating AstraZeneca.

Cynics said that two companies with huge products but obvious future patent problems – '**Losec**' in the case of Astra ('**Prilosec**' in the US) – were putting all their problems in the one basket. Nevertheless, the figures added up well; and the staff were relieved by the obvious shorter term job security the merger promised for most, despite a predicted £660 million of cost savings.

The new AstraZeneca would have £6 billion of annual drug sales, just a whisker behind world leaders Merck and Glaxo Wellcome. Sir David Barnes boldly claimed that: '*Our growth is faster than Glaxo. We will be challenging for number two* [in the world].' The merger was completed the following year.

The new group possessed a 4,000-strong US salesforce, even after 6,000 jobs were axed worldwide. It would be a truly international company: its headquarters would be London, but the research headquarters would be in Sweden and the company's shares would be denominated in dollars. There were to be 14 directors, seven from each side. The first head of R&D, ultimately the man overseeing Mereside's research, would be Claes Wilhelmsson from Astra. A number of key figures departed, particularly on the Zeneca side.

It was not quite an equal marriage. Effectively, Zeneca paid £22 billion for Astra, which incidentally then had to pay out £4 billion to Merck.

The shares soared as more rumours arose. Zeneca investors were pleased; they had quadrupled their money since the group was floated out of ICI. (Zeneca was by then six times bigger than its old partner.)

On 6 April 1999, Sir David and Håkan Mogren announced that all the conditions to the merger offers made by Zeneca for the issued share capital of Astra had been satisfied and that they were therefore unconditional – and that the merger had been completed. Acceptances for a staggering total of 1,580,147,610 Astra shares (96.2% of the total possible) had been received and 797,184,469 new AstraZeneca shares were issued in exchange. Redenomination of Zeneca's share capital into US dollars had taken place the previous day. Dealing started immediately.

On 1 June 1999, ownership of Alderley Park passed formally into the hands of the new company AstraZeneca plc. The new company's sales in 1999 were $14.8 billion (out of a world market estimated at $270 billion) and its total research investment was $2.5 billion.

Serving **ZENECA** Pharmaceuticals

# Scan

Friday
April
30
1999

## 'DELIGHT' AT SHAREHOLDER APPROVAL

"WE ARE delighted that the shareholders of both companies have given such overwhelming support to the creation of AstraZeneca PLC. AstraZeneca is now one of the leading companies in prescription medicines in the world and this confirms the vision behind the merger"

That was the reaction of Sir

**'This new company combines the best of two innovative companies'**

Percy Barnevik, Chairman of AstraZeneca

aged in the merger documentation sent to shareholders in January 1999, was completed on April 5, and the name of Zeneca Group PLC was changed to AstraZeneca PLC with effect from the same date.

A total of 797,184,469 new AstraZeneca shares were issued on

At the end of March, the acceptance by the Federal Trade Commission in the US of the Consent Order in connection with the merger was announced. This fulfilled the only remaining regulatory consent required to effect the merger, following the previously

– Acceptance
ASTRAZENECA ARCHIVES

Sir Tom McKillop was knighted for services to the pharmaceutical industry in 2002 and made Fellow of the Royal Society in 2005.

### Old ICI

A later final footnote was the demise of the old and historic company of ICI itself, the last residue of the company from which Zeneca parted in 1993. On 2 January 2008, in an £8 billion takeover, ICI, now reduced to just the adhesives and paints businesses after many more sales of parts of its portfolio, was swallowed by the Dutch giant Akzo Nobel. It does, however, live on as a name on Dulux™ paint.

At its height, the announcement of ICI's results to the London Stock Exchange would bring the City to a halt, so important was its performance to British industry. Five days after the last takeover, ICI's shares were delisted from the London Stock Exchange altogether.

## LIFE IN THE PARK

### Works for ICI?

The demise of ICI ended not only a company but also a folk culture. Classic one-liners were suddenly no more, such as 'How many people work for ICI? Oh, about half of them!'

And the question that one was incessantly asked, from Land's End to John O'Groats and even further afield, was always 'Do you know So-and-so? He (or she) works for ICI.' Really – one of the 180,000?

'Er – no, I don't know them.' In sixteen years, I knew only once the person referred to. Camping alone on a small island in the remote Shetlands, I went to an isolated cottage for a bucket of water. Where was I from? What was my job? 'Oh – do you know Dave Whalley? He works for ICI …' Dave worked in the next lab to me. The exception proves the rule.

## Rational drug design & molecular modelling

Rational drug design is one of two ways of discovering new chemical 'leads' towards a new drug.

The other, described in the next section, is using robots to do very many chemical reactions, in the hope that just one or two might give useful products. But scarce and expensive resources are far better used in making *only* the chemical compound that the research actually points to – if one knows what it is!

## Fitting into the active site

The aim is to find a small organic molecule that has an effect on the body. But the body is made up of a huge number of cells! So the chemical can only work by influencing a protein or other large molecule, one that in turn influences a whole cell and in turn the whole body. The large 'target' molecule will typically be one specific to a particular disease or condition, or essential to a microbe or virus. The small molecule must affect (and preferably must only affect) the large one.

This happens when a small molecule is complementary in shape and electrical charge to its larger target – so that it fits into the 'active site' on the protein neatly, like a key into a lock, as described earlier. The 'computer-aided drug design' that achieves this relies on computer modelling techniques. But the three-dimensional structure of the protein or other target must be known – or learnt – in advance. And then we must handle a simulated electronic version of it, which in turn requires immense amounts of calculation: hence 'computational' chemistry.

## Drug structure and activity

The Physical Methods function at Mereside originally studied only individual chemicals and their structure. But in the late 1960s physical chemist Peter Taylor, later made a top ICI Research Associate, began to look at comparing the structure with the biological activity of drugs. This was a further step jump following the work of Alf Spinks half a generation before on comparing drug structure and absorption.

Taylor's interest was initially in the study of reaction mechanisms. This became important when the identification of the impurities in drugs – 'fingerprinting' them – proved to be a way of identifying counterfeit drugs and their sources, so tackling this dangerous and costly crime.

In the 1970s, Taylor moved on to the first attempts to match structure against activity mathematically (measuring the 'lipophilicity' or fatty character of each drug molecule) in a Physical Concepts Project led by Ralph Howe. Taylor introduced calculation packages enabling drug properties to be predicted; many chemists later designed drug molecules with the required properties using Hansch analysis. This was

Peter Taylor
*SCAN*, July 1988

research that gained much attention across the pharmaceutical industry and ICI scientists gave many talks on an area in which they were leading research. An important development of his work was the use of complex drug kinetics to investigate drug stability and issues in manufacture.

## Molecular modelling

All this focused interest on how molecules of exact shapes, leading to high biological activity, could be rationally designed with the right properties. The result was the new technique of 'molecular modelling', pioneered in Pharmaceuticals Division by Robin Davies, originally using a computer at Blackley as none was available in the Park. This involved creating a virtual equivalent of the simple physical models that chemists had long made of their molecules. The first major physical model of an enzyme (thrombin) at Mereside was built by Dave Timms in 1977, but 'virtual' molecules could be far more useful.

Molecular modelling nowadays starts with work that discovers the 3D shape of the target protein, commonly by X-ray crystallography of a crystal of the protein. This gives a simulation that can be projected onto a screen and viewed with 3D spectacles.

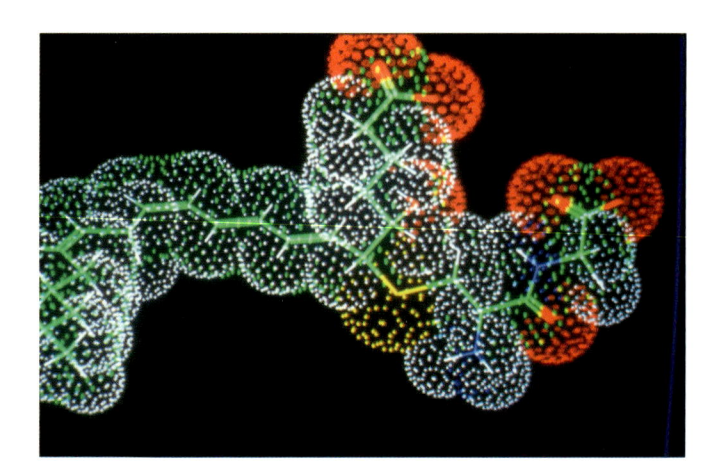

An early computer
simulation of a
molecule
COPYRIGHT
ASTRAZENECA

In recent times, the computing power available in the Park reached astonishing levels. It became possible to create entire virtual enzyme molecules, consisting of many hundreds of atoms in exact positions. Such a construct could then have added to it a virtual model of a potential drug molecule (also known as a ligand). This could be put into the computer projection in order to see how the drug would probably interact with the enzyme. For instance, the drug might present one of its atoms in a position where it could bind to the enzyme irreversibly, rendering it useless. If all of the enzyme present were thus disabled by the drug, this might kill the cancer cell that needed the enzyme. Thousands of similar virtual molecules can be checked for fit in a very short time.

In the Park, this eventually became a seriously high-tech business, as staff realised when the giant screen for the new unit had to be lowered into its position through a hole in the roof! But the ultimate aim was to achieve something immensely valuable: to predict whether a compound will be a potential drug without spending (sometimes huge) time and resources on making it. When this works well, it can save great resources in chemistry and avoid much needless biological testing, as well as speeding up research to a rate that past scientists would never have dreamed of.

## The hardware

Earlier at Mereside, computing was done on VAX mainframes, to use which scientists had to negotiate with computer staff for use of computer time. Then the arrival of PCs equipped with graphic input devices brought in the modern world. The first of these required computers so large that they acted as radiators in winter and were nearly unbearable in summer! But drug design work accelerated in the 1990s to a rate quite incredible to older staff.

## LIFE IN THE PARK

### Oh, sorry!

It does not always work, though! On one occasion, I spent seven months making a single, supposedly critical (and difficult-to-make) product.

When my struggle was finished, I sent the compound to be tested 'against the enzyme' by my bioscience colleagues. It proved to be wholly ineffective.

Two weeks later, my supervisor came up to me and said cheerfully:

'Oh, that structure we gave you to make is pointless. The molecular modellers have obtained a better focused image. There's another bit of protein that we hadn't noticed before. It fills the hole your compound was supposed to occupy. The gap you were trying to fill isn't there at all ...'

SEVEN months!

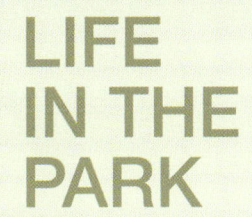

**LIFE IN THE PARK**

# Robots and technology

One thing that bewildered even those behind it was the incessant pace of change in the business and the laboratories. New technology came along so fast that a few systems were out of date and on their way out of the lab (usually donated to universities) almost before some staff had learnt to use them.

The threat to jobs from robots was always a heated topic for discussion. In fact, however, areas of novel research such as compound synthesis proved only partly compatible with robotic methods. Occasionally, new technology was tried with great enthusiasm from managers, then quietly sidelined. And there was still a quiet understanding that critical and difficult areas of research – the sort that could lead to breakthroughs into whole new areas – were unlikely to come from repetition, however facile.

## Chemical technology and Labstock

But the Chemical Technology lab in Block 19, opened in 1999, brought a dramatic change, making large numbers of (fairly similar) compounds simultaneously, and was highly influential in policy planning. It was equipped with seven bespoke Zymark robotic systems alongside a whole range of automated parallel processing systems.

Along with the chemistry, other things had to be automated or accelerated. The chemistry Labstock store, holding over 10,000 bottles, which had been created in the 1990s to bring together many individual stocks and had become a major operation by around 2001, had to be made still more efficient. It eventually became a high-tech lab in the new RA building (Block 33). And the software became very sophisticated, albeit to the regret of one or two older staff members!

## Potential

The potential of robotics in chemistry was soon obvious. At Mereside and everywhere in the pharmaceutical industry, the advent of high-tech ways to make chemicals (using robots or 'multiple parallel synthesis') was an immense advance. It did, however, bring two linked problems: when compounds are easy to make in large numbers, it becomes less wasteful of resources to make 'unnecessary' compounds than *not* to. Also, less thought goes into selecting them and asking what point each one has.

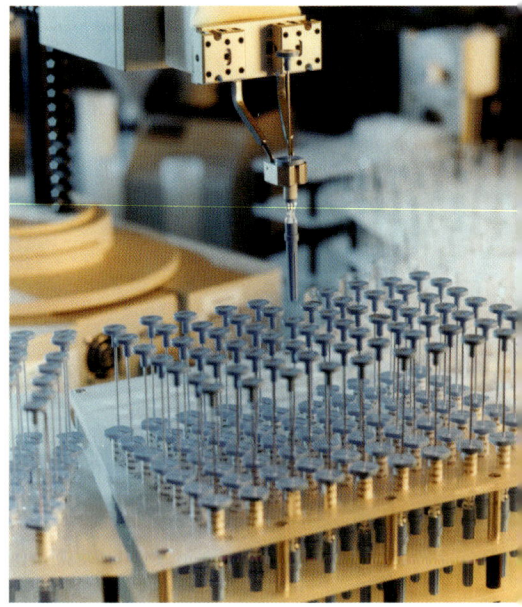

Robotic testing
COPYRIGHT
ASTRAZENECA

Robotics used for
analysis
COPYRIGHT
ASTRAZENECA

## Robots in testing

Robots in bioscience were already part of ordinary lab work, dispensing materials for multitudinous *in vitro* tests. High Throughput Screening has already been mentioned. But the pressure on staff to learn how to operate in what was unendingly a 'new world' in the labs was continuous, exciting, and at times stressful.

Explaining the technology to visitors was also interesting. The less technically minded ones were easy to awe. But the old adage of 'Don't work with animals and children' also applies to robots!

## LIFE IN THE PARK

### Playing with the remote!

One embarrassment over technology involved the company's new chemical Compound Library, a huge automated store containing up to 2 million bar-coded bottles. The new computerised retrieval system was a quite amazing piece of multimillion pound robotic technology, fascinating to watch. On its completion, a prestigious group of journalists from the *Wall Street Journal* were invited to see it work.

Unfortunately, it wasn't working. The main computer was not quite ready. So, unseen by the journalists, a hidden engineer lay underneath the window they were looking through, pressing random buttons on a remote to make the robot do anything that looked impressive!

Roboticised sample store
PHOTO ASTRAZENECA

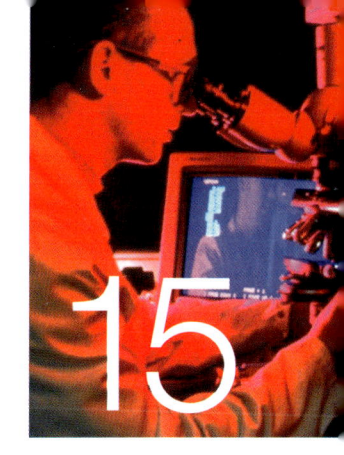

# Targets & hormones

## Choosing targets

How are targets for research chosen? This is an enormous topic. The choice is driven by medical need; by opportunity (arising from discoveries from academia or in-house pure science research); and by the potential for financial return that would pay for the research itself.

### Breakthrough?

Challenging research is risky. As Craig Thornber puts it:

> In 'breakthrough' research, the company seeks a totally novel form of therapy. It may be that there is no current treatment for the disease in question or one which is quite inadequate. The objective is to find a new approach and to find a drug that ultilises this approach. This type of programme involves the highest risks and the highest rewards.

Less difficult is finding a drug that is merely better than a similar one already on the market; but there will be more competition. And such so-called 'me-too' drugs need to make at least some small improvement. The decision on whether to go for a 'breakthrough' product or modify a known one is made at the top of the business, as the research director talks to medical and commercial directors. They decide the 'What'.

### How?

The 'How' is a science question. If a biochemical 'mode of action' can be imagined and a biological test devised, the task falls to medic-inal chemists to plan the work and testing, to synthetic organic chemists to make the compounds, and to the whole team to select a successful product and decide how to use it. But can it usefully be sold?

Choosing targets
COPYRIGHT
ASTRAZENECA

### Small molecules and blockbusters

During the heyday of 'small molecule' research, when biological rather than chemical drugs were still curiosities, big firms aimed to grow by launching so-called 'blockbuster' drugs – comprising molecules small enough to be made from simple materials by traditional methods. Importantly, these needed to be treatments needed by patients for long periods – perhaps even for life. So targets of this sort were favoured.

But there are now many more large and small pharmaceutical companies. An analogy with the fishing industry might help: the pharmaceutical 'ocean' is now full of trawlers large and small. But all the easy-to-catch 'fish' have already been plucked from the fishing grounds. The cost of catching those left is much higher; and nearly all the big fish have probably gone.

So there is a wide view that the day of blockbusters is over. Which begs the question what can we afford to research into now?

## Target types

Different disease types have different issues. Drugs for terminal illnesses are precious from a human point of view, but compared with conditions like heart disease or arthritis the annual number of patients treatable is small. Drugs for infectious diseases may be taken by many, but patients will only take an antibiotic for a few days – although long-term side-effects are then less worrying. New drugs for diseases of the developing world cannot be financed heavily because patients cannot afford them, but may come from related research. Cost is becoming prohibitive even in some diseases of the 'rich' … but is that a bad thing? Should we develop a new treatment for heart disease or asthma when the drugs we already have are frankly very good, and only a slight improvement would cost both researchers and providers a king's ransom?

## Cancer

According to the AstraZeneca 2013 Annual Report, every year more than 14 million people are diagnosed with cancer and more than 8 million die of it. By 2035 annual cases are projected to rise to 35 million. The cancer area was one that became, in the end, the prime research area in Alderley Park. This was, however, the result of a long and reluctant process, for returns on cancer drugs were known to be lower than those on more profitable cardiovascular or gastrointestinal ones. (However, respect across the industry was gained for ICI's courage when it committed itself to the area.)

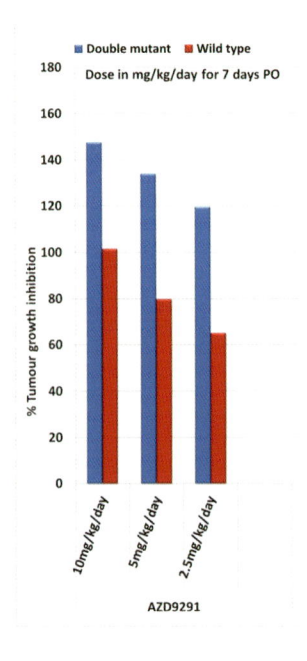

Arthur Walpole's pioneering work was continued by many. The first leaders to follow in his footsteps were Barry Furr, Mike Dukes and Alan Wakeling, who inspired the cancer biology capability at Mereside that led to the company's oncology successes (oncology is the study of tumours).

The company was one of the first in the world to approach cancer not as a disease ultimately to be cured, but as a chronic condition with which patients could learn to live. From 1995, Zeneca's approach was parallel to the drug cocktails that have revolutionised the treatment of AIDS, prolonging the patient's survival rather than curing him or her. Perceiving the potential of this, those who directed Alderley Park's work slowly turned it more and more into a cancer centre. In 1998, the business spent £160 million on cancer R&D. By 1999, AstraZeneca was the second largest company worldwide for anticancer products and it was said to have the second biggest industrial cancer research facility in the world.

The first and most famous cancer drug discovered at Alderley Park was, of course, 'Nolvadex' (tamoxifen). After it came seven more …

# Prostate cancer 1: 'Zoladex'

Two hugely important prostate cancer drugs were discovered in Alderley Park. The first was one of the most problematic medicines invented in the Park.

## Hormones and the prostate

Charles Huggins at the University of Chicago worked out how to measure the effect hormone changes have on the prostate gland. He found that castration (or another technique, suppressing male sex hormone levels by adding female ones) caused it to shrink. In 1941, he and Clarence Hodges showed that this effect, by reducing the level of androgens (male hormones) in the bloodstream, could slow prostate cancer. As the first to demonstrate that hormones could control the spread of some cancers he was awarded the Nobel Prize for Medicine in 1966.

## LHRH (GnRH)

In 1966, Arthur Walpole, who was a regular conference speaker, met Professor Geoffrey Harris of Oxford University. Harris asked for Walpole's help to track down LHRH, an agent he thought was produced by the brain's hypothalamus to stimulate the nearby pituitary gland. (This then produced gonadotrophin, which in turn caused the reproductive organs to generate sex hormones.)

It was a testament to Walpole's persuasiveness that Garnet Davey allowed him to ship huge quantities of sheep hypothalami from Australia! Peptide chemist Harry Gregory carried out months of sterling laborious work (today it would take an hour) isolating the agent. In 1968 it was shown to be a peptide (a small protein-like molecule). He had identified 80% of the structure when two Americans (with a large team!) beat him and were later awarded the Nobel Prize.

Though overtaken, their work at Mereside put ICI in a leading position to study LHRH (also called GnRH, **G**onadotrophin **R**eleasing **H**ormone). There were several research possibilities. Analogues that acted like LHRH could have uses in infertility (or in animal husbandry). LHRH antagonists – drugs which *blocked* its effect – were even more appealing (if they could be found) as potential contraceptives and, more significantly, against hormonal cancers. They could offer non-surgical alternatives to ovary removal in breast cancer and castration in prostate cancer.

## 'Zoladex' (goserelin)

Following publication of LHRH's structure, the Fertility Project Team (as it was then known) made samples of LHRH to test its effects. Rosemary Chester in Mike Dukes' team investigated whether it

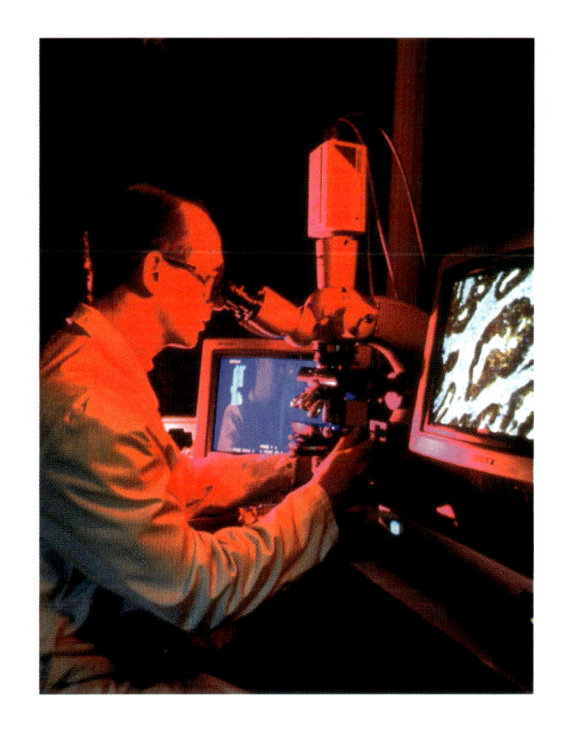

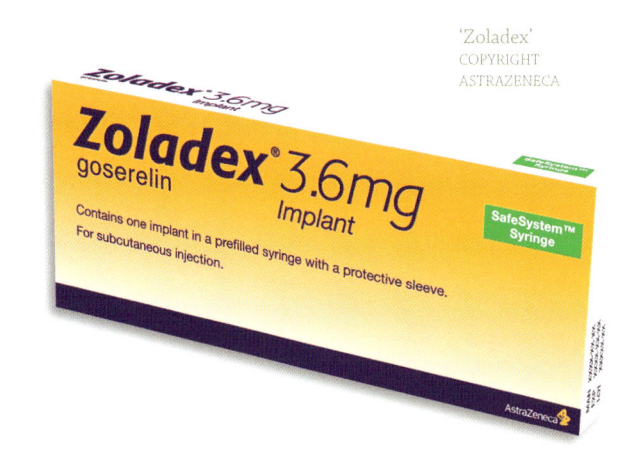

could restore ovulation previously blocked (with 'Nolvadex') in rats. This became an important test for new compounds (and also later during drug manufacture). Barry Furr tested LHRH induction of ovulation in cows.

The Peptide Section immediately set about synthesising analogues. The chemistry team was led by Anand Dutta. In 1976, Mike Giles in Dutta's team synthesised ICI 118,630. This was a decapeptide, similarly to LHRH, but with at least 100 times its potency. A patent application was filed with them and Furr as inventors in April 1977. In that year, Furr took the lead on ICI 118,630 after Arthur Walpole's retirement. The compound was later named **'Zoladex' (goserelin acetate implant)** (and it is actually the acetate salt of the peptide.)

## Strange news

Then around 1978, a tantalising rumour about LHRH surfaced. Scientists in Canada and the US hinted that on *repeated* dosing, LHRH (and its analogues) had a remarkable and unpredicted effect: it ceased to be an agonist! It did not continue to stimulate gonadotrophin secretion, but *inhibited* it. It blocked it, as if it had changed into an *antagonist*.

The first news of this reached the team from research director Brian Newbould, on his return from a US trip. An immediate flurry of studies with ICI 118,630 confirmed ICI's drug had the same effect.

'Zoladex' first causes the body's system to flood with hormone (in cancer treatment, this unsurprisingly causes an initial flare of the tumour). Nevertheless, it then works as a real antagonist of LHRH would, as it persuades the body's pituitary gland to *shut down* production indefinitely. This sort of feedback mechanism is a form of 'downregulation'.

The future was obvious: 'ICI need look no further than the potent LHRH agonist that it already had … to take forward into the treatment of breast and prostate cancer.'

## Peptide Section

'Zoladex' was the leading success of ICI's Peptide Section. This had been set up in the early 1960s when chemist 'Jack' (J. S.) Morley (later a great supporter of 'Zoladex') and Alan Laird had been sent to Liverpool University to learn about peptides and start the section. It was disbanded some forty years later.

Before it closed, some remarkable experiments were done on making peptide 'libraries' (large sets of mixed peptides that were studied *without* the mixtures being separated). One such library that peptide chemistry experts helped me to build myself in only four weeks, during antibiotic research for Neil Hales in the 1990s, contained 2.4 million compounds – several times the number of *real* compounds made by the entire company to that date! It was a taste of the new world to come.

## How to develop it?

The case for developing ICI 118,630 was compiled and energetically put forward by Barry Furr. But there were formidable issues. The synthesis and purification of 'Zoladex' would be very challenging, involving 15 chemical steps, followed by purification.

Anand Dutta
ASTRAZENECA ARCHIVES

'Zoladex' structure
GEORGE B. HILL

**'Zoladex'***
(goserelin acetate)

**\*'Zoladex' is the acetate salt as an implant**

Moreover, 'Zoladex' proved to be a very difficult drug to deliver into the body. The initial clinical trials used an aqueous solution for injection. At least two other LHRH super-agonists that were being developed by other companies did the same. So clinical and commercial ICI colleagues were eager for a means to distinguish 'Zoladex' from these and give it a marketing advantage.

Could the drug be dosed at *long* intervals but released *slowly* (this is called sustained release)? Many were sceptical. Yet Brian Newbould was keen that it should be investigated. He had recently helped create a Formulations Research Group. This came under Formulation Department at Macclesfield but, critically, was based at Mereside. A new member of the group, Frank Hutchinson, a polymer materials scientist who had just moved to Mereside from ICI's Corporate Lab, took on the task.

Brian Newbould
ASTRAZENECA ARCHIVES

## A new road!

**Cometh the hour, cometh the man! Frank Hutchinson's previous projects had included a new road-surfacing material, but he was a deep thinker outside the box. He worked largely alone on the 'Zoladex' project – though this was partly because his new biology colleagues were disconcerted, as his conversation was sprinkled with words like 'entropy' and 'enthalpy' – physical science terms most of them had gratefully left behind at university!**

# LIFE IN THE PARK

## Technology success

In a remarkably short time, Hutchinson devised a unique and biodegradable polymer matrix in which 'Zoladex' could be injected. It released the drug at exactly the required rate. Hutchinson later commented: 'They said we were living in cloud-cuckoo-land … It was the most badly-designed experiment you could think of – but it worked.'

(Frank Hutchinson shortly afterwards developed a serious heart condition, necessitating a heart transplant. His career was curtailed, though not his zest for life; he remains one of the longest surviving transplant recipients.)

The new formulation worked well in man; and provided the desired marketing advantage, so the product development team led by Sandy Todd applied for and were granted permission to drop the aqueous formula. Jega Iswaran and Hugh Adam steered safety assessment work and Robert Donnelly, along with medics Bob (R.A.V.) Milsted and Ian Jackson, coordinated clinical trials in 1982. Alex Pleuvry covered the global marketing side (as he did for 'Nolvadex', 'Casodex' and 'Arimidex'). Les Hughes comments that it was it was Alex 'who persuaded the company to invest. Without him, we would never have made it.'

Using the formulation, 'Zoladex' was launched in the UK in 1987 and approved in 1989 in the US for the treatment of prostate cancer. For patients it required an injection under

the skin only every 28 or 90 days in the form of a tiny rod of biodegradable polymer containing the drug. About the size of a grain of rice, it releases the drug itself very slowly.

The formulation was as important as the drug in making 'Zoladex' the commercial success it has been. In 1991, it won the Queen's Award for Technology and Dutta, Furr and Hutchinson won the Society for Medicines Research Award for Drug Discovery for it. In 2000, they (and 'Zoladex') were featured in a display of British scientific achievements at the Millennium Dome.

More importantly from a commercial perspective, no generic manufacturer has yet been able to reproduce 'Zoladex' and manufacture a generic verson of it, presumably due to its complex synthesis and formulation. Consequently in 2014 'Zoladex' was still AstraZeneca's sixth biggest selling product, at $924million long after the substance patent expired.

But the road had been rocky. Sandy Todd records that:

> At almost every development team meeting either the Process Development chemist or the Pharmaceutical Development representative – or both – reported that he had met a major problem which might threaten the supply of clinical trial material. However, we prevailed in the end …

### In women

'Zoladex' is also effective as cytotoxic chemotherapy against breast cancer (and a combination of it with **tamoxifen** is better than either agent alone). This is because 'Zoladex' also causes oestrogen levels to drop in the same way as it lowers testosterone levels. 'Zoladex' also proved to have much wider medical uses than first thought; it became valuable in endometriosis and other benign gynaecological disorders.

### More needed

The first effect of Zoladex appears to be flooding of the body with androgens. Because this can produce a flare of growth in a cancer before the tumour starts to be suppressed – and to control the cancer over the longer term with reduced side-effects – a different agent (an 'anti-androgen') is also needed.

That was already the target of a different project.

## Prostate cancer 2: 'Casodex'

Reducing androgen levels (using a drug that achieved the equivalent of surgery or 'chemical castration') was not enough. The key effect driving the growth of over two-thirds of prostate cancers is the binding of male hormones to the 'androgen receptor' (AR) in the prostate gland itself. Blocking this effect with an anti-androgen was also a key research target.

## Anti-androgens

The first effective anti-androgen used clinically was **cyproterone acetate**, developed by Neumann and colleagues at Schering AG after a search of chemical collections. First used in 1964, it had a broad range of side-effects including liver toxicity and loss of libido.

Cyproterone acetate was a steroid, a class of molecules that can have many biological effects, which often cause problems. Drug companies hunted for much simpler, non-steroidal molecules. In the 1970s, Neri and others at Schering-Plough discovered **flutamide** (probably a development of a series of anilides identified as having antibacterial activity in 1966/7 by Monsanto). The biologically active 'entity', produced in the human body on dosing the drug, was its derivative **hydroxyflutamide**.

## 'Casodex' (bicalutamide)

In the late 1970s, following the success of **'Nolvadex'** and before the action of **'Zoladex'** was proven, Alan Wakeling proposed a hunt (building on cyproterone and flutamide as leads) for anti-androgens. At the same time Barry Furr proposed a search for a different agent class, anti-progestins. Cyproterone acted as both of these, so it was clear new compounds should also be tested as both. Furr took responsibility for the animal work rather than Wakeling (who was a biochemist). Following a growing emphasis on 'Zoladex' and prostate cancer, Furr later focused on anti-androgens alone.

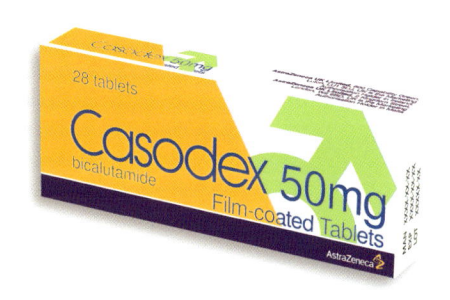

Flutamide was of growing interest, as it had been found to block androgens from the adrenal glands as well as the testes (unlike castration). It was thus likely the action of 'Zoladex' would be enhanced by dosing alongside an anti-androgen. But flutamide, although causing far fewer side-effects than cyproterone acetate, led towards liver toxicity and diarrhea. Also it had a short 'duration of action', requiring a dose every eight hours.

Furr therefore urgently sought a compound as potent as fluta-mide without these shortcomings. The result was Alderley Park's next cancer drug, the anti-androgen **'Casodex'** (**bicalutamide**). This compound was a relative of hydroxyflutamide.

The chemistry was led by Howard Tucker with Les Hughes; and Glynne Chesterson in Tucker's team was asked to make the first sample in November 1982 of what became ICI 176,334 and even-tually 'Casodex', with the patent application filed with Tucker as inventor in the following July.

Barry Furr received the OBE in 2000 for his part in the teams for both 'Zoladex' and 'Casodex'. Gerard Costello steered the latter's development and Bob Milsted its clinical trials. 'Casodex' was launched in 1995 for advanced prostate cancer, and for milder forms from 2002.

It had few side-effects and rapidly overtook flutamide (which was itself approved for use in 1989) in clinical use. As hoped, 'Casodex' became popular because it needed dosing (orally) only every 24 hours. It soon became the agent of choice and proved as effective

'Casodex'
(bicalutamide)

as castration, but with little undesirable effect, in many prostate cancer patients. It also has effect in female cancers. Again as hoped, 'Casodex' became commonly used alongside 'Zoladex'.

'Casodex' was also shown to have two more advantages. It accelerates the degradation of the androgen receptor, preventing the cancer from being stimulated by androgen; and it is effective in early prostate cancer cases.

Sadly most advanced prostate cancer patients eventually appear to become resistant to anti-androgen treatments like 'Casodex', but only after a significant extension of life.

In 1995, Schering-Plough tried to block Zeneca's 'Casodex' entry into the US market using a combination patent they held, which claimed the combined administration of an anti-androgen with an LHRH agonist. However, Schering-Plough were only co-owners of this patent – and Zeneca had already acquired a licence from the other owner, Roussel. Following a legal dispute looking at ownership and infringement, Zeneca was eventually found to have a valid license to the patent and their right to continue marketing 'Casodex' was upheld.

# Breast cancer again 1: 'Faslodex'

Two new breast cancer drugs were also discovered in Alderley Park, of different sorts. One was what the company was looking for; but the first was something it was never convinced it would find!

### A pure anti-oestrogen?

Throughout the story of **'Nolvadex'** (**tamoxifen**), there was an awareness that the latter was an anti-oestrogen but one that had a small degree of oestrogenicity itself. That is, it blocked the action of the female hormones but had a slight hormone-like action. It was believed this might account for the few undesirable side-effects of tamoxifen. A project was begun, to search for a pure anti-oestrogen with no oestrogen effect at all. About 80 compounds were initially investigated.

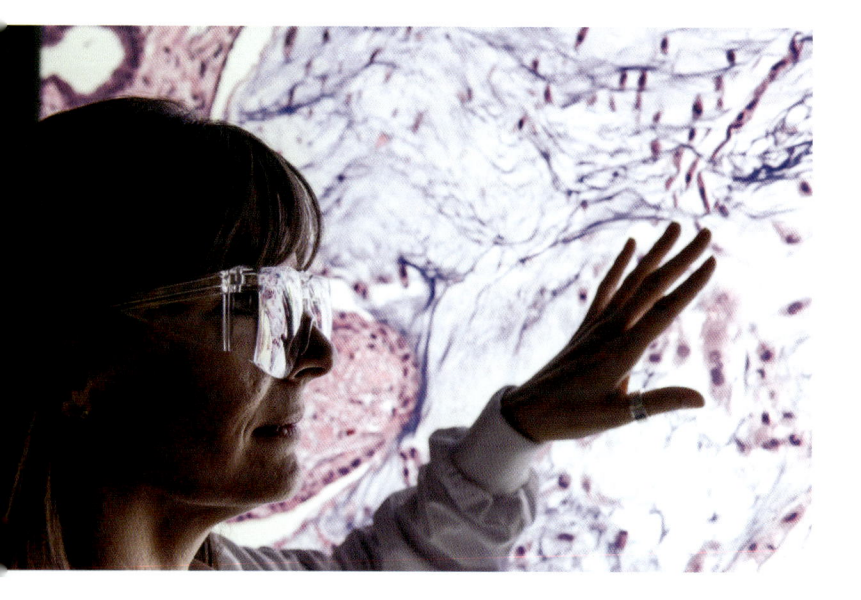

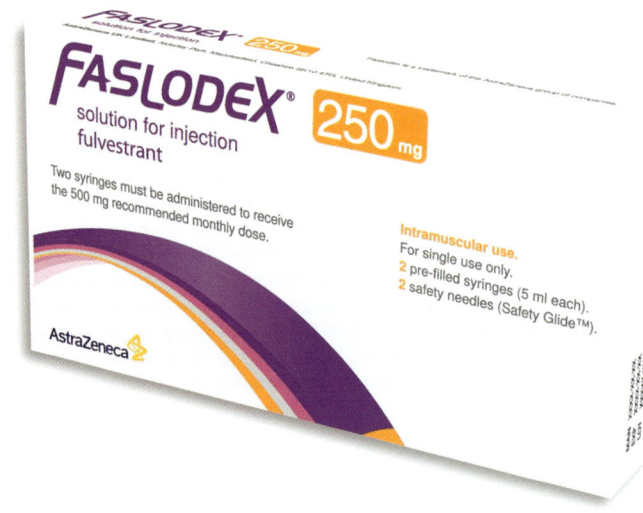

Chemist Jean Bowler recalled chemistry that her team had done previously on steroidal compounds, to create a compound not intended as a drug at all, but rather as an 'affinity probe' for the oestrogen receptor. This had required a potent steroid molecule (one able to bind to the body's receptor) to which a chemical chain had been attached that could be linked to an inert material.

Twelve steps of chemistry, starting with the natural hormone oestradiol, led to a steroid with an 11-carbon alkyl chain attached, the final carbon being a carboxylic acid. Some speculative amides were made to be tested as drug molecules from this hard-won acid and one of them, ICI 160,325, registered by Bowler on the 23 January 1981, had the unique distinction of being the first pure anti-oestrogen ever made.

The news startled both the project team and their senior chemistry manager, Tom McKillop. I was actually in that laboratory as Tom walked in with a beaming smile on his face.

'Congratulations! I was going to shut you down next week!'

By such margins are some remarkable deeds done.

The first compound was not very potent, so Jean and her team set about making improved molecules. One amide made by Tim Lilley, ICI 164,384, became an exciting lead. But there were some concerns that such amides might not be sufficiently stable. In addition they were, in general, poorly crystalline. So Jean's team also replaced the amide with other polar groups such as amines, sulfoxides and sulfones – and found the pure anti-oestrogenic activity could be retained. (Other research group members showed that some non-steroidal oestrogenic compounds, when modified to carry the same types of long chain, were also pure anti-oestrogens.)

## 'Faslodex' (fulvestrant)

But concerns about metabolic instability remained. One way to reduce this is to incorporate fluorine atoms into a molecule. This may also make the molecule more crystalline. Jean, well aware of this, asked a new member of her team to make compounds with chains that were fluorinated. With help from Alan Breeze at Brunner Mond, the hazardous reagent sulfur hexafluoride was used successfully for this by Robert Stevenson. He was asked to

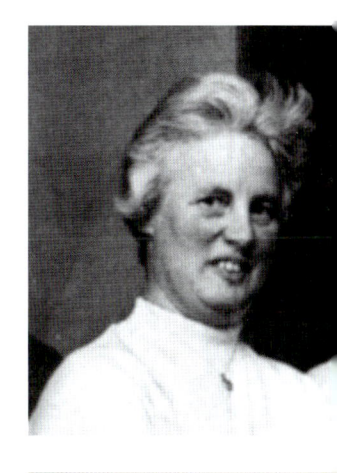

Jean Bowler
PHOTO ASTRAZENECA

Tom McKillop in 1997
PHOTO ASTRAZENECA

Key steps in the original synthesis of 'Faslodex'

'Faslodex' structure
GEORGE B. HILL

'Faslodex'
(fulvestrant)

(Philip J. Hogan, Lyn Powell*, Graham E. Robinson. *Org. Process Res. Dev.*, **2010**, *14*, p1188–1193)

'Faslodex'
(fulvestrant)

make a sulfoxide derivative, which proved to be crystalline and so could be purified more easily. It was August 1983; the compound was ICI 182,780, which became ZD9238. The patent application was filed in October 1984 with Bowler and Brian Tait named as inventors. **'Faslodex'** (**fulvestrant**), the most revolutionary breast cancer drug of its generation – and still the only pure anti-oestrogen ever to reach the market – had been born.

The lead biologist was Alan Wakeling. 'Faslodex' inhibited MCF-7 breast cancer cells by 80% under conditions where 4-hydroxytamoxifen, the active metabolite of 'Nolvadex', achieved only 50%. Technically speaking, the new drug was a selective estrogen receptor down-regulator (a SERD). That is, it works in breast cancer that is still oestrogen-dependent (breast cancer can cease to be hormone-dependent, in which case hormone treatments cease to work). Nevertheless, 'Faslodex' is a unique agent which, in addition, speeds the breakdown of the estrogen receptor itself, just as **'Casodex'** does with the male androgen receptor in prostate cancer (thus preventing the hormones from stimulating the cancers).

## Development

In 1988, 'Faslodex' entered development for breast cancer as well as for benign gynae-cological problems. Clinical studies proved positive, but a decision was later taken to halt clinical development, in order to allow a focus on **'Arimidex'** for breast cancer. The commercial view was that the present injectable ('intramuscular') formulation would not compete well with **'Nolvadex'** tablets. 'Faslodex' was put on the list of drugs for potential out-licensing (to another company).

## We'll keep it!

In 1994, another big pharma company made an approach. They thought they might be able to develop an oral version of 'Faslodex'. So AstraZeneca decided to keep it! Development was restarted to see what could be done. A formulation in soft gelatin capsules (nicknamed 'pina colada'!) worked in animals but showed drug metabolism in man and was stopped in 1996.

## Intramuscular

By 1996, the commercial view had changed; work on an intramuscular formulation was restarted, and this time was successful. Ray Evans and Rosalind Grundy invented the crucial formulation that allowed 'Faslodex' to be injected once a month. The formulation patent application was filed in January 2001.

## Synthesis

A different problem from the beginning had been making the drug at all. As a steroid it was a very expensive drug to make. When 'Faslodex' was first discovered, it was said to be impossible for the company to develop and market it straight away, because the amount needed of **oestradiol** (the starting material) was calculated to exceed the entire world supply at the time. (The main supply was from mare urine and ICI decided it was not in the horse-farming business.) Foreseeing this, the research team had earlier tried hard to find a (far cheaper) non-steroidal compound with similar efficacy.

Later, however, much work by many dedicated experts, and the availability of synthetic sources of steroids, made it possible to manufacture and develop 'Faslodex' itself.

## Dose and approval

Clinical trials showed 'Faslodex' to be efficacious at a dose of 250mg monthly and to have a low incidence of problems. It was launched in the US in 2002 and in the EU in 2004.

By this time, its substance patent life was only short, although a patent term extension was granted in the US (and several other countries) after its approval. Despite the lack of substance patent protection, 'Faslodex' has meaningful patent protection until 2021 provided by a patent on the Evans and Grundy intramuscular formulation. To date, no third party (or indeed AstraZeneca, and goodness knows, we tried!) has come up with an alternative formulation for a drug that has all the solubility of ear wax!

In the late 1990s, further clinical development work showed that doubling the dose to 500mg monthly was more efficacious. This has now replaced the previously registered dose and expanded the use of 'Faslodex' into more markets, such as Japan. In 2014 it held on to the last place in AstraZeneca's list of top ten selling medicines, at $720million, a rise of 6% on the year. Its potential is still being investigated in ongoing clinical trials to see if its benefits can be brought to a wider group of breast cancer patients.

# Breast cancer again 2: 'Arimidex'

In the meantime, the research into 'aromatase inhibitors' had also been successful. Aromatase is a unique enzyme inside cells. It belongs to a very important family of enzymes called 'cytochrome-P450'. Named after the 450nm wavelength at which these absorb light, they are produced by a gene on chromosome 15 in the human genome with the name of CYP19A1.

Aromatase converts 'male' hormones (androgens), such as testosterone, into 'female' oestrogens. As described earlier, manipulation of these 'female' hormones, chiefly oestrone and oestradiol, had long been seen to produce benefits for some breast cancer patients.

## 'Arimidex' (anastrozole)

The target of ICI's aromatase programme (which it achieved) was a selective, long-acting aromatase inhibitor. During it, many chemical types were considered for their potential as aromatase inhibitors, using published data from academic groups in Britain and abroad, especially in the US. Enzyme modelling by Phil Edwards also gave important clues.

The ICI team soon found that discovering potential drugs with adequate *in vitro* potency was not a problem; a great many compounds, even some very simple ones, were very potent. Their $IC_{50}$ (a standard measure of potency) often reached sub-nanomolar, a level quite adequate for an effective drug. This made the project very attractive compared with the rather weak starting points from which many projects begin.

Promising compounds mostly contained imidazole, triazole or pyridine rings. (Compounds in the first chemical series developed had an imidazole ring, but this was later replaced by a triazole.)

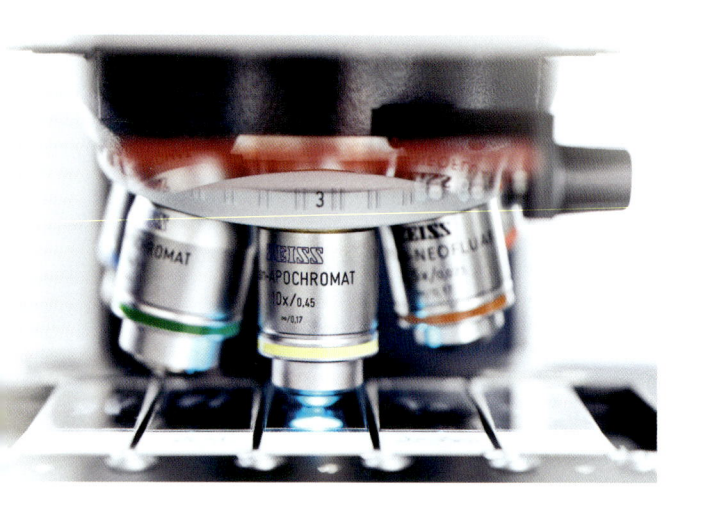

## Getting the balance right

But two important things were needed; and in biological testing (led by Mike Dukes) it proved very difficult to find a compound that had both at the same time.

The potential drug must be potent and long-lasting (when tested in laboratory rats, when dosed orally). But it must also be highly *selective*: it must not have much effect on other, similar and important cytochrome-P450 enzymes in the body. This had to be demonstrated both *in vitro* and (in both laboratory rat and dog) *in vivo*. Without this property, the cancer drug would also attack healthy systems in the body. But the required combination proved very rare.

This is perhaps the biggest challenge in drug design: making sure the new medicine acts only as intended, against the enzyme the cancer cell depends on, while not affecting other enzymes. The first compound chosen by the project team (managed by Dave Julian) was ICI 204,208, a lead from anti-fungal research previously investigated by Tom Boyle. It looked promising, but on long-term animal testing proved to be not quite selective enough. It affected cholesterol synthesis, an important process in the body. This is especially important in the eye, which has no blood flow and so cannot receive cholesterol made by the liver; so the compound caused cataracts, and thus was sadly dropped in 1999.

The scientists had chosen a backup, ICI 207,658, which had been first synthesised in August 1986 by Mike Large in Phil Edwards' team. In early clinical studies this also proved to be a potent, selective, long-acting, oral inhibitor of aromatase. By the time the first lead compound 'fell', Mike Dukes' team 'had already completed the full pharmacology package on its possible replacement and rapidly showed it did not have the same toxicity problem.'

The patent application was filed in June 1988 with Edwards and Large named as inventors. Safety assessment led by Nigel Barrass confirmed its potential and it was named '**Arimidex**' (**anastrozole**). It remains fundamental to breast cancer treatment to this day.

## Competition

Aromatase research was intensely competitive at the time. At least three other companies (Janssen, Novartis and Pharmacia) were ahead of ICI. The latter's licensing group had

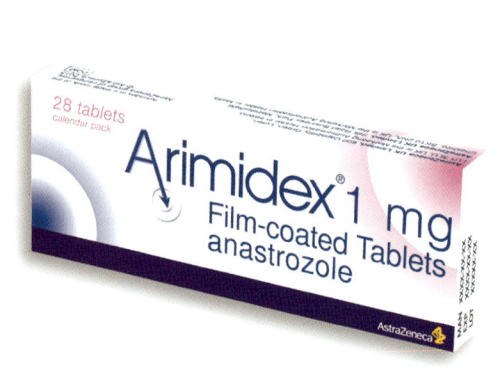

earlier visited Janssen to propose a co-development deal on their compound, **vorazole**. Fortunately they said no! Had they agreed, 'Arimidex' would have been dropped – and vorazole was later abandoned. In the event, ICI overtook the pack and gained approval for 'Arimidex' first. Brent Vose drove the development campaign while US clinician Paul Plourde brilliantly short cut the FDA approval process.

## Launch

'Arimidex' was launched in 1995 for the treatment of advanced disease; approvals for early breast cancer treatment followed in 2002. It was a huge achievement for Alderley Park to launch a second world-leading breast cancer treatment to follow 'Nolvadex'. Its patent expired in 2010.

A major factor in the growth of use of 'Arimidex' was the ATAC study, the largest study of its kind. This was a clinical trial involving almost 10,000 women with localised breast cancer (from all over the world) which compared tamoxifen and 'Arimidex', alone or dosed together. This showed in 2005 that, after five years, the group taking 'Arimidex' had significantly less recurrence of cancer than any of the tamoxifen groups. ATAC's results are still being followed up. Another study showed that patients who were switched to 'Arimidex' after they had been taking tamoxifen for two years post diagnosis had, when analysed across all patients, a 40% decreased risk of breast cancer returning. An even greater reduction was evident when patients whose cancers were not hormone-dependent were excluded from the analysis.

Aromatase inhibitors are among the leading treatments for postmenopausal women, though they are not suitable for premenopausal women. Because they reduce oestrogen synthesis everywhere in the body, patients also may need bone density to be monitored (unlike tamoxifen, which maintains bone density).

One use of anastrozole is something AstraZeneca does not condone. It has become a popular adjunct to the use of anabolic androgenic steroids (AAS) by bodybuilders, who refer to it as just 'adex' for short. When bodybuilders use AAS to build muscle tissue, co-administration of the drug was found to prevent certain AAS compounds becoming aromatised into oestrogens – which could lead to decidedly unwanted consequences! This use of (now generic) anastrozole is certainly not supported by AstraZeneca.

'Arimidex' (anastrozole)

Route to 'Arimidex'
GEORGE B. HILL

'Arimidex' logo 2015
COPYRIGHT
ASTRAZENECA

# Wider horizons still

## Widening the target: 'Tomudex'

Hormonal treatments for breast and prostate cancer have a weakness; if the cancer changes over time and stops depending on hormones, the drug will stop working. Moreover, cancers that do not depend on hormones to grow will not be touched.

This was of concern to the company's investors and marketing department as well as to the rest of humanity. Everyone wanted cancer drugs that did not have horrible side-effects but might work on a range of different cancers that could be treated by any sort of drug. Treatment of cancer that had nothing to do with hormones was an obvious further target.

Structure of 'Tomudex'
GEORGE B. HILL

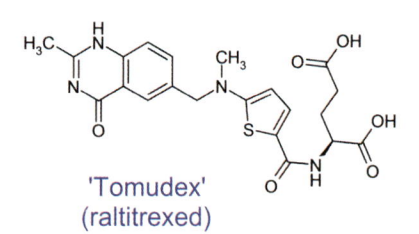

'Tomudex' (raltitrexed)

'Tomudex'
COPYRIGHT ASTRAZENECA

Les Hughes (right) in 2003 with Royal Society of Chemistry President Sir Harry Kroto (invited by Ian Wilson) at the Park
PARK LIFE, November 2003

### 'Tomudex'

An early example was **'Tomudex'** (**raltitrexed**), the first sample of which was made in the team of Les Hughes (by synthetic chemist John Oldfield). This project was a collaboration with the Institute of Cancer Research (ICR) which had been working on the problem for ten years. 'Tomudex', a treatment for colon cancer, is from a class of drug known as anti-metabolites, which inhibit (block) the work of a chemical carrying out a natural process of metabolism in the body.

The area of cytotoxic drugs was then new to the company and Trevor Stephens from the ICR came to ICI to lead the bioscience that led to 'Tomudex'. Safety assessment for it was led by Nigel Barrass. The patent application was filed in 1987 with Hughes named as inventor.

Anti-metabolites are used in cancer treatment to interfere with the production of DNA – and therefore of cell reproduction and of the growth of tumours. In the case of 'Tomudex', this works because 'Tomudex' is an inhibitor of an enzyme called thymidylate synthase (and possibly of other similar ones). The body uses this in a process leading to the organic base called thymidine, one of the building blocks of DNA.

'Tomudex' works because it is chemically similar to the important vitamin folic acid. Because of this, it inhibits these enzymes that use folic acid derivatives. This prevents the formation of DNA which is required for the growth and survival of both normal cells and cancer cells.

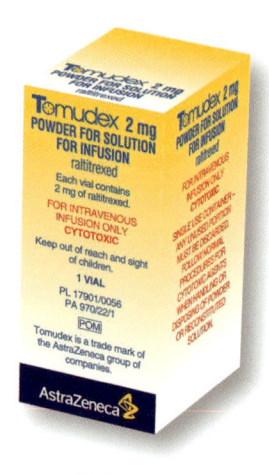

'Tomudex' was approved for sale and launched in Europe in 1995 but not in the US, which limited its use. It worked but was not better than previous similar drugs. A related medicine discovered by Lilly, **'Alimta' (permetrexed)**, achieved more success after a combination with the known cancer treatment **cisplatin** was shown to be better.

# The Big C: 'Iressa'

The biggest non-hormonal target, and the one against which new drugs are commonly aimed, is non-small cell lung cancer (NSCLC). This accounts for ~80% of lung cancers; and lung cancer is the biggest cause of cancer-related deaths, accounting for 13% of all cancers in 2012 and 1.6 million new patients annually.

## Kinases

Roughly speaking, in biochemistry a 'kinase' (pronounced kynase) is a type of enzyme that transfers a phosphate group (a process known as phosphorylation) from energy-rich phosphate molecules to other molecules. This means that kinases act effectively as a form of 'switch' in cells; and their action makes them key to regulating many of the cell's, and thus the body's, systems.

The first kinase was discovered by Gene Kennedy in 1954. Many hundreds are now known. Because kinases are important controllers of cell growth in abnormal cells they can sometimes cause uncontrolled growth, forming tumours. So they are immensely important in the understanding and treatment of cancer.

An important aspect of what kinases do is described by the term 'signalling' e.g., passing on information from receptor molecules on (or inside) a cell, by the triggering of a biochemical chain of events inside the cell. Depending on the cell, this response may alter the cell's metabolism, shape, or some other property. The chain of events is called a signalling 'cascade'. A cascade can involve *a whole series of different kinases, working one after another in order*. This means that cells can respond and adapt to the many different stimuli via these signalling pathways. A reasonable analogy is of the cells being the engine, and kinases acting as the accelerator and brakes, thus controlling such things as how fast cells grow and die.

Kinases are important in healthy cells. However, if the function of one is damaged or hijacked, the result could be a problem for the cell (e.g., causing the brakes to fail, following our car analogy), and this could result in unpleasant side-effects in the body. One such effect could be the misused kinase causing a cancer cell to grow out of control. Therefore, interfering with it will damage the *cancer* cell. If there is a series of kinases acting as part of a cascade, that series fuelling the cancer cell's growth might have many points which could be interfered with. Therefore. a 'spanner in the works' at *any* point in the cell's signalling machinery, e.g., using a drug that could block the kinase activity, could stop or slow the whole machine of the cancer cell!

An obvious possibility was to identify a sort of kinase that was *very* important to a cancer cell, but *not* very important to healthy cells. If that could be done, a modest side-effect in the healthy cells would be a price well worth paying for a big effect on the cancer.

But kinases are all fairly similar. How could a drug be found to affect just *one* kinase?

## In the Park

At Alderley Park, a programme to study kinases was originally proposed by Alan Wakeling. It was not acted on at first, because it was not thought possible chemically to distinguish one kinase from another. It was assumed an action that affected one kinase would affect (even if to a lesser extent) all the other 500 or so kinases in the human body.

A paper in the journal *Science* from a research group in Israel changed that thinking dramatically. Its authors studied the difference between the kinases IGFR (Insulin-like Growth Factor Receptor) and EGFR (Epidermal Growth Factor Receptor). (A growth factor is a protein or steroid in the body that stimulates growth). Importantly, they identified a compound (not a drug) that could show a 100-fold difference in effect between IGFR and EGFR.

This created much interest and a project was sanctioned. A chemistry programme was begun and molecular modelling (using a virtual model of the EGFR enzyme) was used to determine the shape that a drug would need to have. Compounds containing a pyrimidine ring in their structure were predicted to be useful.

The company collection of previously made compounds was examined; and a structure related to pyrimidines, an anilinoquinazoline, proved interesting. This became the project's first lead. ICI was the first company to discover anilinoquinazolines as building blocks for kinase inhibitors, which later became the subjects of research in many pharmaceutical companies around the world – especially ICI's competitors!

## EGFR

Blocking the effect of EGFR became a major target for cancer research.

In 1953, American biochemist Stanley Cohen joined a research group at St Louis investigating a different agent, nerve growth factor (NGF). Later in that work, in 1962, he unexpectedly observed the effect of an unknown growth factor at work, in the skin and cornea cells of mice, and termed it epidermal (skin) growth factor. In the 1970s he was able to isolate human EGF from urine and went on to discover the site to which EGF normally attached itself in the body. This was christened the EGF receptor, or EGFR. Cohen was awarded the Nobel Prize in 1986 for his work.

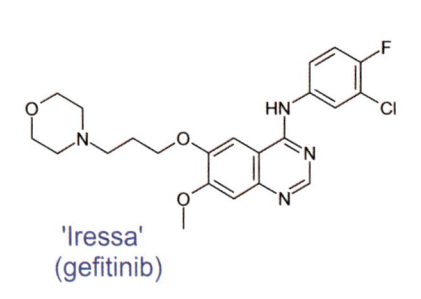

'Iressa'
(gefitinib)

Structure of 'Iressa'
GEORGE B. HILL

In the early 1980s, Texas cancer expert John Mendelsohn proposed that EGFR could be a target for cancer therapy. Along with Gordon Sato at the University of California, he put forward the idea that blocking EGFR would *prevent cancer growth* by blocking the effect of an enzyme, a tyrosine kinase, which was part of the receptor. Later they gained further proof for their theory by developing an antibody that stopped EGF binding to EGFR, and showed that this stopped tyrosine kinase from being 'activated'. They demonstrated the effect in a cell culture in 1983.

In the early 1990s, the first clinical trial of a 'monoclonal antibody' (mAb) agent with an anti-EGFR effect proved that the theory could work medically. This led to the antibody treatment **'Erbitux'** (**cetuximab**) which is used for the treatment of cancer of the colon, lung, head and neck, though it has significant side-effects. But by this time evidence was also accumulating that small molecules could achieve the same result.

### 'Iressa' (gefitinib)

Alderley Park's first lung cancer drug was **'Iressa'** (**gefitinib**).

In 1994, a new class of EGFR tyrosine kinase inhibitors (EGFR-TKIs) was discovered. Following up on the discovery of anilinoquinazolines as kinase inhibitors by Andy Barker (patent application filed in 1993), work at Alderley Park led that year to the first synthesis of a potential drug.

The chemistry was ahead of the biology and there was at first no animal model to test a cancer drug; but the compound was eventually put into pan-tumour Phase I clinical trials in 1998. ('Iressa' was not originally aimed just at lung cancer.) The first physical sample of the drug was synthesised by Mark Healy, a sandwich student (who then took a job with Glaxo!) in a chemistry project led by Keith Gibson under Huw Davies. The patent application for 'Iressa' was filed in 1996 with Gibson named as inventor.

### From lab to patient

The project's lead bioscientist was Jim Woodburn. Gerard Costello steered the transition into the development phase. The drug was tolerated well; but its story would soon take an unexpected and disconcerting turn.

The future of 'Iressa' became the focus of many eyes. In 2002 when 'Iressa' achieved its first marketing authorisation (which was in Japan), its sales were predicted to reach $1 billion by 2006. In May 2003 it was approved in the US under an 'accelerated approval' scheme based on Phase II data. That is, it was given special treatment because it was thought medically valuable, as the first EGFR-TKI to be used in non-small cell lung cancer.

### Strange problem

Unfortunately, the subsequent Phase III data (for the two big INTACT trials), when published in March 2004, failed to confirm the original Phase II findings.

This was a puzzling and massive blow for AstraZeneca, which was already struggling with high profile problems with other products. In patients where it *did* work, the drug worked better in women (and in patients who had never smoked), and was clearly better than existing chemotherapy (with a **platinum** agent and a **taxol** derivative). But it was ineffective in too many of the INTACT patients (who had not been specially selected).

It was noted at the time that – curiously – there were significant differences between the clinical trial results in the US (which were not encouraging) and those done in Asia (which were much better). But the reason for this was unknown.

At this point, its competitor, **Tarceva™ (erlotinib)**, overtook 'Iressa' in the race in the US. In the key Phase III trials, the developers of Tarceva™ elected for the maximum possible dose (compared with one third of it for 'Iressa'). Had this factor, and perhaps a different pattern of patient selection, helped to tip their results in favour of approval?

That did not help 'Iressa'. The American FDA was now doubtful about AstraZeneca's drug. In June 2005, 'the label' for the original US approval was restricted to patients *already*

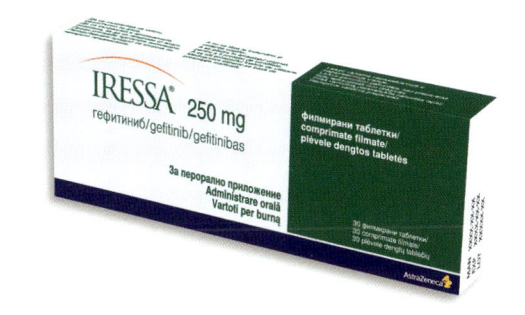

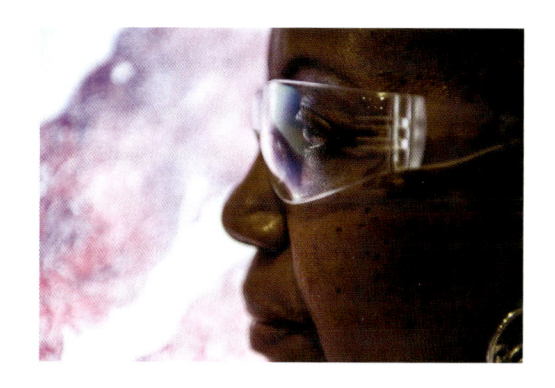

'Iressa'
COPYRIGHT
ASTRAZENECA

Scientist examining tumour biopsy on a digitised image
COPYRIGHT
ASTRAZENECA

receiving benefit from 'Iressa', and the FDA withdrew approval altogether for use of 'Iressa' in new patients, due to lack of evidence that it extended life (in enough cases). (In Japan, by comparison, no restrictions were applied and Japan accounted for $139 million of Iressa's total sales in 2004.)

### Clues already

It is still not clear why Tarceva™ was able to achieve full FDA approval in the US then, and not 'Iressa'; not least, because the hunt to explain the unexpected 'Iressa' trial results was already hot.

In June 2004, two important reports had already been published (by Thomas Lynch and others in the *New England Journal of Medicine* and, on the same day, by Guillermo Paez and others in *Science*) showing that 'Iressa' worked well in a particular class of patients – ones whose EGF receptors were genetically *mutated*. (The successful international Human Genome Project in 2003 helped researchers to work out how.) Moreover, the *Science* paper revealed that such EGFR mutations were more than ten times commoner in the trial's Japanese patients than in its American ones (and suggested that the striking differences proved population diversity was important in cancer clinical trials).

Could *this* explain the original 'Iressa' INTACT trial data?

### Puzzle solved

It was some years before new evidence from further clinical trials, the first of which was IPASS, conclusively demonstrated that 'Iressa' *was* effective in this EGFRm+ patient population – the one that is more common in Asia than in the USA (EGFR mutations can be identified by a number of diagnostic tests).

This evidence enabled 'Iressa' also to be approved and launched successfully in territories where these mutations were less numerous. It became fully available everywhere globally except (until recently) in the US.

### Europe

In Europe, the journey was quite different. Unlike the US, the European authorities declined to accept the original Phase II data, and 'Iressa' was not approved in Europe until 2009 for patients 'with activating mutations of EGFR-TKI receptor'. However, Switzerland *had* accepted the Phase II data, and Swissmedic approved 'Iressa' in 2004. This led to a curious scenario.

### Liechtenstein!

As noted above, 'Iressa' was not approved by the EMEA until 2009, and the European patent term extension for 'Iressa' *should* have been calculated using that date. But there was a problem. European patent term extensions (awards of up to 5 years additional patent term depending on time taken to obtain regulatory approval) are calculated using the *first* approval in any state in the European Economic Area (EEA).

Because the Phase II data had been accepted in Switzerland – although Switzerland was not in the EEA – the Swiss 2004 approval (unhappily for AstraZeneca) automatically triggered an approval in its tiny neighbour Liechtenstein, whose interests Switzerland safeguarded. And Liechtenstein *was* in the EEA!

A legal battle later ensued. Because the 2009 EEA-wide approval was for EGFRm+ patients, unlike the earlier Swiss one, AstraZeneca argued across Europe that their European patent term extension should be calculated from the EMEA 2009 date (not the 2004 Liechtenstein date). The Alderley Park IP team, consisting of Laurence Gainey, Allen Giles and Lucy Padget, ultimately took the UK Patent Office to court, a case that was then referred to the Central Court of Justice of the European Union; but in the end the Lichtenstein date prevailed and AZ 'lost' several years of valuable potential patent life in Europe.

## Reward

'Iressa' was, in any case, recognised as a great achievement. In 2011 Andy Barker and Keith Gibson were awarded the prestigious 'Heroes of Chemistry' award by the American Chemical Society in recognition; and a huge team shared their pride.

Keith Gibson and Andy Barker
ACS PHOTO BY PETER CUTTS

## Approval in the US

Despite its turbulent history, 'Iressa' had sales of $623 million in 2014. And more are likely. As this story was being written, news came in July 2015 that 'Iressa' had (at last!) been approved by the US FDA, ten years after the license was restricted, for only those patients receiving benefit. That is, for first-line treatment of all patients with advanced EGFR mutation-positive non-small cell lung cancer.

American doctors immediately began prescribing 'Iressa' again; and new patients are now joining the 100-plus patients who have remained on 'Iressa' treatment in the US for over 10 years. (These are themselves, of course, wonderful evidence of the effect of 'Iressa'.)

## Personalised medicine

The Iressa and EGFR-TKI story became a paradigm of *targeted* therapy – a pointer to the future.

It revealed the need and opportunity for sub-dividing tumours based on their molecular profile, and then identifying the best drugs to treat the different subsets – e.g. 'Iressa' for the subset of NSCLC patients with EGFRm+ disease. Along with the new prospects in view, there will be the challenge of identifying smaller disease subsets and finding treatments to treat these rarer tumours.

The powerful example of 'Iressa' has helped to start the new era of 'personalized medicine', one that continues to transform how cancers are viewed and treated.

# Multiple pathways: 'Caprelsa'

Drugs that act in more than one way are not generally popular with pharmaceutical companies or regulators – because it means more things can go wrong! It can be more difficult to predict and identify which patients might benefit from the drug, and there is greater potential for side-effects related to 'off-target' activity. But it can be useful to attack cancer cells by more than one mechanism.

## 'Caprelsa' (vandetanib)

One drug conceived and planned in the Park – although the first chemical sample was actually made at the Reims site – was '**Caprelsa**' (**vandetanib**), discovered in 1999. The patent application was filed in November 2000 with Andy Thomas, Elaine Stokes and Laurent Hennequin as named inventors. (Hennequin was later director of the respected AstraZeneca Reims research site from 2004 until it closed abruptly in 2012.)

This is a slightly more general sort of 'kinase inhibitor'. Although not originally designed for that, it has been approved for treating tough cases of an often inherited form of thyroid cancer. 'Caprelsa' acts in three different ways.

Its main designed action is against a different kinase receptor from that of 'Iressa': the VEGF one. The effect of this is to block the development of a blood supply to a growing tumour. It also inhibits the growth and survival of the tumour through two other kinases, EGFR and RET. It is its activity against RET kinase that is understood to be the primary action behind the activity of 'Caprelsa' against medullary thyroid cancer. Donald Ogilvie and Steve Wedge were key bioscientists.

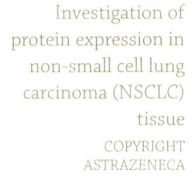

'Caprelsa'
(vandetanib)

Vandetanib was originally designed to be another non-small cell lung cancer drug, and was to be marketed under the name 'Zactima'. Sadly, though it did show tumour shrinkage in some patients, it failed to prove effective at prolonging life and EU regulatory submissions were withdrawn in 2009.

In April 2011 under its new name 'Caprelsa' became the first drug approved in the US for treatment of medullary thyroid cancer in adult patients who cannot have surgery, and the EU gave similar approval in 2012.

By July 2015, 'Caprelsa' was already available in 28 countries when it was sold to Sanofi subsidiary Genzyme in a deal worth up to $300million to AstraZeneca. It was then already in Phase III development for a form of thyroid cancer.

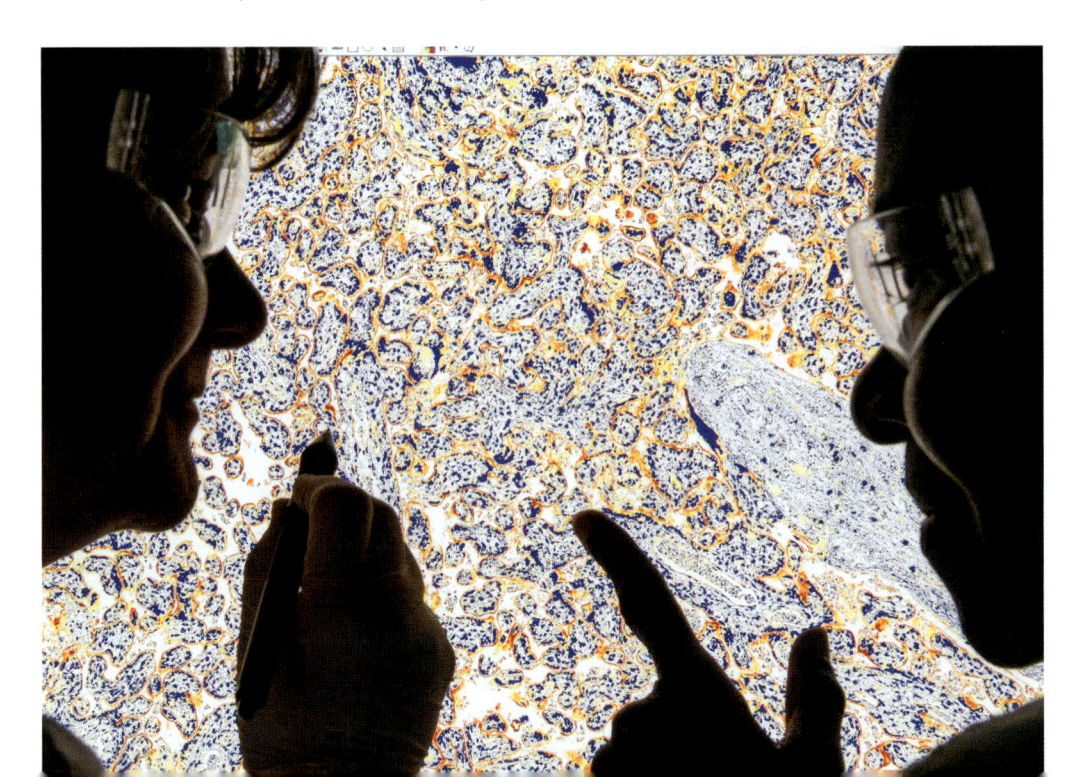

# A perfect project: 'Tagrisso'

The challenges faced by **'Iressa'** became the clues that have unlocked what we all hope is a new major advance against lung cancer.

Drugs like 'Iressa' have allowed lung cancer patients to reach a potential average life from diagnosis of perhaps three years. ('Iressa' typically ceases to be effective after much less than half that time.) Could we make it *five* years? That would be a real breakthrough in lung cancer; it could feel like half a patient's family lifetime – and might buy time to allow a further new invention …

This is the story of **'Tagrisso' (osimertinib)**, known, before it was named, as AZD9291. ('Tagrisso' is actually the mesylate salt of osimertinib.) As the most recent success story from Alderley Park, and as a showcase of what can be done in pharmaceutical research, it is described in some detail. Why did it succeed? In a word, 'Iressa'. Collective knowledge: we had done something rather like it before!

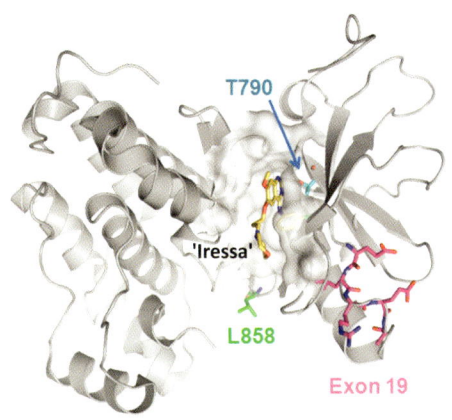

## Sensitising mutations

Patients with non-small cell lung cancer (NSCLC) with mutant EGFR that can be treated with 'Iressa' or related therapies often respond rapidly. In such tumours, the EGF receptors carry one of what are called the activating or 'sensitizing' mutations, most commonly those known as **L858R** and **exon-19 deletions**. (At least 20 different mutations in the EGF receptors in human lung cancers are known). Patients whose cancers do not have such sensitising mutations are not expected to benefit from 'Iressa'.

But sadly, 'Iressa' can stop working even in patients that *do* have them, when the cancer cells become resistant to the drugs. (Just about all tumours eventually develop resistance within 10 to 12 months.) In around 60% of cases the EGF receptor in the tumour (over time, as the 'first-line' treatment acts) develops a secondary mutation called **T790M**. This, unlike those named above, works to *prevent* the drug having an effect. This drug resistance prevents further treatment with 'Iressa' from being very useful at that point against the cancers of patients harbouring such 'double mutant' EGF receptors.

## Selectivity

As described in the 'Iressa' story EGF (Epidermal Growth Factor) is, of course, a natural product of the body and is designed by nature to activate its own receptor (EGFR). Unlike the mutated form of the EGF receptor that many non-small cell lung cancers contain, EGFR elsewhere in the body is *not* mutated and performs useful bodily functions – which the doctor does not want to disturb if possible.

What if this natural or 'wild-type' form of the EGF receptor (wt EGFR) were blocked by a drug? Earlier, it was hoped that this would affect cancer cells more than normal cells. But regrettably, unwelcome side-effects such as skin rash and diarrhea then occur (because 'wild-type' EGFR is common and used by the body in the skin and digestive tract).

Also, because EGF receptor contains a kinase (strictly speaking, a 'kinase domain') in part of its protein, this gives another challenge. Kinases are a class of biological catalyst of which the body has many – and many of them are very similar!

So what was needed was a *selective* drug; one which would:

a) Block the EGF receptor in L858R and/or exon-19 deletion cancers in patients;
b) NOT block the natural 'wild-type' receptor elsewhere in the body of all patients, away from the cancer;
c) NOT interfere with the body's many other kinases to any significant degree;
d) *In some way*, prevent the T790M mutation (if and when it appeared) from interfering and stopping the drug from working.

## EGFRM

The project, which began in May 2009 was named EGFRM (M for mutant) or EGFRm, and was seen from the start as a good prospect and a project that ticked all the boxes. Often in pharmaceutical research it can be difficult to match the research with the need. A remarkable scientific discovery may be made … but what can we do with it? Or there may be a desperate need … but what can we do about it?

In this case, however, the need (the patients who were being successfully treated with 'Iressa' or a similar drug until they developed resistance) and the means (a lung cancer programme already equipped with a great deal of information) were both present. And, of course, if a drug of the right sort could be found then the new drug might sometimes be used straight away instead of the old drug, 'Iressa' itself (unless a new problem arose).

It was very clear which patients would benefit – and there were many of them. If a good working hypothesis could be put forward, this would look like a perfectly aimed project.

## Irreversible inhibitor

So the researchers then had to ask, what *sort* of drug was wanted?

If a receptor in the body needs to be blocked – using, as described in the **'Nolvadex'** story, the equivalent of the 'wrong key' for the 'lock' – then one could imagine two sorts of key. One sort might block the real key from entering, but then fall out again itself (and keep having to be pushed back in). But another sort might jam in the lock, bond to it in some way, and so block it forever. This is the principle behind the sort of drug called an '*irreversible inhibitor*'. This should not just block the cancer cells' EGFR for a short period of time, but disable it for a long period – potentially indefinitely, though this has not yet been proven with 'Tagrisso' – so that the tumour could not change and escape the drug's restraint.

Importantly, this would be a much more potent sort of medicine; and it would make it much easier to find a promising candidate drug. It is that potency that is needed here. Why?

## T790M – the powerful gatekeeper

The troublesome mutation T790M gains its name by being a mutation of a very significant unit or 'residue' of the EGF receptor protein, a residue known as Threonine 790. This is known as the 'Gatekeeper' residue; and that name tells us how important T790 is in controlling what molecules may 'dock' with the receptor – that is, what keys may enter the lock! – or which drugs may work. A mutation in T790 may therefore have a big effect.

And T790M also has, unfortunately, *another* bad effect. To grow, a cell needs energy provided by the body's principal energy source, **adenosine triphosphate** (ATP); and the T790M mutation in a 'double mutant' cell, unhappily, increases that cell's ability to use ATP (its 'ATP affinity') considerably.

It does this so powerfully that a more ordinary sort of drug – such as a *reversible* inhibitor

– will be almost useless. This is because 'competitive' kinase inhibitors generally act in a similar way to ATP (they are 'ATP mimetics'). To work, they must do better than ATP! – they must *outcompete* it. Only an irreversible inhibitor would be able to do this.

So the scientists knew that an irreversible inhibitor was the only sort of medicine they could predict should be powerful enough to kill 'double mutant'-containing cancer cells – their main target.

## Acrylamides

But the chemists faced a problem.

An irreversible inhibitor is, in effect, a chemical on a knife-edge. It is a drug that could potentially attack whatever it is near. When the correct drug reaches the EGF receptor in the cancer cell, it will stay there and damage the cancer's growth.

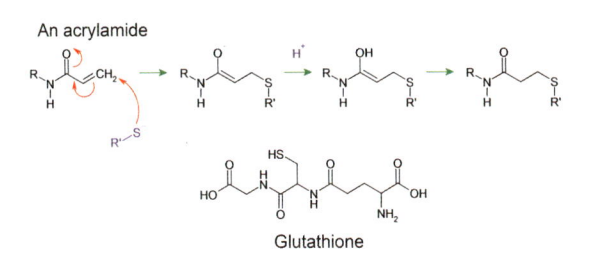

An acrylamide

Glutathione

But what happens before that? It will be a powerful sort of molecule! If it is too powerful, it will not wait to attack the cancer, but will attack another part of the body as soon as it is dosed. On the other hand, if its 'warhead' is too weak, it will reach the cancer cell, but not work until it has stayed there for a long time.

Addition reactions of model compounds with Glutathione
GEORGE B. HILL

So it must have both a carefully calculated (and low) power to attack, and a very high attraction for the EGFR receptor – a high 'reversible affinity' for it. And very many compounds would need testing to find the precise, perfect balance. Unsurprisingly, an irreversible inhibitor does not often become a good drug, and few examples were known. It requires some courage to kill a deadly snake by holding a scorpion to it. Scorpions sting, too!

Many experiments were needed even to prepare the way. To test their proposals, the chemistry team decided to focus on a type of chemical called an **acrylamide**. A range of acrylamide model compounds were compared by reacting each with the simple compound **glutathione**, which acted in the same way the receptor would. This showed how powerful each sort was.

Then acrylamides were made of the potential drug structures themselves. But these acrylamides turned out to be very difficult compounds to make and purify, for they decomposed into useless orange gums unless treated with peculiar care and urgency (I remember them all too well). Numerous variations all had problems. The research team's route followed multiple twists and turns in the evolution of better compounds over some years, with many dead ends. The bioscientists tested immense numbers of compounds.

Scientist looking at 'Tagrisso'
COPYRIGHT ASTRAZENECA

## The basic nitrogen

The chemists eventually found that a nitrogen atom, one that would act as a base, was needed *near* to the acrylamide 'group' (in chemistry a base is the opposite of an acid). To start with, the chemists attached this to the acrylamide itself. But then they found that it also worked when positioned adjacent to the acrylamide. As the team recorded in the *Journal of Medicinal Chemistry* when publishing their account in 2014, 'Our interest was piqued by this architectural modification for a number of reasons.'

One of the biggest reasons was potency. A series of compounds whose molecules held a nitrogen-containing 'piperazine' ring fell short and acted too weakly. Other series in which a nitrogen was replaced with a carbon or oxygen atom were tried, but proved particularly disappointing. But when a small basic side-chain was used to replace the piperazine, potency was achieved, which the team described as 'exquisite', leading eventually (after a great deal more work) to **'Tagrisso' (osimertinib).**

## Standing on shoulders

Around twenty-five each of chemists and bioscientists worked on the early research phase of 'Tagrisso'. Chemistry was led in the later stages of this by Ray (M. R. V.) Finlay and Richard Ward, and bioscience was led by Darren Cross and Susan Ashton, under project leader Teresa Klinowska. The patent application, eventually listing eight inventors, with a 'priority date' of July 2011, was filed in July 2012.

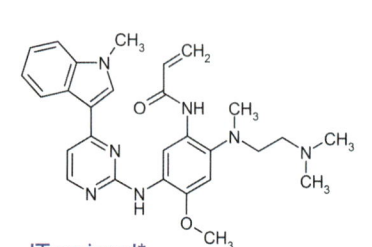

'Tagrisso'*
(osimertinib)

*'Tagrisso' is the mesylate salt

In such a huge project, nothing is done except by standing on the shoulders of others. The final assembly of the 'Tagrisso' molecule is an example of this. Thus, when synthetic chemist Gail Wrigley in Ray Finlay's team was asked to synthesise the first physical sample of the drug itself, she combined chemical precursor 'halves' made earlier by colleague Matt Box and by the project's experienced and knowledgeable lead synthetic chemist, Peter Smith. I record this because tragically in the following year Pete died very unexpectedly, only weeks before the patent was formally filed (and in the same week that I left both the same team and the company). Pete could hardly have left a more precious legacy; this image of him therefore stands for all the unseen workers of the huge 'Tagrisso' team.

In the hands of his determined colleagues, the story of AZD9291 is, hopefully, just beginning – certainly for patients.

## Lack of side-effects

All drugs have side-effects. In particular, drugs that affect kinases can upset very many of the body's mechanisms. A specific problem with compounds sharing similar chemical structures to 'Tagrisso' was the potential to disturb the levels of glucose and insulin in the bloodstream (by affecting a different receptor, the insulin receptor).

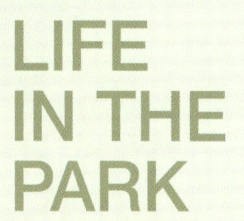
**Mere immortality fails!**

The first generic name proposed for the new drug was 'mereletinib' – derived from Radnor Mere. This was to have revived a tradition of Park-derived names that started with 'Alderlin' in the 1960s, but the name was rejected for technical reasons. But hopefully, the new drug will still become a successful late (hopefully not the last) comment on the historic research done by AstraZeneca and its predecessor companies in the Park.

So avoiding this was critical – the new drug had to have the right 'selectivity profile', affecting EGF only. This was a key project aim and it was very pleasing to confirm that 'Tagrisso' (unlike several very similar compounds tested) had no effect on blood sugar levels.

A different problem is one that has affected all drug research since it was identified. The phenomenon known as 'QT prolongation' is where the heart spends longer than normal in part of its cycle. (The part between the Q and T peak of the so-called 'PQRST' pattern detected by an ECG.) People or families in whom it occurs naturally have a very small risk of sudden death (so must be monitored).

But 'long QT' can also be caused by certain types of drugs, generally any that is chemically a strong base. So all new medicines are now checked rigorously to show that they have no such effect. When 'Tagrisso' was tested, the team found it had a 'very desirable profile' as a potential medicine in this and all other important probable side-effects.

## But would it work?

It was a project that had followed the science, toward a defined patient set and an acceptable side-effect profile. But would it *work*? Yes, it would! 'Proof of principle' was obtained in only four months. 'Tagrisso' did indeed prove to be a highly selective, irreversible inhibitor of both the activating sensitising EGFR mutation and the resistance mutation T790M, which demonstrated sustained tumour shrinkage in preclinical models representing these mutations.

And it attracted support! The early development team that began rapidly to build up included Graham Richmond, Paula Daunt and Anne Galer, with Andrew Mortlock plus Mark Anderton on toxicology and John Freeman with clinical development. A veritable army of AstraZeneca staff then joined in, as the new drug attracted more and more interest and investment.

## FTIM

One of the most crucial moments in the development of a new drug is what is called FTIM, or 'First Time In Man'. 'Tagrisso' was dosed in patients for the first time in March 2013. The scientific papers reporting the results on two of the first patients (both women, as it happened) showed that their Stage IV cancers shrank dramatically, by over 50%.

# TAGRISSO™
## osimertinib

Excitement mounted – and so did apprehension. When Pfizer made a bid for AstraZeneca in January 2014, as described later, AZ's CEO Pascal Soriot grimly warned MPs, at a Parliamentary Business Select Committee hearing, that progress of the company's new cancer treatments could be disrupted by the distraction of a mega-merger: *'What will we tell the person whose father died from lung cancer because one of our medicines was delayed – and essentially was delayed because in the meantime our two companies were involved in saving tax and saving costs?'* He had mainly 'Tagrisso' (and also MedImmune's upcoming **durvalumab**) in mind.

Many more remarkable results, on patients with cancers of the relevant type were being reported. The Federal Drug Authority in the USA noted them and granted 'Tagrisso' in April 2014 a coveted 'breakthrough therapy' designation, as well as 'fast track' status. This exceptional support allowed AstraZeneca to submit the drug for approval by US and EU regulators in the second quarter of 2015 and for a priority review in Japan in the third quarter.

## AURA and approval!

The two pivotal Phase II trials that provided the data for submission to the FDA were known as AURA. They were designed to 'demonstrate efficacy' in a total of 411 patients whose cancers had ceased to respond to older EGFR TKI drugs. They showed a response rate reported as 59% – a stunning result. A supporting Phase I trial showed a median 'duration of response' reported at the time as 12.4 months.

But it was a dramatic surprise when approval by the FDA, expected in February 2016, was announced three months early on 13 November 2015 (very satisfyingly, just as this book's text was nearing its final changes!). The excitement was tangible; within 9 hours of the company's press release, the newly released name 'Tagrisso' found for me over 5,000 hits in a Google search and within 3 days it was 100,000. And more! In Europe 'Tagrisso' received, on 3 February 2016, the EU's first ever new drug (conditional) 'expedited approval'.

## Output and market

Even faster was the medical momentum; the first shipment of the drug to US patients went out within 6 hours of the US approval, and three 'real' patients were treated within the first day!

The marketing effort with Sam Comley now swung into top gear. This followed considerable work by Simon Collett and others in Pharmaceutical Development group (who were responsible for deciding on its mesylate salt as the final form of the drug). An original planned output of 70,000 dose units had been accelerated by an amazing hundred-fold by Pharm Dev to 7 million by the end of 2013; the original launch plan had been aimed at 2019!

## Ongoing trials

Yet immense work lies ahead. Phase III trials are ongoing: the FDA's accelerated approval, based on tumour response rate and duration of response, requires these much larger trials to confirm that the drug gives proven survival benefit.

## Diagnostics

The FDA also simultaneously approved a new companion diagnostic test for use with 'Tagrisso'. This was the result of an agreement made in November 2013 with Roche Diagnostics guided by Rose McCormack and Suzanne Jenkins. The test was developed by Roche in association with AZ, to detect the T790M mutation. This clearly assisted both the approval process and likely patient benefit. The approval was for patients thus identified, with metastatic EGFR, T790M mutation-positive non-small cell lung cancer (NSCLC), whose cancer had progressed despite previous 'Iressa' type therapy.

## Business impact

The patient benefit was, of course, paramount; but a conservative revenue prediction at launch by analysts of $1.1 billion annually by 2020 was excellent news for the business, which itself had estimated $3 billion. The latter may now be plausible, as the lower figure had allowed for an expected competitor's product which then ran into problems. A further boost was the news that 'Tagrisso' (unusually for a drug of this kind) appears to cross the 'blood-brain barrier' sufficiently to give some activity against secondary tumours with EGFR mutations arising from 'metastasis' into the brain. 'Tagrisso' is already available to UK patients under an Early Access to Medicines Scheme. We watch its future with great hope.

But the even bigger impact on the company came from the simple existence of the project. In the context of the failed Pfizer takeover, oncology iMed leader Susan Galbraith put it bluntly: 'Tagrisso' (at the time of writing) looks like it is the drug that has *'helped to save the company'* (and thus at least *some* of the UK pharmaceutical research base that this book has described).

And to one of several patients who have already chosen to record their story and thanks personally, it has simply been: *'Magic – as good as gold dust!'*

## Record breaker!

Finally, at two years and eight months from FTIM to approval, 'Tagrisso' achieved yet one more first, as it became the fastest ever launched drug in the pharmaceutical industry, bar none, to pass between those points. (And it started a year behind the competition …) This was an extraordinary achievement, ensuring that AstraZeneca's great history of research and development at Alderley Park would certainly end in style.

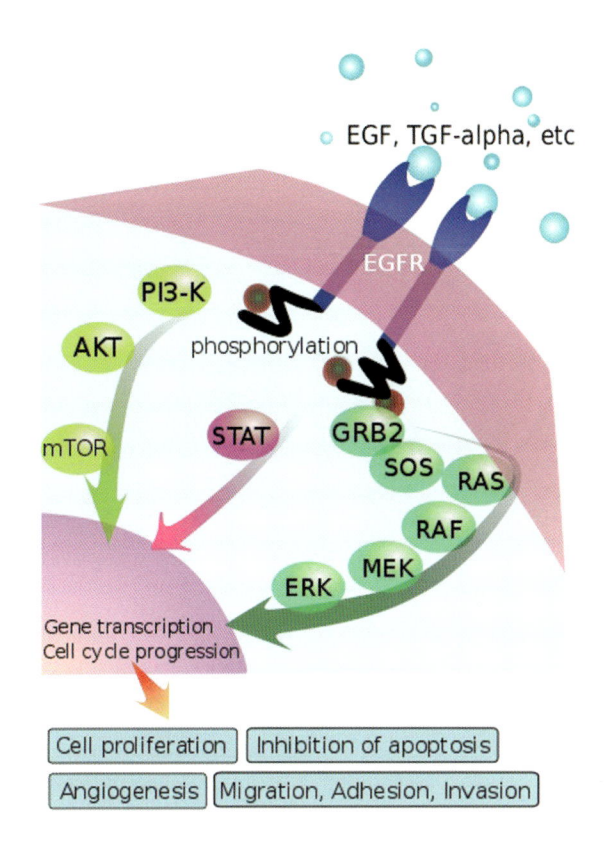

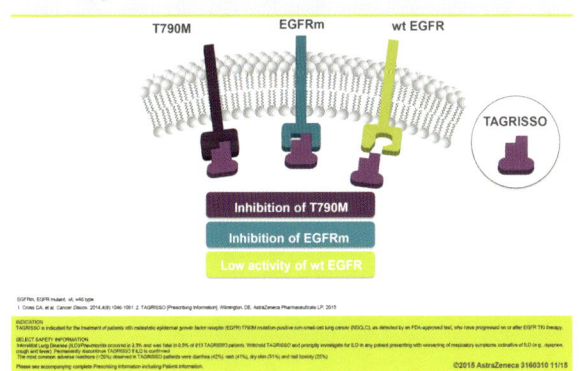

A 'cascade' of kinase enzymes used in cancer cells which are past, present and future targets for research
PUBLIC DOMAIN

'Tagrisso' mode of action
COPYRIGHT ASTRAZENECA

# A site in constant growth: 1993–2008

## Zeneca plc builds on

By September of 1993 the site population had reached 3500, including 375 at CTL. The new laboratories of Block 3 at Mereside opened in October at a cost of £19 million. New animal units also opened that year.

In 1996 Mulberry's was opened, and Alderley Park's woodland was surveyed and a regeneration programme begun. Birdwatchers on the estate were delighted when the new Charles Tunnicliffe bird hide was erected beside Radnor Mere in November 1997.

In 1998, the staff on site first totalled 4000. In the same year, Alderley House's £3.9 million Stage IV was completed; during its building, an interesting discovery was that of a sealed well from the old Stanley mansion.

## AstraZeneca plc builds on, too

On 1 June 1999, ownership of Alderley Park passed into the hands of the new company AstraZeneca plc, formed by the merger of Astra and Zeneca.

At the time, some remarkable statistics about the new company included:

- More than 400,000 sandwiches and 30,000 kg of chips were prepared in the restaurants each year.
- The energy used on site was equivalent to 6000 households, or the same size as Congleton.
- 33,000 visitors came to Alderley Park each year.
- 1.5 million pieces of glassware were cleaned each year.
- Over 150 km of cabling were installed during 1998.
- 14,000 square metres of floor-space required cleaning – equivalent to 30 football pitches.
- 1,400 lambs were born on the estate farm that year.

The merger led to a new wave of development. In 1999, the Block 21 Bioinformatics facility was opened for study of molecular biology. In 2001 a new site development plan was approved, including a series of new buildings and £35 million spent on major capital investment into the utilities infrastructure over the next nine years. Blocks 50 and 52 were approved (and opened three years later at a cost of £58 million; the fully automated compound store alone cost £6.2 million). These were the first buildings in Alderley Park to show an obvious Swedish influence on their design following the merger. Also in 2001 came the opening of Alderley Park's first tiered car park at Mereside, a capacious if curiously disorienting building because of its various levels, which staff were strictly instructed *not* to describe as a 'multi-storey'. At the height of change, no fewer than five tower cranes were in use around Mereside.

## Parklands

One of the biggest projects had become possible only because of another. In 2001 new Alderley Park Farm buildings were built in the parkland south of Church Lodge. This left vacant the site of the old farm, demolished the previous year, upon whose site was approved

### David Beckham's curtains!

Parklands achieved fame in legend as the building which helped to persuade David Beckham to leave the country! The renowned Manchester United and England footballer lived at that time less than a mile away. This was within sight (in winter) of Parklands, a building that its architects had chosen to illuminate with unusual purple lighting.

Reputedly, the driver of the Alderley Park site bus, a shuttle vehicle that transported people between Mereside and Alderley House, was filling his tank with diesel at a local filling station when David drove in. Seeing the AZ logo on the bus, David stomped across. He demanded to know whether it was our building that filled his bedroom at night with a purple glow. And what were we going to *do* about it?

The startled bus driver shrugged. 'Well, have you thought of drawing your curtains?'

History does not record David's reply; nonetheless, it does report that shortly afterwards he and Posh were on their way to Real Madrid. And Parklands still has a purple glow at night!

# LIFE IN THE PARK

Parklands @ night
PHOTO ASTRAZENECA

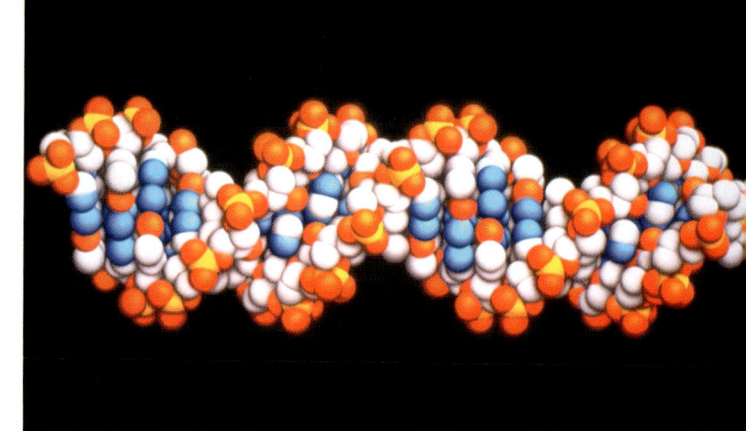

the building of the four-square five storey Parklands office building, designed by LMP. It was opened in 2003 at a cost of £34 million and contained 800 workstations, 30 km of cable and 4,500 square metres of external glass. With its revolutionary chilled ceilings and beams instead of air-conditioning, it is now considered one of the best office buildings in the country – not least because it has good parking!

### Blocks 17, 23 and 50–52

Block 17, opened in 2002, was built to hold the company's DNA archive and genotyping facility, designed to underpin global projects with leading-edge genetic research as a brand-new concept in high throughput screening. The Genome Group itself had been established in 1994 under Rakesh Anand to build on advances in research on the human genome.

The huge new Safety Assessment building, Block 23, opened in April 2003, showed the great importance attached by the company and the industry in general to that subject. The £27 million building, opened by Lord Sainsbury, was built in only 28 months but was designed to be the core of Mereside's safety evaluation of all new products.

The Centre for Advanced Lead Discovery, Blocks 50–52, was opened at a cost of £58 million in 2004, becoming the building on site perhaps more full of high technology than anything yet.

### Radnor Reception and the RA building

Of more public impact in 2005 was the completion of the massive new Cancer Research Area development at Mereside. Approached by a wide apron, landscaped by mature trees, the sandstone and glass Radnor Reception with its adjoining buildings designed by AMEC and Stevenson Bell represented an investment of no less than £60.9 million, the largest in the Park to that date.

It included the 75-foot-high atrium of Block 30 with its many breakout areas, together with the two state-of-the-art flanking laboratory blocks of 33 and 35 to take 265 scientists, plus a link building to the Mereview Restaurant. Radnor Reception opened to visitors and staff in October 2005.

Also completed in 2005 was a new Logistics Centre (Block 109).

## High water

The year 2007 was the high water mark for the research facilities of Alderley Park, when staff peaked at around six thousand.

It also saw the 50th anniversary celebration of the opening of Mereside, which was attended by over 140 guests who heard speakers including CEO David Brennan. The Government guest speaker was the Rt Hon Alistair Darling MP, Secretary of State for Trade and Industry. The keynote speech was given by Sir James Black, who had worked at Alderley Park between 1958 and 1964.

Another historic event that year was the vacating of CTL as Syngenta left the site. But an entertaining appendix to the 50th anniversary was the news that, in Mereside's 50th year, the herons in the growing colony on Radnor Mere's island had also for the first time occupied more than 50 nests!

## Conference Centre and bypass

In 2008 Block 22 at Mereside was opened, as was, much more interestingly to most people, the long delayed and finally sanctioned new £10 million Alderley Park Conference Centre and lecture theatre, this replacing the old cramped one in Block 8. The Centre, designed by Alderson Design, completed the design set of buildings around the Radnor Reception at Mereside and was a remarkable, state-of-the art structure designed to be used by external visitors as well as business conferences. It included a large raked 233-seat auditorium and three combinable meeting rooms.

In the same year, another (even longer delayed) project that finally came to pass was the start of the building of Melrose Way, the three-mile long A34 Alderley Edge north–south bypass.

The last link in the Manchester southward traffic flow system, it was built in the hope of relieving congestion caused by the 26,000 vehicles that previously passed through Alderley Edge daily. It was finished over 6 months early.

The bypass roundabout was sited half way between the two main Alderley Park lodges, Church Lodge and Bollington Lodge, about 400 yards north of the latter. It had been planned for many years and the company had been invited to contribute to its £56 million cost (which would have advanced its date) but had declined. Its route crossed no company land, but would massively affect both traffic flow and environment around the Park.

Nevertheless, it was not the biggest change on the horizon for the Park.

# Value judgement

## Approaching the cliff

Multinational companies only exist because of their investors. This is not a popular fact; during the merger, when rumours were driving up the share price, fund managers Capital and Mercury Asset Management, each owning 5 per cent of Zeneca, made £150 million in two days.

But one of the biggest events of the company year at Alderley Park was the day the City analysts came for their gritty annual briefing on the company's prospects. (It was also the one day of the year when the staff of Mereside were, to their disgust, banned from Mereview Restaurant, the scene of the big event!)

### The pipeline and the patent cliff

The two big issues were, of course, what drug the company might be able to launch next; and how fast its sales would decline on old drugs whose patents had just expired – as the company approached the dreaded 'patent cliff' where its profits would suddenly fall away.

The company's pipeline became a key issue in the years after the merger. AstraZeneca's executives publicly acknowledged that its pipeline became too dependent on precious but risky new first-in-class medicines. Over the course of a few years, a series of major drug failures decimated AstraZeneca's late-stage pipeline and knocked investor confidence. **'Iressa'** entered in 2004 its most traumatic phase; then in February 2006 pending marketing applications for the anti-clotting treatment **'Exanta'** (**ximelagatran**) were finally withdrawn due to concerns about hepatotoxicity; the diabetes drug **'Galida'** (**tesaglitazar**) was scrapped; and the stroke drug **'Cerovive'** (**disufenton sodium**) abandoned.

In 2004 John Patterson, who after piloting the trials for 'Nolvadex' trials had transferred to the commercial side of the company, was recalled to R&D as development director to begin a major review.

By 2007 the whole research process had been accelerated, while staff numbers were beginning to decline. The company, along coincidentally with the global economy, was approaching a peak that concealed a decline already underway.

The first major recent case of a big pharmaceutical company falling over a really dramatic patent cliff came in 2011 when Pfizer lost exclusivity to their blockbuster drug 'Lipitor'. In 2005, 'Lipitor' comprised about forty percent of total profits for Pfizer. After the patent for the drug expired in 2011, Pfizer instantly suffered as global sales of 'Lipitor' fell by forty-two percent and Pfizer's total

Beware of the cliff

profit declined by nineteen percent just in the first quarter. While analysts predicted drops in revenue, nobody had expected Pfizer's income to fall so massively.

AstraZeneca's 'Nexium' produced over $4.9billion in sales in 2010. Its patent expiration in 2014 soon became, in the City's view, a train coming the other way in the tunnel (although the train was briefly delayed by a bureaucratic tangle in the US).

# Ethics, testing and public support

This book is not the place for a study of the ethics of pharmaceutical research; but the questions do not need experts to ask them. What should we research *on?* Where is development work best conducted? The ethical complexities of clinical trials have already been described. How much is it right to spend on development of new medicines? How fast should progress be? Ethical judgements are made upon the basis of society's normal expectation of what is right and proper.

Laboratory testing
COPYRIGHT
ASTRAZENECA

## Environment

What do we research *with?* Chemicals and the environment: the company's detailed care of the natural Alderley Park environment has been described but biodiversity loss is only one of the 'big four' green dangers. Waste of resources, pollution and global warming all demand our attention, too.

## Concerns

The deeply-felt humane concerns are even more obvious: the testing of medicines for humans on animals; the law and standards that regulate such tests; research on human material for which consent can never be sought (e.g. embryonic stem cells) or involving genetic manipulation; and the directing of investment in medical research toward the diseases of those who can pay or not.

## The burden

Research for whom? The question of which diseases (and which humans) to research treatments for is one of economics. We all knew that AZ scientists would not have jobs, and no research would be possible, if we worked on diseases for whose treatment most patients could not pay.

As a young chemist in 1980, I once challenged my section manager, Tom McKillop, about why the company did not do more research into diseases of the developing world. He was friendly, but blunt and painfully rational. I heard no more. However, many years later, the decision fell to the (by then) Sir Tom as to whether to close the Bangalore labs in India, which were doing unique research into tuberculosis, a terrible developing world killer. He decided to keep them open. I took pleasure in writing to him to thank him for that decision, and received a friendly reply; and the Bangalore labs continued doing valuable research, though sadly only for another decade.

## Humans and animals

On this, I can offer only a personal perspective. I have already described in my preface how my wife faced surgery for breast cancer when we had a young family; and how she was treated with tamoxifen, a drug from a project on whose follow-up I worked early in my career (in a large team whose work later led to 'Faslodex'). This involved testing on very many animals, particularly rats.

Within weeks of my wife's surgery, my supervisor asked me to move back into cancer research (I had spent some years working on drugs for rheumatoid arthritis). My first reaction was to feel I had been slapped across the face. Then I came to see that I was standing where few people ever have the privilege to stand. Millions raise funds for the fight against cancer; I could be part of that fight. My wife started radiotherapy, followed later by cytotoxic chemotherapy and then tamoxifen, on the same day I moved back into the cancer labs.

It is illegal in every country of the world to sell drugs not tested on animals. And all long-term testing must be done on two species, only one of which is a rodent. It would be utterly reckless to use on a human a drug that had not been confirmed safe at a much higher dose in animals. All possible steps are taken; the number of animals used in research by AstraZeneca dropped by over 25% between 2013 and 2014. Yet, while I do not pretend in any way to be competent to write about animal testing, I have lived with the knowledge of its consequences … from both sides.

## Humans and humans

The issues around use of human tissue and embryonic material are even more complex. But scientists have always had their personal principles taken into account. On one occasion, I refused to work on anti-cancer drugs related to the 'morning-after' pill. My principles were respected by the company. But science is never black and white. Tamoxifen itself had, two decades earlier, originally been intended for that purpose.

## Research lives on

And the possibilities of research can never wholly be predicted. Tropical disease expert John Ryley once told a meeting of the Alderley Park Christian Fellowship how he had worked over many years at Mereside on chemotherapy against all the major tropical diseases areas. He gained expertise and made many contacts among scientists in them all. All of those areas eventually failed to deliver newer treatments; and ran, one by one, out of time and money. At the end of his career, John wondered what it had all been for.

Then, he realised that his experience was unique. The ICI collection of hundreds of thousands of chemicals could now be tested (as they could not previously) on an immense scale; and the World Health Organisation was keen for industry collections to be tested – and to find people who knew what they were doing and had contacts and knowledge right across tropical medicine. After he had retired, John started a new work — and found what his old career had been for. So even when we think we have failed, our science may yet surprise us …

COSHH symbol
for acids

# Safe at work

The safety of staff was, of course, the company's first priority of all. For example, it was repeatedly pointed out that for all the hundreds of millions of pounds invested in the Mereside complex, it contained nothing irreplaceable – except its people.

Thus the SHE department was a foundational part under one name or other of all that happened at Mereside. The first meeting of the Laboratory Safety Committee at Mereside was held on 19 September 1957, with Bert Crowther as chairman, even before the labs were formally opened, by which time the first of a legion of safety inspections had already taken place. This began the process that led to the huge Global SHE department of AstraZeneca – and to the careful oversight of all of Mereside today.

Behavioural SHE grew constantly in importance. The safe handling of materials under COSHH regulations (Control of Substances Hazardous to Health) became central to all lab work, with the danger of allergic reactions to materials always a concern.

But the more immediate dangers of sudden events was never forgotten. An explosion can kill you now, not next year!

In some workplaces, safety regulations and procedures are regarded as boring. In science labs, the culture is very different. And that applies to the management too, when it is made up of former scientists.

## Fast reaction

I had this strikingly illustrated to me one day. I was in charge of a potentially hazardous hydrogen facility when I realised a serious problem might have occurred (it had not). I sent an immediate e-mail to my section manager, Andy Barker.

Senior managers of big companies can get hundreds of e-mails per day. Andy was no exception. So I was impressed when Andy appeared at my desk well under five minutes later. He had seen the e-mail, dropped everything, dashed across one block, climbed two flights of stairs and hurried the length of another. Because safety came first.

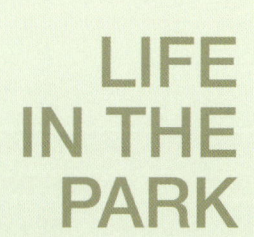

# LIFE IN THE PARK

### Fire, or don't fire?

Not all safety decisions, of course, are that clear-cut. On another occasion, one cheerful colleague of mine gave the whole lab a dilemma.

He was using the lab's best weighing balance. Unfortunately, he spilt underneath it a very flammable chemical. Unwilling to disturb the sensitive balance by moving it, he sprayed some alcohol underneath the valuable instrument, to try and destroy the spilt residue. But the mixture caught fire!

He yelled, and all the chemists in the lab gathered round the precious weighing balance in horror. The plastic base of the balance was slowly shrivelling, as if it were sitting on a gas burner. Consternation ensued. Half of his colleagues wanted to discharge a fire extinguisher at the balance; and one was levelled in readiness.

But the other half of the lab shouted in dismay that *that* would destroy the lab's best weighing balance! As we argued, Rome burnt.

Happily, someone thought of water. A well-aimed jet solved the problem … only just in time!

# Alderley Park's 15 drugs (so far!)

Twelve successful international products for human medicine were actually made for the first time in Alderley Park. New development work on an earlier launched product, 'Hibitane', was also done here. A fourteenth, 'Atromid-S', was recognized here and was described earlier; and while the first chemical sample of a fifteenth, 'Caprelsa', was actually made in France, the project was an Alderley Park one. The dates are of first chemical synthesis unless stated:

Research by Radnor Mere
PHOTO ASTRAZENECA

| Brand name | Generic name/identity | Purpose | Made (or launched) |
|---|---|---|---|
| 'Hibitane' | chlorhexidine | antiseptic | 1950 |
| 'Atromid-S' | clofibrate | lipid lowering | 1958 (date launched) |
| 'Inderal' | propranolol | hypertension | 1962 |
| 'Nolvadex' | tamoxifen | breast cancer | 1962 |
| 'Tenormin' | atenolol | hypertension | 1970 |
| 'Diprivan' | propofol | anaesthetic | 1975 |
| 'Zoladex' | goserelin | prostate cancer | 1977 |
| 'Casodex' | bicalutamide | prostate cancer | 1984 |
| 'Faslodex' | fulvestrant | breast cancer | 1985 |
| 'Tomudex' | raltitrexed | bowel cancer | 1987 |
| 'Arimidex' | anastrozole | breast cancer | 1988 |
| Invanz™ (Merck) | ertapenem | antibiotic | 1993 |
| 'Iressa' | gefitinib | lung cancer | 1996 |
| 'Caprelsa' | vandetanib | thyroid cancer | 1999 |
| 'Tagrisso' | osimertinib | lung cancer | 2011 |

## The firsts!

Another perspective on the company's historic work in (or before) Alderley Park is given by a different table of world firsts; the dates this time are of launch:

| | |
|---|---|
| First low toxicity prophylactic against malaria | 1946 |
| First skin-persistent antiseptic | 1954 |
| First modern inhalation anaesthetic | 1956 |
| First oral antifungal agent | 1959 |
| First beta-blocker | 1965 |
| First anti-oestrogen for use in cancer therapy | 1973 |
| First cardioselective and hydrophilic beta-blocker | 1976 |
| First long-acting GnRH antagonist for prostate cancer | 1989 |
| First long-acting low toxicity safe oral anti-androgen for prostate cancer | 1995 |
| First selective aromatase inhibitor for breast cancer | 1995 |
| First pure anti-oestrogen for breast cancer | 2002 |
| First selective EGFR inhibitor for lung cancer | 2003 |
| First personalised cancer treatment following rebiopsy and molecular test | 2015 |

Not all ICI's medical advances came from within Pharmaceuticals Division; the first visible light cured dental restorative, **'Fotofil'**, came from the corporate research labs at Runcorn in 1978.

## Serendipity

Garnet Davey
*SCAN* April 1997

The words were quoted earlier of Archie Sexton, research director in 1954:

> … unexpected observations can lead to results of considerable moment … scientific work must be so conceived that an <u>opportunist</u> policy can be pursued when necessary.

Reflection on the origin of the major medicines discovered in Alderley Park could hardly illustrate this better. The drug discovery business must be one of the most startling enterprises in human history in terms of its unpredictability. Just consider the following:

- Another company had already made 'Inderal' and discarded it. And its main medical value, against hypertension, was described by Garnet Davey as 'absolute luck'.
- The first sample of 'Inderal' was made as a substitute because the chemist could not obtain the starting chemical he planned to use.
- 'Tenormin' was the only active compound in its series – and was made first. And it unexpectedly achieved three clinical aims at the same time.
- 'Nolvadex' was a failed contraceptive marketed reluctantly and never expected to make a profit.
- 'Diprivan' was an old compound made many years before, which had long been forgotten.
- Invanz™ was made by a brand new employee merely as a first compound to give her initial practice at chemistry.
- The project that led to 'Faslodex' was to be shut down the next week.

Need we say more?

# Five more worked on in the Park

Four more important drugs were 'licensed in' from other companies but were the subject of large or small amounts of development work in Alderley Park before actual launch. A fifth, 'Lynparza', was acquired and developed after the company itself (KuDOS) was bought. These were:

| Brand name | Generic name | Purpose | Source | Launched |
|---|---|---|---|---|
| 'Meronem' | meropenem | antibiotic | Sumitomo | 1996 launch in US |
| 'Zestril' | lisinopril | hypertension | Merck (1986) | 1988 launch |
| 'Zomig' | zolmitriptan | migraine | GSK (1996) | 1997 launch in US |
| 'Crestor' | rosuvastatin | lipid lowering | Shionogi (1998) | 2002 launch in EU |
| 'Lynparza' | olaparib | cancer | KuDOS (bought 2005) | 2014 launch in US |

## Licensing

Licensing involves the buying of some or all of the rights to a drug discovered by another company. In simple cases this may be a matter merely of selling another company's drug, often in a market which that company is not interested in – perhaps because your company has a much better sales team in that 'territory' than theirs.

'Meronem' was described earlier along with other antibiotics. Two of the remaining three feature below:

## 'Zestril' (lisinopril) and 'Statil' (ponalrestat)

The ACE inhibitor heart drug **'Zestril'** (**lisinopril**) was a good news story, boosted by some clever thinking in the Park and by ICI Pharmaceutical Divisions American subsidiary Stuart, but with a peculiar twist.

ACE inhibitors are drugs related to a peptide found in the venom of the jararaca, a Brazilian pit viper. This one came to ICI as the result of a deal with lisinopril's discoverer, Merck, whereby ICI would gain shared worldwide marketing rights to lisinopril, a drug already close to launch, in exchange for an interest in ICI's new Alderley Park diabetes treatment **'Statil' (ponalrestat)** (ICI 128,436). This was an aldose reductase inhibitor, a revolutionary new drug class, which was less advanced in the clinical trial cascade but which had generated some excitement.

## ICI play hardball

Roy Vagelos, Merck's CEO, recorded the corporate story in his book *Medicine, Science and Merck*: 'Merck wasn't working on diabetes; and our laboratories urged me to negotiate for the right to sell the ICI product.' Merck also looked at a different firm's diabetes drug but decided ICI's was better. On the other hand, Merck's marketing group was unwilling to share lisinopril, though it already had a similar drug of its own.

But ICI declared in turn that they were considering another company's ACE inhibitor. ICI's management 'were playing hardball, and I flinched. I wanted their potential breakthrough drug and was now convinced we could beat them in marketing our own product.' The deal was struck. ICI's 'Zestril', originally ZD3286, was first approved in September 1987 in New

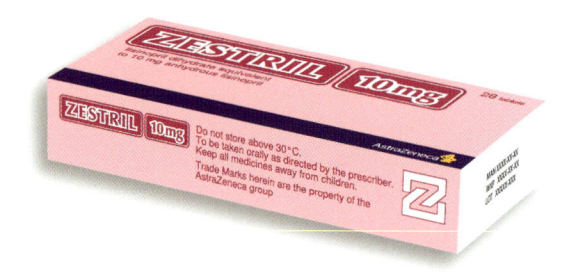

'Zestril'

'Zestril' and other
structures

GEORGE B. HILL

'Zestril'
(lisinopril)

'Zomig'
(zolmitriptan)

'Crestor'
(rosuvastatin)

'Lynparza'
(olaparib)

Zealand. (As mentioned earlier, Merck marketed it under the different brand name of Prinivil™.)

But ICI then sprang ahead of Merck with some smart thinking by Alderley Park's medical department, which invented an information programme for doctors called 'Resolutions'. From February 1988, this pushed the (undoubtedly correct) view that 'Zestril', which needing taking only once a day, improved on earlier drugs by making sure patients kept taking the medicine. The same was true of Merck's brand, of course, but when ICI launched 'Zestril' this probably helped ICI's cardio-vascular sales team (which was a strong one as it was already selling 'Tenormin').

Also ICI, used to the bulk chemical trade where price completion was a way of life, undercut Merck's price for their version of the same drug (thus attracting bulk buyers of it), and beat Merck at their own game. The drug was launched by both companies; 'Zestril' became a huge seller for ICI.

Merck were further embarrassed (as was ICI) when ICI's drug 'Statil' failed in clinical tests. 'We never learned whether the theory behind 'Statil' was wrong, or whether it was simply not good enough to block the complications from diabetes.' But ICI still retained 'Zestril'.

'Later,' as Vagelos described it: 'the fair-minded chairman at ICI agreed that the deal was unbalanced and that his company owed Merck a product candidate. But when I retired as CEO, nothing had come of that promise.' The demise of 'Statil' was announced in January 1990.

You lose some, you win some.

## 'Zomig' (zolmitriptan)

The acute migraine drug '**Zomig**' (**zolmitriptan**) works partly by blocking the serotonin receptor in the brain. It was discovered by Wellcome and bought from GSK in 1996 when the American authorities insisted on its sale for competition reasons as a condition of the Glaxo merger. In March 1997 it was approved in the UK and launched. Triptans are used for treating migraine both with and without cluster headaches. 'Zomig' became the world's leading second-generation triptan by 2001, with 2.7 million prescriptions having been dispensed by then in the US alone. It actually worked in two ways, both in blood vessels and in the brain itself. Its prospects looked (and were) good, as the world market for migraine drugs was approaching $2 billion by the turn of the millennium.

Curiously, Wellcome nearly dropped the drug when a very small early clinical trial appeared poor. But then the Wellcome scientists discovered that 'there was one patient who had just taken a dose … and suddenly realised that her car was parked outside and she hadn't paid the parking fee. She looked out the window to see it being towed away and flipped into a major migraine crisis. Analysis without that patient did show adequate significance … had we not known … we probably would have killed the project!'

### 'Crestor' (rosuvastatin)

The huge cholesterol-lowering statin story for Alderley Park began in 1997, when AstraZeneca's licensing group led by Chris Henderson with Martin Carlisle started talks with the drug's discoverers, Shionogi of Japan. The successful commercial deal brought **'Crestor'** (**rosuvastatin**) to AstraZeneca and to Alderley Park in April 1998. Only the marketing rights for Japan remained with Shionogi. They had carried out the early stages of clinical evaluation, which looked very promising.

Development work was then done in Alderley Park, led by Fergus McTaggart, which was crucial in positioning the drug for market. 'Crestor' was first approved in the EU in November 2002 and launched that year, and in the US in 2003. It had been used to treat one million patients by April 2004, and, to cut a very long story short, by 2014 was the fifth-highest-selling prescription medicine in the United States, at $5.8 billion. Patents are now close to expiry.

'Crestor'
COPYRIGHT
ASTRAZENECA

'Crestor' logo
COPYRIGHT
ASTRAZENECA

### 'Lynparza' (olaparib)

This was the one that paused for breath!

An inhibitor of a DNA repair enzyme called PARP, **'Lynparza'** (**olaparib**) was discovered by KuDOS Pharmaceuticals of Cambridge, a privately-owned biotechnology company in Cambridge, UK, founded by Stephen Jackson and bought by AstraZeneca in 2006. The drug was originally designated KU 59436 and targeted mutations that change the way DNA is repaired.

It was designed selectively to kill cancer cells whose repair mechanism (known as homologous recombination or HR) is deficient compared to normal cells. In particular, these cancer cells occur in patients who have inherited certain gene mutations (such as BRCA1 and BRCA2).

Mark O'Connor was a lead bioscientist. Olaparib went into Phase II trials, where early results showed promise. But later the main trial results did not support continuation into Phase III; and AstraZeneca announced in December 2011 that it would not proceed. Instead olaparib was sent back into the iMed (research department) where intense data analysis showed that the drug was indeed valuable for at least one particular set of patients.

With this knowledge, in 2013, the hiatus was ended. The regulatory authorities were approached and in December 2014 'Lynparza' was approved in the US by the FDA as part of its Accelerated Approval programme, on condition it was put into a Phase III trial for ovarian cancer, the same target as before (other trials are also underway).

In the UK, the National Institute for Health and Care Excellence (NICE) initially declared 'Lynparza' too expensive. Then in October 2015, the UK press announced that olaparib had been effective in what the BBC called a *'milestone'* trial by the Institute of Cancer Research against prostate cancer. Less than a third of patients showed a response  but, strikingly, out of those in whom a relevant mutation of some kind was identifiable, 88% responded strongly; these included both patients born with that mutation and ones in whom it developed inside the tumour. Many were doing well on the drug after more than a year, despite having original life expectancies of less than that. In December, at a reduced price, NICE gave partial approval for its use in a subgroup of patients pending better understanding of the benefits.

## Soon: cediranib?

Also on the cusp of possible launch is the VEGF inhibitor **cediranib** for ovarian cancer, although its intended brand name of **'Recentin'** is now to be changed. A combination of it with 'Lynparza' has been reported in one clinical trial to be better against certain cancers than 'Lynparza' alone. Early exploration and its first synthesis were achieved by the excellent chemists at AstraZeneca's Reims labs, although the actual project and all elements of its molecular biology, cell and in vivo testing were developed and conducted at Mereside. It has the potential to be the 'best in class' of its kind. In July 2014 it was granted the useful status of 'orphan drug' (an early step towards approval of an apparently important medicine) by the European Commission.

## Drugs not from the Park – blockbusters from elsewhere

Other great medicines, marketed by the company but both discovered and entirely developed at other company sites, are not described in this book and would require another volume. All have complex and vast stories that must be (or have been) told elsewhere. Each can sadly only be named here: impressive adjectives could be added to their globally famous names but would hardly do them justice in such a short space.

**'Seroquel'** (**quetiapine**) for CNS conditions and **'Accolate'** (**zafirlukast**) for asthma were discovered at the ICI Pharmaceuticals Division lab at Wilmington in the US.

From Astra in Sweden came the local anaesthetics **'Xylocaine'** (**lidocaine**) and **EMLA** and later **'Naropin'** (**ropivacaine**). Astra's best-known global product was the gastrointestinal treatment **'Losec'/'Prilosec'** (**omeprazole**), which was followed by its single isomer **'Nexium'** (**esomeprazole**), the latter being launched in the UK in 2000 and remaining in 2014 the fourth biggest selling prescription medicine in the US. **'Vimovo'** combined the latter with the anti-inflammatory **naproxen**. Other Swedish Astra or Astra derived products included the second generation beta-blocker **'Seloken'** (**metoprolol**); different brands containing **budesonide** (**Entocort, Rhinocort and 'Pulmicort'**) variously for asthma and inflammatory bowel disease, along with **'Symbicort'** (budesonide with **formoterol**), also for asthma; and the calcium channel blocker **'Plendil'** (**felodipine**). The heart drug **'Brilinta'**

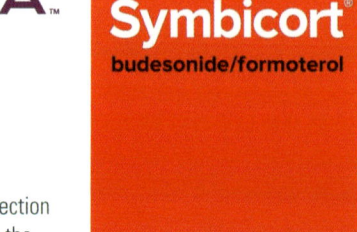

(**ticagrelor**) came from the former Fisons facility at Charnwood in Leicestershire, UK that was bought by Astra in 1995.

The first major 'biologic' owned by AstraZeneca plc was MedImmune's 1998 monoclonal antibody **'Synagis'** (**palivizumab**) for respiratory virus infections, particularly in infants. A different sort of product was MedImmune's nasal flu vaccine **'FluMist'** (**'Fluenz'**).

## Drugs from other companies

Over many years, other drugs that were part of the company's but not of the Park's story were bought in, licensed in, or were the result of swap deals with other pharma companies. Amongst a large list, ones not mentioned earlier included the antibiotic **'Apatef'** (**cefotetan**) discovered by Yamanouchi in Japan; the angiotensin receptor blocker **'Atacand'** (**candesartan**) from Takeda; and the opioid receptor antagonist **'Movantik'** (**naloxegol**) from Nekta. In-house drugs were also combined with external discoveries, such as in the COPD treatment **Duaklir** (formoterol with Forest's **aclidinium bromide**).

'Onglyza' logo
COPYRIGHT
ASTRAZENECA

A major competitor that AstraZeneca partnered for a time was Bristol-Myers Squibb. In 2007, a deal was struck to give AZ access to the diabetes treatments being developed by BMS. In mid-2012 as part of this BMS announced the purchase of Amylin Pharmaceuticals for $5.3billion; but an end to the collaboration was announced in December 2013 when for up to $4billion Amylin passed to AZ, along with all of the BMS or Amylin diabetes drugs, **'Onglyza'** (**saxagliptin**), **'Forxiga'** (**dapaglifozin**) and **'Symlin'** (**pramlintide**). **'Kombiglyze'** combines saxagliptin with **metformin**.

## Comparisons

It is hard to compare earlier and later successes. I could find no good global *historical* comparison of major medicines by numbers of prescriptions both before and after patent expiry, or even a (less informative) inflation-indexed comparison of all best-selling drug sales. Comparable data would certainly be hard to gather.

But there is no doubt that 'Inderal', 'Tenormin', 'Nolvadex' and 'Diprivan', plus their generic equivalents, would be among the leaders. What is known is that, based on sales *without* inflation balancing, 'Crestor', 'Seroquel' and 'Nexium' all stand in today's top 20 all-time sales list. A true comparison between the latter and earlier sets would be a fascinating (but arduous) task for someone …

## Latest success: durvalumab

Both internally and externally, exciting new drugs are constantly being sought, though no more will be developed in the Park. AZ's next rising star from elsewhere is **durvalumab**, a human monoclonal antibody developed at Cambridge and Maryland by MedImmune and designed to counter a tumour's immune-evading tactics by blocking a signal that helps tumours avoid detection. It is hoped it will work in a wide range of cancer types. It has just become the first immuno-oncology drug to be granted 'breakthrough' status in the US (against some bladder cancers). Other 'biologics' lining up include **brodalumab** for psoriasis and **tremelimumab** for mesothelioma. The MEK inhibitor **selumetinib**, licensed in from Array Biopharma, remains in trials for lung cancer. Acquisition of more small companies for the products they have discovered remains big business for AZ; the latest in November 2015 was the £1.8billion purchase of ZS Pharma, owner of promising potassium-binding compound **ZS-9**.

# Therapeutic areas: painful failures

The scale of unproductive research in all science dwarfs the successful. And the pharmaceutical industry is perhaps the supreme example of one where the amount of effort spent is not proportionate to success.

Mereside's work has always been 'therapy area led' and project driven. But it has often been alleged, convincingly, that ICI's biggest mistakes lay in spreading its research too thinly. When I arrived to work at Mereside as a young graduate in 1977, a multitude of projects existed. Most were deeply meaningful but thinly resourced. One or two felt more like hobby explorations by senior scientists whose past achievements allowed them the luxury. They were not all good examples to younger staff (particularly one who always insisted on wearing a 3-piece suit in the lab to do chemistry, with no other protection!)

## Painful near misses

This could be a very long list and only a few examples are mentioned. It should be pointed out that several possible drugs failed to reach the clinic because there were too few patients involved (i.e., compared with other more justifiable projects) even if a drug were launched purely as a service to medicine.

**'Vivalan'** (**viloxazine**), discovered in 1969, was an example of a drug that nearly managed to switch therapeutic area. It resembled a beta-blocker-type molecule but became a promising antidepressant, reaching the market before sickness side-effects caused its withdrawal.

**'Clozic'** (**clobuzarit**, ICI 55,897) was a rheumatoid arthritis treatment that in rare cases produced a severe skin reaction and was dropped against 'tremendous resistance from the physicians who were using the drug'.

The beta-blocker **'Corwin'** (**xamoterol fumarate**, ICI 118,587), was an inotropic agent (for use after a heart attack) discovered in 1977 and launched in 1988, but withdrawn in 1990 when it was found to worsen heart failure in some patients and returned to medical use with severe restrictions in 1993. (The earlier failure in this area, **'Eraldin'**, is mentioned elsewhere.)

Diabetes was an area in which the company would dearly have loved to find a valuable treatment. After the fall of **'Statil'**, mentioned earlier, 'licensing in' of a drug from a competitor was in the end the only way to get something on the market.

Not all failures failed! ICI 63,197, made by Mike Dukes, was intended to treat asthma but caused severe vomiting. So, ICI Agrochemicals added it to 'Paraquat' weedkiller (which itself had been invented by William Boon, one of the original 'Pharms' chemists). It saved many people who had accidentally ingested 'Paraquat', and was for a time was the Division's biggest product – by tonnage!

I worked for many years on both arthritis and novel antibiotic mechanisms. Both areas made fundamental discoveries from an academic point of view (which were published), but although reaching agonisingly high scientific plateaux were still far from giving useful medicine. Other major areas sadly also

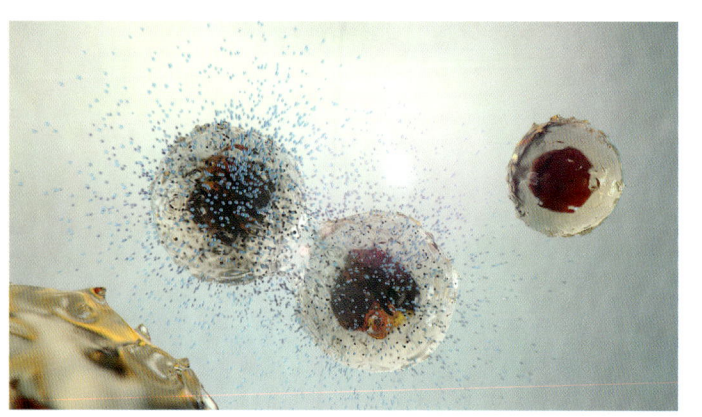

Pancreatic beta cells at different stages of regeneration

## Swallowed!

Not all the failures were scientific ones. Brilliant peptide scientist Harry Gregory set up a project to (for the first time) purify urogastrone, as a putative treatment for ulcers.

As a natural product found in urine, the raw material for urogastrone was obtained without charge, by placing large plastic containers in the gents' toilets at Mereside and Macclesfield! The contents, supplied by hundreds of staff, underwent a long and complex purification process to isolate the urogastrone from it.

Sadly for Harry, a problem occurred with the first batch. A final concentrate of just 50 milligrams was obtained. But then a scientist who was pipetting it (a procedure still then done by mouth) managed to swallow the entire six month's supply of pure material by mistake!

But Harry later succeeded, and made urogastrone famous, when he published proof that it was the same as human EGF, described as part of the 'Iressa' story.

Harry Gregory
*SCAN* April 1989

---

failed, while still others were effective in so few cases that they could not be marketed. Other company sites around the world faced similar struggles, as AstraZeneca's Sweden labs discovered to their chagrin with **'Exanta'**.

The biggest issue was that scientists often made the wrong great discoveries, ones that were ahead of their time, or led to drugs that would cost a fortune to use. Only occasionally did science and need manage to connect. When they did, of course, history was made.

## ICI's cure-all?

ICI's antiseptic 'Hibitane' was nearly developed, after work by Rosemary Chester, as a contraceptive; the *Telegraph* headline was 'Gran's gargle beats the pill'!

# 18

# People, time & chance

## Changing of the guard – CEOs

Sir Tom McKillop was chief executive in turn of Zeneca Pharmaceuticals, of Zeneca plc and then AstraZeneca plc from 1994 to 2005, as already described. He believed strongly in organic growth and mostly resisted big and undoubtedly traumatic corporate adventures. The company under him trod paths of caution both in target and appetite.

### Below the top

After Claes Wilhelmsson retired in 2002, the research and development managements were split, between Jan Lundberg and Martin Nicklasson respectively, with Barrie Thorpe over Operations. After stints by John Patterson and Anders Ekblom in development, the research and development roles were later recombined under Martin Mackay in 2010; when he left Mene Pangalos took over research and early development from 2013. Thorpe was followed by David Smith.

### Top of the tree

But what about the big choice? David Brennan, who replaced Sir Tom as CEO, was from a sales rather than research background. He worked to grow the company much more by acquisitions than his predecessor.

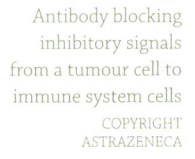

Antibody blocking inhibitory signals from a tumour cell to immune system cells
COPYRIGHT ASTRAZENECA

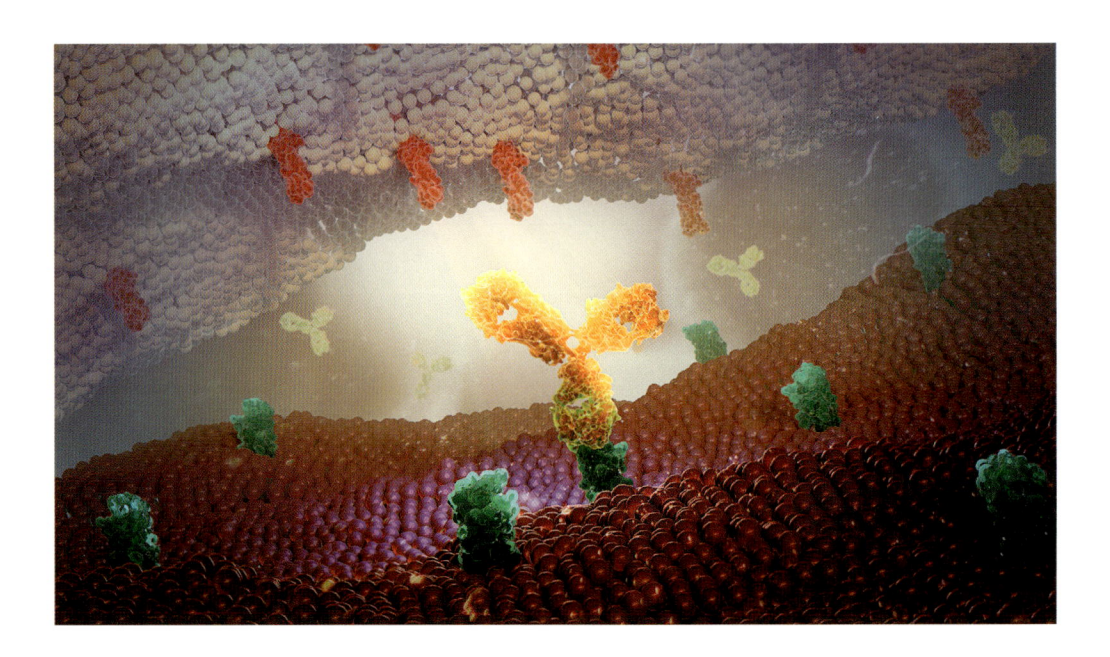

Cambridge Antibody Technology (CAT) was bought for £702 million in 2006. Then came the decision to purchase MedImmune LLC in April 2007 for £15.2 billion, a price widely considered high, if fairly defensible from a strategic point of view. This was a huge step that gave the company major access into 'biologics', specifically into existing products in flu vaccine and anti-viral products. CAT and MedImmune LLC were combined and MedImmune became the new name of AstraZeneca's entire dedicated biologics division. Sadly, and surprisingly to some though not all, the earlier results failed to justify the investment. David Brennan left in April 2012, with newspaper reports suggesting he was under pressure, although there was no public evidence of this and the company board refused to comment.

The new CEO was biologist Pascal Soriot. Former Chief Operating Officer of Roche Pharma AG, he was named on 28 August 2012 as the new head of AstraZeneca plc. He took up the post on 1 October, aged 53.

As measured by 2012 prescription drug sales, his company was the world's seventh largest pharmaceutical company (after Pfizer, Sanofi, Novartis, Merck & Co., Roche and GlaxoSmithKline), with operations in over 100 companies. By 2014 it was eighth, one behind GlaxoSmithKline.

In August 2014 AstraZeneca plc had a market capitalisation of approximately £54.8 billion, the sixth largest of any company with a primary listing on the London Stock Exchange.

## Not as planned …

Amid the huge machinery of modern industry, there are always a few moments when nuts fall harmlessly off bolts, people decided to be people, team spirit is tested, and truth becomes stranger than fiction.

Here are two unplanned moments from Alderley Park …

### Nothing left?

At one time we used large ten litre hydrogen balloons for certain experiments. One colleague was walking up the lab toward me, past a fumehood holding an especially full balloon. Momentarily, he disappeared from my sight behind some shelves. At that instant, the balloon beside him, perhaps disturbed by the draught of his passage, exploded with the peculiarly vicious crack that hydrogen balloons produce.

All that appeared beyond the shelves, at first, were my (deafened) friend's flying spectacles. My first stunned thought was, 'Is that all that is left of him …?'

### A typical lab

Then there was the moment when our section manager opened our laboratory door, to usher an important visitor into the lab. 'This is a typical chemistry laboratory–' BANG!

LIFE
IN THE
PARK

# Another great sale

## The downsize begins

In 2008 the proximity of the patent cliff that AstraZeneca had been approaching began to undermine budgets deeply. The last building to be completed on site was the new £42million Bioscience Block 41 in 2011. An interesting historical note is that it stands on the sites of the old Blocks 4 and 7 of the original Mereside complex; that is, on the exact site once occupied by the eighteenth-century Deer House of Sir Edward Stanley.

Many parts of buildings around the site were slowly vacated as the workforce was downsized heavily in a major redundancy programme (much of it thankfully voluntary). In 2012, company profits slumped by 34%.

## R&D closure announced

It still came as a shock to almost everyone, nevertheless, when on 21 March 2013 the closure of all R&D in Alderley Park was announced, with a move to a new £515 million new facility in Cambridge. (By 2015 when planning permission was granted this was being quoted as £330 million.) 1600 jobs were to move to the new Cambridge Biomedical Campus where around 2,000 should finally be employed, 550 were to be lost and 700 were to stay in the Park, comprising IT, HR, communications and finance. The economic

The Mereside
research complex
PHOTO ASTRAZENECA

## Titanic complaint

Announcements of large-scale redundancies were made to whole departments. Questions were always invited, but most people sat grim-faced in silence.

When my own chemistry department heard the sad news, announced from the platform by a brave and equally grim Susan Galbraith, only a few technical queries were made from the floor. Some obviously wished for a more plangent protest, but could find nothing worth saying. Eventually, as the oldest in the department, I posed a subdued question that I thought might put the session quietly out of its misery. I repeat my question here by popular demand.

'Ah, Susan? We appreciate that the management has seen an iceberg ahead. And we can all see why they have put lots of people on the lookout, and on the steering.

'But what we don't quite understand is why they have decided to throw a lot of the stokers and the coal overboard? When the ship slows nearly to a halt in mid-ocean … er, what then …?'

My question was described to me afterwards as a 'curve ball' (whatever that is). To my embarrassment, it was met with a universal roar of black hilarity (as well as a patient and gracious reply from Susan); and I received a round of applause, and a small final place in departmental legend.

I left the Park in June 2012.

**LIFE IN THE PARK**

cost to Cheshire and the North West was estimated at 3,000 jobs and £245 million, with much wider implications for the North West science base as a whole. The future of the site's international science status, with its 856,807 square feet of bioscience facilities, hung in the balance.

Equally distressed were the small army of caterers and cleaners who served the Park, as well as the dozen taxi firms that serviced it. An estimated 1,000 people working at Alderley Park as contractors were likely to be affected.

The move would have been more drastic and Alderley Park would have closed entirely, according to reports, if the Park had not, by chance, lain within the parliamentary constituency boundary (of Tatton) of the Chancellor of the Exchequer, the Rt Hon. George Osborne, MP. He, along with former Minister for Universities and Science David Willetts, helped persuade AstraZeneca to retain a presence within the Park.

Mr Osborne declared it:

> … very difficult news for people directly affected by the decisions around Alderley Park. I have worked hard with AstraZeneca over the last few weeks to make sure that a substantial number of jobs are kept there and will work closely with Cheshire East council and the government taskforce we are creating to bring new companies to the site. We are all determined that Alderley Park shall remain a success story and at the heart of our local community.

His private reaction is not hard to guess, considering that he had only months before helped AstraZeneca secure a £5million government grant to develop Alderley Park's R&D facilities, a loan subsequently announced by the company to be on hold, though it later became an important contribution to the site's new incarnation.

Government, council, university and business figures rapidly convened a high-powered taskforce on 26 April to develop a clear vision for the site. The Unite union said Astra's relocation was a 'massive blow' for north-west England. The GMB union called it 'disastrous news'. Pascal Soriot, AstraZeneca's new chief executive, said the job cuts and restructuring represented an 'exciting and important opportunity' that would allow the company to 'accelerate innovation'.

The company announced that it would now focus globally on three areas: Respiratory, Inflammation & Autoimmunity; Cardiovascular & Metabolic Disease; and Oncology.

## The decision?

The decision received support only from commentators in the south east of England. The BBC's 'Mind the Gap' web page of TV presenter Evan Davis was an example:

> It becomes clear talented people do better working next to each other. This helps Evan explain the story behind one of the scariest business announcements since the crash – drug giant AstraZeneca's decision to move its scientists from a rural research park in the North West to the buzzing biomedical hub of Cambridge.

This hardly seems a rational perspective. For example, in recent times AstraZeneca's core research (and nearly all its successful product range) has become increasingly focused on cancer; and The Christie, the largest cancer centre in Europe, treating more than 40,000 patients a year, is less than 15 miles and only 30 minutes' drive from the Park.

And then there is the near-permanent gridlock on the Cambridge ring road …

The decision was one that many felt defied logic. Was it a hard business decision? Or was it a new man's new broom? *Surely* no issue of mere prestige clouded the minds of anyone involved? Again, time will tell. But it marked the end of an era. And there was no going back. Both Park and company are in uncharted waters now.

## Sale brochure

The sale included of all of Alderley Park's 400 or so acres (including 39 acres of buildings, 11 of them in the Mereside complex and at the time employing just under 3000 people). The sale brochure advertising the site included many glossy photographs and features including:

- Car parking for approximately 4000 vehicles.
- Mereview (actually described as Mereside) Restaurant seating 400, and the Water Garden Restaurant for 250
- In excess of £330 million recently invested in new facilities since 1997, in addition to over £250 million in the ongoing improvement and enhancement of the older assets and infrastucture on the site
- Just over five million square feet (165,000 m$^2$) of research and office space
- Block 52 with its high-density solid and liquid storage facilities capable of holding 5 million samples

A list of the Alderley Park's site's vital statistics in the brochure was impressive: an electricity supply capacity of 50,000 volt-amperes; a gas supply of 3000 square centimetres per hour; a water supply of 100 cubic metres per hour; a fire fighting capacity of 230 cubic metres of water per hour; a water reservoir with a capacity of 20,000 gallons; a high pressure hot water system capacity of 65 megawatts; and a steam heating output of 24 tonnes per hour. But the Park's fine old buildings were also included in the sale.

## Buyer

A buyer was sought for the Park who would take on ownership, and from whom AstraZeneca would lease back such premises as it still required. Bids under tender were reported as ranging from £30 to £50 million.

The sale was subsequently announced, for around the latter figure, to Manchester Science Parks (MSP). It included the entire Alderley Park property (except for Little Clays field). The main investor behind MSP is Bruntwood Ltd, and the others are Manchester University; Manchester Metropolitan University; Central Manchester University Hospitals NHS Foundation Trust; and Manchester and Salford City, and Cheshire East, councils.

## The future (and past) of AstraZeneca

Some 700 AstraZeneca staff will stay in Alderley Park. The new research labs at Cambridge, when built, will now be the research base of the company in the UK (though much work is now contracted out to China or India).

Interestingly, a historical narrative is also possible in Cambridge, where archaeology prior to the start of building in April 2015 revealed Neolithic, Bronze Age, Iron Age and Roman occupation of the site, and even a rare glass bottle that would have held an unguent.

## Not Pfizer!

But other problems can arise! In November 2013, the cash-rich pharma mega-giant Pfizer informally approached AZ, chairman to chairman. On 5 January 2014, Pfizer formally offered £60 billion for its rival.

A week later, the AstraZeneca board rejected the bid, saying that it 'very significantly' undervalued the company. The story did not break in the press until published by the *Sunday Times* on 20 April. On 26 April, Pfizer made a second bid, also rebuffed, then on 17 May two further offers, eventually offering £55 per share, valuing its rival at £69 billion.

There was huge public concern, on the basis of Pfizer's past record of takeover and research closure, and the scientific community in the UK was stentoriously opposed to the merger. It was pointed out grimly that, as a share of its overall revenue, in 2013 AstraZeneca spent almost 50 per cent more on research and development than Pfizer. Even more embittering was the revelation that Pfizer's bid was driven by a US tax loophole, something that infuriated both US politicians and UK industry.

In the end, Pfizer effectively ran out of time under the strict rules for takeovers. Their bid was abandoned on 26 May; and before another could be launched, the US government ensured that the tax 'flip' Pfizer had planned became much harder.

So that was that. For now …

However, in November 2015 Pfizer bought Dublin-based Allergan in a monster friendly merger – now with the *new* US tax rules in their sights. But at least we were now off their radar.

# PART V
# Living Alderley Park

*'We are too young to realize that certain things are impossible; so we will do them anyway.'*
*William Wilberforce (or else to him, by his friend William Pitt the Younger)*
*The first Lord Stanley of Alderley was a friend of Wilberforce.*

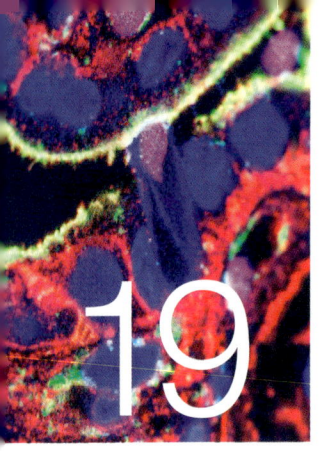

# 19

# And now, the future

## Manchester Science Partnerships

What could come beyond? And could it be achieved, whatever it was?

The closure of AstraZeneca plc's R&D laboratories in Alderley Park by 2016 (in theory) was announced on 21 March 2013. The Park passed into Manchester Science Park's hands on 1 April 2014. On completion of the sale, AstraZeneca immediately signed a lease to keep such control of the property as it required, including responsibility for the Park's security.

Chancellor of the Exchequer the Rt Hon. George Osborne MP (centre) was present as ownership was formally handed over on Friday 9 May by Mene Pangalos, Executive Vice President of Innovative Medicines and Early Development (right), to Chris Oglesby, Bruntwood CEO and MSP Chairman (left), who promised: '… With our partners and the community of exceptional scientists who choose to remain at the Park, we will build a flag-ship for a vibrant life sciences industry in the North of England'.

In October 2014 it was announced that MSP's name was being changed to Manchester Science Partnerships.

### BioHub

The £600 million Mereside research facility is still a world-class set of research facilities – and is far better than many small scientific firms and enterprises in the area would be able to provide for themselves otherwise. MSP itself began by committing £35 million to capital investment over a ten year period, mostly towards reconfiguring space previously used by a single company to multi-user occupancy, but also to creating a campus style communal environment, and to maintaining facilities near state-of-the-art level while the new occu-pants began to generate revenue.

Handover of
the Park:
Chris Oglesby;
Rt Hon. George
Osborne MP;
Mene Pangalos
PHOTO MSP &
ASTRAZENECA

Radnor Reception approach
PHOTO ASTRAZENECA

The December 2015 map of MSP's proposed development of Alderley Park
COPYRIGHT MSP

The overall project to turn Mereside into a new science park was led for AstraZeneca by Graeme Bristow, Alderley Park Site Manager, and for MSP by Chris Doherty, Site Director for the new owners. They started working out in autumn 2012 how to bring new people and innovation to the site. Later, the vision became how to transform the site from a single occupier into a life science cluster – a 'science for life park'. Regional and national support were harnessed, plus £5 million from AstraZeneca that was 'matched funded' by government. Advice was taken from similar teams at other science parks at Sandwich, Colworth and Harwell.

Tho leading partner was the pioneering group BioCity, who were asked in late 2012 'to look into the feasibility of creating a specialist centre for the next generation of life science companies'. Following the R&D closure announcement in March, BioCity was engaged by AstraZeneca in May 2013 to develop the new BioHub, designed as an innovative 'business support ecosystem'.

Discussions were led with AstraZeneca by Dr Glenn Crocker of BioCity. Initially the plan was to provide 36,000 sq. ft. of high-end laboratory and office facilities for early-stage and growing firms engaged in innovative drug discovery and development. In addition to providing the BioHub laboratory and office accommodation, AstraZeneca provided tenant firms with access to restaurants, meeting rooms and conference facilities. Firms would also be able to rent highly specialised technical equipment and services.

Demand for accommodation was expected to be competitive, and so it proved.

By October 2014, eighteen months after its conception, the BioHub 'incubator' had grown from nurturing three to 81 tenant companies, employing 310 staff, covering a full spectrum of life sciences including the first US arrival of a Chicago-based company. The total passed 100 companies in 2015, ahead of target. In addition, over 35 groups of former AstraZeneca staff undertook a 'business bootcamp', leading to 5 new companies with 25 more in the pipeline.

Nevertheless, a key aim is the securing of a scientific 'anchor tenant'. This would assist with Alderley Park's current operating costs of some £25million p.a. – no small challenge!

## First plans for the Park

Following the sale, the wider consequences for the whole Park began to become clear. Initial plans were made then, in January 2015, Cheshire East Council's Cabinet approved a planning draft and public consultation. (It was the latest of over 200 planning applications for the Park since 1957!) On 9 April 2015, the tenancy of all the buildings in the southern Alderley House part of the Park passed out of the hands of AstraZeneca. The consultation led to changes in June, including the dropping of initial housing development plans near to the A34 (for scenic green belt reasons) and of an environmentally unhelpful proposal to build on the Radnor Car Park beside the Mere.

Old and beautiful
PHOTO GEORGE B. HILL

If the plans proceed, some things will remain familiar. The science park ('campus') itself will remain a secure area with the biggest change being the ambitious refurbishment of the former CTL, following demolition of obsolete parts of that complex in late 2015, possibly with an academic institution tenant in mind. The 1 million square feet or so of life sciences facility in the Park being retained would be increased by over another 400,000 for laboratories, offices and light manufacturing, which could, in time, support a total workforce of up to 7,000, overtaking AstraZeneca at its peak.

## Housing and sports

But a number of buildings will be demolished (and some replaced), including at Mereside, Blocks 2, 9, 10, 13, 17 and 21, and in the south of the Park the Bollington Lodge gatehouse and the Water Garden restaurant. Alderley House was vacated entirely by the spring of 2014 and preparation for its demolition (admittedly of a building well past its sell-by date) began in November 2015 – a curious historical link between this book and Nancy Mitford's books of letters in 1938 which recorded the demolition of the first Alderley House mansion.

The rear of the House site and all of the Southbank portakabin complex, if the plans published in December 2015 go ahead, are destined to become high quality housing housing. Also expected to be replaced by housing are both of the existing football pitches; the old cricket pitch and the tennis courts in the walled garden; some of the old Matthews site; and, surprisingly, the Mulberry's sports building itself which will also be demolished. There will be a total of around 275 new homes, some of them retirement homes. Perhaps they might attract old employees?

The new cricket pitch will remain and a new football pitch will be laid out near Bollington Lodge. But the main area in the 'new' Park for brand new sports facilities and pitches will be to the east of the site road opposite Parklands, an area presently occupied by disused buildings and a large car park. There, a new sports centre of similar size to Mulberry's with indoor sports hall, dance studios, courts, changing, fitness studio, spin studio, café and spa, will separate new football and tennis facilities from another small housing zone.

### New village hub

Alongside the main housing area, a new village hub will offer complementary uses such as a shop, cafe, restaurant and pub. These developments will embrace the listed buildings in the south of the Park which have, of course, protected status. The former Sir James Black Conference Centre is expected to return to its original name of Tenants' Hall, with a future incarnation perhaps as a restaurant and possibly a wedding venue. A hotel with up to 100 bedrooms may stand on the site of the frontage of the former mansion and (later) company offices, with the same fine view north across the Park. But the Water Garden itself and the adjacent Arboretum will be preserved.

### Access and response

Most interesting to many readers, though of some environmental concern, will be the opening of much of the Park to public access, particularly to the local community. The site's lovely wildlife and heritage will be watched with care and concern by me and many others. Various new footpaths are likely; one good suggestion is of a path down to the historic Nether Alderley Mill.

The plans have aroused great local interest. Chief Executive Officer of MSP, Rowena Burns's statement published in the local press held a pleased note: 'It is encouraging that the response the proposals have received from the local community has been very positive. People want to see a successful future for Alderley Park, which builds on its achievements and international reputation.'

# Faith and future in the Park

The pharmaceutical industry is often viewed with antipathy. For differing reasons, some outsiders resent its failings more than they appreciate its achievements. Perhaps the most wounding insinuation is that those working in the industry inevitably have poor motives. This is not merely unfair; it is a reversal of the truth for many. I chose pharmaceutical research myself in the full knowledge that career prospects were more limited there for most chemists than in, say, the energy industries. Very many others have also valued the privilege of working in a business addressing some of society's major needs.

And many of them, including me, do this from particular ethical or credal worldviews. They believe and find hope beyond themselves, or have a vision of some desired and less painful far country.

### Belief

In any case, belief in the wide sense is also something that those outside the Park have always invested in it. Throughout my seventeen years in cancer research, I was often moved to realise there were millions of people out there who were campaigning, giving, hoping and sometimes praying for the battle against cancer, for a start. On tiring days, the energy that came from recalling this hidden hinterland of hopefulness always gave me a huge sense of purpose, as well as of concern that so much goodwill was directed toward us all.

On a personal note, I wore one other cap during my time in Alderley Park: that of a founder member of the Alderley Park Christian Fellowship, a section of ClubAZ which was started by NMR spectroscopist David Smith in 1979 and still meets at the time of writing. One of the Fellowship's chosen tasks was to pray regularly, from its start, for the company's leadership, work, staff and purpose. (One guest speaker at our Fellowship, Rod Badams, once challenged us: 'Well, think about it. If people like you don't pray for your company, who will?') As a token of our interest, when Sir Tom McKillop retired as CEO, our Fellowship offered him (and he warmly accepted) the gift of a Bible – a New International Version, chosen to match the global character of the company's purpose.

Work and faith can grow together. I chose to spend my career at the bench; so did Jesus. Last year, I walked in Nazareth near where his workbench presumably stood, as he began his work on earth with a plain day job, according to a later writer 'working as a carpenter among men, making ploughs and yokes; by which he taught the symbols of right-eousness and an active life'. A bench is a good place to share and appreciate day-to-day life, no matter who you are. For some scientists in this book, starting with Frank Curd and Frank Rose, bench work has evidently gone alongside believing.

In recent years a quiet room has existed at Mereside for individual devotions. But probably very few employed at the Park have not brought sincere convictions of some sort with them. Those who malign multinational giants often forget that their employees possess their own views, and generally have families and friends whom they love and want to trust in them. The business itself needs honest commitment for success. (And it recognizes things beyond itself; in the aftermath of events such as 9/11 and the Boxing Day tsunami, it encouraged many staff to attend reflective meetings.)

## Commitment

In any case, very many showed their practical conviction at times, by putting in longer and harder hours than they were paid for, when important campaigns were underway. *Everyone* who works harder than they need to for a creditable end shows they believe in something good. This is the contribution which Alderley Park has made to science, vision, and the drive to achieve significant outcomes for the public good. Great achievements came through people with personal values, determination, altruistic motives, and differing world-views. Commitment was the name of the game that led to goals scored.

And the match report is also worth reading! In the Bible, the apostle Paul advised the Colossians: 'Whatever you do, work at it with all your heart'. This is advice for every one of us, whatever worldview we draw our wisdom from. A secular analogue might be my favourite motto, Gandalf's words to Pippin in *Lord of the Rings* before the siege of Gondor: 'Generous deed should not be checked by cold counsel.' These two quotations particularly spoke to my decision to write this book; there were those who felt such a project would attract far too many brickbats to be worthwhile. Yet you are reading it.

## Reminding us

Why did I write it? If I were to put the thing I find most meaningful in life into two words, they would be these: I like *reminding us*. Reminding us of what we *did* at Alderley Park, of what we all (together) achieved there. Recalling how we *did* discover that new drug, did change that result. Reminding us of what we *learnt*; of what we found was true, and valuable, and worth fighting for, even if we sometimes lost heart; of what we believed in and knew then, even if we grew cynical or world-weary later. Reminding us of mistakes or lost insights we might still redeem or rediscover in some way, or encourage others who can. Retelling the

great explorations we made at the Park, the voyages to the medicines that helped the world – and the routes that may still; for duty calls.

## Responsibility

Rod's words above may not strike many people as they did me. But his wider sentiment should. If those inside the organization do not take responsibility for it, then who else can possibly be expected to? If those who actually create the science do not direct their discoveries toward helping humanity (insofar as it lies within their power), what sort of world will this become?

Scientists, more than perhaps any other people, are very good at excavating holes which they slowly dig deeper in search of their own little research treasure … and in which they become more and more isolated from those around them. Perhaps the biggest problem in science today, even (or even more?) in this modern e-world, is not lack of inspiration but lack of *connection*. We know much; but do we share enough? And do we see the *whole* picture? We should emulate the scientific and intellectual polymaths of earlier centuries (perhaps the first Lord Stanley might even be considered one of the latter?).

## Connected and motivated?

Can we stay connected? Can we link not only to what Alderley Park was but what it might be? Are we connected to the responsibilities of science to humanity and the future?

At this pivotal moment in the Park's history, are we motivated by its legacy in our own life journeys, inside it or beyond? Are we fired not only by the energy of past discoveries but of enthusiasm for new ones? Are we motivated not only by what has been done but what still *can* be done? Can we take the drive which led to the triumphs, history and heritage of the past – and the legacy of the gifted men and women of the Alderley Park family who then and now have served so well – into a meaningful future? It *is* possible to believe that we can. We must all hope for that; and some of us will try.

# The question of research

One of my chemistry supervisors, Andrew, once told me – to the amusement of us both – that the Chemistry department was the least popular of all with the Human Resources department. HR had a constant complaint:

> 'When we issue a new policy, every other department conforms to it. But the chemists always cause trouble! They always want to know <u>why</u> they must do it. They say to us, 'What's the reason? Why should we? Why?''

The answer is simple. Research scientists are trained to do research! They are trained to question everything; to *ask*. Why should they not? If not, what's the point of having a research department?

Scientists persist by asking questions. This is not easy. And no belief system, of any form – or none – can be allowed to signpost the road. In a 1967 introduction to the *Communist Manifesto*, historian A. J. P. Taylor commented pungently of Karl Marx that:

> Marx never made a discovery in the scientific sense. He never had an illumination which turned his ideas upside down. He decided beforehand what he wanted to discover and then sure enough discovered it.

Needless to say, the research scientist must not do this. Why? Science does not start at the laboratory bench. Before a scientist can begin an experiment, he or she has to make plans. In research those plans must, by definition, be novel ones. In making them, the scientist may have to ask questions that no-one has asked or perhaps even thought of before. Socrates suffered for less. He or she might need to grasp concepts no-one has previously imagined, or even realised *were* concepts, and try to predict what problems will arise, whilst navigating a land no-one has heard of before, let alone seen. This can be a remarkably hard thing to do. In science as a whole, it can be a labour of weeks, years, or even centuries.

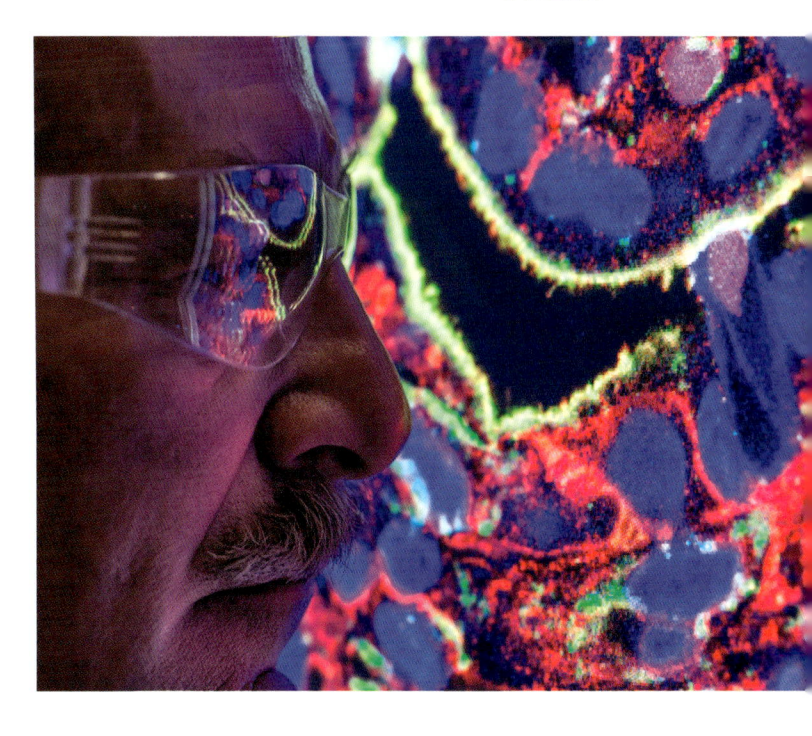

## Persistence

Asking is a crucial part of learning. When I first started work, I was advised to ask lots of questions. It was wise advice, and I learnt fast because of it. Over the years I saw many 'new starters' arrive in new jobs. Some asked many questions. Others did not, for a variety of reasons. Some were genuinely brilliant and quick on the uptake, or had had good on-the-job training previously. Some were too nervous. Some were too proud, for asking questions is never an ego-boosting process, and people of all intellectual levels can find it difficult to seek help, even when in the right place to ask.

Yet, Mereside has often been blessed with senior and very clever scientists of deep experience who have themselves asked questions very loudly! Research is a team business, and there is nothing better for getting the team 'going' than making them aware of a question for which they have no answer.

## Tension

Yet questioning creates tension. It is hard to face down clever opposition. And science has often had to progress against *resistance.*

For, truth can easily be opposed – even by great scientists. Galileo attacked dogma in his church; but he also derided other science more visionary than his own (such as Kepler's suggestion that the moon affected tides on the earth, which Galileo called one of his *'puer ilities'*). Others ridiculed Harvey, who discovered blood circulation; Pasteur, who explained fermentation; Semmelweis, who discovered the cause of childbed fever; the opponents of the false chemical theories of phlogiston and vitalism; and Wegener, who claimed that the continents must once have been elsewhere. Raymond P. Ahlquist, on whose theories Alderley Park's great early medicines were built, was rejected when he first tried to publish his theory.

Scientists must value humility. We must all be prepared to see our convictions go the way of the phoenix, if necessary. Max Planck, the inventor of quantum theory, wrote words in 1949 that those slow to move would do well to bear in mind: 'A new scientific truth does not triumph by convincing its opponents and making them see the light, but rather because its opponents eventually die, and a new generation grows up that is familiar with it.' Yet even he later struggled with the implications of his own discoveries! In short, no problem can be safely approached except from first principles.

Magnolia in the Water Garden
PHOTO GEORGE B. HILL

## Growing up

And the future? Can we still start from the beginning? Can we still be growing up? Have we more truth to discover? Is Alderley Park still young enough to go on finding the new? Is AstraZeneca? The BioHub certainly could be. Or have we aged and, though still apparently vital, lost our curious edge? (Old chemists never die; they merely fail to react …) Were 'Inderal', 'Nolvadex', 'Zoladex', 'Diprivan', and the rest, all heights that no-one in the Park can rise to again? Surely not. Research of all kinds will continue, both by those inside the Park and by those who have moved away. It will continue to be the intense, urgent, meaningful, intellectually scintillating, beautiful world that has attracted so many fine minds and challenged so many to offer so much over so many years. It will continue to be *work*; for, behind my trivial words lies the untold labour, the patience, the hard-won experience and insights, the grind. The search will advance.

Alderley Park will also continue. It has been here for centuries. It has seen the birth of world-class treatments that have changed the face of medicine and the life expectancies of millions. It has seen multitudes come and go: lords and ladies; cabinet ministers; a prime minister and a greater one to be; artists, writers, naturalists and explorers; and visitors from scouts to an emperor – and to Prince Andrew, Duke of York, the Park's very latest notable guest, who visited Alderley Park in February 2016 to speak at the Conference Centre, in support of biotech entrepreneurs aiming to create new businesses. (By then, the BioHub was already home to over 130 start-up companies).

The 'Park' has provided meaningful careers for thousands of staff in hundreds of different specialties in a rainbow of departments, for employees from students, bench scientists and office workers to world experts, clinicians and physicians, science heroes and heroines, inventors who have stirred the globe, at least one Nobel Prize winner and authors of a whole library of eye-opening publications. And it has provided, most of all, for grateful patients.

For them and the world, the work certainly must go on.

# Afterword

Sir James Black – Fellow of the
Royal Society and Nobel Laureate
KENNEDY, *ICI, THE COMPANY THAT CHANGED
OUR LIVES*

Dr Arthur Walpole – surely might
have shared a Nobel Prize, had he lived ...
V. CRAIG JORDAN

# Appendix 1: Historical tables

## A) Ownership of Radnor Mere and Beech Wood (part of the Nether Alderley manor lands):

| | |
|---|---|
| Godwin | before Domesday |
| Bigot | Domesday |
| Hugh | ? |
| the Lords of Aldford | –1220 |
| the Arderne family | 1220–1413 |
| various Stanleys | 1413–1495 |
| William Brereton | 1495–1536 |
| Edmund Peckham | 1536–1556 |
| Margaret Moreton | 1556–1557 |
| Sir Edward Fitton | 1557–1602 |
| Sir Thomas Stanley, Bart. | 1602–1672 |
| five more Stanley Baronets | 1672–1807 |
| six Lords Stanley of Alderley | 1807–1938 |
| Hambling, Crundall & Co. Ltd* | 1938–1950 |
| Imperial Chemical Industries Ltd | 1950–1993 |
| Zeneca plc | 1993–1999 |
| AstraZeneca plc | 1999–2014 |
| Manchester Science Partnerships** | 2014– |

\* = originally Mr A. E. Becheley Crundall
\*\* = initially Manchester Science Parks

## B) Ownership of the manor of Over Alderley:

| | |
|---|---|
| Brun | before Domesday |
| William Fitz-Nigel | Domesday–? |
| The Barons of Halton | ? |
| The Barons of Montalt | ? |
| The Arderne family | ? – before 1360 |
| The Weever family | before 1360 & between 1446 and 1461 |
| Sir John Stanley | between 1446 and 1461; then united with other Stanley lands at Alderley |

## C) History of the Park House / Alderley House site (originally part of the monastic lands):

| | |
|---|---|
| part of the estates of Dieulacres Monastery | before dissolution |
| two brothers Sheldon | after dissolution |
| Greene family | ? |
| Ralph Bentley | ? |
| Thomas Bentley/Deane de Park | until 1629 |
| Thomas Deane de Park (son) | 1629–1695 |

*(the latter built the Park House, 1668, on the site of an older Park Tenement)*

| | |
|---|---|
| William Stanley | from 1695 |
| William Stanley de Park | |
| Thomas Stanley de Park | until 1748 |
| purchased by the main Stanley family | 1748–1779 |
| became the principal Stanley residence | 1779–1930s |

*(the Park House was largely demolished in 1818 and a new mansion built on the same site)*

*(the Alderley Park mansion was damaged by fire in 1931 and demolished in 1933)*

| | |
|---|---|
| Hambling, Crundall & Co. Ltd | 1938–1950 |
| Imperial Chemical Industries | 1950–1993 |

*(the present Alderley House was built on the same site in 1963/64)*

| | |
|---|---|
| Zeneca plc | 1993–1999 |
| AstraZeneca plc | 1999–2014 |
| Manchester Science Partnerships | 2014– |

## C) THE STANLEY BARONETS

| | |
|---|---|
| Sir Thomas Stanley | 1597–1672, created baronet in 1660 |
| Sir Peter Stanley | 1626–1683, second baronet |
| Sir Thomas Stanley | 1652–1721, third baronet |
| Sir James Stanley | 1680?–1746, fourth baronet |
| Sir Edward Stanley | 1690?–1755, fifth baronet, brother of James |
| Sir John Thomas Stanley | 1735–1807, sixth baronet |

## D) THE LORDS STANLEY OF ALDERLEY

| | |
|---|---|
| John Thomas Stanley | 1766–1850 |
| Edward John Stanley | 1802–1869 |
| Henry Edward John Stanley | 1827–1903 |
| Edward Lyulph Stanley ('Lyulph') | 1839–1925 |
| Arthur Lyulph Stanley | 1875–1931 |
| Edward John Stanley | 1907–1971 |
| Lyulph Henry Victor Owen Stanley | 1915–1971 |
| Thomas Henry Oliver Stanley | 1927–2013 |
| Richard Oliver Stanley | 1956– |

# Appendix 2: Wildlife checklists

Records cover the years 1791–2015. Other wildlife classes (e.g., flora) are not listed but checklists are available from the author.

† = breeds or has bred recently

†† = non-recent breeding recorded or suspected

(100 in 2011) = largest flock ever recorded

* = rarely reported

[3] = number of records if very rare (or year given)

** = pre-1950 records only

## BIRDS

### Waterfowl

| | |
|---|---|
| Coot† | (532 in 1977) |
| Cormorant | |
| Diver, great northern** | [1] |
| Gadwall* | |
| Gannet* | [1] |
| Goosander | (36 in 2005) |
| Goose, barnacle*†† | |
| Goose, Canada† | (525 in 2011) |
| Goose, pink-footed* | |
| Goose, white-fronted* | |
| Goldeneye* | |
| Grebe, black-necked* | [1] |
| Grebe, great-crested† | |
| Grebe, little*†† | |
| Grebe, Slavonian* | [1] |
| Goose, greylag | (132 in 2008) |
| Mallard† | (648 in 1982) |
| Mandarin† | (154 in 2012) |
| Moorhen† | |
| Pintail* | |
| Pochard | (139 in 2008) |
| Ruddy duck*†† | (47 in 1982) |
| Scaup* | [1] |
| Scoter, common* | [3] |

| | |
|---|---|
| Scoter, velvet* | [1] |
| Shag* | [1] |
| Shelduck* | |
| Shoveler | (97 in 1979) |
| Smew* | |
| Teal | (313 in 1979) |
| Tufted duck† | (180 in 1997) |
| Swan, Bewick's* | [2] |
| Swan, mute† | (21 in 2002) |
| Swan, whooper* | [4] |
| Water rail* | |
| Wigeon | (80 in 1923) |

### Waterside birds, waders, gulls and terns

| | |
|---|---|
| Avocet | [1] |
| Bittern* | [2] |
| Curlew* | |
| Dunlin* | [1] |
| Egret, little* | [3] |
| Greenshank* | [1] |
| Heron, grey† | |
| Gull, black-headed | |
| Gull, common* | |
| Gull, greater black-backed | |
| Gull, herring | |
| Gull, Iceland* | [1] |
| Gull, lesser black-backed | |

| | |
|---|---|
| Gull, little* | [1] |
| Kittiwake* | [1] |
| Lapwing | |
| Oystercatcher | |
| Petrel, Leach's* | [1] |
| Plover, golden* | |
| Redshank* | [1] |
| Sandpiper, common | |
| Tern, black* | |
| Tern, common* | |
| Tern, sandwich* | [1] |
| Snipe, common* | |
| Snipe, jack* | [1] |
| Woodcock† | |

### Birds of prey and larger ground birds

| | |
|---|---|
| Buzzard, common† | |
| Corncrake*†† | [2] |
| Harrier, Hen* | [2] |
| Hobby* | |
| Kestrel† | |
| Kite, Red*†† | |
| Merlin* | [2] |
| Osprey* | [5] |
| Peregrine* | |
| Partridge, grey* | [1] |
| Partridge, red-legged† | |
| Pheasant† | |
| Sparrowhawk† | |

### Corvids, doves, owls, woodpeckers, etc.

| | |
|---|---|
| Crow, carrion† | |
| Cuckoo* | |
| Dove, collared† | |
| Dove, stock† | |
| Dove, turtle* | [3] |
| Jackdaw† | |
| Jay† | |
| Kingfisher*† | |
| Magpie† | |
| Martin, house† | |
| Martin, sand | |
| Owl, barn* | |
| Owl, little*† | |
| Owl, tawny† | |

| | |
|---|---|
| Pigeon, feral | |
| Pigeon, wood† | |
| Raven*†† | |
| Rook† | |
| Swallow, barn† | |
| Swift | |
| Woodpecker, great spotted† | |
| Woodpecker, green* | |
| Woodpecker, lesser spotted*†† | |

### Songbirds of the Park

| | |
|---|---|
| Blackbird† | |
| Blackcap† | |
| Brambling | |
| Bullfinch† | |
| Bunting, reed† | |
| Chaffinch† | |
| Chiffchaff† | |
| Crossbill* | [2] |
| Dipper* | [1] |
| Dunnock† | |
| Fieldfare | |
| Flycatcher, pied*† | |
| Flycatcher, spotted*† | |
| Goldcrest† | |
| Goldfinch† | |
| Greenfinch† | |
| Hawfinch*†† | [3] |
| Linnet*† | |
| Nightingale* | [1] |
| Nuthatch† | |
| Pipit, meadow* | |
| Pipit, tree*†† | until 1982 |
| Redpoll, lesser | |
| Redstart, black* | [3] |
| Redstart, common* | [3] |
| Redwing | |
| Robin† | |
| Siskin | |
| Skylark*†† | |
| Sparrow, house*†† | |
| Sparrow, tree* | [3] |
| Starling† | |
| Thrush, mistle† | |
| Thrush, song† | |

Tit, blue†
Tit, coal†
Tit, great†
Tit, long-tailed†
Tit, marsh*                until 1987
Tit, willow*††          until 1997
Treecreeper†
Warbler, garden*†
Warbler, reed*††          [1]
Warbler, sedge*††
Warbler, willow†
Warbler, wood*††         [2]
Wagtail, grey†
Wagtail, white*
Wagtail, pied†
Wagtail, yellow*
Waxwing*              [2]
Wheatear*          [1]
Whitethroat, common*††
Whitethroat, lesser**
Wren†
Yellowhammer*

## Escapes

Budgerigar
Cockatoo, sulfur-crested
Goose, bar-headed
Goose, Egyptian
Goose, lesser white-fronted
Goose, snow
Ibis, sacred
Muscovy duck
Peafowl, common
Pheasant, Lady Amherst's
Pochard, red-crested
Shelduck, ruddy**
Stork, white

## MAMMALS EXCEPT BATS

(breeding not shown)
Badger
Cat (feral)
Deer, fallow (herd)**
Deer, red*         2007
Deer, roe*        2006

Fox, red
Hare*            1980s
Hedgehog
Mink*
Mole
Mouse, house*
Mouse, wood
Muntjac*        2014
Polecat*
Rabbit
Rat, brown
Shrew, common
Shrew, pigmy
Shrew, water*
Squirrel, grey
Squirrel, red**     1920s
Stoat*
Vole, bank or field
Vole, short-tailed
Weasel*

## BATS

Brandt's bat*      2012
Brown long-eared bat
Common or Alto bat
Daubenton's bat
Natterer's bat
Noctule
Soprano bat
Whiskered bat

## FISH

Barbel
Bream
Carp
Chub
Crucian carp
Ghost carp
Golden Orfe
Goldfish
Koi carp
Pike
Roach
Rudd
Silver Ide

Tench
Three-spined Stickleback
Rainbow trout*                        gone
Wels catfish

## REPTILES AND AMPHIBIANS

Adder*                                1960s & 70s
Frog, common
Lizard, common
Newt, great-crested
Newt, smooth
Slowworm**                            1920s
Snake, grass*                         1960s
Toad

## LADYBIRDS

Cream-spot ladybird*
Eleven-spot ladybird*                 2002
Eyed ladybird*                        [3]
Fourteen-spot ladybird
Harlequin ladybird                    from 2006
Hieroglyphic ladybird*                [2]
Seven-spot ladybird
Ten-spot ladybird
Two-spot ladybird

## BUTTERFLIES

Brimstone
Clouded yellow*
Comma                                 from 1982
Common blue*
Dark green fritillary*                1990
Gatekeeper
Green-veined white
Holly blue
Large skipper
Large white

Meadow brown
Orange tip
Painted lady
Peacock
Purple hairstreak*                    1990
Red admiral
Small copper
Small skipper                         from 1992
Small tortoiseshell
Small white
Speckled wood                         from 1986
Wall*                                 until 1980s

## DAMSELFLIES AND DRAGONFLIES

### Damselflies

Azure damselfly
Blue-tailed damselfly
Common blue damselfly
Emerald damselfly*
Large red damselfly
Red-eyed damselfly                    from 1992

### Dragonflies

Banded demoiselle*
Black darter*                         1997
Black-tailed skimmer*                 2000 & 2009
Broad-bodied chaser*
Brown hawker
Common darter
Common hawker*
Emperor                               from 1998
Four-spotted chaser
Migrant hawker*                       2009
Ruddy darter*
Southern hawker

# Appendix 3: What and who

This book is an account only of some successes and advances, not of the whole enterprise arising from Alderley Park. It is certainly *not* a balanced academic summary, either of what was done or who did it. Perhaps one should be written; but not by me, though my record should assist a future author of one. The drug stories differ considerably in length; this purely reflects a reader's likely interest from what I have learnt of them myself.

Nor is this a corporate history taken from official minutes. Rather, it is the *story* of achievements in the Park – and what led to them.

Nor is this even an *attempt* to explain *all* that happened. The later drug development stories lie outside all possible readable compass. Published sources (such as Craig Thornber's overview listed in the Bibliography) should be consulted to learn about the pharmaceutical industry's *real* overall work. Rather, I cover in some detail only the early phases of discovery, and merely hint at what came after. This is, of course, a simulacrum of the truth, since the scale of development dwarfs that of research, in all of scale, expense and range of skills needed.

Nor do I attempt to list fully even the pivotal players who helped to make our medicines global, not in the early stages and most certainly not later. I could do no justice to the immense effort carried out in Alderley Park by many thousands of people over several decades.

Naming people is a problem in all achievement publications. *Everyone* ought to be named, like the team sheet at the end of the football match report. Since this is impossible, some would leave it all nameless. Yet history without personality is cold and unmemorable; and outsiders may write it then, with even less accuracy. No story of global discovery would have room for all the heroes and heroines of Antarctic adventure; yet no-one would be enthused to read that 'people discovered Antarctica'. But who to name? The captain and navigator who first found the new land? The financier and shipbuilder that supported them? The cook that fed them? Their lookout's sharp eyes, just two of many that might have spotted land? Those on the adjacent ship that was stopped by impassable pack-ice? Those who rose deservedly to high rank over others? The famous that followed – Davis, Ross, Shackleton, Amundsen, Scott, Fuchs, Byrd – and their teams? The thousands who built massive enterprises after them? But how to respect them all?

Even worse, what if key people are missed out? And how can anyone be *sure*, long after the event? Memories differ. What part was *really* played by her, or him? Or was it neither? How did their peers see it? And all those unsung and unsingable others who worked under them?

So I dedicate this account firstly to (from a different triumph) the Tom Bourdillons and Charles Evanses of the Park. In writing this I recalled *their* names without resorting to Google – would you have? They failed, three days before Edmund Hillary and Tenzing Norgay succeeded in their footsteps; and in any case it was John Hunt whose *team* first conquered Everest (and the leaders were honoured and knighted for all). The unremembered should come first, like the Unknown Warrior. And where someone did 'lead' successfully, it should be obvious that theirs were not the only footsteps in the snow, wherever it lay.

The criteria I then used were simply those of history and story. For the former, my record is comparable to the Mereside reception's new 'Patent Wall' display, which lists only the key new product filings and their authors. This defines only the first march of each journey – but a march that *stirs*. And, for the latter, I have identified mainly those having the finest adventures and taking the epochal steps.

It seemed profoundly important to do at least this little. To do more was never possible – not least, because of the rapid dispersal of old records, and because of the passage of time (not least, mine).

# Appendix 4: Companies leadership

## Chairmen:
### ICI Ltd (ICI plc from 1974)

| | |
|---|---|
| Sir Alfred Mond (later Lord Melchett) | 1926–1930 |
| Sir Harry McGowan (later Lord McGowan) | 1930–1950 |
| Mr John Rogers | 1950–1953 |
| Sir Alexander Fleck (later Lord Fleck) | 1953–1960 |
| Sir Paul Chambers | 1960–1968 |
| Sir Peter Allen | 1968–1971 |
| Sir Jack Callard | 1971–1975 |
| Sir Rowland Wright | 1975–1978 |
| Sir Maurice Hodgson | 1978–1982 |
| Sir John Harvey-Jones | 1982–1987 |
| Sir Denys Henderson | 1987–1993 demerger |

## Non-executive Chairs:
### Zeneca plc

| | |
|---|---|
| Sir Sidney Lipworth | 1993–1999 merger |

### AstraZeneca plc

| | |
|---|---|
| Percy Barnevik | 1999–2004 |
| Louis Schweitzer | 2004–2012 |
| Leif Johansson | 2012– |

## Chief Executive Officers:
### Zeneca plc

| | |
|---|---|
| Sir David Barnes | 1993–1999 merger |

### AstraZeneca plc

| | |
|---|---|
| Sir Tom McKillop | 1999–2005 |
| David Brennan | 2005–2012 |
| Pascal Soriot | 2012– |

### Bruntwood

| | |
|---|---|
| Chris Oglesby | current |

### MSP

| | |
|---|---|
| Rowena Burns | current |

# Appendix 5: The Park's famous scientific papers & patents

## PATENTS (for human medicines invented by the company; & approved in the Park's era)

FI = Filing date for the complete specification; FA = First approval received. The initial 'priority date' (of precedence) for the patent is NOT shown but was, for every substance patent, just under twelve months prior to the FI date.

Hibitane™ (chlorhexidine): **New bactericidal substances**. Francis Leslie Rose, Geoffrey Swain. GB 705,838. FI: 4 Jan 1952; FA: unknown.

Atromid-S™ (clofibrate): **Pharmaceutical Compositions comprising α-aryloxy aliphatic carboxylic acids etc.** GB 860,303. William Glynne Moss Jones, Jeffrey Meyrick Thorp, Wilson Shaw Waring. FI: 22 May 1959; FA: unknown. [US 3262850].

Inderal™ (propranolol): **3-Napthyloxy-2-hydroxypropylamines.** Albert Frederick Crowther, Leslie Harold Smith. ICI Ltd GB 994,918. FI: 28 Oct 1963; FA: 18 Jun 1965 (France). [US 3337628].

Nolvadex™ (tamoxifen): **Alkene derivatives.** Arthur Leonard Walpole, Dora Nellie Richardson, Michael John Kennedy Harper. ICI Ltd GB 1,013,907. FI: 21 Aug 1963; FA: 30 Aug 1973 (UK).

Tenormin™ (atenolol): **Alkanolamine derivatives.** Arthur M. Barrett, John Carter, Roy Hull, David J. LeCount, Christopher J. Squire. ICI Ltd GB 1,285,038. FI: 28 Jan 1970; FA 26 Apr 1976 (UK). [US 3663607].

Diprivan™ (propofol): **2,6-Diisopropylphenol as an anaesthetic agent.** John Baird Glen, Roger James. ICI Ltd GB 1,472,793. FI: 17 Mar 1975; FA: 11 Dec 1985 (New Zealand). [US 4056635].

Diprivan™ formulation patent: **Oil in water emulsions containing propofol and disodium edetate.** Christopher Buchan Jones, John Henry Platt. EP 814787. FI: 17 Mar 1999.

Zoladex™ (goserelin): **Polypeptide.** Anand Swaroop Dutta, Barrington John Albert Furr, Michael Brian Giles. ICI Ltd GB Patent 1,524,747. FI: 21 Apr 1977; FA: 1 Dec 1986 (UK). [US 4100274].

Zoladex™ formulation patent: **Continuous release pharmaceutical compositions.** Francis Gowland Hutchinson. EP 0058481. FI: 27 Jan 1982.

Casodex™ (bicalutamide): **Amide derivatives.** Tucker, Howard. ICI plc. EP 0100172. FI: 08 Jul 1983; FA: 18 Jan 1995 (South Africa).

Faslodex™ (fulvestrant): **Steroid derivatives.** Jean Bowler, Brian Steele Tait. ICI plc. EP 0138504. FI: 2 Oct 1984; FA: 25 Apr 2002 (US).

Faslodex™ formulation patent: **Fulvestrant formulation.** Rosalind Ursula Grundy, John Raymond Evans. EP 1,250,138. FI: 08 Jan 2001.

Arimidex™ (anastrozole): **(Substituted aralkyl) heterocyclic compounds.** Philip N. Edwards, Michael S. Large. ICI plc. EP 0296749. FI: 14 Jun 1988. FA: 11 Aug 1995.

Tomudex™ (raltitrexed): **Anti-tumor agents.** Leslie Richard Hughes. ICI plc. GB 2,188,319. FI: 24 Mar 1987; FA: 11 Aug 1995 (UK). [EP 0239362].

Invanz™ (ertapenem): **Carbapenems containing a carboxy substituted phenyl group, processes for their preparation, intermediates and use as antibiotics.** Michael John Betts, Gareth Morse Davies, Michael Lingard Swain. Zeneca Limited. EP 0579826. FI: 2 Feb 1993; FA: 21 Nov 2001 (US, by Merck; compound sold to Merck).

Iressa™ (gefitinib): **Quinazoline derivatives**. Keith Hopkinson Gibson. Zeneca Limited. EP 0823900. FI: 23 Apr 1996; FA 5 Jul 2002 (Japan). **Tyrosine kinase inhibitors.** Andrew John Barker. Zeneca Limited. US 5942514. Same dates.

Caprelsa™ (vandetanib): **Quinazoline derivatives as VEGF inhibitors.** AstraZeneca Ab. Andrew Peter Thomas, Elaine Sophie Elizabeth Stokes, Laurent François André Hennequin. EP 1876545. FI: 1 Nov 2000; AD: 6 Apr 2011 (US).

Tagrisso™ (osmertinib): **2-(2,4,5-substituted-anilino) pyrimidine derivatives as EGFR modulators useful for treating cancer.** Sam Butterworth, Maurice Raymond Verschoyle Finlay, Richard Andrew Ward, Vasantha Krishna Kadambar, Reddy C. Chandrashekar, Andiappan Murugan, Heather Marie Redfearn, Claudio Edmundo Chuaqui. AstraZeneca AB. WO 2013014448. FI: 25 Jul 2012; FA: 13 Nov 2015.

## PATENTS (for externally created compounds worked on in the Park)

Meronem™ (meropenem): **Carboxylic thio-pyrrolidinyl beta-lactam compounds and production thereof.** [Sumitomo]. EP 0126587. FI: 9 May 1984; FA: 17 Aug 1994 (Italy).

Zestril™ (lisinopril): **Carboxyalkyl dipeptide derivatives, process for preparing them and pharmaceutical composition containing them.** [Merck]. EP 0012401. FI: 10 Dec 1979; FA: 24 Sep 1987 (New Zealand).

Zomig™ (zolmitriptan): **Therapeutic Heterocyclic Compounds.** [GSK]. EP 0486666. FI: 6 Jun 1991; FA: 7 Mar 1997 (UK).

Crestor™ (rosuvastatin): **Dyslipidemia.** [Shionogi]. EP 0,521,471, FI: 30 Jun 1992; EP 1,155,015, FI: 15 Feb 2000; FA: 6 Nov 2002 (Netherlands).

Lynparza™ (olaparib): **Phthalazinone derivatives.** [Kudos]. EP 1633724. FI: 12 Mar 2004; FA: 16 Dec 2014 (EU).

## SOME KEY INITIAL PAPERS

Hibitane™ (chlorhexidine): **1:6-di-4'-chlorophenyldiguanidohexane ('Hibitane'). Laboratory investigation of a new antibacterial agent of high potency.** Davies, G. E.; Francis, J. Martin, A. R. Rose, F. L. Swain, G. *Brit. J. Pharmacol.* 1954, *9*, 192–6.

Atromid-S™ (clofibrate): **Modification of metabolism and distribution of lipids by ethyl chlorophenoxyisobutyrate.** Thorp, J. M.; Waring, W. S. *Nature*. **1962**, *194*, 948–949.

Inderal™ (propranolol): (1) **β-Adrenergic blocking agents. II. Propranolol and related 3-amino-1-naphthoxy-2-propanols.** Crowther, A. F.; Smith, L. H.[†] *J. Med. Chem.* **1968**, *11*, 1009–13. (2) **A new adrenergic beta-receptor antagonist.** Black, J. W.; Crowther, A. F.; Shanks, R. G.; Smith, L. H.; Dornhorst, A. C. *The Lancet* **1964**, *I*, 1080–81. (3) **The Discovery of the First Beta-Adrenergic Blocking Agents.** Crowther, A. F. *Drug Design and Delivery*, **1990**, *6*, 149–156.

Nolvadex™ (tamoxifen): (1) **Preparation and identification of *cis* and *trans* isomers of a substituted triarylethylene.** Bedford, G. R. & Richardson, D. N.[†] *Nature*, **1966**, *212, 733–734.* (2) **The history of Nolvadex.** Richardson, D. N. *Drug Design and Delivery*, **1988**, *3*, 1–14. (3) **Tamoxifen: a most unlikely pioneering medicine.** Jordan, V. C. *Nature Reviews: Drug Discovery*, **2003**, *2*, 213.

Tenormin™ (atenolol): (1) **A new type of cardioselective adrenoceptive blocking drug.** Barrett, A. M.; Carter, J.; Fitzgerald, J. D.; Hull, R.; LeCount, D. *Br. J. Pharmacol.* **1973**, *48*, 340P. [†] = Squire, C. J. (2) **'Atenolol'.** LeCount, D. in J. S. Bindra, J. S.; Lednicer, D. (eds.) *Chronicles of Drug Discovery*, Vol 1. (Wiley, 1982).

Diprivan™ (propofol): (1) **Synthesis, biological evaluation and preliminary structure-activity considerations of a series of alkylphenols as intravenous anaesthetic agents.** James, R; Glen, J. B. *J. Med. Chem.* **1980**, *23*, 1350–1357. (2) **ICI 35868 (Diprivan™): a new intravenous anaesthetic. A comparison with Althesin.** Kay B., Stephenson D. K. *Anaesthesia* **1980**, *35*, 1182–7.

Zoladex™ (goserelin): (1) **Synthesis and biological activity of highly active α-aza analogues of luliberin.** Dutta, A. S.; Furr, B. J. A; Giles, M. B.[†]; Valcaccia, B. *J. Med. Chem.* **1978**, *21*, 1018. (2) **The discovery and development of goserelin (Zoladex™).** Dutta, A. S.; Furr, B. J. A; Hutchinson, F. G. *Pharm. Med.* **1993**, *7*, 9–28.

Casodex™ (bicalutamide): (1) **Nonsteroidal**

antiandrogens. Synthesis and structure-activity relationships of 3-substituted derivatives of 2-hydroxypropionanilides. Tucker, H.; Crook, J. W.; Chesterson, G. J.† *J. Med. Chem.* **1988**, *31*, 954–9. (2) **ICI 176,334: a novel nonsteroidal peripherally selective antiandrogen.** Furr, B. J. A.; Valcaccia, B.; Curry, B.; Woodburn, J. R.; Chesterson G. J.; Tucker, H. *Journal of Endocrinology*, **1987**, *113*, R7–9. (3) **The preclinical development of bicalutamide: Pharmacodynamics and mechanism of action** *[also rationale and comparison with similar agents].* Furr, B. J. A.; Tucker, H. *Urology* **1996**, *47*(Supplement 1A), 13–25.

**Faslodex™ (fulvestrant): (1) Novel antioestrogens without partial agonist activity.** Wakeling, A. E., Bowler, J. *J. Steroid Biochem.* **1988**, *31*(4B), 645–53. (2) **A Potent Specific Pure Antiestrogen with Clinical Potential.** Wakeling, A. E.; Dukes, M.; Bowler, J. *Canc. Res.* **1991**, *51*, 3867–3873. † = Stevenson, R.

**Tomudex™ (raltitrexed): (1) Quinazoline antifolate thymidylate synthase inhibitors: heterocyclic benzoyl ring modifications.** Marsham, P. R.; Hughes, L. R.; Jackman, A. L.; Hayter, A. J.; Oldfield, J.†; Wardleworth, J. M.; Bishop, J. A.; O'Connor, B. M.; Calvert, A. H. *J. Med Chem.* **1991**, *34*, 594–605. (2) **ICI D1694, a novel inhibitor of thymidylate synthase for clinical study.** Jackman, A. L.; Newell, D. R.; Taylor, G. A.; Jodrell, D. I.; Gibson, W.; Stephens, T. C.; Calvert, A. H.; Hughes, L. R. *Br. J. Canc.* **1991**, *63*, 23.

**Arimidex™ (anastrozole): (1) The preclinical pharmacology of 'Arimidex' (anastrozole; ZD1033) – a potent, selective aromatase inhibitor.** Dukes, M.; Edwards, P. N.; Large, M.†; Smith, I. K.; Boyle, T. *J. Steroid Biochem. Mol. Biol.* **1996**, *58* (4), 439–445. (2) **From Programme Sanction to Clinical Trials: A Partial View of the Quest for Arimidex™, a Potent, Selective Inhibitor of Aromatase.** Edwards, P. N. In *Smith and Williams' Introduction to the Principles of Drug Design and Action, 3rd ed.*; Smith H. J., Ed.; Harwood Academic Publishers: Amsterdam, 1998; pp208–235. 4th Edition, CRC Press LCC: Boca Raton, 2006; pp233–256.

**Iressa™ (gefitinib): Studies Leading to the Identification of ZD1839 (Iressa™): An Orally Active, Selective Epidermal Growth Factor Receptor Tyrosine Kinase Inhibitor Targeted to the Treatment of Cancer.** Barker, A. J.; Gibson, K. H.; Grundy, W.; Godfrey, A. A.; Barlow, J. J.; Healy, M. P.†; Woodburn, J. R.; Ashton, S. E.; Curry, B. J.; Scarlett, L.; Henthorn, L.; Richards, L. *Bioorg. Med. Chem. Lett.* **2001**, *11*, 1911–1914.

**Caprelsa™ (vandetanib): (1) Novel 4-Anilinoquinazolines with C-7 Basic Side Chains: Design and Structure Activity Relationship of a Series of Potent, Orally Active, VEGF Receptor Tyrosine Kinase Inhibitors.** Hennequin, L. F.; Stokes, E. S. E.; Thomas, A. P.; Johnstone, C.; Plé, P. A.; Ogilvie, D. J.; Dukes, M.; Wedge, S. R.; Kendrew, J.; Curwen, J. O. *J. Med. Chem.* **2002**, *45* (6), 1300–1312. (2) **ZD6474 inhibits vascular endothelial growth factor signaling, angiogenesis, and tumor growth following oral administration** Wedge, S. R.; Ogilvie, D. J.; Dukes, M.; Kendrew, J.; Chester, R.; Jackson, J. A.; Boffey, S. J.; Valentine, P. J.; Curwen, J. O.; Musgrove, H. L.; Graham, G. A.; Hughes, G. D.; Thomas, A. P.; Stokes, E. S. E.; Curry, B.; Richmond, G. H. P.; Wadsworth, P. F.; Bigley, A. L.; Hennequin, L. F. *Canc. Res.* **2002**, *62* (16), 4645–4655. † = Pasquet, G.

**Invanz™ (ertapenem): Ertapenem: a Group 1 carbapenem with distinct antibacterial and pharmacological properties.** Hammond, M. L. *J. Antimicrob. Chemother.* **2004**, *53* Suppl 2, ii7–9. † = Darbyshire, C.

**Tagrisso™ (osimertinib): The Discovery of a potent and selective EGFR inhibitor (AZD9291) of both sensitising and T790M resistance mutations that spares the wild type form of the receptor.** Finlay, M. R. V.; Anderton, M.; Ashton, S.; Ballard, P.; Bethel, P.; Box, M.; Bradbury, R.; Brown, S.; Butterworth, S.; Campbell, A.; Chorley, C.; Colclough, N.; Cross, D.; Currie, G.; Grist, M.; Hassall, L.; Hill, G.; James, D.; James, M.; Kemmitt, P.; Klinowska, T.; Lamont, G.; Lamont, S.; Martin, N.; McFarland, H.; Mellor, M.; Orme, J.; Perkins, D.; Perkins, P.; Richmond, G.; Smith, Pete; Ward, R.; Waring, M.; Whittaker, D.; Wells, S.; Wrigley, G.† *J. Med. Chem.* **2014**, *57*, 8249–8267.

† = handled the first physical sample, if known.

# References

## INTRODUCTION

'this fairytale place'       Kennedy, Carol (1986), p136.

## CHAPTER 1 – THE EARLY PARK & LAND

'the park … Alderley'       Earwaker, J. P. (1890), Vol 2, p609.
Nether Alderley history, etc.       Summerfield, Phyllis & Massey, Eleanor (1951)
Aerial survey:       Report by members of the Alderley Edge Landscape Project Team, c.2000.
'… the same … waste.'       Domesday Folios. (1087–88). 266, 266b (reproduced in Earwaker).
'William son … shillings.'       *Loc.cit.*
Medieval deer park:       Harrison, W. (1902), *20*, p26
Family history generally:       Stanley, Peter. (1998).
'Much of the family's'       *Op cit.,* p6.
'The Stanleys managed to'       *Op cit.,* p6.
'Some very interesting'       *Op cit.*, p172.
Repton style landscape:       From ICI notes on visit of Editors to Alderley Park, September 1974.
'… bewailed … park …'       Adeane, Jane H. (ed.). (1899), p229.
Views       *Note: the formula giving the distance across the earth's curvature from which a particular high point may be seen (without allowing for refraction) is: DISTANCE (in km) = 3.5678 x root [HEIGHT of hill + HEIGHT of observer (in m)]*

## CHAPTER 2 – BARONETS TO BARON

'new mansion house'       Clare Pye, through *pers. comm.*
'the house stood in'       Stanley, Louisa (1843), p56.
'probably because the king'       Clare Pye, in chapter 19 of Prag, John (ed.) (2016).
'A fair man'       Adeane, Jane H. (ed.) (1899), p9.
'A house stood before'       Notebook, c.1820, Chester Records Office Ref. DSA 3752/1.
'newly fronted and in'       Lysons, Daniel. (1810).
'a contemporary picture'       Mitford, Nancy (1938), p. xxii.
'consisted of three sides'       Stanley, Louisa (1843), p56.
'returners from the Alehouse at 10'       Notebook, c.1820, Chester Records Office Ref. DSA 3752/1.
'from some Wood work'       Bethell, David (1979), p156.
'It is said that Sir John'       Mitford, Nancy (1938), p. xxii.
'In the summer of'       Adeane, Jane H. (ed.) (1899), p4.
'Alderley Hall was burnt'       *Op. cit.,* p8.
Beeches planted by first baronet       Stanley, Louisa (1843), p57.
'A thick venerable'       Adeane, Jane H. (ed.) (1899), p100.
'thanks will not be paid'       *Op. cit.,* p92.
'the Road which was'       Cited in Bethell, David (1979).
Blanche Stanley wedding       *Alderley & Wilmslow Advertiser*, Friday July 19, 1912, p19.
'… formerly a part'       Stanley, Louisa (1843), p53.

| | |
|---|---|
| 'occupied what was' | *Loc. cit.* |
| 'Mary Stanley, the' | *Op. cit.,* p55. |
| 'a respectable gable' | Bethell, David (1979), p157. |
| 'A fair specimen' | Adeane, Jane H. (ed.) (1899), p10. |
| 'By degrees, however,' | Mitford, Nancy (1938), p. xxii. |
| 'continued enlarging' | *Loc. cit.* |
| 'I have made the house' | Adeane, Jane H. (ed.) (1899). |
| 1872 Ordnance Survey map | First Series 1:2500, sheet XXXVI, 5. (N part of Park on sheet 1). |
| 'The first indication' | St Mary's Church, Nether Alderley (CD); Volume 2 – *The Stanleys in Alderley*, p263 (newspaper report from November 10th, 1939). |
| 'dignified, high-minded' | Adeane, Jane H. (ed.). (1899), p. viii. |
| 'in 1789 he sailed' | *Loc. cit.* |
| 'Maria is happy and placid' | *Op. cit.,* p147. |
| 'Gibbon used to lament' | *Op. cit.,* p414. |
| 'I wish I knew' | *Op. cit.,* p253. |
| '… we did not want liborty' | Russell, Bertrand (1937) Vol 1, p02. |
| 'The year after the burning' | Adeane, Jane H. (ed.) (1899), p30. |
| 'The dear old Man is' | *Op. cit.,* p147. |
| 'We are as busy' | *Op. cit.,* p159. |
| 'the sluggish blood' | *Ibid.* |
| 'How I gained it' | Adeane, Jane H. (ed.) (1899), p6. |
| 'We have just' | *Op. cit.,* p175. |
| 'We are buried in snow' | *Op. cit.,* p190. |
| 'You and Lady S.' | *Op. cit.,* p228. |
| 'Since I wrote' | *Op. cit.,* p195. |
| 'whose Liberal principles' | Mitford, Nancy (1938) p59. |
| 'As we were at' | *Op. cit.,* p60. |
| 'I have heard of' | *Op. cit.,* p61. |
| 'I had a talk with' | *Op. cit.,* p81. |

## CHAPTER 3 – ELEGANCE AND WATER

| | |
|---|---|
| 'In the park just' | *Alderley & Wilmslow Advertiser*, Friday June 12, 1931, p9. |
| 'The kitchen garden' | *Op. cit.,* Friday September 11, 1931, p3. |
| 'To be sure, I do like' | Adeane, Jane H. (ed.). (1899), p330. |
| 'Winnington is' | *Op. cit.,* p308. |
| 'went through the' | *Notes on the History of Alderley*, Chester Records Office DSA3752/1. |
| 'ivy-mantled parish church,' | Stanley, Edward (1835), vol 2, p2. |
| 'An address to my parishioners' | Summerfield, Phyllis & Massey, Eleanor (1951) |
| 'The purplish brown of' | Stanley, Arthur P. (1880) p133. |
| Herons nesting in nineteenth century | *Proc. Chester Soc. Nat. Sci. and Lit.*, iv, pp226–243. |
| A 10lb pike | Mitford, Nancy (1938) p49. |
| 'Fishing is the' | *Op. cit.,* p215. |
| 'Any boy who falls into' | *SCAN*, 25 October 1968. |
| Underground crypt found | *Manchester Evening News*, Feb 21, 2008. |
| 'Benjamin Backbite' | Russell, Bertrand (1937) Vol 1, p14. |
| 'Yesterday I assisted' | Mitford, Nancy (1939), p64. |
| 'She loved him' | Mitford, Nancy (1938), p. xx. |
| 'Nothing could have' | Mitford, Nancy (1939), p94. |
| 'Did I tell you there was' | *Op. cit.,* p222. |

| | |
|---|---|
| 'It is quite hard to' | Mitford, Nancy (1938) p. xxiv. |
| 'When he died,' | Mitford, Nancy (1939), p. xx. |
| 'a woman of vigorous' | Russell, Bertrand (1937) Vol 1, p17. |
| 'by coming down early' | Mitford, Nancy (1939), p. xi. |
| 'definitely stupid' | Russell, Bertrand (1937) Vol 1, p19. |
| 'Bertrand Russell' | *Loc. cit.* |
| 'Lyulph wants to buy' | Mitford, Nancy (1939), p279. |
| Pacifist meeting | Russell, Bertrand (1937), preface. |
| 'Not your hat' | Mitford, Nancy (1939), p. xi. |
| Henry with Mr Haydon | Mitford, Nancy (1938) p31. |
| 'an image so awful' | *Op. cit.,* p. xxv. |
| 'At Xmas time a tea party' | 'A Stately Home', *Fulshaw Times,* **8**(2), Spring 1964, p16. |
| 'There are numberless' | Soames (2001), p27. |
| 'that great Welshman,' | *Daily Telegraph*, 1999. |
| 'We could see a whirlpool' | *Scan*, 1967. |
| 'There is to be' | Estate book, Chester Records Office DSA 5/10 (1802), p1. |

## CHAPTER 4 – REMEMBERING

| | |
|---|---|
| Alderley Mummers Play | Cited in *Alderley Park and Radnor Mere Wildlife Report 1995*, pp35–44. |
| 'the ever-popular and' | Newspaper report, 1939. |
| 'The Tenants Hall' | Newspaper report, 1876. |
| Alderley Mummers Play | Cited in *Alderley Park and Radnor Mere Wildlife* Report 1995, p35–44. |
| 'Mrs Jane Bracegirdle,' Et seq. | Reports in the *Wilmslow & Alderley Advertiser*, 7, 14 & 21/10/1938. |
| 'All is happily over,' | Adeane, Jane H. (ed.) (1899), p247. |
| Ice House reconstruction | *Scan*, 26 January 1973; 27 July 1973. |
| 'filling in the Ice' | Stanleys *Day Book 1877–81*, Chester Records Office Ref. DSA/241/21. |
| 'Earller in the day' | Reports in the *Wilmslow & Alderley Advertiser*, 7, 14 & 21/10/1938. |
| 'The sale itself' | Hyde, Matthew (1999), p52. |
| '… Lord Stanley's dustbin' | *Sunday Dispatch*, 17 July 1938. |
| 'It will ease things' | Letter Books Nov 1920 – Dec 1921, Chester Records Office DSA/245/2. |
| 'We had evacuees like' | Alderley Edge Landscape Project archive Ref. AELP 17008_poyner; also in British Library Sound Archive (Alderley Edge No. 55). |
| '… was a mixed arable' | Mrs Mary Watts (*pers.comm.*) |
| 'heard the whistle' | *Scan*, 24 May 1968. |
| 'situated on the edge' | *Scan*, 30 December 1990. |
| 'We used to have' | *Ibid.* |
| 'We had evacuees' | In http://www.alderleyedge.manchester.museum/downloads/tfd_history.pdf |
| '… was a mixed' | *Pers. comm.* to the author. |

## CHAPTER 5 – NATURE RECORDING IN THE PARK

| | |
|---|---|
| 'In memory of' | Inscription on the Bird Hide. |
| 'On the other side' | Adeane, Jane H. (ed.) (1899), p100. |
| *'It was a favourite* | Stanley, Edward (1835), vol 2, p164. |
| 'pale dove-coloured' | Coward, T. A. *The Birds of Cheshire* (1900). |
| 'At one extremity' | Stanley, Edward (1835), vol 1, p219. |
| '[The mere's] *western margin'* | *Op. cit.,* vol 2, p7. |
| 'It is now a rare' | *Op. cit.,* vol 2, p9. |

## CHAPTER 6 – SPECIES OF THE PARK

| | |
|---|---|
| 'I hoped to approach' | Tunnicliffe, Charles F. (1948), p42. |
| [adders] 'have been observed' | *Fulshaw Times,* **8**(8), Autumn 1965, p9. |
| Bat records | Courtesy of Clare Sefton and Ged Ryan, Cheshire Bat Group. |

## CHAPTER 7 – FARM AND ESTATE

| | |
|---|---|
| Estate and Farm records | Surviving ones are in the Chester Records Office. |
| 'All useful timber' | ICI internal notes, c.1972. |
| Plantation creation & management | Article, *Fulshaw Times,* **5**(8), Spring 1958, p2. |

## CHAPTER 8 – PEOPLE IN THE PARK

| | |
|---|---|
| *Alderley Park Estate History* booklet | Listed in Bibliography. |
| ICI, Zeneca & AstraZeneca trail guides | Ditto. |
| 'A large section' | *Summerfield, Phyllis & Massey, Eleanor (1951) |
| 'In the great days' | Parish History Series CD, Vol 2 – The Stanleys in Alderley (Family History Society of Cheshire) |
| 'Overwhelmingly the' | *Ibid.* |

## CHAPTER 9 – COMPANY BEGINNINGS

| | |
|---|---|
| 'Our aim and hope' | Loshak, David (1994) p16. 213280, P/ALDE in Chester Records Office. |
| 'It was quite a common' | Kennedy, Carol (1986), p125. |
| 'an intention to' | Reader, W.J. (1975), p459. |
| 'We're not going to' | Kennedy, Carol (1986), p136. |
| 'This was magnificent' | Reader, W.J. (1975), p460. |
| 'Almost from that' | Alfred Spinks memoir, *Biogr. Mems Fell. R. Soc.* **1984**, *30,* 566–594. |
| 'Because of the all-prevalent' | *Ibid.* |
| Charles Bunn penicillin account | See Clarke, H.T.; Johnson, J.R.; Robinson, R. (ed). *The Chemistry of Penicillin* (1949). |
| 'In the last war' | Loshak, David (1994) p20. 213280, P/ALDE in Chester Records Office. |
| ICI film story | Referenced in Tansey and Reynolds (2000), pp3–4. |
| 'The sulfonamide' | Thornber, Craig (1994), p283. |
| 'Hail chemotherapy – set' | *Op. cit.,* p23. |
| J. Yule Bogue US & Canada tour | Letter to P.W. Cunliffe, 12 March, 1987. |
| 'pursuit of unexpected' | Lecture, 1954. |
| 'Absolutely filthy – all' | Kennedy, Carol (1986), p121. |
| Search for a new Site | J. Yule Bogue, letter to P.W. Cunliffe, 12 March, 1987. |
| Outline planning permission | Ref 5/5/469, 12 October 1950. |
| 'The problems were' | ICI internal history, c.1972. |
| Mereside Opening | *Alderley Park Review*, published for the opening. |
| Ditto. | *Manchester Guardian*, pp7–10, 1 October 1957. |
| The griseofulvin story | Davenport-Hines, R.P.T. (2012), p209. |
| French farm workers story | Tansey & Reynolds (2005), p72. |
| Rhinoceros story | Ford, Hilary (née Curd), *pers. comm.* |
| Memories of Frank Curd | Ford, Hilary (née Curd), *pers. comm.* |
| 'was always keen' | Francis Leslie Rose memoir, *Biogr. Mems Fell. R. Soc.* **1990**, *36,* 491–524. |

## CHAPTER 10 – HISTORIC MILESTONES

| | |
|---|---|
| 'They who are' | Quoted in Stapleton, Melanie (1997). |
| Pharms early history | Quirke, Viviane (2005) & refs cited. |
| Ditto | Holland, K. (1985). |
| Ditto | Francis Leslie Rose memoir, *Biogr. Mems Fell. R. Soc.* **1990**, *36*, 490–524. |
| Beta-blocker story | Quirke, Viviane (2006) & refs cited. |
| 'Ahlquist's theory' | Kennedy, Carol (1986), p137. |
| Sir James Black and propanolol | Stapleton, Melanie (1997). |
| Bogue employed Black | J. Yule Bogue, *ibid*. |
| Stephenson brilliant | Quirke, Viviane (2006). |
| No pressure on Black | Kennedy, Carol (1986), p138. |
| 'We were astonished' | Black, James (1988). Nobel lecture. |
| 'We've done it' | Crowther (1990), |
| Propranolol previously made elsewhere | Hara, Tajuki (2003), p45. |
| '*Since then, propranolol*' | Nicolaou, K.C., Montagnon, T. *Molecules that Changed the World* (Wiley, 2008), p315. |
| 'absolute luck' | Hara, Tajuki (2003), p46. |
| 'When a company has' | Interview quoted in Hara, Tajuki (2003), p44. |
| 'After Crowther's' | J. Yule Bogue, *ibid*. |
| 'the most important drug' | *The New York Times*, Nov 16, 2004. |
| Early oestrogens found, 1930s | Maximov, McDaniel & Jordan (2013), pp1–6. |
| 'Tamoxifen was born' | Jordan, V. Craig. *The Lancet Oncology* **2000**, *1*, p43–49. |
| Haddow and Patterson paper | *British Medical Journal*, **1944**, *2*, 393–7 (23 September 1944). |
| Arthur Walpole tribute | 'The development of tamoxifen for breast cancer therapy: a tribute to the late Arthur L. Walpole.' Jordan, V. Craig, *Breast Cancer Research and Treatment*, **1988**, *11*, 197–209. |
| 'He quickly asked' | Dukes (2015), unpublished. |
| Tamoxifen isomer assignment | Richardson, D.N. (1988) p4. |
| 'never considering that' | Maximov, McDaniel & Jordan (2013), pp1–6. |
| '… almost a caricature' | Sandy Todd, *pers. comm.* |
| 'our work on the study of cancer' | ICI Ltd (1957), p69. |
| 'The particular advantage' | Cole, M.P.; Jones, C.T.A.; Todd, I.D.H. *Br. J. Cancer* **1971**, *25, 270–275.* |
| 'The commercial folk' | Mike Dukes, *pers. comm.* |
| April 1972 decision | Jordan, V. Craig (2003) p207. |
| 'In 1972 … no-one' | Jordan, V. Craig. *British Journal of Pharmacology* **2006**, *147*, S269–S276. |
| 'I was lucky. I walked' | *The Pharmafocus* interview, 19 December 2007. |
| Tamoxifen patent problems | Jordan, V. Craig (2003) p209. |
| 'solely due to the' | John Patterson, *pers.comm*. |
| Meropenem Japanese origin | http://www.ds-pharma.com/rd/new_value/meropen.html |
| 'highly fluorinated compounds' | Kennedy, Carol (1986), p131. |
| Ferguson review | Sneader, Walter (2005), p85. |
| 'They either flew or they didn't.' | Kennedy, Carol (1986), p132. |
| 'the course of' | *Scan*, 14 January 1983. |
| 'Right, what are you going to' | *Op. cit.,* p135. |
| 'single most changing' | '50 years of Alderley Park' Heritage DVD (2007). |

## CHAPTER 11 – SUPPORTING THE WORK

| | |
|---|---|
| Costs of drug discovery | From a review published in *Nature Reviews: Drug Discovery* (2010). |

## CHAPTER 12 – ROUTE TO THE PATIENT

| | |
|---|---|
| 'I … arrived for' | http://www.chemheritage.org/research/institute-for-research/oral-history-program/projects/critical-mass/index.aspx |
| '… competition from other' | Ellingworth, S. (1957), p209. |
| 'Zestril' & Prinivil™ story | Vagelos, P. Roy; Galambos, L. *Medicine, Science and Merck* (Cambridge University Press, 2004), p196. |

## CHAPTER 13 – GROWING AND TELLING

| | |
|---|---|
| 'The mileage covered' | Letter from Dr T. Leigh to Miss Angela Haygarth-Jackson, 26/7/1973. |
| 'done a superb job with the estate' | *Scan*, 9th September 1988. |
| 'If I attempted to answer' | The Duke, to the Secretary of State for War during the Peninsula Campaign. Also quoted in Vivian Fuchs & Edmund Hillary, *The Crossing of Antarctica* (Penguin, 1958), p28, in which I came across it. |
| 'the philosophy of' | Professor Nigel Baber in Tansey and Reynolds (2008), p4. |

## CHAPTER 14 – THE BIG NEW WORLD

| | |
|---|---|
| 'ICI is widely' | Ellingworth, S. (1957), p201. |
| Demerger story | A good external overview is Owen, G.; Harrison, T. *Harvard Business Review* (March–April 1995), pp133–142. |
| ICI Bio | Thompson & Martin (2010), p717–8. |
| 'had a weakness' | *Management Today*, December 1995, p58. |
| 'Our growth is faster than Glaxo.' | *Daily Mail*, 10 December 1988. |
| 'AstraZeneca re-organised' | AstraZeneca 1999 Annual Report, p21. |

## CHAPTER 15 – TARGETS & HORMONES

| | |
|---|---|
| 'In 'breakthrough' research,' | Thornber, Craig (1994), p282. |
| 'ICI need look no' | Dukes (2015), unpublished. |
| 'They said we were' | *SCAN*, 18 December 1992, p4. |
| 'who persuaded the' | Les Hughes, *pers. comm.* |
| 'At every development' | Sandy Todd*, pers. comm.* |
| 'So called 'me-too'' | *Op. cit.* p283. |
| 'One does not discover' | Gide, André *Les faux-monnayeurs* [The Counterfeiters], (1925). |

## CHAPTER 16 – WIDER HORIZONS STILL

| | |
|---|---|
| Paez 'Iressa' paper: | '*EGFR* Mutations in Lung Cancer: Correlation with Clinical Response to Gefitinib Therapy'. J Guillermo Paez *et al. Science*, 4 June 2004, pp1497–1500. |
| 'Iressa' vs 'Tarceva' & doses | 'The Current Situation: Erlotinib (Tarceva®) and Gefitinib (Iressa®) in Non-Small Cell Lung Cancer', R. L. Comis, *The Oncologist*, **2005**, *10*(7), pp467–70. |
| 'Our interest was piqued by' | Finlay, M. R. V. *et al.* (2014) (*J. Med. Chem.*) in the KEY PAPERS list |
| 'exquisite' | Ditto. |
| 'a very desirable profile' | Ditto. |
| 'What will we' | *The Guardian*, 13 May 2014. |
| 'Tagrisso' duration of response | AstraZeneca *Late-stage pipeline conference call*, 2 December 2015, https://www.astrazeneca.com/content/dam/az/our-company/investor-relations/20151202%20Call%20L-Sp%20Wsh%20prsn%20sent(1).pdf |

| | |
|---|---|
| 'Tagrisso' and blood-brain barrier | *AZD9291 Shows Clinical Activity in Non-small Cell Lung Cancer Patients With Leptomeningeal Disease*, American Association for Cancer Research, 7 Nov. 2015, http://www.aacr.org/Newsroom/Pages/News-Release-Detail.aspx?ItemID=787#.VnMEtvmLRD8 |
| 'helped to save the company' | Quoted by kind permission of Susan Galbraith. |

## CHAPTER 17 – VALUE JUDGEMENT

| | |
|---|---|
| Lisinopril Merck story | Vagelos, P. Roy, Galambos, L. *ibid*. |
| 'there was one' | Tansey and Reynolds (2005), p72. |
| 'tremendous resistance' | Professor Nigel Baber in Tansey and Reynolds (2008), p7. |

## CHAPTER 18 – PEOPLE, TIME & CHANCE

| | |
|---|---|
| '… very difficult news' | *The Guardian*, Monday 18 March 2013. |

## CHAPTER 19 – AND NOW, THE FUTURE

| | |
|---|---|
| Alderley Park plans | Cheshire East Council planning ref. 15/5401M. |
| 'He was in the' | *Dialogue with Trypho*, Justin Martyr (c.150AD), ch. 88. |
| 'Marx never made' | Taylor, A.J.P. in introduction to Marx, K., *The Communist Manifesto* p10 (Penguin Books, 1967). |
| 'No great discovery was' | Quoted in *The Palladium: a monthly journal*, **1850**. I, p151. |
| 'If we knew what we were' | Source unknown. |
| 'La République' | Probably an urban myth. |
| 'The most exciting phrase' | Source unknown. |
| 'Puerilities' | Galilei, Galileo. *Dialogue on the Two World Systems* (1632) (University of California Press translation by S. Drake, 1970), p462; cited in Forster, Roger and Marston, Paul, *Reason & Faith* (Monarch Publications Ltd, 1909), p101. |
| 'A new scientific truth' | Planck, Max. *Scientific autobiography* **1950**, p33. |
| 'to be kind, because' | Dylan, Bob. *Chronicles*, vol 1. (Simon & Schuster, 2004). |

# Bibliography

*Alderley Park Estates sale brochure.* 1938.

Adeane, Jane H. (ed.). *The early married life of Maria Josepha Lady Stanley, with extracts from Sir John Stanley's Praeterita, 1766–1816.* Longmans, Green & Co.,1899 (republished BiblioLife, 2010).

'Alderley Park bought.' *The Times*, 22nd June 1950: 3.

Anglesey Antiquarian Society. 'An Alderley Park Anthology.' Unpublished, 1988.

Angus-Butterworth, L. M. *Old Cheshire Families and their Seals.* 1932.

AstraZeneca plc. '50 years of Alderley Park.' *Heritage DVD,* internal, 2007.

— *Alderley Park Estate History.* Internal, c.2006.

— *Ice House & Trees Walk – Alderley House Trail Circuit GOLD ROUTE.* Internal, 2003.

— *Parkland & Old Buildings – Alderley House Trail Circuit PURPLE ROUTE.* Internal, 2003.

— *Radnor Mere and Beech Wood – Mereside Trail Circuit RED ROUTE.* Internal, 2005.

— *Radnor Mere and Beech Wood – Mereside Trail Circuit RED ROUTE.* Internal, 2005.

— *Water Garden & Serpentine – Alderley House Trail Circuit BLUE ROUTE.* Internal, 2003.

Bethell, David. *A Portrait of Cheshire.* Robert Hale Ltd, 1979.

Black, James. 'Drugs from emasculated hormones: the principles of syntopic antagonism.' Nobel lecture, 1988.

Croston, James. *County Families of Lancashire and Cheshire.* 1887.

— *Historic Sites of Lancashire and Cheshire.* 1883.

Crowther, A. F. 'The discovery of the first beta-adrenergic blocking agents.' *Drug Design and Delivery*, 1990: 149–156 (152).

Davenport-Hines, R. P. T. and Slinn, Judy. *Glaxo: A History to 1962.* Cambridge University Press, 2012.

*Domesday Book*, Folios 266, 266b, c.1087–88.

Driver, George W. *Anecdotes about Alderley Park.* Internal, 1988.

Driver, George W. *The Stanleys of Alderley.* Internal, 1989.

Dukes, Mike. 'Arthur Walpole' (unpublished notes on his career). Pers. comm., 2015.

Earwaker, J. P. *East Cheshire, Past & Present.* 1890.

Ellingworth, S. 'The Evolution of the Pharmaceuticals Division of ICI.' (227 pages). Internal, 1957.

Frost, Victoria. *A Brief History of Alderley Park.* Internal, 1996.

Hara, Tajuki. *Innovation in the Pharmaceutical Industry: The Process of Drug Discovery and Development.* Edward Elgar Publishing, 2003.

Harding, John R. 'From ICI to AstraZeneca: the development of synthetic isotope chemistry at Alderley Park, UK.' *J Label Compd Radiopharm*, 2007: 927–932.

Harrison, W. 'Ancient Forests, Chases and Deer Parks in Cheshire.' *Transactions of the Lancashire and Cheshire Antiquarian Society*, 1902.

Haworth, D., and Comber, W. M. (eds.). *Cheshire Village Memories.* Cheshire Federation of Women's Institutes, 1952.

Hignett, Norman H. 'The Stanleys of Alderley.' *Cheshire Life*, 1936: 6.

Hill, George B. (ed.). *Alderley Park and Radnor Mere Bird Report.* Private publ., for the years 1978–1989.

Hill, George B. (ed.). *Alderley Park and Radnor Mere Wildlife Report*, see Natural History Section entry.

Hill, George B. 'Alderley Park's History to 1779.' *Alderley Park and Radnor Mere Wildlife Report*, internal, 1993.

Holland, K. 'IC Pharmaceuticals.' *Pharmaceutical Journal*, 1985: 286.

Houghton, Helen L. *The Changing Landscape of Alderley Park.* Manchester University thesis, 1986.

Hyde, Matthew. *The Villas of Alderley Edge*. The Silk Press, 1999.

ICI Ltd *Pharmaceutical Research in ICI, 1936–57*. Birmingham: The Kynoch Press, 1957.

ICI plc. *Alderley Park History & Nature Trail – BLUE ROUTE: Water Garden & Alderley House* (blue cover). Internal, 1990.

— *Alderley Park History & Nature Trail – RED ROUTE: Radnor Mere and Beech Wood* (grey cover). Internal, 1988.

Jordan, V. Craig. *Tamoxifen: A Guide for Clinicians and Patients*, PRR, 1996.

Jordan, V. Craig. 'Tamoxifen: a most unlikely pioneering medicine.' *Nature Reviews: Drug Discovery*, 2003: 205–213.

Jordan, V. Craig, and Furr, Barry J.A. (eds.) *Hormone Therapy in Breast and Prostate Cancer.* Humana Press, 2002.

Kennedy, Carol. *ICI, The Company That Changed Our Lives.* Hutchinson, 1986.

Keyworth, W. Wykeham. *Sketches in the Neighbourhood of Alderley Edge*. George Falkner & Sons, late 1800s.

Kollewe, J. and Wearden, G. 'ICI: from Perspex to Paints.' *The Guardian*, 18th June 2007.

Lee, Frank. 'Sans Changer.' *ICI Magazine*, September 1978: 212–7.

Loshak, David. *From ICI to Zeneca: A Pharmacological Revolution.* Zeneca Pharmaceuticals, 1994.

Lysons, Daniel. *Magna Britannia, Volume II – The County Palatine of Chester.* 1810.

Maximov, Phillipp Y., McDaniel, Russell E., and Jordan, V. Craig. *Tamoxifen: Pioneering Medicine in Breast Cancer.* Springer Basel, 2013.

Mitford, Nancy (ed.). *The Ladies of Alderley, 1841–50.* 1938; republ. Hamish Hamilton Ltd, 1967.

— *The Stanleys of Alderley, letters 1851–65.* 1939; republ. Hamish Hamilton Ltd, 1968.

Moss, Fletcher. *Pilgrimages to Old Homes.* 1903.

Natural History Section (successively of Icicals/Zeneca Leisure/ClubAZ) (eds. George B. Hill & A. Hugh Pulsford). *Alderley Park and Radnor Mere Wildlife Report.* Internal, for the years 1990–2011.

Ormerod, George. *History of Cheshire, Volume III.* 1882.

Owen, G. and Harrison, T. 'Why ICI Chose to Demerge.' *Harvard Business Review,* March 1995. https://hbr.org/1995/03/why-ici-chose-to-demerge.

Powell, Jane. 'History of Alderley Park' (3 articles). *Scan*, November-December 1971.

Prag, A. John N.W. (ed.). *The Story of Alderley: Living with the Edge.* Manchester University Press, 2016.

Pulsford, A. Hugh (ed.). *Alderley Park and Radnor Mere Wildlife Report*, see Natural History Section entry.

Quirke, Viviane. 'From evidence to market: Alfred Spinks' 1953 survey of new fields for pharmacological research, and the origins of ICI's cardiovascular programme.' In *Medicine, the market and the mass media: producing health in the twentieth century*, by V. Berridge and K. Loughlin (eds.). Routledge, 2005: 144–169.

Quirke, Viviane. 'Putting Theory into Practice: James Black, Receptor Theory and the Development of the Beta-Blockers at ICI, 1958–1978.' *Medical History*, 2006: 69–92.

Reader, W.J. *Imperial Chemical Industries: A History (Volume II, 1926–52).* OUP, 1975.

Richardson, D.N. 'The History of Nolvadex.' *Drug Design and Delivery*, 1988: 1–14.

Russell, Bertrand and Patricia. *The Amberley Papers.* 1937.

Soames, Mary. *Winston and Clementine: The Personal Letters of the Churchills.* Houghton Mifflin Harcourt, 2001.

St Mary's Church, Nether Alderley. 'Volume 2 – The Stanleys in Alderley.' (CD). No date.

Stanley, Arthur P. *Memoirs of Edward and Catherine Stanley.* 1879.

Stanley, Edward. *A Familiar History of Birds.* John W. Parker, 1835.

Stanley, Louisa Dorothea. *Alderley Edge and its Neighbourhood.* James Swinnerton, 1843.

Stanley, Peter. *The House of Stanley from the 12th century.* Pentland Press, 1998.

Stanley, Thomas (Lord Stanley of Alderley). *The Stanleys of Alderley 1927–2001*, AMCD (Publishers) Ltd, 2004.

Stapleton, Melanie P. 'Sir James Black and Propranolol.' *Texas Heart Institute Journal*, 1997: 336–342.

Storey, Harold. 'Heraldry of the Tenants' Hall.' *Alderley Park and Radnor Mere Wildlife Report*, internal, 1996: 38–45.

Summerfield, Phyllis, and Eleanor Massey (eds.). *Looking Back: memories of the villages of Nether Alderley and Warford.* Private publication; abridged version in Haworth, D. & Comber, W.M. (eds.), 1951.

Syngenta plc. *60 Years of Scientific Excellence: Memories of the Industrial Hygiene Research Laboratory (1948–2008).*

Tansey, L.A., and Overy, C. (eds.). 'Migraine: Diagnosis, Treatment and Understanding, c.1960–2000.' *Wellcome Witness Volumes, Volume 49; http://www.histmodbiomed.org/sites/default/files/133771.pdf*, 2014: 39.

Tansey, L.A., and Reynolds, E.M. (eds.). 'Cholesterol, atherosclerosis and coronary disease in the UK, 1950–2000.'

*Wellcome Witness Volumes Volume 27; http://www.histmodbiomed.org/sites/default/files/44850.pdf*, 2006: 72.

Tansey, L. A., and Reynolds, E. M. (eds.). 'Clinical Pharmacology in the UK c.1950–2000; Industry and Regulation.' *Wellcome Witness Volumes, Volume 34; http://discovery.ucl.ac.uk/14888/1/14888.pdf*, 2008: 4–7.

Tansey, L. A., and Reynolds, E. M. (eds.). 'Post penicillin antibiotics: from acceptance to resistance?' *Wellcome Witness Volumes, Volume 6; http://www.histmodbiomed.org/sites/default/files/44828.pdf*, 2000: 3.

Thompson, John L, and Martin, Frank. *Strategic Management: Awareness & Change.* 2010.

Thornber, Craig. 'The Pharmaceutical Industry.' In *The Chemical Industry*, by Heaton, C. A. (ed.), Dordrecht: Springer Science+Business Media (2nd Ed.), 1994: 272–305.

Tunnicliffe, Charles F. *Mereside Chronicle.* Country Life Ltd, 1948.

Zeneca plc. 'Zeneca Group PLC – Company Profile, Information, Business Description, History, Background Information.' *Reference For Business – Company History Index.* c.1997. http://www.referenceforbusiness.com/history2/63/Zeneca-Group-PLC.html.

— *Alderley Park History & Nature Trail – BLUE ROUTE: Water Garden & Alderley House* (blue cover). Internal, 1997.

— *Alderley Park History & Nature Trail – RED ROUTE: Radnor Mere and Beech Wood* (now red cover). Internal, 1996.

— *The Water Garden Throughout the Year* by Hill, George B., Stafford, Lesley E. and Tinston, David J. Internal, c.1998.

# Index

Not all names of medicines or people in the text are included in the Index.
Page numbers in bold refer to images (not all accompanied by text).
* = incomplete list
** = and see Appendix 5
† = chemical structure

marketing 173, 183, 193, 222, 242, 244–5, 321 and in drug stories

Markham, Sir Alex 257–8

mass spectrometry 230–1

Matthews Garden Centre **133**

Matthews, Fred 46–7, 132–3, 169

*Mayflower* and beeches 29

Mayo, John 273

McKillop, Sir Tom xvi, 144, 261, **274**–8, **291**, 316, 328, 341

Meadow Cottage 85

media (glassware, etc.) 210–**11**

Medical Centre 258

medical agents

anaesthetics 154, 172, 176, 203–7, 319–20, 324

anti-androgens 288–290, 320

antibiotics 8, 124, 153, 173, 200–202, 217, 320–1, 325

anti-oestrogens (antiestrogens) xiii, 189–198, 290–292, 320

aromatase inhibitors 293–5, 320

beta-blockers xvi, 3, 179–186, 252, 320, 324

carbapenems 201

irreversible inhibitors and acrylamides 304–5

penicillin and penicillins 155–6, 160–**1**, 200

prostaglandins 177

medicines and medicinal science (see biological materials and targets; formulation; medical agents; pharmaceuticals, etc.)

MedImmune 271, 308, 325, 329

Medinger, Till **245**

meeting rooms **214**

Morck 201–3, 236, 272, 277, 319–22

mere, see Radnor Mere

Mereside designed and opened **163**, **166**

Mereside Library (original) **165**, **166**, **200**, 249

Mereside Library (recent) 249, **269**–70

Mereside Restaurant (old) 3, 80, 164, **211**

Mereview Restaurant (new) xiii, 252–**3** and see food

merger 260, 271, 276–7, 310, 315

metabolism (see pharmacokinetics)

military training (Alderley Volunteers) **6–7**, 27, 37, 51

mill pool 21, 23, 29, 55, 90, 92

Mill, Nether Alderley ix, **10**, 55, 81, **90**, 124, 340

Mitford, Nancy xx, 17, 26, 28, 60, 339

molecular modelling 278–**280**

motorway not built 132

MSP, see Manchester Science Partnerships (formerly Parks)

mulberries 48, **146**, 169

Mulberry's (ClubAZ) 3, **145**, 310, 339

Mummers **75–6**

mural in old Restaurant **80**

mutations 300, 303–9, 323

nature trails, see guidebooks

Natural History Section of ClubAZ x, 102, 105, 141

Nether Alderley (and see buildings by name) 9–11, 14–15, 18, 23, 62, 90–1, 147

New Year Honours List, Mereside 261–2

Newbould, Brian 166, **177**, 206, **252**, 286–**7**

NMR, see Nuclear Magnetic Resonance

Nobel Prize xv, 1, 3, 72, 180, **183**–4, 252, 346

noise levels 92, 236, 247

notebooks (see also computing – software) 217–8

Nuclear Magnetic Resonance (NMR) 191, **231**–2

office supply 214

offices and office life **215**–6

Oglesby, Chris xvi, **336**, 355

Old Drive or Chestnut Avenue xiii, 30–**1**

Old Hall, see Alderley Old Hall

online security 220

ornithology 52, 98, 106

Osborne, Rt Hon George, MP xvii, 331, **336**

Over Alderley 9, 10, 15, 17

Owen, Margaret 28, 77

Pangalos, Mene 328, **336**

Park House 18, 25, 28, 32, **33**, 34, 38–9, 52, 348

*Park Life* 25**5**–6

Parklands **126**, 211–**2**, **240**, **311**–2

patent cliff **315**

patents xi–xiii, 153, 156, 160, 182, 190–193, 197–8, 217, 235–7, 242, 300–1, 315 and in drug stories

patent wall **xxii**, 354

patient (the author's wife, Christine) **246**

patient safety 237, 239, 244, **246**–7

Patterson, John xvii, **144**, 196–8, **197**, 245, 315, 328

Pension Fund, ICI 260, 273

pensioner's association, see Amazen

Peptide Section 286

performance assessment 259, 261

Perkin's Mauve 152–3

personalized medicine 301

Personnel (see Human Resources)

Pfizer 308–9, 315–6, 333

pharmacodynamics, drug 159

pharmacokinetics, drug and DMPK 159, 233

pharmaceuticals for animals*

'Aimax' (methallibure) 175–6, 190

'Antrycide' (quinapyramine) 175–6

'Estrumate' (cloprostenol) **177**†

pharmaceuticals (human) from early ICI research before the Park*

'Fluothane' (halothane) 172, 203–**4**, **205**†

'Fulcin' (griseofulvin) 172–3

'Gynosone' **188**, 190

'Hibitane' (chlorhexidine) (also see 'Corsodyl') 172–3†, 319**

# List of subscribers

Cynthia Abbott
David Ainscough
Andrew and Jackie Aldridge
Dr Sandra Allen
AMAZEN Association
J. David Atkinson
Paul T. Atkinson
John Bailey
Mrs Alison Bamford
Sarah Barber
Alan C. Barker
Jeff Barlow
David John Barnfield
Emma Barton
Sally Basford
Albert Beattie
Vicky Beattie
Julie Bell
Ian Berry
Maria Berry
Roger Bolton
Rob Bradbury
Rob Bramley
Ingrid Brotherhood
Elaine Brown
Steve Brown
Jennifer Bull
Wendy Burke
   (at AP 1972–2015)
Heather Burrows
Dr Roger J. Butlin
Peter W. R. Caulkett
David Checkley
Mr Glynne Chesterson
Ian Cockshott
Paul N. Coffey
Tim Conry
Michael J. Cooper
Graham Crawley
David Martyn Crewe
Janet Culshaw
Sheena Cummins
Janet Dahlstrom
Valerie Dann
Julia A. Davenport
Alan Davies
Mr Terry Davies
Denny and Bernard
Timothy Dobson

J. Bryan Doherty
Mike Dukes
Dr Anand S. Dutta
Jonathan Mark Eden
Dr Jeff Edwards,
   Infection R & D, 1968–2002
Phil Edwards
Philip Edwards
Pat Ellison
Paul Elvin
Richard J. Ernill
Graham Evans
J. Ray Evans
Mary Evans
Graham Fawcett
Hilary Fine
Dr Ray Finlay
Eric Fisher
Anne & Paul Flackett
Roger Foden
Olwen Foley
Hilary R. S. Ford née CURD
Dr Steve Foster
Margaret Fowell
Laurence Gainey
Carole Garlland
Paul Gellert
John Gillett
Alasdair Glen
Dr John Gonzalez
Eileen Goodwin
Reg Goodwin
Mark A. Graham
Beverley Greenway
Dr Malcolm Haddrick
Ian Hardern
Sylvia Harman
Elaine Hatton
David Hawkins
Barry Hayter
David Heard
Lorraine Heath
Ian J. Heathcote
D. W. Heaton
Phil Hickson
Mark Hindley
Ken Hobbs
Vince Hobday
Nick Howe

Dr Ralph Howe
David Hudson
Les Hughes
Roger Hume
Dr Jonathan Hutton
Anne Jeater
John Johnson
David V. Jones
John Jones
Ann Jonynas née MANLEY
Joyce and Hugh
Dave Jude
David Julian
Adam Kalinowski
Larry Keenan
John V. Kemp
Susan Kennedy
Teresa Klinowska
Cathy Kynaston
Roger Langley,
   Public Affairs Department
Hilary Lewis
Linda Littler
Ann Llewellyn
Dr E. A. (Ted) Lock
Pam Lowe
Richard Luke
Dr Jim Lynch
Phill Marley
Peter Marsham
Pat Mart
Graham Mason
Ian Robert Mason
Rachel Mayers
Elizabeth C. McCrea
Trevor McKelvey
Bernard Joseph McLoughlin
Karina Meachin
Lawrence and Margorie Metcalf
Roger Metcalf
Colin Miller
Margaret Mirrlees
Michael Morton
Andrea Moss
Shirley Moss
MSP Alderley Park
Dr John A. Murdoch
Lyndon and Sue Murgatroyd
Keith Nicklin

Gary Nunn
Alex Oldham
Terry Orton
Lucy Padget
Graham & Dorothy Palmer
Clare Pattenden
Dave and Paula Perkins
Stephen Peters
Alan E. Pickup
Kurt Pike
Dr David Priaulx
Dr Joher Raniwalla
John Rayner
Andrea Reeve
Peter Reeves
Mr Alan Reid
Paul Reynolds
Dr David T. Robinson
Gail Robinson
Kerrian W. Rogers
Kevin Rogers
David Rudge
Ken Russell
Thelma Rutherford
Bill Rutter
Ged Ryan
Dr John F. Ryley
Susan Sadler
Olusola Sangojinmi

Brenda Scott
George Sependa
Robert Shaw
Andrew Shore
Barry Smith
Dennis Smith
Gaynor Smith
Michael Smith
Steve Smith
Dr Mike Smithers
Dr Joe de Sousa
Marie South
Will Spinks
Laura Stanley
Dr Trevor Stephens
Robert Stevenson
Joanna Stiles
Mike Stonard
Colin Strawson
Mike Swain
Angela Sykes
John Taylor
Michael Taylor
Zoe Taylor
Ian Thompson
Dr Tom Thompson
Dr Craig W. Thornber
S. M. Thornton
Dave Timms

David Tipney
Sandy Todd
Catherine Trigwell
Eric Trueman
Howard Tucker
Dr Julie Ann Tucker
David Tuffin
Jonathan Tugwood
Kim Vaughan
Ian Waddell
Andrew Walding
Ken Ward
Dr Richard A. Ward
Ken Warren
Dr Paul Watkins
Roy Welsh
David Whalley
Sue Whalley
Mary Whitehead
Paul Whittamore
Barry Wienholdt
John Wilkinson,
    AH Building Services
Bruce M. Williams
Lisa Williams
Hilary Wood
Paul Woodcock
Pauline Yearsley